Distributed Parameter Systems

CONTROL AND SYSTEMS THEORY

A Series of Monographs and Textbooks

Additional Volumes in Preparation

Distributed Parameter Systems

Identification, Estimation, and Control

edited by

W. Harmon Ray
Department of Chemical Engineering
University of Wisconsin–Madison
Madison, Wisconsin

Demetrios G. Lainiotis
Department of Electrical Engineering
State University of New York at Buffalo
Buffalo, New York

MARCEL DEKKER, INC. New York and Basel

Library of Congress Cataloging in Publication Data

Main entry under title:
Distributed parameter systems.

(Control and systems theory ; v. 6)
Includes bibliographical references and index.
1. System analysis. 2. Control theory. I. Ray, Willis Harmon II. Lainiotis, Demetrios G.
QA402.D57 003 77-26050
ISBN 0-8247-6601-6

MARCEL DEKKER, INC.

270 Madison Avenue, New York, New York 10016

Current printing (last digit):
10 9 8 7 6 5 4 3 2 1

PRINTED IN THE UNITED STATES OF AMERICA

PREFACE

This volume is the result of a feeling on the part of the editors that the literature on system identification, state estimation, and control as applied to distributed parameter systems had reached a sufficient state of maturity that a coherent exposition of the results in this field was needed. To ensure the best possible coverage, we asked the leading people in the field to write chapters over their area of expertise. To our delight, this very distinguished group shared our perceptions of such a need, and each has contributed a survey of his own area of specialization. Because of the stature of the contributors and the breadth of the subject matter it is our hope that this volume may serve as a basic reference for workers in the field while providing inspiration and material for expanding the scope of graduate courses in control systems design.

Because of the great diversities in the background of the contributors, and the large number of authors for whom English is not their mother tongue, no attempt was made to enforce a uniform style or even a uniform format for presentation. Rather, the flavor of the material runs from a style comfortable for the applied mathematician to something more familiar to the palate of the engineer. We hope that the reader, from whatever discipline, shares our belief that this diversity of presentation enhances the value of the volume.

W. Harmon Ray
Demetrios G. Lainiotis

CONTENTS

CONTRIBUTORS

ALAIN BENSOUSSAN Paris-IX University, Paris, France, and IRIA-LABORIA, Rocquencourt, France

KUN SOO CHANG Department of Chemical Engineering, University of Waterloo, Waterloo, Ontario, Canada

WEN H. CHEN* Gulf Research and Development Company, Pittsburgh, Pennsylvania

R. EUGENE GOODSON† A. A. Potter Engineering Center, Purdue University, West Lafayette, Indiana

M. KÖHNE Institut für Systemdynamik und Regelungstechnik, Universität Stuttgart, Stuttgart, West Germany

A. J. KOIVO Department of Electrical Engineering, Purdue University, West Lafayette, Indiana

H. N. KOIVO‡ Department of Electrical Engineering, Tampere University of Technology, Tampere, Finland

JACQUES-LOUIS LIONS Collège de France, Paris, France, and IRIA-LABORIA, Rocquencourt, France

MICHAEL P. POLIS Department of Electrical Engineering, Ecole Polytechnique of Montreal, Montreal, Canada

G. RODRIGUEZ Jet Propulsion Laboratory, California Institute of Technology, Pasadena, California

---

*Current affiliation: Production Research Department, Gulf Science and Technology Company, Pittsburgh, Pennsylvania.

†Current affiliation: Institute for Interdisciplinary Engineering Studies, Purdue University, West Lafayette, Indiana.

‡Part of the chapter was written while the author was with the Department of Electrical Engineering, University of Toronto, Toronto, Canada.

JOHN H. SEINFELD Department of Chemical Engineering, California Institute of Technology, Pasadena, California

SPYROS G. TZAFESTAS* Control Systems Laboratory, Electrical Engineering Department, University of Patras, Patras, Greece

BRUNO A. J. VAN DEN BOSCH Institut voor Chemie-ingenieurstechniek, Katholieke Universiteit te Leuven, Leuven, Belgium

P. K. C. WANG Department of Systems Science, University of California at Los Angeles, Los Angeles, California

---

*Also with the Department of Reactors, NRC Democritos, Aghia Paraskevi Attikis, Athens, Greece.

Distributed Parameter Systems

INTRODUCTION

W. Harmon Ray

Department of Chemical Engineering
University of Wisconsin
Madison, Wisconsin

The field of distributed parameter system theory, which began in earnest some 15 years ago, is concerned with the dynamic behavior of processes distributed in space as well as evolving in time. The state space models for these systems usually take the form of partial differential equations or integral equations while the Laplace domain representations result in transcendental transfer functions. In addition to this increased complexity of the modeling equations over lumped systems, there are qualitative differences in system structure which give rise to new considerations in satisfying controllability, observability, and optimality conditions. The simplest illustration of this fact can be seen in Figure 1 where the domain of a one-dimensional distributed process is sketched. The state $\underline{x}(z,t)$ will respond to control actions $\underline{u}(z,t)$ applied in the interior of the domain, but will also respond to control actions $\underline{u}_0(t)$ and $\underline{u}_1(t)$ taken at the two boundaries $z = 0$ and $z = 1$, respectively. Thus <u>controllability</u> and <u>optimality</u> must concern itself with both surface and volume controls. Similarly, sensors of the process output can only be placed at certain locations in space. Figure 1 shows the trajectories of three sensors at z_1^*, z_2^*, z_3^* <u>fixed in space.</u>

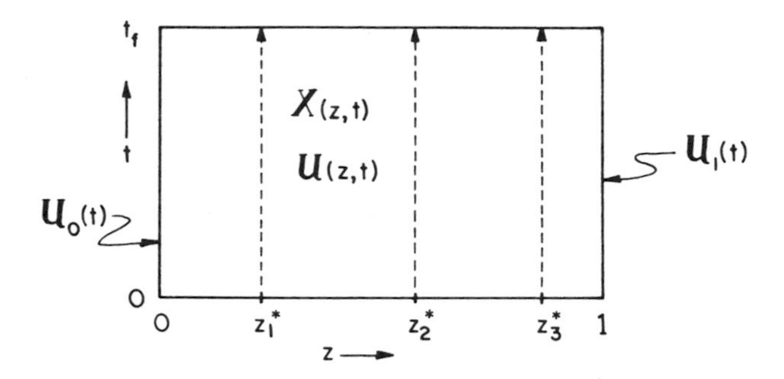

FIG. 1. Domain of a one-dimensional distributed process.

Clearly one may also envision sensors which change position with time in a distributed process. Therefore, the observability of a distributed process must include considerations of sensor location as well as which state combinations are measured.

STRUCTURE OF THE PRESENT VOLUME

In the chapters which follow, the fundamentals of the theory will be developed in Part 1, and then several important specific applications will be discussed in detail in Part 2. Part 1 begins with a discussion of distributed parameter control theory. Several fine monographs [1-3] and survey articles [4,5] are available for background reading in this area. Building on this foundation, Professor Lions in Chapter 1 of this volume discusses recent work in the optimal control of deterministic distributed parameter systems, covering both the steady-state and dynamic conditions.

Chapter 2 is concerned with the identification of parameters in distributed system models. Professors Goodson and Polis present a very comprehensive literature review and discussion of the several approaches which have been used in tackling the identification problem. This chapter is particularly rich in examples which illustrate the algorithms discussed. Distributed parameter state estimation is the topic of Chapter 3. In this chapter, Professor Tzafestas presents a detailed discussion of the existing theory for both linear and nonlinear systems and follows this by the presentation of several application case studies. Both Chapters 2 and 3 deal extensively with the questions of observability and optimal sensor placement, crucial considerations in the design of schemes for parameter or state estimation.

In Chapter 4, Professor Bensoussan presents theoretical results important for the control of stochastic distributed parameter systems, building upon the ideas of state and parameter estimation established in Chapters 2 and 3.

The control and estimation of systems having time delays is treated in Chapter 5 by Professors H. N. Koivo and A. J. Koivo, brothers, who themselves have contributed a great deal to our knowledge about this special class of distributed parameter processes. Although the formulation of time delay problems may somethimes result in models having the form of differential-difference equations or integrodifferential equations, the most general representation of time delay problems gives rise to coupled ordinary differential equations and first-order hyperbolic partial differential equations. Thus, a clear understanding of the behavior of systems having time delays can only come through recognizing their distributed nature.

The last chapter of the fundamentals section, Chapter 6, deals with computational methods for calculating optimal control policies. Professor Chang, who has accumulated a great deal of computational experience with these techniques, first surveys the techniques available and then concentrates his discussion on second-order methods. Through several examples, he illustrates the relative advantages of these methods.

Part 2 of the volume, which is concerned with practical applications, begins with a treatment of the control of vibrating elastic systems in Chapter 7. Dr. Köhne presents a detailed discussion of the basic problem for these systems modeled by fourth-order partial differential equations and concludes his chapter with an analysis of the control of hauling pipes used in deep sea mining. Of additional interest in this chapter is the section dealing with distributed parameter observers, a relatively undeveloped area for distributed parameter systems.

Chapters 8 and 9 are devoted to energy applications of distributed parameter systems theory. In Chapter 8, Professor Wang and Dr. Rodriguez present a detailed discussion of the feedback control of high energy plasmas through manipulation of the external current density surrounding the plasma. This type of control application is of great importance in the development of controlled fusion energy devices. The identification of the properties of underground oil reservoirs is the topic of Chapter 9. In this chapter, Professor Seinfeld and Dr. Chen first develop mathematical models for the reservoir and then describe parameter estimation schemes especially suited to this complex problem. Inasmuch as current oil recovery methods leave more than two-thirds of the discovered oil in the ground, research of this type, directed toward improved modeling and control of these distributed parameter processes, would seem to have great economic importance.

In the final chapter of the volume, Dr. Van den Bosch discusses the problem of parameter estimation in the most complex distributed parameter system found in the process industries: the distributed chemical reactor. An individual who has developed efficient computational algorithms for treating such problems, he describes the most important problems and presents several computational approaches for tackling these problems.

SOME PERCEIVED FUTURE NEEDS

In discussing future needs, one must first make an assessment of the present state of the field of distributed parameter systems theory and practice. As one may judge from the following chapters, the theory is in a fairly advanced state. Powerful results are available for controllability, observability, optimality, and for the estimation and control of stochastic distributed parameter systems. As in all systems theory, the results for linear systems are finely developed while the treatment of nonlinear systems requires more formal and approximate methods.

In contrast, the areas of computational algorithm development and their experimental testing are not very far advanced. The very complex partial differential and integral equations arising from the system modeling and associated control and estimation schemes can be extremely difficult to solve in real time in the absence of clever and efficient numerical schemes. Therefore, it would be very useful to see a larger research effort in this direction. The very few reports of real time implementation (e.g., [6,7]) give encouragement that even very complex distributed parameter systems can be controlled in real time, particularly with the rapid improvements in performance and cost of available minicomputers.

Another broad area in which more emphasis is needed is in specific practical applications. Although a number of these areas are treated in the present volume, many other promising applications exist. Some problem areas of current economic and social importance would include nuclear reactor control and safety, in situ combustion of coal, oil shale, or tar sands, state and parameter estimation for weather prediction, the monitoring and control of environmental air and water quality, etc. Because the practitioner having the problem usually does not understand the potential usefulness of distributed parameter systems methods, while the control scientist or engineer often does not grasp all the practical constraints of the problem, it is through joint efforts that meaningful solutions to important problems will be found. It is to be hoped that a substantial increase in such joint work will be forthcoming, and that the present volume will contribute to narrowing the gap between distributed parameter systems theory and practice.

REFERENCES

1. A. G. Butkovskiy, Distributed Control Systems, American Elsevier, New York, 1969 (translation from the Russian, 1965).

2. J. L. Lions, Optimal Control of Systems Governed by Partial Differential Equations, Springer-Verlag, Berlin, 1971 (translation from the French, 1968).

3. E. D. Gilles, in Systeme mit Verteilten Parametern (R. Oldenbourg, Verlag, Munich, 1973 (in German).

4. A. G. Butkovskiy, A. I. Egorov, and K. A. Lurie, Optimal control of distributed systems, SIAM J. Contr., 6: 437 (1968).

5. A. C. Robinson, A survey of optimal control of distributed parameter systems, Automatica, 7: 371 (1971).

6. M. B. Ajinkya, M. Köhne, H. F. Mäder, and W. H. Ray, The experimental implementation of a distributed parameter filter, Automatica, 11: 571 (1975).

7. H. F. Mäder, Zeitoptimale Steuerung und Modale Regelung eines Praktische Realisierten Warmeleitsystems, Ph.D. thesis, Univ. of Stuttgart, 1975.

PART 1

FUNDAMENTALS

Chapter 1

OPTIMAL CONTROL OF DETERMINISTIC DISTRIBUTED PARAMETER SYSTEMS

Jacques-Louis Lions

Collège de France
Paris, France
and
IRIA-LABORIA
Rocquencourt, France

In this chapter we present some ideas on the problems connected with the control of distributed systems in the stationary case (Section I) and in the evolution case (Section II). In each case we consider linear and nonlinear systems (but always deterministic; for the stochastic case, the reader is referred to Chapter 4).

I. STATIONARY SYSTEMS

A. Setting of the Problem and General Remarks

To fix ideas let us start with a very simple example. Let Ω be a bounded open set in R^n ($n = 2$ or 3 in the applications) with a smooth boundary Γ; let A be a second-order elliptic operator in Ω, which is supposed to be linear. To simplify even further, let us take

$$A = -\Delta + I, \qquad \Delta = \frac{\partial^2}{\partial x_1^2} + \cdots + \frac{\partial^2}{\partial x_n^2}, \qquad I = \text{identity} \tag{1}$$

The state equation is given by

$$Ay = f \quad \text{in } \Omega \tag{2}$$

$$\frac{\partial y}{\partial \nu} = v \quad \text{on } \Gamma \tag{3}$$

where in (2), f is given in some functional space and in (3), $\partial/\partial\nu$ denotes the normal derivative taken toward the exterior of Ω. The above formulation is not sufficiently precise; indeed "solving" a partial differential equation (PDE) makes precise sense only after defining exactly in which class one looks for a "solution." There are indeed very many function spaces where one can solve, in a unique fashion, problem (2), (3). The choice of the function spaces where one wants to solve the state equation (2), (3) depends on the nature of the cost function. Let us give some (simple) examples.

Example 1. Let us assume for the time being that (2), (3) admits a unique solution $y(v)$ and suppose that the cost function is given by

$$J(v) = \int_\Omega [y(v) - z_d]^2 \, dx + N \int_\Gamma v^2 \, d\Gamma \tag{4}$$

where z_d is a given function in $L^2(\Omega)$ and N is a given constant > 0 (which corresponds to the cost of the control). Then v should be taken in $L^2(\Gamma)$ [in order for (4) to make sense], and therefore one has to solve (if possible) (2), (3) for $v \in L^2(\Gamma)$. This is indeed possible by using Sobolev spaces and projection theory in Hilbert spaces.

We introduce

$$H^1(\Omega) = \{\varphi \mid \varphi, \frac{\partial\varphi}{\partial x_1}, \cdots, \frac{\partial\varphi}{\partial x_n} \in L^2(\Omega)\} \tag{5}$$

for $\varphi,\Psi \in H^1(\Omega)$, we define their scalar product

$$(\varphi,\Psi)_{H^1(\Omega)} = (\varphi,\Psi) + \sum_{i=1}^{n} (\frac{\partial\varphi}{\partial x_i}, \frac{\partial\Psi}{\partial x_i}) \tag{6}$$

where $(f,g) = \int_\Omega f(x)g(x)\, dx$ (all functions we consider are real valued).

It is very simple to verify that, provided with the scalar product (6), $H^1(\Omega)$ is a (real) <u>Hilbert space</u>. For $\varphi,\Psi \in H^1(\Omega)$, we also set

$$a(\varphi,\Psi) = (\varphi,\Psi)_{H^1(\Omega)} \tag{7}$$

and we verify, by a formal computation, that if y satisfies (2), (3), then it satisfies, if $y \in H^1(\Omega)$,

$$a(y,\Psi) = (f,\Psi) + \int_\Gamma v\Psi\, d\Gamma \qquad \forall\Psi \in H^1(\Omega) \tag{8}$$

in (8) we have to remark that

$$\Psi \to \int_\Gamma v\Psi\, d\Gamma$$

is a continuous linear form on $H^1(\Omega)$, whenever $v \in L^2(\Gamma)$ (this uses a trace theorem in Sobolev spaces (see Sobolev [1] and, for instance, Lions and Magenes [2]).

One verifies by a straightforward application of the projection theorem in Hilbert spaces that (8) admits a unique solution; we therefore define the state of our system as the element $y = y(v) \in H^1(\Omega)$ which is the solution of the variational equation of elliptic type:

$$a(y(v),\Psi) = (f,\Psi) + \int_\Gamma v\Psi\, d\Gamma \qquad \forall\Psi \in H^1(\Omega) \tag{9}$$

The mapping $v \to y(v)$ is (affine) continuous from $L^2(\Gamma) \to H^1(\Omega)$, and therefore (4) defines a function $v \to J(v)$ which is continuous from $L^2(\Gamma) \to R$.[†]

Example 2. Suppose now that $J(v)$ is of the form

$$J(v) = \int_\Omega |y(v) - z_d| \, dx + N \int_\Gamma |v| \, d\Gamma \tag{10}$$

In order for (10) to make sense, v should be in $L^1(\Gamma)$, and the "natural" space to choose in (3) is $v \in L^1(\Gamma)$. The corresponding space to be chosen for the y's is then much more complicated than in Example 1. The reader is referred to Bidaut [3].

Remark 1. If the cost function is given by

$$J(v) = \int_\Omega |y(v) - z_d|^p \, dx + N \int_\Gamma |v|^p \, d\Gamma, \qquad 1 < p < \infty \tag{11}$$

then, of course, one has to consider problem (2), (3) with $v \in L^p(\Gamma)$. One then finds $y(v) \in W^{1,p}(\Omega)$[‡], i.e.,

$$y(v) \in L^p(\Omega), \qquad \frac{\partial y(v)}{\partial x_i} \in L^p(\Omega) \qquad \forall i$$

$[$If $p = 2$, $W^{1,2}(\Omega) = H^1(\Omega)]$

The Problem of Optimal Control

We now consider the case when J is given by (4). Let $\mathcal{U}_{ad}$ be a closed convex subset of $L^2(\Gamma)$ ($\mathcal{U}_{ad}$ = set of admissible controls). The problem is to find

$$\inf J(v), \qquad v \in \mathcal{U}_{ad} \tag{12}$$

If $\mathcal{U}_{ad} = L^2(\Gamma)$, we say we are in the "no constraint" case.

Since $v \to J(v)$ is a continuous function which is strictly convex, and which is such that

$$J(v) \to +\infty \qquad \text{if } \|v\|_{L^2(\Gamma)} \to \infty \tag{13}$$

we have:

---

† In fact one can prove more: if Γ is of class C^1, then $y(v)$ is in a Sobolev space $H^{3/2}(\Omega)$ [1]; it means that fractional derivatives of order 3/2 of $y(v)$ belong to $L^2(\Omega)$.

‡ One can improve this result by using "fractional" derivatives.

$$\underline{\text{There exists a unique element } u \text{ in } \mathcal{U}_{ad} \text{ such that}} \qquad J(u) = \inf J(v), \quad v \in \mathcal{U}_{ad} \tag{14}$$

The problem now is to see how one can characterize the unique element u in (14), which is called the optimal control.

Remark 2. In Section I.B we give the optimality conditions on the particular example introduced above; but the methods are quite general. The reader is referred to Lions [4] for the general case.

B. Optimality Conditions

The function $v \to J(v)$ is differentiable; it is simple to verify (see Lions [4]) that the optimal control u is characterized by

$$(J'(u), v - u) \geq 0 \quad \forall v \in \mathcal{U}_{ad}, \qquad u \in \mathcal{U}_{ad} \tag{15}$$

Remark 3. If $\mathcal{U}_{ad} = L^2(\Gamma)$ (case without constraints), then (15) reduces to the Euler equation

$$J'(u) = 0 \tag{16}$$

By a simple computation, one verifies that (15) applied to (4) gives (after dividing by 2):

$$\int_\Omega [y(u) - z_d][y(v) - y(u)]\, dx + N \int_\Gamma u(v - u)\, d\Gamma \geq 0 \qquad \forall v \in \mathcal{U}_{ad}, \quad u \in \mathcal{U}_{ad} \tag{17}$$

We now use a classical idea of the calculus of variations, namely the idea of introducing an adjoint state. We set

$$y(u) = y \tag{18}$$

and define the adjoint state p as the solution of

$$A^*p = y - z_d{}^{\dagger} \quad \text{in } \Omega \tag{19}$$

---

† A^* denotes the adjoint of A; in this particular case $A^* = A$, but the present remarks apply to the general case when $A^* \neq A$.

$$\frac{\partial p}{\partial \nu^*} = 0 \qquad \text{on } \Gamma^\dagger \tag{20}$$

Then, an application of Green's formula, which is justified by general results in PDE (see Lions and Magenes [2]) leads to

$$\begin{aligned}\int_\Omega (y - z_d)[y(v) - y]\,dx &= \int_\Omega A^*p[y(v) - y]\,dx \\ &= \int_\Gamma p\left[\frac{\partial y(v)}{\partial \nu} - \frac{\partial y}{\partial \nu}\right] d\Gamma + \int_\Omega p[Ay(v) \\ &\quad - Ay]\,dx \\ &= \int_\Gamma p(v - u)\,d\Gamma\end{aligned}$$

so that (17) reduces to

$$\int_\Gamma (p + Nu)(v - u)\,d\Gamma \ge 0 \qquad \forall v \in \mathcal{U}_{ad}, \qquad u \in \mathcal{U}_{ad} \tag{21}$$

We can summarize: the unique optimal control u [for the cost function (4)] is given by the unique solution $\{y,p,u\}$ of the optimality system:

$$\begin{aligned}&Ay = f, \qquad \frac{\partial y}{\partial \nu} = u \qquad \text{on } \Gamma \\ &A^*p = y - z_d, \qquad \frac{\partial p}{\partial \nu^*} = 0 \qquad \text{on } \Gamma \\ &\int_\Gamma (p + Nu)(v - u)\,d\Gamma \ge 0 \qquad \forall v \in \mathcal{U}_{ad} \\ &u \in \mathcal{U}_{ad}\end{aligned} \tag{22}$$

<u>Example 3</u>. (Case without constraints.) If $\mathcal{U}_{ad} = L^2(\Gamma)$, then $p + Nu = 0$, and one solves first the system of PDE of elliptic type

$$\begin{aligned}&Ay = f, \qquad A^*p = y - z_d \qquad \text{in } \Omega \\ &\frac{\partial y}{\partial \nu} + \frac{1}{N}p = 0, \qquad \frac{\partial p}{\partial \nu^*} = 0 \qquad \text{on } \Gamma\end{aligned} \tag{23}$$

---

† $\partial/\partial\nu^*$ coincides with $\partial/\partial\nu$ in this particular case.

then

$$u = -N^{-1}p \tag{24}$$

Example 4. Let us consider the case

$$\mathcal{U}_{ad} = \{v | v \in L^2(\Gamma),\ v \geq 0 \text{ a.e. on } \Gamma\} \tag{25}$$

Then (21) is equivalent to

$$p + Nu \geq 0, \qquad u \geq 0, \qquad (p + Nu)u = 0 \qquad \text{on } \Gamma \tag{26}$$

One sees therefore that there are on Γ two regions:

$$\Gamma_0 = \{x \mid u(x) = 0\},\ \Gamma_1 = \{x \mid u(x) = -N^{-1}p(x)\} \tag{27}$$

These two regions are not given but are, on the contrary, an essential unknown of the problem. This kind of problem fits in the framework of free boundary problems such as those arising in mathematical physics in evaporation, welding, infiltration, etc.

Remark 4. The set of inequalities (15) is called a "variational inequality," after Lions and Stampacchia [5].

C. Examples of Physical Interest

Examples arising in physics which can fit into a theory extremely similar to that of Section I.B are given in Lions [4,6,7]. We would like to discuss an example which can be treated along similar lines (with, of course, many new technical points) and which arises in the equilibrium of a plasma (see Boujot, Morere, and Temam [8][†], and similar examples in Colli-Franzone [9] and Colli-Franzone and Gazzarniga [10].

Remark 5. The method introduced in Section I.B readily extends to situations where the state equation is given by a higher-order PDE or by a system of PDE.

Remark 6. For the case when N = 0 in the example of Section I.B, the reader is referred to Chapter 1, Section III of Lions [7].

[†] See also Chapter 8 in this volume.

D. Control in the Coefficients[†]

In a number of problems arising in applications, the state equation is a linear PDE of elliptic type, but with the control variable appearing in the coefficients of the operator, which makes the problem highly nonlinear.

Suppose the state y is given by

$$-\Sigma \frac{\partial}{\partial x_i}\left[v_{i,j}(x)\,\frac{\partial y}{\partial x_j}\right] = f \quad \text{in } \Omega$$
$$y = 0 \quad \text{on } \Gamma \tag{28}$$

where the control variable is

$$v = \{v_{i,j},\ i,j = 1, \ldots, n\} \tag{29}$$

subject to the constraints

$$\mathcal{U}_{ad} = \{v \mid v_{i,j} \in L^\infty(\Omega)$$
$$\alpha|\xi|^2 \le \Sigma\, v_{i,j}(x)\,\xi_i\xi_j \le \beta|\xi|^2 \quad \text{a.e. in } \Omega \tag{30}$$
$$0 < \alpha < \beta < \infty\}$$

For every $v \in \mathcal{U}_{ad}$, problem (28) admits a unique solution

$$y = y(v) \tag{31}$$

one has

$$y(v) \in H_0^1(\Omega)$$

where $H_0^1(\Omega)$ denotes the (closed) subspace of $H^1(\Omega)$ of those functions $\varphi \in H^1(\Omega)$ which are 0 on Γ.

Let the cost function be given by, for instance,

$$J(v) = \int_\Omega |y(v) - z_d|^2\, dx \tag{32}$$

Examples of this type arise in applications (see [11-13]).

[†] The content of this section is not indispensable for the reading of the other sections in this chapter.

Remark 7. Although the state equation (28) is linear for a given v, the mapping $v \to y(v)$ is of course nonlinear.

The problem of optimal control is now to find

$$\inf J(v), \quad v \in \mathcal{U}_{ad} \tag{33}$$

We are faced here with a major difficulty, leading to very interesting mathematical developments.

Let u_n be a minimizing sequence; then if $u_n = \{u_{n,i,j}\}$, one has, by virtue of (30), $\|u_{n,i,j}\|_{L^\infty(\Omega)} \leq$ constant; therefore one can extract a subsequence, still denoted by u_n, such that

$$u_{n,i,j} \to u_{i,j} \quad \text{in } L^\infty(\Omega) \text{ weak star}, \quad \forall i,j \tag{34}$$

Let us set

$$y(u_n) = y_n \tag{35}$$

by virtue of (30), y_n remains in a bounded set of $H_0^1(\Omega)$, and therefore we can assume (by a new extraction of a subsequence) that

$$y_n \to y \quad \text{in } H_0^1(\Omega) \text{ weakly} \tag{36}$$

But results (34) and (36) are generally not sufficient to conclude that

$$y = y(u) \quad \text{if } u = \{u_{i,j}\}$$

See a counter example, for the case when the state equation is given by

$$-\frac{d}{dx}\left[v(x)\frac{dy}{dx}\right] + v(x)y = 0 \tag{37}$$

with nonhomogeneous boundary conditions in Murat [14].

One can obtain positive results by choosing a set $\mathcal{U}_{ad}$ contained in (30) and having a particular structure.

Example 5. Let us define

$$\mathcal{U}_{ad}^1 = \{v \mid v \in \mathcal{U}_{ad} \text{ [defined in (30)], and such that } v_{i,j} \in \text{compact set of } L^\infty(\Omega)\} \tag{38}$$

Then it is immediate to obtain the existence of an optimal control in $\mathcal{U}^1_{ad}$.

Example 6. The following example is much more delicate. It is a direct application of results relative to systems with rapidly varying coefficients [15-21].

Suppose we are given functions $a_{i,j}$ which belong to $L^\infty(R^n)$ which are periodic with period 1 in all variables and which satisfy

$$\Sigma\, a_{i,j}(y)\xi_i\xi_j \geq \alpha|\xi|^2 \qquad \text{for a.e. } y \in R^n \tag{39}$$

Let us then set

$$\mathcal{U}^2_{ad} = \{v \mid v_{i,j}(x) = a_{i,j}(\tfrac{x}{\epsilon}) \quad \forall i,j, \quad \forall \epsilon > 0\} \tag{40}$$

We define a subset of $\mathcal{U}_{ad}$ in this manner (if β is chosen so that $\Sigma\, a_{i,j}(y)\xi_i\xi_j \leq \beta|\xi|^2$). A natural question to ask in this framework is then to study the behavior, as $\epsilon \to 0$, of the solution y_ϵ of

$$-\Sigma \frac{\partial}{\partial x_i}\left[a_{i,j}\left(\frac{x}{\epsilon}\right)\frac{\partial y_\epsilon}{\partial x_j}\right] = f$$
$$y_\epsilon = 0 \qquad \text{on } \Gamma \tag{41}$$

One can show [15-21] that, when $\epsilon \to 0$,

$$y_\epsilon \to y \qquad \text{in } H^1_0(\Omega) \text{ weakly} \tag{42}$$

where y is the solution of

$$\mathcal{A}y = f, \qquad y = 0 \qquad \text{on } \Gamma \tag{43}$$

and where $\mathcal{A}$ is a second-order elliptic operator with constant coefficients constructed in the following manner: Let W be the space of functions defined on $]0,1[^n = \Pi$, which belong to $H^1(\Pi)$ and are periodic (i.e., their values on opposite faces of Π are equal). For $\varphi,\Psi \in W$, let us set

$$\alpha(\varphi,\psi) = \Sigma \int_\Pi a_{i,j}(y)\,\frac{\partial\varphi}{\partial y_j}\,\frac{\partial\Psi}{\partial y_i}\,dy \tag{44}$$

there exists a unique element $\chi^i \in W/R$ such that

$$\alpha(\chi^i,\Psi) = \alpha(y_i,\Psi) \qquad \forall\Psi \in W \tag{45}$$

we define

$$q_{i,j} = \alpha(\chi^i - y_i,\ \chi^j - y_j) \tag{46}$$

then

$$\mathcal{A} = -\Sigma\, q_{i,j} \frac{\partial^2}{\partial x_i \partial x_j} \tag{47}$$

Remark 8. The preceding construction shows the "noncompact" structure of sets such that (30): in order to obtain existence theorems, one has to add the "G-limits" to this set, in the sense of de Giorgi and Spagnolo [16], of the operators

$$-\Sigma \frac{\partial}{\partial x_i}\left[v_{i,j}(x) \frac{\partial}{\partial x_j}\right]$$

Remark 9. For the case when the coefficients are step-functions, the reader is referred to Tartar [22].

E. Constraints on the State-Duality Methods

Let us now return to the "linear quadratic" case considered in Section I.B, but with constraints on the state. Let $\mathcal{L}$ be a "curve" given on Γ ($\mathcal{L}$ is a variety of dimension n-2 contained in Γ). Let the state be given by (2), (3) with $f = 0$[†]; let $\mathcal{U}_1$ be a closed convex subset of $L^2(\Gamma)$, and let g be a given function on $\mathcal{L}$. We define

$$\mathcal{U}_{ad} = \{v \mid v \in \mathcal{U}_1,\ y(v) = g \text{ on } \mathcal{L}\} \tag{48}$$

We suppose that $\mathcal{U}_{ad}$ is not empty. It is indeed the case if, for instance, $\mathcal{U}_1 = L^2(\Gamma)$ and g is smooth enough (see Lions [7] for further details). Let the cost function be given by:

$$J(v) = \int_\Gamma |y(v) - z_d|^2\, d\Gamma + N \int_\Gamma v^2\, d\Gamma \tag{49}$$

[†] This is in order to simplify slightly. This does not restrict the generality.

Exactly as in Section I.B, one immediately sees that there exists a unique u such that

$$u \in \mathcal{U}_{ad}, \qquad J(u) = \inf J(v), \quad \forall v \in \mathcal{U}_{ad} \tag{50}$$

This optimal control u is characterized by

$$\begin{aligned} &\int_\Gamma (y - z_d)[y(v) - y]\, d\Gamma + N \int_\Gamma u(v - u)\, d\Gamma \geq 0 \qquad \forall v \in \mathcal{U}_{ad} \\ &y = y(u) \end{aligned} \tag{51}$$

As in Section I.B, we introduce the adjoint state p given by

$$A^*p = 0, \qquad \frac{\partial p}{\partial \nu} = y - z_d \qquad \text{on } \Gamma \tag{52}$$

and (51) is seen to be equivalent to

$$\int_\Gamma (p + Nu)(v - u)\, d\Gamma \geq 0 \qquad \forall v \in \mathcal{U}_{ad} \tag{53}$$

But it is difficult to interpret (53) because of the implicit character of the constraints in $\mathcal{U}_{ad}$ [because of the condition $y(v) = g$ in $\mathcal{L}$].

We will show that this difficulty can be "removed" by using a general duality theorem in convex analysis (see Rockafellar [23] and Ekeland and Temam [24]). The idea of using duality arguments in case of "constraints on the state" is from Mossino [25]; the example we present here is taken from Lions [7]; other examples can be found in Mossino [25].

We define

$$V = L^2(\Gamma), \qquad Q = L^2(\Gamma) \times L^2(\mathcal{L})$$

$$F(v) = \frac{1}{2} N \int_\Gamma v^2\, d\Gamma, \qquad v \in V$$

$$G(q) = G_1(q_1) + G_2(q_2), \qquad q = \{q_1, q_2\} \in Q$$

$$G_1(q_1) = \frac{1}{2} \int_\Gamma |q_1 - z_d|^2\, d\Gamma$$

$$G_2(q_2) = \{0 \text{ if } q_2 = g, \ +\infty \text{ if } q_2 \neq g\}$$

$\Lambda v = \{y(v)|_\Gamma, y(v)|_\mathcal{L}\}$, which defines Λ as a continuous linear map from V to Q. With these notations, (50) is equivalent to

$$\inf\,[F(v) + G(\Lambda v)], \qquad v \in V \tag{54}$$

We now explain in a formal manner how to transform (54) by duality (see Rockafellar [23] and Ekeland and Temam [24] for a precise analysis). We set

$$\Phi(v,q) = F(v) + G(q) \tag{55}$$

and we observe that

$$\inf_v F(v) + G(\Lambda v) = \inf_v \inf_q \sup_{q^*} [\Phi(v, \Lambda v - q) - \langle q^*,q\rangle] \tag{56}$$

where q^* spans the dual space of Q and where $\langle q^*,q\rangle$ denotes the scalar product between Q and its dual. Formula (56) is immediate, since $\sup \langle -q^*,q\rangle = +\infty$ unless $q = 0$, and since

$$\Phi(v,\Lambda v) = F(v) + G(\Lambda v)$$

We now commute in a formal manner (which can be justified in the present situation) the inf and sup in (56):

$$\inf F(v) + G(\Lambda v) = \sup_{q^*} \inf_{v,q} [-\langle q^*,q\rangle + \Phi(v, \Lambda v - q)]$$

If we set

$$\Lambda v - q = \rho \tag{57}$$

we obtain

$$\inf F(v) + G(\Lambda v) = \sup_{q^*} [-\sup_{v,q} \langle q^*, \Lambda v - \rho\rangle - \Phi(v,\rho)] \tag{58}$$

We have to compute

$$\sup_{v,q} [\langle q^*,\Lambda v\rangle - \langle q^*,\rho\rangle - \Phi(v,\rho)] = \sup [\langle \Lambda^* q^*,v\rangle - F(v)] + \sup [\langle -q^*,\rho\rangle - G(\rho)] \tag{59}$$

But if Ψ is a given convex function on a space X, its conjugate function Ψ^* on X^* is given by

$$\Psi^*(x^*) = \sup_{x \in X} [\langle x^*, x\rangle - \Psi(x)] \tag{60}$$

Then (59) gives

$$\sup [\langle q^*, \Lambda v\rangle - \langle q^*, \rho\rangle - \Phi(v,\rho)] = F^*(\Lambda^* q^*) + G^*(-q^*)$$

and therefore (58) gives

$$\inf_{v} F(v) + G(\Lambda v) = \sup_{q^*} [-F^*(\Lambda^* q^*) - G^*(-q^*)] \tag{61}$$

We now compute Λ^*, F^*, and G^*.

We identify V and Q with their dual.

Given a function $q_2 \in L^2(\mathcal{L})$, we define $q_2^{\mathcal{L}}$ as a distribution on Γ by

$$\langle q_2^{\mathcal{L}}, \Psi\rangle_\Gamma = \int_{\mathcal{L}} q_2 \Psi \, d\mathcal{L} \tag{62}$$

We consider the problem

$$\begin{aligned} A^*\Phi &= 0 \\ \frac{\partial \Phi}{\partial \nu} &= q_1 + q_2^{\mathcal{L}} \quad \text{on } \Gamma \end{aligned} \tag{63}$$

One can show that this problem admits a unique solution that we denote by $\Phi(q)$, such that $\Phi(q)|_\Gamma \in L^2(\Gamma)$ (using the technique of Lions and Magenes [2]) and that

$$\Lambda^* q = \Phi(q)|_\Gamma \tag{64}$$

It is a simple matter to verify that

$$F^*(v) = \frac{1}{2N} \int_\Gamma v^2 \, d\Gamma \tag{65}$$

$$\begin{aligned} G_1^*(q_1) &= \frac{1}{2} \int_\Gamma q_1^2 \, d\Gamma + \int_\Gamma q_1 z_d \, d\Gamma \\ G_2^*(q_2) &= \int_{\mathcal{L}} q_2 g \, d\mathcal{L} \end{aligned} \tag{66}$$

Then (61) shows that

$$\inf F(v) + G(\Lambda v) = \sup_q \left[-\frac{1}{2N} \int_\Gamma \Phi(q)^2 \, d\Gamma - \frac{1}{2} \int_\Gamma q_1^2 \, d\Gamma + \int_\Gamma q_1 z_d \, d\Gamma + \int_{\mathcal{L}} q_2 g \, d\mathcal{L} \right] \tag{67}$$

Then the initial original problem $\inf J(v)$ is equivalent to the dual problem

$$\inf \left[\frac{1}{2N} \int_\Gamma \Phi(q)^2 \, d\Gamma + \frac{1}{2} \int_\Gamma q_1^2 \, d\Gamma - \int_\Gamma q_1 z_d \, d\Gamma - \int_{\mathcal{L}} q_r g \, d\mathcal{L} \right] = \inf Y(q) \tag{68}$$

where the state of the system is now given by (63). There are no constraints. The price paid here lies in the fact that problem (68) does not necessarily admit a solution. But (68) is nevertheless useful in particular for numerical computations (see Mossino [25]).

F. Nonlinear State Equations

We have already seen an example in Section I.D where the mapping $v \to y(v)$ is nonlinear, although the state equation (28) was linear. In a large number of applications, the state equation itself is nonlinear. Let us give two examples.

Example 7. Let the state $y = y(v)$ be given by the solution of

$$-\Delta y + y^3 = f \quad \text{in } \Omega \tag{69}$$

$$\frac{\partial y}{\partial \nu} = v \quad \text{on } \Gamma \tag{70}$$

where we assume that $f \in L^2(\Omega)$ and $v \in L^2(\Gamma)$. This problem is equivalent, in its variational form, to find y such that

$$\int_\Omega (\text{grad } y \text{ grad } \Psi + y^3 \Psi) \, dx = \int_\Omega f\Psi \, dx + \int_\Gamma v\Psi \, d\Gamma \qquad \forall \Psi \in H^1(\Omega), \quad y \in H^1(\Omega) \tag{71}$$

One can show by standard methods such as in Lions [26] that problem (71) admits a unique solution which belongs to $H^1(\Omega) \cap L^4$ [if $n \le 4$, $H^1(\Omega) \subset L^4(\Omega)$]. Let us suppose that the cost function $J(v)$ is given by

$$J(v) = \int_\Omega |y(v) - z_d|^2\, dx + N \int_\Gamma v^2\, d\Gamma \tag{72}$$

The problem of optimal control is now to find inf $J(v)$, $v \in \mathcal{U}_{ad}$, $\mathcal{U}_{ad}$ = closed convex subset of $L^2(\Gamma)$.

One can show without great difficulty that there exists $u \in \mathcal{U}_{ad}$ such that

$$J(u) = \inf J(v), \qquad v \in \mathcal{U}_{ad} \tag{73}$$

There is no reason to expect uniqueness of u in such a situation. The function $v \to J(v)$ is differentiable, and a necessary condition for u to be an optimal control is:

$$(J'(u), v - u) \geq 0 \qquad \forall v \in \mathcal{U}_{ad} \tag{74}$$

If we define $\hat{y}(v)$ by

$$\hat{y}(v) = \frac{d}{d\lambda} y(u + \lambda v) \Big|_{\lambda=0} \tag{75}$$

we have

$$\begin{aligned} -\Delta\hat{y}(v) + 3y^2\hat{y}(v) &= 0 \quad \text{in } \Omega \\ \frac{\partial}{\partial\nu}\hat{y}(v) &= v \quad \text{on } \Gamma \end{aligned} \tag{76}$$

where $y = y(u)$. With these notations, (74) can be written (after dividing by 2):

$$\int_\Omega (y - z_d)\hat{y}(v - u)\, dx + N \int_\Gamma u(v - u)\, d\Gamma \geq 0 \qquad \forall v \in \mathcal{U}_{ad} \tag{77}$$

We now introduce the adjoint state p defined as the solution of

$$\begin{aligned} -\Delta p + 3y^2 p &= y - z_d \quad \text{in } \Omega \\ \frac{\partial p}{\partial\nu} &= 0 \quad \text{on } \Gamma \end{aligned} \tag{78}$$

by applying Green's formula, one finds that

$$\int_\Omega (y - z_d)\hat{y}(v - u)\, dx = \int_\Gamma p(v - u)\, d\Gamma$$

so that (77) is equivalent to

$$\int_\Gamma (p + Nu)(v - u)\, d\Gamma \geq 0 \tag{79}$$

We can summarize: a necessary condition for u to be an optimal control is that $\{y,p,u\}$ is a solution of the optimality system:

$$\begin{gathered} -\Delta y + y^3 = f, \qquad -\Delta p + 3y^2 p = y - z_d \quad \text{in } \Omega \\ \frac{\partial v}{\partial \nu} = u, \qquad \frac{\partial p}{\partial \nu} = 0 \quad \text{on } \Gamma \\ \int_\Gamma (p + Nu)(v - u)\, d\Gamma \geq 0 \quad \forall v \in \mathcal{U}_{ad}, \quad u \in \mathcal{U}_{ad} \end{gathered} \tag{80}$$

Let us remark that, from what has been said, we know the existence of a solution $\{y,p,u\}$ of (80), but the question of uniqueness is open.

Example 8. Let us now consider a situation where the state equation corresponds to a free boundary problem. We suppose that the state $y = y(v)$ is the solution of

$$Ay(v) = f \quad \text{in } \Omega, \qquad A = -\Delta + I \tag{81}$$

with the unilateral boundary conditions

$$\frac{\partial y(v)}{\partial \nu} - v \geq 0, \qquad y(v) \geq 0, \qquad y(v)\left[\frac{\partial y}{\partial \nu}(v) - v\right] = 0 \quad \text{on } \Gamma \tag{82}$$

Remark 10. Problem (81), (82) is a "model" problem of questions arising in mechanics: unilateral constraints, friction, plasticity, etc. (see Duvaut and Lions [27]).

Remark 11. Problem (81), (82) can be formulated as a variational inequality (VI) in the sense of Lions and Stampacchia [5] and similar to what we have already met in this chapter; indeed if we define

$$K = \{\varphi \mid \varphi \in H^1(\Omega),\ \varphi \geq 0 \text{ on } \Gamma\} \tag{83}$$

then one easily checks that problem (81), (82) is equivalent to finding $y(v)$ such that

$$\begin{gathered} y(v) \in K \\ a(y(v), \varphi - y(v)) \geq (f, \varphi - y(v)) + \int_\Gamma v[\varphi - y(v)]\, d\Gamma \quad \forall \varphi \in K \end{gathered} \tag{84}$$

The existence and uniqueness of $y(v)$, solution of (84), follows from a general result of Lions and Stampacchia [5].

Remark 12. In this particular situation where A is symmetric, it is easy to verify that (84) is equivalent to minimizing

$$\frac{1}{2} a(\varphi,\varphi) - (f,\varphi) - \int_\Gamma v\varphi \, d\Gamma = \mathcal{E}_v(\varphi) \tag{85}$$

over K, hence the existence and uniqueness of $y(v)$ follows. Let the cost function be given by

$$J(v) = \int_\Gamma |y(v) - z_d|^2 \, d\Gamma + N \int_\Gamma v^2 \, d\Gamma \tag{86}$$

Let us consider the case without constraints: we look for

$$\inf J(v), \qquad v \in L^2(\Gamma)$$

One can show the existence and uniqueness of an optimal control, characterized in the following fashion (this is a highly nontrivial result, due to Mignot [28]). Let us first introduce the following notation: if $\{y,p\}$ is given in $H^1(\Omega) \times H^1(\Omega)$, we define

$$S(y,p) = \{\varphi \mid \varphi \in H^1(\Omega), \quad \varphi \geq 0 \text{ a.e. on the set of } \Gamma \tag{87}$$

$$\text{where } y = 0^{\dagger}\}, \; a(y,\varphi) = \int_\Omega f\varphi \, dx - \frac{1}{N}\int_\Omega p\varphi \, d\Gamma$$

One then has

$$u = \text{optimal control} = -\frac{1}{N} p \tag{88}$$

where $\{y,p\}$ is the unique solution of the "optimality system"

$$\begin{aligned} &a(y,\varphi - y) \geq (f,\varphi - y) - \frac{1}{N}\int_\Gamma p(\varphi - y) \, d\Gamma \qquad \forall\varphi \in K \\ &y \in K \end{aligned} \tag{89}$$

---

† This set is defined up to a set of capacity 0 on Γ.

$$a^*(p,\varphi) \le \int_\Gamma (y - z)\varphi \, d\Gamma \qquad \forall\varphi \in S(y,p)$$
$$p \in S(y,p) \tag{90}$$

Remark 13. (89), (90) enters into the family of the so-called "quasi variational inequalities" introduced in Bensoussan and Lions [29] in order to solve problems of impulse controls.

Remark 14. It follows from (82) that there are two regions on Γ; $y(v) = 0$ on Γ_0, and $\partial y/\partial\nu(v) = v$ on Γ_1; these regions are not given a priori, so that this problem is indeed in the class of free boundary problems.

Remark 15. Remark 12 naturally leads to the following general problem: Let $\mathcal{E}(\varphi,v)$ be a function given on $K \times \mathcal{U}_{ad}$, where K = a closed convex set of a Hilbert (or a reflexive Banach) space. Let us assume that the problem

$$\inf_{\varphi \in K} \mathcal{E}(\varphi,v) \tag{91}$$

admits a unique solution $y(v)$ in K; let $J(v)$ be the cost function:

$$J(v) = \Phi[y(v)] + N \|v\|_{\mathcal{U}} \tag{92}$$

where Φ is a functional on K and where $\| \ \|_{\mathcal{U}}$ denotes the norm in a Hilbert (or Banach) space containing $\mathcal{U}_{ad}$.

Question: In case we have the existence of a solution u of $J(u) = \inf J(v)$, $v \in \mathcal{U}_{ad}$, how can we obtain the necessary conditions of optimality? The Mignot's theorem, reported above, gives an answer in a particular case; however, the general question is open.

Remark 16. The main difficulty in this kind of problem lies in the nondifferentiability of the mapping $v \to y(v)$.

Example 9. A very important class of problems of optimal control where the mapping $v \to y(v)$ is neither linear nor differentiable is when the control variable is a domain. These problems enter into the family of optimum design problems. The reader is referred to Pironneau [30,31].

Remark 17. The reader is referred to Puel [32] for the study of problems of control when the state equation admits an ordered set of solutions, the state of the system being, for instance, the maximum solution of this set.

Remark 18. The reader is referred to Mignot [33], Saguez [34], and Van de Wiele [35] for the study of problems of optimal control where the state equation is defined by the first eigenvalue of a second-order elliptic problem.

III. EVOLUTION SYSTEMS

A. Optimality Conditions

Let Ω be an open set in R^n again; we assume that Ω is bounded to fix ideas, with a smooth boundary Γ. We consider in Ω a second-order elliptic operator given by

$$A\varphi = - \sum_{i,j=1}^{n} \frac{\partial}{\partial x_i} \left[a_{i,j}(x) \frac{\partial \varphi}{\partial x_j} \right] \tag{93}$$

where we assume that

$$\begin{aligned} & a_{i,j} \in L^{\infty}(\Omega) \\ & \Sigma\, a_{i,j}(x)\xi_i\xi_j \geq \alpha|\xi|^2, \qquad \alpha > 0, \qquad \text{a.e. in } \Omega \end{aligned} \tag{94}$$

Remark 19. We do not assume symmetry; $a_{i,j}$ can be different from $a_{j,i}$. Some of the computations which follow (in particular in Section II.C), are somewhat clearer when $A^* \neq A$.

Remark 20. The results of Sections II.A and II.B immediately extend to the case when the coefficients $a_{i,j}$ depend on t : $a_{i,j}(x,t)$. However, this is not the case for the results of Section II.C.

The state equation is now given by the parabolic PDE:

$$\frac{\partial y}{\partial t} + Ay = f + v \qquad \text{in } Q = \Omega \times]0,T[\tag{95}$$

$$y(x,t) = 0 \qquad \text{if } x \in \Gamma \tag{96}$$

$$y(x,0) = y_0(x), \qquad y_0 \text{ given in } \Omega \tag{97}$$

This boundary value problem is a classical one; it is well known that it admits a unique solution, under various hypotheses. Let us assume here that the control function v satisfies

$$v \in \mathcal{U}_{ad}$$
$$\mathcal{U}_{ad} = \text{a closed convex subset of } \mathcal{U} = L^2(Q) \tag{98}$$

Let us assume that f and y_0 are given satisfying

$$f \in L^2(Q) \tag{99}$$

$$y_0 \in L^2(\Omega) \tag{100}$$

Then (see for instance Lions and Magenes [2]) it is known that the problem (95) to (97) admits a unique solution $y = y(v)$ in the class $W(0,T)$:

$$W(0,T) = \left\{\varphi \mid \varphi, \frac{\partial\varphi}{\partial x_i} \in L^2(Q) \quad \forall i \right.$$
$$\frac{\partial\varphi}{\partial t} \in L^2[0,T;H^{-1}(\Omega)]^{\dagger}, \qquad \varphi = 0 \tag{101}$$
$$\left. \text{in } \Sigma = \Gamma \times]0,T[\right\}$$

We provide $W(0,T)$ with the Hilbertian norm $\|\varphi\|_{W(0,T)}$ given by

$$\|\varphi\|^2_{W(0,T)} = \int_0^T \left[\|\varphi(t)\|^2_{H^1_0(\Omega)} + \|\frac{\partial\varphi}{\partial t}(t)\|^2_{H^{-1}(\Omega)}\right] dt$$

Then the mapping $v \to y(v)$ is (affine) continuous from $L^2(Q) \to W(0,T)$.

Remark 21. In the above problem we have taken a distributed control. In most of the applications, the control appears like a boundary control or as a ponctual control. But the methods we are going to introduce are general: for specific applications to boundary control and to ponctual control, the reader is referred to Lions [4-7, 36] and to the bibliography therein.

The cost function is given by

$$J(v) = \int_0^T (y(v) - z_d)^2 \, dt + N \int_0^T |v|^2 \, dt \tag{102}$$

where in (102), $| \ |$ denotes the norm in $L^2(Q)$; therefore, more explicitly, (102) reads

---

$^{\dagger}H^{-1}(\Omega)$ denotes the dual space of $H^1_0(\Omega)$.

$$J(v) = \int_0^T \int_\Omega |y(x,t;v) - z_d(x,t)|^2 \, dx \, dt + N \int_0^T \int_\Omega v(x,t)^2 \, dx \, dt \tag{103}$$

in these formulas, z_d is given in $L^2(Q)$ and N is given > 0.

The problem of optimal control is now to find

$$\inf J(v), \qquad v \in \mathcal{U}_{ad} \tag{104}$$

By virtue of the continuity of the mapping $v \to y(v)$ from $L^2(Q) \to L^2(Q)$ (in particular), we see that the function $v \to J(v)$ is continuous, strictly convex, and such that $J(v) \to +\infty$ as $\|v\|_{\mathcal{U}} \to \infty$. Therefore, exactly as in Section I, there exists a unique element u of $\mathcal{U}_{ad}$ such that

$$J(u) = \inf J(v), \qquad v \in \mathcal{U}_{ad} \tag{105}$$

The optimal control u can be characterized, as in Section I, by the variational inequality (since J is differentiable)

$$\begin{aligned} &(J'(u), v - u) \geq 0 \qquad \forall v \in \mathcal{U}_{ad} \\ &u \in \mathcal{U}_{ad} \end{aligned} \tag{106}$$

If we explicitly compute (106) we obtain (after dividing by 2):

$$\begin{aligned} &\int_0^T (y - z_d, y(v) - y) \, dt + N \int_0^T (u, v - u) \, dt \geq 0 \qquad \forall v \in \mathcal{U}_{ad} \\ &u \in \mathcal{U}_{ad} \end{aligned} \tag{107}$$

where in (107) we have set: $y(u) = y$ and where (f,g) denotes the scalar product in $L^2(Q)$.

We now transform (107) by introducing the adjoint state p defined by:

$$\begin{aligned} &-\frac{\partial p}{\partial t} + A^* p = y - z_d \qquad \text{in } Q \\ &p = 0 \qquad \text{on } \Sigma = \Gamma \times \,]0,T[\\ &p(x,T) = 0 \end{aligned} \tag{108}$$

Problem (108) admits a unique solution in the class $W(0,T)$ (it is of course the same type of problem as (95) to (97), after changing the time orientation). A simple application of the Green's formula shows that

$$\int_0^T (y - z_d, y(v) - y)\, dt = \int_0^T (p, v - u)\, dt$$

so that (107) is equivalent to:

$$\int_0^T (p + Nu, v - u)\, dt \geq 0 \qquad \forall v \in \mathcal{U}_{ad}, \qquad u \in \mathcal{U}_{ad} \tag{109}$$

We can summarize: problem (104) admits a unique solution u which is given by the unique solution $\{y,p,u\}$ of the optimality system:

$$\begin{aligned} &\frac{\partial y}{\partial t} + Ay = f + u \\ &y = 0 \text{ on } \Sigma, \qquad y(x,0) = y_0(x) \end{aligned} \tag{110}$$

together with (108) and (109).

Example 10 (case without constraints). Suppose that $\mathcal{U}_{ad} = L^2(Q)$. Then (109) reduces to

$$p + Nu = 0 \tag{111}$$

One can then eliminate u; one first solves the system

$$\begin{aligned} &\frac{\partial y}{\partial t} + Ay + \frac{1}{N} p = f \\ &-\frac{\partial p}{\partial t} + A^* p = y - z_d \qquad \text{in } Q \\ &y = 0, \qquad p = 0 \qquad \text{on } \Sigma \\ &y(x,0) = y_0(x), \qquad p(x,T) = 0 \qquad \text{in } \Omega \end{aligned} \tag{112}$$

then u is given by (111).

Remark 22. It follows from what has been said above that system (112) admits a unique solution; for a direct approach to this problem, the reader is referred to Yebra [37].

Remark 23. We shall return in Sections II.B and II.C to study (112).

Example 11. Let us now suppose that

$$\mathcal{U}_{ad} = \{v \mid v \geq 0 \quad \text{a.e. in } Q\} \tag{113}$$

Then (as in Example 4,) (109) is equivalent to

$$p + Nu \geq 0, \quad u \geq 0, \quad u(p + Nu) = 0 \tag{114}$$

We see that there are two regions in the cylinder Q:

$$Q_0 = \{x,t \mid u(x,t) = 0\}, \quad Q_1 = \{x,t \mid u(x,t) = -N^{-1}p(x,t)\} \tag{115}$$

The interface between Q_0 and Q_1 plays the role of a free surface somewhat similar to the free surface appearing, for instance, in the Stefan's problem in mathematical physics.†

Remark 24. The reader is referred to Lions [4,6,7] for the optimality system in many other situations.

Remark 25. The same approach is valid for hyperbolic state equations. If the state equation is given by

$$\begin{aligned} &\frac{\partial^2 y}{\partial t^2} + Ay = f + v \quad \text{in } Q \\ &y(x,t) = 0 \quad \text{on } \Sigma \\ &y(x,0) = y_0(x), \quad \frac{\partial y}{\partial t}(x,0) = y_1(x) \quad \text{in } \Omega \end{aligned} \tag{116}$$

where A is given by (93) with (94) and $a_{i,j} = a_{j,i} \quad \forall i,j$, methods entirely similar to what has been said before apply [4,6,7]. If the state equation is given by a first-order hyperbolic system, there are technical differences due to methods one has to use for the solution of the corresponding PDE (the reader is referred to Refs. 38-42, and to the bibliographies therein).

---

†The free surface problems of mathematical physics are also closely related to problems of optimal stopping time and optimal impulse control. The reader is referred to Bensoussan and Lions [29, 43-45].

B. Riccati's Integrodifferential Equation

We now return to the study of (112). There is a way to "transform" (112) (which is a boundary value problem in x and in t) into a boundary value problem in x only and into a Cauchy problem in t. One can show directly (see Lions [4], chapter 3) that there exists an operator P(t) from $H_0^1(\Omega)$ into itself, and a function r(t) with values in $H_0^1(r)$ such that

$$p(t) = P(t)y(t) + r(t) \tag{117}$$

where in (117), we denote by y(t) the function $x \to y(x,t)$, etc.

One can derive equations which characterize P and r by an identification; taking (117) into the second equation (112) one obtains (after verifying that $t \to P(t)$ is differentiable),

$$-\frac{\partial P}{\partial t} y - P \frac{\partial y}{\partial t} - \frac{\partial r}{\partial t} + A^*Py + A^*r = y - z_d$$

we replace $\partial y/\partial t$ by its value obtained from the first equation (112); we obtain

$$-\frac{\partial P}{\partial t} y + P[Ay + \frac{1}{N}(Py + r)] - Pf - \frac{\partial r}{\partial t} + A^*Py + A^*r = y - z_d$$

But this is an identity in y (due to the fact that t = 0 does not play a particular role and that y_0 is arbitrary); therefore

$$-\frac{\partial P}{\partial t} + PA + A^*P + \frac{1}{N} P.P = I \tag{118}$$

and

$$-\frac{\partial r}{\partial t} + A^*r + \frac{1}{N} Pr = Pf - z_d \tag{119}$$

Since p(T) = 0, we have the conditions

$$P(T) = 0 \tag{120}$$

$$r(T) = 0 \tag{121}$$

We have to remember that in (118) the P(t)'s are operators which satisfy

$$P(t) \in \mathcal{L}[H_0^1(\Omega);H_0^1(\Omega)]^{\dagger} \tag{122}$$

$$\begin{aligned} &P(t) \in \mathcal{L}[L^2(\Omega);L^2(\Omega)], \quad P(t) \text{ being symmetric, i.e.,} \\ &[P(t)f,g] = [f,P(t)g] \quad \forall f,g \in L^2(\Omega) \end{aligned} \tag{123}$$

and that [since $r(t) \in H_0^1(\Omega)$]:

$$r(x,t) = 0 \quad \text{on } \Sigma \tag{124}$$

The conditions (118), (120), (122), (123) characterize $P(t)$; this is an evolution equation with the Cauchy data $P(T) = 0$. Next we define $r(t)$ by (119), (121), (124), and this is again an evolution equation with Cauchy data $r(x,T) = 0$. Then $p(t)$ is given by (117).

One can give a much more concrete interpretation of (118) by using the kernel of the mapping $P(t)$; since $P(t)$ satisfies (122) it follows from Schwartz's kernel theorem [46] that, $\forall\varphi \in H_0^1(\Omega)$

$$P(t)\varphi(x) = \int_\Omega P(x,\xi,t)\varphi(\xi)\, d\xi \tag{125}$$

where $P(x,\xi,t)$ is a uniquely defined distribution on $\Omega \times \Omega$, such that $P(t)$ satisfies (122), (123); the symmetry of $P(t)$ is equivalent to

$$P(x,\xi,t) = P(\xi,x,t) \tag{126}$$

One then verifies that (118) is equivalent to

$$-\frac{\partial P}{\partial t} + (A_x^* + A_\xi^*)P + \frac{1}{N}\int_\Omega P(x,\xi,t)P(\zeta,\xi,t)\, d\zeta = \delta(x - \xi) \tag{127}$$

where $\delta(x - \zeta)$ is the distribution given by

$$\int_\Omega \delta(x - \xi)\varphi(\xi)\, d\xi = \varphi(x)$$

with P subject to

$$P(x,\xi,t) = 0 \quad \text{if } x \in \Gamma \text{ or if } \xi \in \Gamma \tag{128}$$

$$P(x,\xi,T) = 0 \quad \text{for } x,\xi \in \Omega \times \Omega \tag{129}$$

$^{\dagger}\mathcal{L}(X;Y)$ denotes the space of continuous linear maps from X into Y.

Therefore P is characterized by a parabolic nonlinear PDE (with a quadratic nonlinearity).

Remark 26. The obvious difficulty of this approach lies in the increase in the dimension of the problem. We return to that in Section I.C below.

Remark 27. The existence and uniqueness of a solution of (127) to (129) follows from the "indirect" approach followed above; a direct solution of this problem has been given by Tartar [47] (see also Temam [48]).

Remark 28. For a study of the similar problems for hyperbolic systems, the reader is referred to Curtain and Pritchard [49] and the bibliography therein.

Remark 29. A numerical analysis study of the nonlinear PDE of Riccati's type has been made by Nedelec [50] using "splitting-up methods" of the type used in more classical PDEs by Marchuk [51], Yanenko [52], and Temam [53]. Some information on this method is given in Lions [6], chapter 6 (see also Nedelec [54]).

C. Reduction of Riccati's PDE

We give now a method of reduction of the PDE satisfied by P(t). For the algebraic manipulations which follow we return for a moment to Eq. (118). We introduce

$$\frac{\partial P}{\partial t} = M \tag{130}$$

and we take the t-derivative of (118) (which can be justified at the end of the computation). We obtain the equation

$$-\frac{\partial M}{\partial t} + MA + A^*M + \frac{1}{N} MP + \frac{1}{N} PM = 0 \tag{131}$$

$$M(T) = -I \tag{132}$$

If we introduce

$$\mathcal{A} = A + \frac{1}{N} P \tag{133}$$

(which depends on t), then (131) can be written

$$-\frac{\partial M}{\partial t} + M\mathcal{A} + \mathcal{A}^*M = 0 \tag{134}$$

let us assume for a moment that we can find L satisfying

$$\begin{aligned} &-\frac{\partial L}{\partial t} + L\mathcal{A} = 0 \\ &L(T) = I \end{aligned} \tag{135}$$

Then

$$M = -L^*L \tag{136}$$

Indeed, if we set $\tilde{M} = -L^*L$, then $\partial\tilde{M}/\partial t$ is shown to satisfy [by using (135)] $+ \partial\tilde{M}/\partial t = \tilde{M}\mathcal{A} + \mathcal{A}^*\tilde{M}$; and since $\tilde{M}(T) = -I$, we obtain (136), since (131), (132) admit a unique solution; for this last point observe that one can think of (131) as the linear parabolic equation

$$-\frac{\partial M}{\partial t} + (A_x^* + A_\xi^*)M(x,\xi,t) + \frac{1}{N}\int_\Omega M(x,\xi,t)P(\zeta,\xi,t)\,d\zeta + \frac{1}{N}\int_\Omega P(x,\zeta,t)M(\zeta,\xi,t)\,d\zeta = 0 \tag{137}$$

subject to the boundary conditions

$$M(x,\xi,t) = 0 \qquad \text{for } x \in \Gamma \text{ or } \xi \in \Gamma \tag{138}$$

and to

$$M(x,\xi,T) = -\delta(x - \xi) \tag{139}$$

We now observe that (135) does admit a solution; if $L(x,\xi,t)$ denotes the kernel of L, then (135) is equivalent to

$$-\frac{\partial L}{\partial t} + A_\xi^* L(x,\xi,t) + \frac{1}{N}\int_\Omega L(x,\xi,t)P(\zeta,\xi,t)\,d\zeta = 0 \tag{140}$$

with

$$L(x,\xi,t) = 0 \qquad \text{for } \xi \in \Gamma \tag{141}$$

$$L(x,\xi,T) = \delta(x - \xi) \tag{142}$$

This is a linear parabolic problem which admits a unique solution. Since the kernel of L^* is $L(\xi,x,t)$, (136) gives

$$M(x,\xi,t) = -\int_{\Omega} L(\zeta,x,t)L(\zeta,\xi,t)\, d\zeta \tag{143}$$

i.e., using (130),

$$\frac{\partial P}{\partial t}(x,\xi,t) + \int_{\Omega} L(\zeta,x,t)L(\zeta,\xi,t)\, d\zeta = 0$$
$$P(x,\xi,T) = 0 \tag{144}$$

Summarizing: one can compute P by solving the system (140), (141), (142), (144). This implies a considerable reduction of the "amount of computation" involved, since in (140), x plays the role of a parameter. See a detailed study by Sorine [98].

Remark 30. Computations have been made by using this technique in a number of different examples in Leroy [55].

Remark 31. If, even after the reduction indicated above†, the computation of P is too complicated one can use the method of "a priori feedback" as indicated in Lions [7], Bermudez [56], and Bermudez, Sorine, and Yvon [57].

D. Nonlinear Systems

In a large number of applications the state equation is nonlinear. Let us mention Yvon [58] for a system governed by a nonlinear heat equation and Kernevez [59], Kernevez and Thomas [60], and Brauner and Penel [61,62] for systems arising in biochemistry. These systems are governed by nonlinear PDEs of parabolic type. One can use techniques entirely similar to those of Section II.A together with those of Section II.F. The reader is also referred to Chapter 8, and to Yvon [63] and Lions and Yvon [64].

E. Controllability

Let us consider here controllability by boundary controls. There is a significant difference between the parabolic and the hyperbolic cases. We first consider a parabolic system:

---

†The above method fails when the coefficients of A depend on t. (See Casti [65,66], Casti and Ljung [67], and Casti and Yvon [68].)

$$\frac{\partial y}{\partial t} + Ay = 0 \quad \text{in } Q = \Omega \times]0,T[\tag{145}$$

$$\frac{\partial y}{\partial \nu} = v \quad \text{on } \Sigma = \Gamma \times]0,T[\tag{146}$$

$$y(x,0) = 0 \quad \text{in } \Omega \tag{147}$$

In (146), $\partial/\partial\nu$ denotes the conormal derivative associated to the operator A, still given by (93) with (94).

There is approximate controllability in $L^2(\Omega)$ for every T positive, i.e.,

$$\begin{aligned} &\forall T > 0, \quad y(x,T) \text{ spans a dense subset of } L^2(\Omega) \text{ when} \\ &v \text{ spans } L^2(\Gamma \times]0,T[) \end{aligned} \tag{148}$$

Indeed, let $\Psi \in L^2(\Omega)$ be such that

$$\int_\Omega y(x,T;v)\Psi(x)\, dx = 0 \qquad \forall v \in L^2(\Gamma \times]0,T[\tag{149}$$

Let us define ζ as the solution of

$$\begin{aligned} &-\frac{\partial \zeta}{\partial t} + A^*\zeta = 0 \quad \text{in } \Omega \times]0,T[\\ &\frac{\partial \zeta}{\partial \nu_*} = 0 \quad \text{on } \Gamma \times]0,T[\\ &\zeta(x,T) = \varphi(x) \quad \text{in } \Omega \end{aligned} \tag{150}$$

Then

$$\int_Q \left(-\frac{\partial \zeta}{\partial t} + A^*\zeta\right) y(v)\, dx\, dt = 0 = \int_\Sigma \zeta \frac{\partial y(v)}{\partial \nu}\, d\Sigma - \int_\Omega \Psi(x)y(x,T;v)\, dx$$

and therefore, by virtue of (149) we have

$$\int_\Sigma \zeta v\, d\Sigma = 0 \qquad \forall v \in L^2(\Sigma)$$

i.e.,

$$\zeta = 0 \quad \text{on } \Sigma \tag{151}$$

But then the Cauchy data of ζ are 0 on Σ, and by the uniqueness theorem of Mizohata [69], one has $\zeta = 0$, i.e., $\Psi = 0$.

Remark 32. The space spanned by $y(x,T;v)$ is not closed in $L^2(\Omega)$.

The study of controllability for hyperbolic systems has been mainly studied by Russell in a series of papers [70-73]. Let us present a result of Graham and Russell [74]. We consider for the state equation the <u>wave equation</u>

$$\frac{\partial^2 y}{\partial t^2} - \Delta y = 0 \qquad \text{in } \Omega \times]0,T[\tag{152}$$

where Ω is the <u>open unit ball</u>, with the boundary control

$$\frac{\partial y}{\partial \nu} = v \qquad \text{on } \Sigma = \Gamma \times]0,T[\tag{153}$$

and the zero initial data:

$$y(x,0) = y_0, \qquad \frac{\partial y}{\partial t}(x,0) = y_1 \tag{154}$$

where

$$y_0 \in H^1(\Omega), \qquad y_1 \in L^2(\Omega) \tag{155}$$

Then, we say that the system is controllable at time T if there exists $v \in L^2(\Sigma)$ such that

$$y(x,T;v) = 0 \qquad \frac{\partial y}{\partial t}(x,T;v) = 0 \tag{156}$$

It is proved in Graham and Russell [74] that

The system (152) to (154) is controllable (resp. not controllable) if $T > 2$ (resp. $T < 2$) (157)

<u>Remark 33</u>. For other aspects of controllability theory, the reader is referred to Fattorini [75,76], Fattorini and Russell [77], Saint Jean Paulin [78] (where a system partly parabolic and partly hyperbolic is studied), Triggiani [79,80], Sakawa [81], Tsujioka [82], and Henry [83].

F. Additional Remarks

<u>Remark 34</u>. One meets in applications problems of <u>control of variational inequalities of evolution</u> (see in particular problems arising in biochemistry with semiimpervious membranes; consult Kernevez and Thomas [84] and, for the numerical analysis of such problems, Yvon [85]. One can conjecture that one has for evolution

problems results similar to those of Mignot, reported in Section I.F, Example 8, but this remains to be proven (Yvon [86]). Results along these lines have been obtained by Saguez [95].

Remark 35. The main difficulty one is faced with in the end is the dimensionality of the problem. The methods one can think of in order to "diminish the complexity" of the problems are as follows:

1. Decomposition methods: The reader is referred to Bensoussan, Lions, and Temam [87] and the bibliography therein. We would like to draw attention to the numerical results of Lemonnier [88] obtained by using this kind of technique.
2. One can try to "simplify" the state equations: One of the main tools to do that is the method of perturbations. In particular, if one wants to decrease the order of the system, one can use the method of singular perturbations when some of the parameters are "small." The reader is referred to Lions [89,90,96], and Brauner [91] for examples and applications. Another possibility is to use some averaging methods, or methods of "homogenization" similar to those reported in Example 6.

Remark 36. One also encounters problems where the coefficient N appearing in the cost function [see for instance formula (4)] is "small"; since N is related to the "cost of the control," one says that one has then a problem of cheap control; this kind of problem is related to problems of singular perturbations; the reader is referred to Lions [7,89], Jameson and O'Malley [92], and the References in this chapter.

Remark 37. For the problem of sensitivity and the search for the implementation of a "low sensitivity" control, the reader is referred to Stravroulakis [93], Abu El Ata [94], and the bibliography therein.

Remark 38. A problem arising in nuclear engineering is of the following type: The state of the system is given by

$$\begin{aligned} &\frac{\partial y}{\partial t} - \Delta y = f + \sum_{i=1}^{q} v_i(t)\delta(x - b_i), \quad \text{in } Q = \Omega \times]0,T[\\ &y = 0 \quad \text{on } \Sigma = \Gamma \times]0,T[\\ &y(x,0) = y_0(x) \quad \text{in } \Omega \end{aligned} \tag{158}$$

where the b_i's are subject to stay in E = closed subset of Ω. In (158), the control functions and variables are the v_i's *and* the b_i's (and possibly the number q itself of Dirac masses). The corresponding problem, with a quadratic cost function, has been studied by Saguez [34], and with moving B_i by Bamberger [97].

REFERENCES

1. S. L. Sobolev, *Applications of Functional Analysis to Equations of Mathematical Physics*, Leningrad, 1950 (in Russian); (English translation: American Mathematical Society).

2. J. L. Lions and E. Magenes, eds., *Problèmes non Homogènes et Applications*, vols. 1-3, Dunod, Paris, 1970, 1972, 1973 (English translation: vols. 181-183, Springer, Grundlehren).

3. M. F. Bidaut, *Un Problème de Contrôle Optimal à Fonction Coût en Norme L^1*, C.R.A.S., Paris, 1975.

4. J. L. Lions, *Contrôle Optimal de Systèmes Gouvernés par des Equations aux Dérivées Partielles*, Dunod, Gauthier-Villars, Paris, 1968, (English translation: vol. 170, Springer, Grundlehren, 1970).

5. J. L. Lions and G. Stampacchia, *Variational Inequalities*, vol. XX, C.P.A.M., 1967, pp. 493-519.

6. J. L. Lions, Some aspects of the optimal control of distributed parameter systems, in *Regional Conference Series in Applied Mathematics*, Nb. 6, (SIAM, ed.), 1972.

7. J. L. Lions, Various topics in the theory of optimal control of distributed systems, London, Canada, in *Lecture Notes in Economics and Mathematical Systems*, (B. J. Kirby, ed.), Springer, New York, 1973.

8. J. P. Boujot, J. P. Morera, and R. Temam, An optimal control problem related to the equilibrium of a plasma in a cavity, *Appl. Math. Optim.*, *2*: (1975).

9. P. Colli-Franzone, *Problemi di Controllo Ottimale Relativi a Sistemi Governati da Problemi ai Limiti Misti*, Istituto Lombardo, Accad. di Scienze e Lettere, A.108, 1974, pp. 557-584.

10. P. Colli-Franzone and G. Gazzaniga, *Approssimazione Numerica di Alcuni Problemi*, Istituto Lombardo, Accad. di Scienze e Lettere, A.108, 1974, pp. 623-651.

11. K. A. Luré, Optimum control of conductivity of a fluid moving in a channel in a magnetic field, *P.M.M.*, *28*: 258-267 (1964).

12. K. A. Luré, *Optimal Control in Problems of Mathematical Physics*, Moscow, 1975 (in Russian).

13. G. T. McAllister and S. M. Rohde, An optimization problem in hydrodynamic lubrication theory, Appl. Math. Optim., 2: (1975).

14. F. Murat, Un contre-exemple pour le problème du contrôle dans les coefficients, C.R.A.S., 273: 708-711 (1971).

15. E. Sanchez-Palencia, Comportement local et macroscopique d'un type de milieux physiques hétérogènes, Int. J. Eng. Sci., 12: 331-351 (1974).

16. E. de Giorgi and S. Spagnolo, Sulla convergenza degli integrali dell'energia per operatori ellittici del secondo ordine, Bull. U.M.I., 3: 391-411 (1973).

17. S. Spagnolo, Sulla convergenza di soluzioni di equazioni paraboliche ed ellittiche, Ann. Scuola Norm. Sup. di Pisa, 22: 571-597 (1968).

18. C. Sbordone, Sulla G-convergenza di equazioni ellittiche e paraboliche, Ric. di Mat. (1975).

19. N. S. Bahbalov, Doklady Akad. Nauk, vol. 218, no. 5, 1046-1048 (1914).

20. I. Babuska, Solution of the interface problem by homogenization, in Reports, Institute for Fluid Dynamics and Applied Mathematics, 1974.

21. A. Bensoussan, J. L. Lions, and G. Papanicolaou, Sur Quelques Phénomènes Asymptotiques Stationnaires, C.R.A.S., Paris, 1975.

22. L. Tartar, Problèmes de contrôle des coefficients dans des équations aux dérivées partielles, in Lecture Notes in Economics and Mathematical Systems, vol. 107, Springer, New York, 1975, pp. 420-426.

23. T. Rockafellar, Duality and stability in extremum problems involving convex functions, Pac. J. Math., 21: 167-187 (1967).

24. I. Ekeland and R. Temam, Analyse Convexe et Problèmes Variationnels, Dunod, Paris, 1974 (English translation: North Holland, 1976).

25. J. Mossino, An application of duality to distributed optimal control problems with constraints on the control and on the state, J.M.A.A., 50: 223-242 (1975).

26. J. L. Lions, Quelques Méthodes de Résolution des Problèmes aux Limites Nonlinéaires, Dunod-Gauthier-Villars, Paris, 1969.

27. G. Duvaut and J. L. Lions, Les Inéquations en Mécanique et en Physique, Dunod, Paris, 1972 (English translation: Springer, New York, 1975).

28. F. Mignot, "Contrôle dans les inéquations," to be published.

29. A. Bensoussan and J. L. Lions, Nouvelle Formulation de Problèmes de Contrôle Impulsionnel et Applications, C.R.A.S., Paris, 1973, 276 pp., pp. 1189-1192.

30. O. Pironneau, On optimum profiles in Stokes flow, J. Fluid Mech., 59:117-128 (1973).

31. O. Pironneau, On optimum design in fluid mechanics, J. Fluid Mech., 64: 97-110 (1974).

32. J. P. Puel, "Contrôle optimal sur les solutions minimales de certains problèmes quasi linéaires," to be published.

33. F. Mignot, "Un problème de contrôle en fonction propre," to be published.

34. C. Saguez, Contrôle ponctuel et contrôle en nombres entiers de systèmes distribués, Rapport LABORIA, No. 82 (1974).

35. J. P. Van de Wiele, "Résolution numérique d'un problème de contrôle optimal de valeurs propres et de vecteurs propre," thesis, Paris VI Univ., 1974.

36. J. L. Lions, Sur le contrôle optimal de systèmes distribués, IMU Lectures in L'Enseignement Mathématique, vol. XIX, 1973, pp. 125-166.

37. Yebra, work to be published.

38. D. L. Russell, Quadratic performance criteria in boundary control of linear symmetric hyperbolic systems, SIAM J. Contr., vol. 11, no. 3: 475-509 (1973).

39. T. L. Johnson, "Optimal control of first order distributed systems," M.I.T. Electronic Systems Laboratory, Report 482, 1972.

40. R. B. Vinter and T. L. Johnson, "Optimal control of non-symmetric hyperbolic systems in n-variables on the half space," to be published.

41. C. Bardos, Problèmes aux limites pour les équations aux dérivées partielles du 1er ordre, Annales E.N.S., 3: 185-233 (1970).

42. P. Krée, Journal de Mathématiques Pures et Appliquées (1973).

43. A. Bensoussan and J. L. Lions, Nouvelles Méthodes en contrôle impulsionnel, Appl. Math. Optim., vol. 1, no. 4: 289-312 (1975).

44. A. Bensoussan and J. L. Lions, Problèmes de temps d'arrêt optimal et inequations variationnelles paraboliques, Applicable Anal., 3: 267-294 (1973).

45. A. Bensoussan and J. L. Lions, Sur Quelques Aspects du Contrôle Optimal, vol. 1, Temps d'arrêt, Hermann, Paris, 1976.

46. L. Schwartz, Théorie des noyaux, Proc. I.C.M., 1: 220-230 (1950).

47. L. Tartar, Sur l'étude directe d'équations non linéaires intervenant en théorie du contrôle optimal, J. Funct. Anal. (1975).

48. R. Temam, Sur l'équation de Riccati associée à des opérateurs non bornés en dimension infinie, J. Funct. Anal., 7:85-115 (1971).

49. R. Curtain and A. J. Pritchard, The infinite dimensional Riccati equation for systems defined by evolution operators, SIAM J. Contr.

50. J. C. Nedelec, thesis, Paris, 1970.

51. G. I. Marchouk, Numerical Methods for Meteorological Forecast, Leningrad, 1967 (in Russian) (French translation: A. Colin, Paris, 1969).

52. M. M. Yanenko, Méthode à Pas Fractionnaires, A. Colin, Paris, 1968 (translated from the Russian).

53. R. Temam, Sur la stabilité et la convergence de la méthode des pas fractionnaires, Ann. Mat. Pura Appl., 79: 191-380 (1968).

54. J. C. Nedelec, "Finite elements for Riccati's P.D.E.," to be published.

55. D. Leroy, Rapport LABORIA (1975), to be published.

56. A. Bermudez, Contrôle de systèmes distribués par feedback a priori, Rapport LABORIA, Nb. 129 (1975).

57. A. Bermudez, M. Sorine, and J. P. Yvon (1975), to be published.

58. J. P. Yvon, Etude de quelques problèmes de contrôle pour des systèmes distribués, Appl. Math. Optim., to be published.

59. J. P. Kernevez, Control of the flux of substrate entering an enzymatic membrane by an inhibition at the boundary, J. Opt. Theory Appl. (1973).

60. J. P. Kernevez and D. Thomas, Numerical analysis and control of some biochemical systems, Appl. Math. Optim., vol. 1, no. 3, 222-285 (1975).

61. C. M. Brauner and P. Penel, Un problème de contrôle optimal non linéaire en biomathématique, Annali dell'Università di Ferrara, vol. XVIII, no. 7 (1973).

62. C. M. Brauner and P. Penel, Perturbations singulières dans un problème de contrôle optimal intervenant en biomathématique, in Lecture Notes in Economics and Mathematical Systems, vol. 107, 1975, pp. 643-668.

63. J. P. Yvon, Etude de la méthode de bouele ouverte adaptée pour le contrôle de systèmes distribués, in Lecture Notes in Economics and Mathematical Systems, vol. 107, 1975, pp. 427-439.

64. J. L. Lions and J. P. Yvon, in preparation.

65. J. Casti, Reduction of dimensionality for systems of linear two point boundary value problems with constant coefficients, J.M.A.A., 42: 522-531 (1974).

66. J. Casti, Matrix Riccati equations, Dimensionality reduction and generalized X and Y functions, Utilitas Mathematica, 6: 95-110 (1974).

67. J. Casti and L. L. Ljung, Some new analytic and computational results for operator Riccati equations, SIAM J. Contr., (1975).

68. J. Casti and J. P. Yvon, to be published.

69. S. Mizohata, Unicité du prolongement des solutions pour quelques opérateurs différentiels parabolique, Mém. Coll. Sc., Univ. Kyoto, A., 219-239 (1958).

70. D. L. Russell, Non-harmonic Fourier series in the control of distributed parameter systems, J. Math. Anal. Appl., 18: 542-560 (1967).

71. D. L. Russell, Boundary value control of the higher dimensional wave equation, SIAM J. Contr., vol. 1, no. 9, 29-42 (1971); vol. 2, no. 9, 401-419 (1971).

72. D. L. Russell, Control theory of hyperbolic equations related to certain questions in harmonic analysis and spectral theory, J. Math. Anal. Appl., 40: 336-368 (1972).

73. D. L. Russell, A unified boundary controllability theory for hyperbolic and parabolic partial differential equations, Studies Appl. Math., LII: 189-211 (1973).

74. K. D. Graham and D. L. Russell, Boundary value control of the wave equation in a spherical region, SIAM J. Contr., 13: 174-196 (1975).

75. H. O. Fattorini, Control in finite terms of differential equations in Banach space, C.P.A.M., XIX: 17-34 (1966).

76. H. O. Fattorini, The time optimal control problem in Banach spaces, Appl. Math. Optim., vol. 1, no. 2, 163-188 (1974).

77. H. O. Fattorini and D. L. Russell, Exact controllability theorems for linear parabolic equations in one space dimension, Arch. R.M.A., 43: 272-292 (1971).

78. J. Saint Jean Paulin, "Contrôle en cascade dans un problème de transmission parabolique-hyperbolique," to be published.

79. R. Triggiani, A note on complete controllability and stabilizability for linear control systems in Hilbert space, SIAM J. Contr., 12: 500-508 (1974).

80. R. Triggiani, Controllability and observability in Banach space with bounded operators, SIAM J. Contr., 13: 462-491 (1975).

81. Y. Sakawa, Controllability for partial differential equations of parabolic type, SIAM J. Contr., 12: 389-400 (1974).

82. K. Tsujioka, Remarks on controllability of second order evolution equations in Hilbert spaces, SIAM J. Contr., 8: 90-99 (1970).

83. J. Henry, thesis, Paris, 1978.

84. J. P. Kernevez and D. Thomas, in preparation.

85. J. P. Yvon, Optimal control of systems governed by variational inequalities, Fifth Conference on Optimization Techniques, vol. I, Lecture Notes in Computer Sciences, Springer, New York, 1974.

86. J. P. Yvon, Some Optimal Control Problems for Distributed Systems and Their Numerical Solutions, I.F.A.C., Boston, 1975, to be published.

87. A. Bensoussan, J. L. Lions, and R. Temam, Sur les méthodes de décomposition, de décentralisation et de coordination et applications, Cahier de l'IRIA, no. 11, 5-189 (1972). Also in the book by G. I. Marchuk and J. L. Lions (eds.), Dunod, Paris, 1975.

88. P. Lemonnier, Résolution numérique d'équations aux dérivées partielles par décomposition et coordination, Cahier de l'IRIA, 11: 191-239 (1972).

89. J. L. Lions, Perturbations singulières dans les problèmes aux limites et dans le contrôle optimal, in Lecture Notes in Mathematics, vol. 323, Springer, New York, 1973.

90. J. L. Lions, Contrôle optimal de systèmes distribués: propriétés de comparaison et perturbations singulières, in Metodi Valutativi Nella Fisica Matematica, Rome.

91. C. M. Brauner, "Perturbations singulières dans des systèmes non linéaires et applications à la biochimie," thesis, Univ. Paris-Sud, Orsay, 1975.

92. A. Jameson and R. E. O'Malley, Jr., Cheap control of the time invariant regulator, Appl. Math. Optim., vol. 1, no. 4, 337-354 (1975).

93. P. Stavroulakis, Implementation of a low sensitivity control of distributed parameter systems via an observer, Proc. I.E.E.E. Conference on Decision and Control, Phoenix, Ariz., 1974, pp. 891-896.

94. Abu El Ata, Report L.A.N., no. 189, Univ. Paris VI, 1976.

95. M. Saguez, Report Laboria, 1978.

96. J. L. Lions, Survey of Singular Perturbations in Optical Control of Distributed Systems, IFAC meeting, Warwick, 1977.

97. A. Bamberger, Report Laboria and thesis, Paris, 1978.

98. M. Sorine, Report Laboria, 1978.

Chapter 2

IDENTIFICATION OF PARAMETERS IN DISTRIBUTED SYSTEMS

R. Eugene Goodson[†]
A. A. Potter Engineering Center
Purdue University
West Lafayette, Indiana

Michael P. Polis
Department of Electrical Engineering
Ecole Polytechnique of Montreal
Montreal, Canada

I. INTRODUCTION

In this chapter the identification of parameters in partial differential equations is discussed. Parameter identification is defined as the determination from experimental data of a set of unknown parameters in a mathematical model of a physical process such that over a desired range of operating conditions the model outputs are

---

[†]Current affiliation: Institute for Interdisciplinary Engineering Studies, Purdue University, West Lafayette, Indiana.

close, in some well-defined sense, to the process outputs when the two are subject to analogous inputs. Parameters are defined as functions or constants, other than the dependent variables, which appear explicitly in the mathematical model. Distributed systems are those which require partial differential equations or other models which account explicitly for spatial variations of the dependent variables. Since most authors treat the distributed system parameter identification problem for partial differential equation models, it will be assumed unless specifically noted that the system description is given in terms of a set of partial differential equations. Several abbreviations are appropriate to shorten the exposition. They are:

DS — Distributed system
DSPIP — Distributed system parameter identification problem
PDE('s) — Partial differential equation(s)
BC('s) — Spatial boundary condition(s)
IC('s) — Initial condition(s)
ODE('s) — Ordinary differential equation(s)

It is appropriate to provide motivation for the attention given such problems. All real physical systems are, in fact, distributed. The equations modeling a given process can usually be specified from basic conservation laws along with constitutive relations within which the parameters to be determined appear. Much of the classical and modern science and engineering has been concerned with this fundamental problem. Laboratory and experimental determination of chemical rate constants, heat transfer coefficients, specific heats, electromagnetic properties, gas properties, diffusion constants, elastic moduli, strain properties, etc. is an ongoing effort throughout the scientific world. Traditionally, the measurement of such fundamental parameters has been under rigidly controlled experimental conditions. The problem of identifying parameters in systems from dynamic (time series) data can be traced back to early celestial observation problems. It has been only in the last 10 to 15 years, however, that readily usable tools have been developed which allow the determination of unknown parameters in dynamic systems using time series data. Where laboratory experimental determination of parameters under constant controlled conditions is possible, it is preferred. In modeling operational processes, however, identification methods allowing noisy time series data provide a powerful tool for analysis of complex phenomena not transferable to a laboratory environment.

Examples of applications areas abound. They include structural analysis and design where vibrations and dynamic behavior are central, acoustics problems, design in the basic process industries, where heat and mass transfer and chemical reactions are important, geophysical analysis including underground water and oil exploration, earthquake study, meteorological predictions, agricultural productivity assessments, and demographic analysis.

Although much work on the distributed system parameter identification problem (DSPIP) has been done in order to derive a control strategy, it is equally important to accurately predict process behavior should control not be practical. A case in point is the problem of weather prediction. Since all processes are by nature distributed and since our ability to simulate complex processes is continually increasing, parameter identification should become a basic tool in model building, and the DSPIP will of necessity receive greater attention as time goes on.

In this chapter the DSPIP is separated into a number of largely independent subproblems which allow the problem to be treated in a unified manner via a step-by-step approach. The steps involved in the DSPIP are the following:

1. Write the mathematical description, containing unknown parameters, of the physical system under consideration.
2. Choose a method to solve the mathematical description.
3. Decide on measurement type and location in the spatial domain.
4. Choose a criterion of performance.
5. Perform a sensitivity analysis.
6. Perform the physical experiment to obtain data.

7. Choose an optimization scheme.
8. Perform an error analysis.

The steps outlined above convert the DSPIP into a standard optimization problem where any one of a number of optimization techniques may be used.

The chapter is organized as follows. An analysis of the published literature in the field is given in Section II. Overviews discussing each of the steps outlined above are included in the section. Particular results developed by the authors or their coworkers are summarized in Section III. The list of references closes the chapter.

II. LITERATURE SURVEY AND ANALYSIS

The references cited at the end of the chapter were analyzed to determine both the problem attacked and assumptions made. The eight steps listed in the introduction for the DSPIP were defined so as to facilitate categorization of the references by problem addressed. Each reference was studied to determine the:

Parameter type (constant or variable)
Approximation method
Measurement location in space and observability assumptions
Criterion for parameter estimation
Type of error analysis
Optimization scheme
Type of data (computer generated or measured)

The results of this analysis are given in Table 1. In Table 2 a summary of the detailed analyses of Table 1 is given. A discussion of the status of the field follows.

The first attempts to treat parameter identification in distributed systems involved approximating the system by a lumped model. The method assumed the system to be linear and developed a transfer function approximating the distributed model. A pure time delay, generally accounting for mass transfer time, was introduced into the

TABLE 1

An Analysis of the Contribution of the Existing Literature to Each Step in the Solution of the DSPIP

No.	Parameter type	Approximation method	Measurement type (observability)	Criterion	Sensitivity analysis (error analysis)	Optimization scheme	Type of data
1	Constant	Galerkin's method	1 point measurement on $\partial\Omega$ (--[a])	Least square measurement error	--[a] (--[b])	Nonlinear filtering and Bayesian estimation	Real data from a vinyl acetate pilot reactor
2	Constant	Galerkin's method and finite differences	Same as [1] (--[a])	Same as [1]	--[a] (--[b])	Same as [1]; also "matrix inversion"	Same as [1]; also simulated data
11	Constant	Laplace transforms	v(x,t) (--[a])	The log of the gradient of the probability density	--[a] (--[a])	A Newton-Raphson technique is suggested. This paper is a theoretical development and its utility in practice is not considered	No examples are given
13 14 15	Constant; also F(v) is considered in [13]	Finite differences	Many point measurements on Ω. A method is suggested for choosing measurement locations based on sensitivity considerations (--[a])	Least square measurement error	Both the sensitivity of the dependent variable and the criterion to the parameters are discussed (--[b])	A modified gradient method	Simulated data
19	Constant	Finite differences	Several point measurements (--[a])	Least square measurement error	--[a] (--[b])	A first-order perturbation method is used so that the equations may be written in the form of a Kalman filter	Simulated noisy data
22 23	Constant a(x)	Characteristics	Point measurements (yes: see [60])	Mean square measurement error	--[a] (Biases are estimated, --[b])	Stochastic approximation	Simulated noisy data
24	Constant a(x), a(t)	Finite differences	Several point measurements on Ω (--[a])	Least square equation error	--[a] (--[b])	Matrix inversion and quasilinearization	Simulated deterministic data

TABLE 1 (Continued)

No.	Parameter type	Approximation method	Measurement type (observability)	Criterion	Sensitivity analysis (error analysis)	Optimization scheme	Type of data
25 26 27 28 30	Constant, a(x), a(t), a(v)	Finite differences used on the result of the optimization	Both point measurements and v(x,t) are considered. (Observability is considered for the examples based on uniqueness)	Least square measurement error	The sensitivity is discussed [25-27] (--[b])	A method based on control theory in function spaces is presented in which the adjoint state is introduced facilitating the calculation of the gradients necessary in minimizing the criterion	Simulated data [25-28]; real oil field data [26,27,30]; real aquifer data [25,26]
31	a(x)	--[c]	Functions of v(x,t) or these functions at many points in Ω and t (--[a])	Weighted least square measurement error	--[a] (--[b])	Steepest descent and nonlinear filtering	Simulated noisy data
32	Constant, a(x)	Finite differences	v(x,t) and point measurements (considers uniqueness for the examples)	Least square measurement error	--[a] (Considers errors for various noise variances)	Bayesian estimation and nonlinear filtering	Simulated noisy and deterministic data
34	Constant	Eigenfunctions	Many point measurements (yes)	Least square measurement error	--[a] (--[b])	Least squares and Bayesian estimation	Simulated data
35	Constant	Finite differences	Many point measurements (--[a])	Equation error	--[a] (--[b])	Matrix inversion	Simulated data
38 39	Constant, a(x)	Separation of variables and numerical integration	Resonance frequencies (--[a])	Least square error between the measured and calculated resonance frequencies	--[a] (--[b])	A method is proposed which transforms the problem into an ODE problem with x as the independent variable. Then a least-squares method is used to minimize the criterion	Simulated noisy data. Also real data found in the literature

43	Constant, a(x)	Power series and Chebychev polynomials	It would seem that v(x,t) would be necessary (--[a])	Equation error	--[a] (--[a])	Successive integration to yield a system of algebraic equations in the parameters. Then matrix inversion to find the parameters	No examples
44	Constant	Finite differences of the result of the optimization	Many point measurements in Ω (--[a])	Least square measurement error	--[a] (--[b])	The epsilon method	Simulated noisy data
45	Constant	A modal technique is used and the optimization gives the modes as functions of x and t	Many point measurements in Ω (--[a])	Least square measurement error	--[a] (--[b])	The epsilon method	Simulated noisy data
46 68	a(t) appearing in both the equation and BCs for a specific parabolic PDE	Polynomials on each of K segments of $t \in [0,T]$	Boundary measurements (uniqueness and existence are considered	--[a]	--[a] (Estimates of the error are given)	--[a]	--[a]
50	Constant, a(t)	The method of lines	Several point measurements (--[a])	Equation error	--[a] (--[b])	Integration by parts and multiplication by a weighting function yields an algebraic equation in the parameters	Simulated data
54 55 56	Constants	Fourier transforms	Several point measurements (--[a])	Least square measurement error	--[a] (Yes)	Regression analysis	Real data

TABLE 1 (Continued)

No.	Parameter type	Approximation method	Measurement type (observability)	Criterion	Sensitivity analysis (error analysis)	Optimization scheme	Type of data
63	Constant, a(t)	Finite differences	Many point measurements (--[a])	Weighted equation error	--[a] (Yes)	Matrix inversion and nonlinear programming	Simulated noisy data; also real data from a test beam
65	Constant a(t), c(t)	Finite differences	Many point measurements also functions of v(x,t) (--[a])	Weighted least square measurement + equation + boundary + initial errors	--[a] (--[b])	Nonlinear filtering (see [42])	Simulated noisy data
77	Constants in the PDEs describing the evolution of BO and BOD in polluted streams	See [23]	See [23]	See [23]	--[a] (--[b])	See [23]	Real data taken from the literature
78	Estimation of the Green's function	Chebyshev polynomials	Several point measurements (--[a])	Measurement error	--[a] (--[b])	Integral transforms and regression	Real data from a furnace in a hot strip mill
81 82 83	Constant, states	Finite differences of the filter equation	v(x,t) or many point measurements in Ω	Weighted least square measurement, equation and boundary error	--[a] (--[b])	Nonlinear filtering	Simulated noisy data
85	Constant	Approximate Laplace transforms	Several point measurements in Ω (--[a])	The likelihood function	Sensitivity to model order is discussed (yes)	Maximum likelihood	Real data taken on a test rod. Also simulated data

86	Constant	Finite differences	Several point measurements in Ω (--[a])	See note in column 7	See note in column 7	Note: This paper compares three methods [63,85, 125] which may be used to identify the diffusivity of a test rod	Real data taken on test rod
87	a(x) in a nonlinear elliptic problem	Finite differences	v(x), also boundary measurements. (Establishes uniqueness)	Weighted least square measurement error	--[a] (Studies the effect of noise on the estimates)	See [25,100]	Simulated data
91	Constant	Method of lines	Several point measurements in Ω (--[a])	Equation error	--[a] (Discusses errors and convergence rates)	Newton-Raphson expansions yield a linear ODE boundary value problem	Very accurate simulated data
92 93	Constant	See [22,23]	Boundary measurements (--[a])	Least square measurement error	Discusses convergence of the algorithms as a function of the parameter values (--[b])	Quasilinearization	Simulated data. The examples are of interest
94	Constant, states	The emphasis is on modeling and reducing the PDEs to a set of ODEs using quadratic splines	Several point measurements in Ω (--[a])	--[c]	--[a] (--[b])	A filter based on "mean value linearization"	Simulated data for a catalytic reactor which distinguishes between fluid and solid states
100	Constant, a(x)	Modal approximations	Many point measurements in Ω (--[a])	Least square equation error	--[a] (--[b])	Integration by parts to yield a set of algebraic equations in the measurements and the parameters	Experimental data on a beam. Also simulated data
104 105 106 107	Constant	Galerkin's method	A single point measurement in Ω (N-mode observa= bility, see [57])	Least square measurement error	Considered in examples (--[b])	Nonlinear filtering [104, 105,107]; gradient [104,105,107]; direct search [104,107]	Simulated noisy data. Also real data from a test beam [104, 106]

TABLE 1 (Continued)

No.	Parameter type	Approximation method	Measurement type (observability)	Criterion	Sensitivity analysis (error analysis)	Optimization scheme	Type of data
109	Constant	Finite differences	Several point measurements in Ω (--[a])	Weighted least square measurement error	Yes (considers rates of convergence of the algorithm for various ICs)	An algorithm is given which is based on sensitivity theory	Simulated data
110	States	Finite differences	Many point measurements in Ω (--[a])	Weighted least square measurement + equation + initial error	--[a] (--[b])	Nonlinear filtering similar to [82]	Simulated data
112	Periodic functions of space and time	The use of power series is suggested to solve the result of the optimization over small subdomains of Ω	v(x,t), or many point measurements (--[a])	The mean square frequency domain equation error	--[a] (--[a])	Fourier transforms are used to yield integral equations in the frequency domain which must be solved to yield the parameters	No examples
113 114	Constants in the assumed solution	The dependent variables are assumed to be represented by a set of unknown constants multiplied by orthogonal polynomials in both x and t	Many point measurements (--[a])	Mean square measurement error	--[a] (--[b])	Stochastic approximation	Simulated noisy data
116	Constant	Finite differences	Several point measurements. Also scanned measurements (considers initial state observability)	Least square measurement error	Calculates the sensitivity of v(x,t) to parameter variation (--[b])	Steepest descent	Simulated deterministic data

117	Constant	Finite differences	Measurement of v(x,t) (yes)	Least square measurement + equation error	Yes (--[b])	Nonlinear filtering	Simulated noisy data
118	a(x,t) and parameters which determine $\partial\Omega$	--[c]	Functions of v(x,t) and also these functions at several points in Ω (--[a])	Weighted least square measurement error	--[a] (--[a])	Necessary conditions for optimality are derived for the two types of parameters considered. Steepest descent and conjugate gradient algorithms are suggested. The methods are essentially the same as those employed in [25, 30]	No examples
119	Constant	Finite differences	Many point measurements (yes)	Least square measurement error	--[a] (yes)	Steepest descent, quasilinearization	Simulated noisy data
120	Constant and a(t)	Finite differences in x and an approximate form of the covariance matrix	Many point measurements (--[a])	Least square measurement + equation error	--[a] (--[b])	Nonlinear filtering (see [42])	Simulated noisy data
122	Constants and states	--[c]	Several point measurements at discrete times (--[a])	Weighted least square measurement + equation + initial errors	--[a] (--[a])	Nonlinear filtering	Simulated noisy data
124	Constant	Cubic splines	Many point measurements (--[a])	Least square equation error	--[a] (--[a])	Splines are used to transform the DS into a set of ODEs. Then a recursive least squares procedure is used to determine the parameters	Simulated noisy data

TABLE 1 (Continued)

No.	Parameter type	Approximation method	Measurement type (observability)	Criterion	Sensitivity analysis (error analysis)	Optimization scheme	Type of data
131	Constant	Finite difference in time	$v(x,t)$, spatial derivatives of $v(x,t)$ and the inputs. Also these functions at many measurement points in Ω (--[a])	Least square equation error	--[a] (--[b])	The PDE is multiplied by a spatial weighting function and integrated over the spatial domain yielding a set of difference equations in t. The method is similar to that used in [100]	Simulated noisy data
133	$a(x)$	--[c]	$v(x,t)$ and all spatial derivatives of $v(x,t)$ or only $v(x,t)$ (--[a])	Weighted least square equation error	--[a] (--[a])	For DSs which are continuous in x, discrete in time and can be modeled by a linear regression equation, an algorithm is given in which a gradient method and matrix inversion is used to determine the parameters	No examples
135	States	Finite differences	$v(x,t)$ (--[a])	The likelihood function	--[a] (--[b])	Differential dynamic programming is used to derive the nonlinear filter equations	Simulated noisy data
137	$a(x)$ and constants	Finite differences	Scanned measurements (yes)	Least square measurement error	Yes (yes)	Matrix inversion	Real data
138	$a(\underline{x},v)$, at $\underline{x}_i$ and $\partial\Omega$	Finite differences	Measurements at several points in Ω are used to obtain values at the grid points. (Not specifically considered, however process knowledge is used to reject extraneous solutions)	Least square measurement error	--[a] (--[b])	The maximum principle is used and the adjoint equations are derived. Then steepest descent combined with a heuristic method is used to identify the parameters	Real data. The observed water table data

142 143	Constants in nonlinear BCs	Laplace transforms and infinite product expansions	A few point measurements in Ω (--[a])	Least square measurement error	--[a] (--[b])	Regression analysis is used to yield a set of algebraic equations in the parameters	Simulated noisy data. Also real data from a test beam
144	A pointwise approximation to the Green's function	Finite differences	Many point measurements in Ω (--[a])	Mean square measurement error	--[a] (--[b])	Stochastic approximation	Simulated noisy data
146	Constant	Finite differences	Many point measurements (--[a])	The mean square measurement error	--[a] (--[b])	Stochastic approximation	Simulated noisy data
147	Constant	Finite differences	Many point measurements (--[a])	The likelihood function. Also the mean square measurement error	--[a] (Yes)	Decision theory, maximum likelihood and stochastic approximation	Simulated noisy data

References Which Present Reviews or Summaries of Results Noted in Above

Ref. no.	Remarks
29	A resume of some of the work done by the authors. See [25,26,27,28,32].
33	This paper reviews some of the methods used for parameter identification in vibratory systems. It is basically an overview of the monograph <u>System Identification of Vibrating Structures</u> [103]. See also [34].
58	A review of some of the work done by the author and his coworkers. See [60,105,142].
61	A preliminary version of the present paper.
108	A broadranging review of the work done in the U.S.S.R. on identification. Included is an extensive review of the DSPIP work and a list of references. See [78,109,147].
121	This paper is a summary of much of the DS work done at L.A.A.S. at Toulouse, France. See [1,2].
148	This paper presents a summary of the work done by the authors. See [147,148].

TABLE 1 (Continued)

Ref. no.	Remarks
References Treating Observability or Sensor Location as a Major Topic	
21	This paper gives error bounds on the state estimates for the diffusion equation. A single measurement transducer is used and its location in Ω is based on observability and accuracy requirements in the state estimation.
40 41	Questions of observability are considered for certain classes of DSs.
49	This paper presents a method for choosing measurement locations based on a variational formulation. However, the observability of the system considered is not commented upon.
60	A discussion of observability questions for DSs. Observability definitions are given which do not require recovery of the initial state. Also an N-mode observability definition is given.
66	A method is advanced for the determination of the distance between measurement locations. The method is based on the statistical characteristics of the DS and a desire to estimate the states at points other than sensor locations to a given accuracy. However, observability questions are not considered.
69	The author uses an optimization method to determine the "optimal" heat sensor configuration. The sensor location and control problems are considered together. The approach used does not take into account observability considerations.
73	This paper discusses the relation between observability, sensor location, non well-posed problems, and state estimation for DSs. An example is presented and error bounds are derived.
74	A discussion of observability problems for DSs. See also [60].
75	A discussion of observability of the wave equation.
95	A part of the paper considers the determination of "optimal" inputs for the DSPIP. The information matrix is defined and an optimization method is suggested to yield the inputs which maximize a norm of this matrix. Measurements are assumed available at a number of points in Ω and, although not specially mentioned, it would be necessary to insure observability.
98 99, 115	These papers give an excellent discussion of problems associated with observability and state estimation for diffusion processes.
127	Necessary conditions are derived for initial state observability for a class of hyperbolic DSs. See also [45].
139	It is noted here that the observability definition given in [60] reduces to the usual definition for lumped systems. In the author's reply the inadequacy of initial state observability for DSs is demonstrated through an example.
140 141	These papers discuss the control problem for DSs. In addition an observability definition is proposed for DSs which is an extension of the usual definition for lumped systems. The definition requires the recovery of the initial state from the measurements and BCs.
145	Necessary and sufficient conditions are presented for initial state observability of a restricted class of hyperbolic DSs.

Others (Mathematics, PDEs, etc.)[d]

4	A survey of the parameter identification problem for lumped systems.
5	A discussion of many problems associated with identification, estimation, and control of DSs.
16	This paper discusses the determination of the order of the auto regressive moving average representation of a DS. The results are applied to demographic data.
20	A review of work done in the U.S.S.R. on the control of DSs. An extensive list of references is given and some discussion of the DSPIP is included.
42	A nonlinear filter is derived for lumped systems. This filter has been extended to DSs in [116,105].
47, 48	These papers give identification methods for lumped systems.
52	A review of weighted residual methods for solving PDEs.
53	An excellent reference on finite difference methods for solving PDEs.
57	A review of infinite product expansions which may be used to simulate the dynamic response of DSs.
59	Discusses control and modeling of DSs.
67	It is suggested that the identification problem may be transformed into a network matching problem through the use of Laplace transforms.
70	Gives an observability definition for lumped systems. This definition is now considered to be the "usual" observability definition.
71	This paper uses lumped models to approximate heat processes and discusses the assumptions required and errors introduced.
88	A basic reference for control work in DSs.
89	A review of many results on the control of DSs. See also [88].
90	Gives some results on existence and uniqueness for DSs.
123	Discusses the modeling of DSs by using various interconnections of lumped blocks.
125	Gives an experimental method for determining diffusivity.

---

[a] The subject is not considered.

[b] The subject is not expressly considered, although examples with known parameter values are used to indirectly consider the errors.

[c] Not specifically mentioned.

[d] References [36,62,72,84,128] are not specifically listed since they are basic mathematical references for distributed system work.

TABLE 2

Summary of Analysis of References

The data below are useful only as an indication of the status of the field and the major techniques used. Since many authors use more than one technique the totals are not necessarily consistent.

Major Subject Treated	Totals
1. The DSPIP [1,2,11,13-15,19,22-35,38,39,43-46,50,54-56, 58,61,63,65,68,77,78,81-83,85-87,91-94,100,104-110, 112-114,116-122,124,131,133,135,137,138,142-144,146-148]	78
2. State estimation for linear DS's [3,6-10,12,17,51,64, 76,79,80,96,97,101,102,111,126,129,130,132,134,136]	24
3. Observability and sensor location [18,21,40,41,49,60, 66,69,73-75,95,98,99,115,127,139-141,145]	20
4. Others: PDEs, mathematics, etc. [4,5,16,20,36,37,42, 47,48,52,53,57,59,62,67,70-72,84,88-90,103,123,125, 128]	26
Measurements, Data, and Parameters for References Treating the DSPIP	
1. Experimental data [1,2,25-27,30,38,39,54-56,58,63,77, 78,85,86,100,104,106,137,138,142,143]	24
2. Simulated data [2,13-15,19,22-29,31-35,38,39,44,45,50, 58,63,65,81-83,85,87,91-94,100,104,105,107-110,113, 114,116,117,119,120,122,124,131,135,142-144,146-148]	58
3. Variable parameters [13,22-32,38,39,43,46,50,63,65,68, 87,100,112,118,120,133,137,138]	28
4. Consider observability [22,23,25-30,32,34,46,58,68,77, 104-107,116,117,119,137,138]	23
Solution Techniques	
1. Finite differences [2,13-15,19,24-30,32,33,35,44,58, 63,65,81-83,86,87,108,109-110,116,117,119,120,131,135, 137,138,144,146-148]	40
2. Modal techniques [1,2,33,34,38,39,43,45,46,58,78,100, 104-107,112-114,142,143]	21
3. Characteristics [22,23,77,92,93]	5
4. Laplace or Fourier transforms [11,54-56,85]	5
5. Others [50,91,94,124]	4
Optimization techniques	
1. Nonlinear filtering [1,2,31,32,65,81-83,104,105,107, 110,117,120,122,135]	16
2. Stochastic approximation [22,23,77,113,114,144,146,147]	8
3. Gradient methods [13-15,25-31,50,87,100,104,105,107, 116,118,119,131,133,138	22

TABLE 2 (Continued)

	Totals
4. Bayesian or maximum likelihood [1,2,32,34,85,86,135, 147]	8
5. Least squares [34,38,39,124]	4
6. Direct search [104-107]	4
7. Regression or matrix inversion [2,24,35,43,63,78,86, 142,143]	9
8. Quasilinearization [24,92,93,119]	4
9. Others [11,19,44,45,63,91,94,109,112,118]	10
Consider Sensitivity [13-15,25-27,85,104-107,109,116]	13

transfer function. The form of the transfer function [47] used a gain (K), one or more lags (τ_i) (usually accounting for diffusion), and a time delay (T_d). The following type of model resulted:

$$G(s) = \frac{Ke^{-T_d s}}{(1 + \tau_1 s)(1 + \tau_2 s)} \tag{1}$$

Once such a model had been chosen, the parameters (K, T_d, τ_1, τ_2) could be identified using methods developed for lumped systems. Gilath et al. [55] and Gilath and Stuhl [56] use this technique in applications to environmental problems. Such techniques parallel identification methods for lumped systems described by ordinary differential equations (ODEs).

The major body of literature treating the parameter identification problem for lumped systems is summarized in an excellent review paper by Astrom and Eykhoff [4] and as such will not be treated here.

A wide range of methods have been proposed to identify the parameters in distributed systems (DS's). Douglass and Jones [46] and Jones [68] considered the identification of a time-varying parameter appearing simultaneously in a parabolic partial differential equation (PDE) and in one of the spatial boundary conditions (BCs); the problem effectively reduced to the indentification of the BC. Beck [13-15] used finite differences and an approximate Taylor series expansion to derive an iterative parameter identification procedure for the heat equation. Perdreauville and Goodson [100] used integration by parts to reduce the PDEs to a set of algebraic equations in the measurements and the parameters. Fairman and Shen [50] modified Perdreauville and Goodson's method to avoid the spatial integration; also, time integration was accomplished using Poisson filter chains at the locations dictated by the finite difference scheme used to approximate the spatial derivatives. Zhivoglyadov and Kaipov [146] presented a method using stochastic approximation to identify the parameters. A method of identifying constant parameters appearing linearly in a set of PDEs was presented by Collins and Khatri [35]; the method requires the inversion of a matrix obtained by finite differencing the PDEs. Measurement types and error criteria for the DSPIP were discussed by Seinfeld [116]; also, the concept of sensitivity functions (adjoint variables) for DS's was introduced and used in a steepest descent algorithm [31,119]. In addition, quasilinearization was used for nonsequential parameter estimation in DSs. Chavent [25-28] approached the DSPIP using an optimal control formulation. The introduction of the adjoint variable and the simultaneous solution of the system and adjoint equations allows the computation of the gradient of the performance criterion. A steepest descent algorithm was proposed as an optimization scheme. The approach is particularly advantageous when identifying variable parameters. Vemuri and Karplus [138] considered the identification of variable parameters in a PDE describing ground water flow. The formulation is similar to that given by Chavent [25], and a steepest descent algorithm was proposed for identifying the parameters on a hybrid computer.

Carpenter, Wozny, and Goodson [23] used the method of characteristics and stochastic approximation to yield parameter estimates for a class of hyperbolic systems. Malpani and Donnelly [92,93] used characteristics as suggested in [23] along with quasilinearization to identify parameters in one- and two-phase flow processes.

Koivo and Phillips [77] applied the methods of [23] to identify the parameters in the equations describing the evolution of dissolved oxygen (DO) and biological oxygen demand (BOD) concentrations in polluted streams. Chaudhuri [24] resolved the DSPIP using a method based on differential approximation and quasilinearization. Di Pillo and Grippo [44] have extended the ϵ-method to the DSPIP. Kozhinsky and Rajbman [78] consider the parameters and dependent variables to be components of a vector random field. A set of multidimensional integral equations is derived which must be solved to yield an optimal least squares error estimate of the operator which maps the field into its various components. Leden, Hamza, and Sheirah [86] presented a comparative study of three methods [63,85,125] used in determining the diffusivity of a rod. Ward and Goodson [143] treated the problem of identifying constant parameters appearing linearly in nonlinear BCs for DS's with linear fields.

Collins, Young, and Kiefling [34] discussed identification techniques applied to problems of shock and vibration; particular emphasis was given to a method based on Bayesian estimation. Dale and Cohen [39] considered the problem of identifying constant parameters in linear vibratory systems where the steady-state equations may be reduced to a set of ODEs in the spatial variable. The identification is accomplished using steady-state frequency response data near resonance frequencies.

Saridis and Badavas [114] approximated the solution to a linear DS by a finite series of orthogonal functions. The constant parameters multiplying these functions were then identified using stochastic approximation. Ruban [109] proposed an identification method using sensitivity functions which is similar to that proposed by Seinfeld [116]. Wozny, Carpenter, and Stein [144] identified the Green's function for a class of time-invariant linear DS's. Stochastic approximation was used to yield a pointwise approximation of the Green's function. Polis, Goodson, and Wozny [105] used Galerkin's method [52] to reduce the PDEs describing a distributed system to a set of ODEs. The constant parameters were then identified using standard methods for ODEs. Luckinbill and Childs [91] reduced the PDEs to a set of ODEs by finite differencing in all but one of the independent variables. A Newton-Raphson-Kantorovich expansion was used to solve the resulting linear boundary value problems. Hamza and Sheirah [63] used finite difference methods to approximate the partial derivatives in the system equations. A weighted equation error criterion was minimized using matrix inversion for the case of constant parameters and nonlinear programming for the variable parameter case.

The sequential estimation of states and parameters in nonlinear DS's has been considered by Seinfeld and coworkers [31,65,117,120]. Lamont and Kumar [83] and Sahgal and Webb [110] used invariant imbedding, and filters were derived under various assumptions on the disturbances. Tzafestas and Nightingale [135] used differential dynamic programming to derive their filter for nonlinear DS's. Many papers have appeared (see Table 2) which consider problems of state estimation and construction of optimal filters for linear DS's. These works have been referenced for completeness; however, they are of little utility in treating the DSPIP since in most cases considering the parameters as states makes the problem nonlinear.

Several papers which do not deal directly with parameter identification are pertinent. Athans [5] discusses some of the problems inherent in DS's. Lions [89] notes that the DSPIP may be considered in the context of the more general control problem. Butkovskiy [20] reviews research on the DS control problem carried out in the U.S.S.R. and gives several references which treat the DSPIP.

Several review papers have appeared [29,33,58,108,121,148]; however, they are generally limited in scope and have summarized work done by the authors and their coworkers. These are noted in Table 1. Upon reading the literature it is evident that a wide range of optimization schemes and approximation techniques have been proposed for use in solving the DSPIP. The majority of papers concentrate on the optimization scheme or on the method for solving the equations describing the DS. Application papers are just beginning to become available. Tables 1 and 2 support these conclusions.

In developing methods for solving other than the simplest DSPIP, one is faced sooner or later with the necessity of approximation. Even when analytic solutions are available they are often in the form of an infinite series which must be approximated with a finite number of terms. Two problems are important in this step: The first is at what point in the solution procedure to introduce an approximation, and the second is the choice of approximation method.

Athans [5] suggests that any approximation should be made as late as possible in order to retain the distributed nature of the process until numerical results are required. Retaining the distributed nature of the process usually results in additional sets of PDEs or integral equations, however, which must then be approximated. These equations are generally of such complexity that process knowledge and engineering judgment are of little aid in choosing an approximation method. A major advantage of early approximation is that knowledge of the process behavior may be more readily incorporated into the approximation procedure. Also, early approximating may yield algorithms which are less complex than those derived by retaining the distributed nature of the process. In addition, certain approximation methods lead naturally to specific optimization schemes. Zhivoglyadov and Kaipov [146], for example, use a standard finite difference method which readily lends itself to their recursive stochastic approximation scheme. Of greater importance than when to approximate may be the fact that certain approximations at any stage may yield invalid results. For example, finite difference approximation of hyperbolic equations can completely mask the pure time delays inherent in such systems. Furthermore, in acoustic and vibratory systems, the number of modes which are candidates for excitation may be so great that drastic simplification is required before the problem is at all tractable.

Approximate solutions generally involve computer simulation of the DS, and the majority of the optimization schemes require repeated solution of the system equations to yield parameter estimates. Thus, an important factor in the choice of approximation method is the amount of simulation time required to yield an approximate solution. Even with modern high-speed computers large amounts of simulation time may be required for distributed problems. Several approximation methods are available to the investigator faced with a DSPIP. The various methods used by different authors are given in Table 2.

The most widely used methods for solving the DS model are finite difference methods. The goal of all finite difference methods is to transform the distributed equations into a set of difference equations. A variety of implicit and explicit methods are available which have been discussed in detail in the literature [53]. Although finite difference methods are a powerful tool, they should be used with care. Several points may be brought out concerning their use:

The use of simplistic finite difference schemes may require excessive computer time and storage due to stability requirements imposed on the resulting difference equations.

Higher-order approximations are often necessary at the boundaries in order to maintain a desired accuracy in the solution.

In certain problems, such as those encountered in boundary layer theory, the finite difference approximations do not converge to the solution of the distributed equations.

Care must be taken so that the finite difference approximations do not violate mass, energy, or momentum conservation.

Modal approximation techniques have been used by several authors [1,100,105,113]. These techniques are based on the assumption of an approximate solution to the components of $\underline{v}$ in Eq. (8) of the form [72]:

$$\tilde{v}_i(\underline{x},t) = \varphi_0(\underline{x},t) + \sum_{j=1}^{N} F_j(\underline{x})\varphi_j(t) \tag{2}$$

where $\varphi_0(\underline{x},t)$ is a function whose form may be chosen to satisfy BCs and/or forcing functions, or to use asymptotic solutions to yield better approximations. The $F_j(\underline{x})$'s are a set of "approximating" or basis functions which are chosen by the investigator as known functions of $\underline{x}$. The $\varphi_j(t)$'s are a set of time functions and N is the number of terms in the approximation; both the $\varphi_j(t)$'s and N must be determined. Usually, several sets of approximating functions are admissible, and the set chosen will depend on experience and familiarity with the particular problem being considered. It is often useful to choose the approximating functions so that they satisfy the BCs exactly and are members of a complete orthonormal set of functions on the spatial domain Ω.

Once the functions $\varphi_0(\underline{x},t)$ and $F_j(x)$, $j = 1, \ldots, N$ have been chosen, one of a number of weighted residual criterion [52] may be used to transform the DS into a set of ODEs. Polis, Goodson, and Wozny [105] and Aguilar Martin et al. [2] have used Galerkin's criterion in considering the DSPIP. The method is outlined below for the diffusion equation with a constant parameter. The equation and initial residuals are defined respectively as:

$$
\begin{aligned}
R_E &= \frac{\partial \tilde{v}}{\partial t} - a\frac{\partial^2 \tilde{v}}{\partial x^2} - u(x,t) \\
R_I &= \tilde{v}^0 - v^0(x)
\end{aligned}
\tag{3}
$$

The residual is the amount by which the approximate solution differs from the exact solution, i.e., if $\tilde{v} \equiv v$, then $R_E, R_I \equiv 0$. It is desirable, in some sense, to reduce the residual to zero. To do this the residual is distributed over the spatial domain by using a weighting function $W_i(x)$ such that

$$
\int_\Omega R_E W_i(x)\, dx = 0 \qquad \text{and} \qquad \int_\Omega R_I W_i(x)\, dx = 0 \qquad i = 1, \ldots, N \tag{4}
$$

The weighting functions must now be chosen. There are several methods of picking weighting functions. Galerkin's method, which is equivalent to orthogonalizing the residual with respect to the approximating functions, is to choose:

$$
W_i(x) = F_j(x) \qquad i = j = 1, \ldots, N \tag{5}
$$

The result of a specific choice of approximating functions is a system of coupled first-order ODEs in the $\varphi_j(t)$ and the unknown parameters. A disadvantage of this method is that the choice of approximating functions is not unique, and a poor choice yields poor approximations. Since the choice is left to the investigator, however, process knowledge and engineering judgment may be incorporated. Another problem with Galerkin's method is that increasing N does not insure a better approximation if the $F_j(x)$'s are not well chosen. Despite these drawbacks, results [2, 104-107] have shown that this method may be used to yield satisfactory approximations with relatively few terms in the summation of Eq. (2).

Under appropriate conditions on the matrices in Eq. (8), the Laplace transform may be used to yield the exact solution $\underline{v}(\underline{x},s)$, where s is the Laplace variable. However, taking the inverse Laplace transform may be extremely difficult. Goodson [57] has proposed the use of infinite product expansions to yield approximations $\tilde{\underline{v}}(\underline{x},s)$, which can easily be transformed to give the time response. This method is used by Ward and Goodson [143] in identifying nonlinear boundary conditions in problems where the PDE is linear.

The method of characteristics [36] may be used to find exact solutions for an important class of DS problems, i.e., those of hyperbolic type. To illustrate this method consider the hyperbolic PDE,

$$\frac{\partial v}{\partial t}(x,t) + a_1 \frac{\partial v}{\partial x} + a_2 v(x,t) = 0 \tag{6}$$

The characteristic equations for (6) are:

$$\begin{aligned} \frac{dx}{dt} &= a_1 \\ \frac{dv}{dt} &= -a_2 v \end{aligned} \tag{7}$$

If the value of v is available at two points x_1 and x_2 in the spatial domain, the characteristic Eqs. (7) may be used to extend the data recorded at x_1 to the second point x_2. This extended or "predicted" value may then be compared with the measured value of v at the point x_2, an error formed, and the values of a_1 and a_2 chosen to minimize this error. Carpenter, Wozny, and Goodson [23] use the method for noisy measurement data and for various restrictions on the parameters.

If analytic solutions are available, they may be used to yield exact solutions to the DS problem. Analytic solutions are usually not available for most problems of interest. In addition, the analytic solution is often in the form of an infinite series which must be approximated with a finite number of terms. Green's functions and other similar methods convert a PDE to an integral equation which may be solved with special techniques.

It is necessary to hypothesize an equation model for the DSPIP. The models will generally include a set of BCs, initial conditions (ICs), and perhaps random inputs representing noise or modeling errors. Many different model forms have been proposed for DSs, and each has utility for different tasks. A form using only first-order differentials which arises directly from fundamental physical axioms is as follows:

$$A_0(\underline{v},\underline{x},t,\underline{a}) \frac{\partial \underline{v}}{\partial t}(\underline{x},t) + \sum_{i=1}^{n} A_i(\underline{v},\underline{x},t,\underline{a}) \frac{\partial \underline{v}}{\partial x_i}(\underline{x},t) = \underline{B}(\underline{v},\underline{x},t,\underline{a},\underline{u}) + \underline{\eta}(\underline{x},\underline{t}) \tag{8}$$

where $\underline{v}(\underline{x},t)$ is an r-vector of dependent variables, $\underline{x}$ is an n-vector of spatial variables defined on the domain Ω with components x_i, t is time, $\underline{a}(\underline{x},t)$ is a q-vector of unknown parameters to be determined, A_i $i = 0,1, \ldots ,n$ are matrices, $\underline{B}$ is a vector functional, $\underline{\eta}(\underline{x},t)$ represents unknown inputs or equation errors, and $\underline{u}(\underline{x},t)$ is a vector of controls or known forcing functions. It should be noted that A_0 is not necessarily of full rank, i.e., rank $(A_0) = r_1 \le r$.

The boundary conditions are represented by:

$$\underline{C}(\underline{v},\underline{x},t,\underline{c}) + \underline{\eta}_b = 0 \qquad \underline{x} \in \partial\Omega \tag{9}$$

where $\underline{C}$ is an m-vector defining the relation between the variables on the boundary $\partial\Omega$ of the spatial domain, $c(\underline{x},t)$ is a vector of unknown parameters, and $\underline{\eta}_b$ represents the boundary errors. It should be noted that higher-order PDEs may be expressed in the form of Eq. (8) but that systems of first-order PDEs cannot always be expressed as a single higher-order PDE [36].

In addition to the BCs, ICs (perhaps unknown) are given as:

$$\underline{v}(\underline{x},0) = \underline{v}^{o}(\underline{x}) \tag{10}$$

where only r_1 of the components of $\underline{v}$ need be specified at $t = 0$.

The measurements are of the form:

$$\underline{z}(t) = \int_{\Omega_s} \underline{D}(\underline{v},\underline{x},t)\, d\underline{x} + \underline{\zeta}(t) \tag{11}$$

where $\underline{\zeta}(t)$ represents the measurement errors and $\underline{D}$ is a vector function mapping the dependent variables into the measurement vector $\underline{z}$ through the integration over some time-varying spatial domain $\Omega_s(t)$. The integration is important since every real transducer averages over some portion of the spatial domain.

Once a model is chosen, the type of forcing functions for identification must be decided upon. These functions are usually applied on the boundary or at specific points in the spatial domain. It is also necessary to characterize any errors which may be assumed present in the mathematical description. To complete the system representation any constraints on the parameters, forcing functions or dependent variables should be specified.

A performance criterion for the DSPIP is given by

$$J(\underline{a},\underline{c},t_f) = \int_0^{t_f} \int_{\Omega} \underline{E}(\underline{x},t)^T H(\underline{x},t) \underline{E}(x,t)\, d\underline{x}\, dt \tag{12}$$

where $\underline{E}(x,t)$ is an error vector whose components are the equation, boundary, and measurement errors of Eqs. (8), (9), and (11), respectively. Many authors who derive nonlinear filters for DSs use this criterion [65,81-83,110,120]. However, most researchers who treat the DSPIP consider only the measurement error which is the difference between the computed and observed results. Seinfeld [116] discusses the measurement error criterion in some detail, and Chavent [25-27] derives the adjoint equations for the least square measurement error criterion. Some authors [24,50,63,100] consider only the equation error. The use of the equation error facilitates the use of variational methods since the parameters appear explicitly in the performance criterion.

The choice of the weighting matrix $H(\underline{x},t)$ is dependent upon the assumptions made on the problem under consideration (for example, negligible equation error), the preference of the investigator, and the complexity of the resulting optimization schemes. Care should be taken in the choice of $H(\underline{x},t)$ to insure that one component of the error does not dominate the others.

Two types of sensitivity have been considered in the DSPIP. The first is the sensitivity of the dependent variables to the parameters $[(\partial v/\partial \underline{a}),(\partial v/\partial \underline{c})]$. The second is the sensitivity of the performance criterion to the parameters $[(\partial J/\partial \underline{a}),(\partial J/\partial \underline{c})]$.

The sensitivity of the dependent variables to the parameters is closely related to questions of both observability and parameter identifiability for DSs. Seinfeld [116] has derived the equations describing the evolution of the sensitivity functions. Both Seinfeld [116] and Beck [14] suggest that observability and sensor location might be based on sensitivity considerations. Mehra [95] suggests a method for calculating "optimal" inputs for the DSPIP based on the sensitivity functions.

Hamza and Sheirah [63] identify the parameter $a_2 = 0$ in the diffusion equation

$$\frac{\partial v}{\partial t} - a_1 \frac{\partial^2 v}{\partial x^2} + a_2 \frac{\partial v}{\partial x} = 0$$

The eigenvalues for the solution satisfy

$$\lambda_n^2 = a_1\left[\theta_n^2 + \left(\frac{a_2}{2a_1}\right)^2\right]$$

$$\theta_n + \frac{a_2}{2a_1} \tan \theta_n = 0$$

For small a_2 ($a_2/a_1 < 10$, for example) the solution is very insensitive to a_2 so that very accurate data would be required to successfully determine a_2.

The sensitivity of the performance criterion to the parameters is dependent on the weighting matrix $H(\underline{x},t)$ in Eq. (12). This matrix is chosen by the investigator and judicious choice of $H(\underline{x},t)$ can aid in the optimization since the speed of convergence is related to the sensitivity of the performance criterion to the parameters.

The problem of proper choice of variables for measurement for identification is common to both distributed systems and those governed by ODEs. In DSs, however, the location of the transducers in the spatial domain must be chosen. Practical considerations often place restrictions on where measurements may be taken, but the combination of variables and locations which are feasible allow a wide range of choice for most DSs. For the identification problem an observability theory is required to insure that there is sufficient information in the measurement data to identify the relevant parameters and variables.

The observability question reduces to the question of determining the vector $D(\underline{v},\underline{x},t)$ in Eq. (11). This vector is entirely free except for physical restrictions on transducers and spatial locations. The vector $D(\underline{v},\underline{x},t)$ should be chosen so as to provide sufficient information in $z(t)$ of Eq. (11) to determine uniquely those unknowns desired among the parameters $\underline{a}$ and $\underline{c}$ and the system dependent variables $v(\underline{x},t)$ in a specified space-time domain. When only the parameters $\underline{a}$ and $\underline{c}$ are desired, the observability question is less severe than when the dependent variable is to be identified in some space-time domain. A measurement location could be chosen where the data would not contain all the information necessary to determine the dependent variable but would contain sufficient information to determine the parameters.

General observability results are not available; however, several specific results have been published. Two observability definitions have been proposed. The first, proposed by Wang [140,141], is concerned with the recovery of the initial state. This reduces to the ability to determine r_1 of the components of $\underline{v}^0(\underline{x})$ in Eq. (10). The second, offered by Goodson and Klein [60], is based on uniqueness of solutions. This definition allows an arbitrary space-time domain to be specified over which the vector $\underline{v}(\underline{x},t)$ must be uniquely determined by the measurement $\underline{z}(t)$ and those BCs available. Specific results are also offered in this paper. Yu and Seinfeld [145] and Thowsen and Perkins [127] followed Wang's definition and derived necessary and sufficient conditions for the observability for certain classes of hyperbolic DSs. Mizel and Seidman [98,99] and Seidman [115] discussed uniqueness (observability) problems for diffusion processes based on "the strong inherent smoothing." Delfour and Mitter [40,41] studied observability for a class of DSs defined on a Hilbert space. Cannon and Klein [21] considered diffusion processes and based the choice of measurement locations on observability considerations.

Some authors have attempted to use optimization methods to determine measurement type and location. Kaiser [69] considered the control and sensor configuration problem for thermal processes. Ewing and Higgins [49] formulated the variational problem corresponding to the PDE under consideration and used a minimization procedure to yield the sensor locations. Itskovich [66] determined the distance between sensors based on the statistical characteristics of the DS. These methods have some drawbacks in that there is no assurance that the system will be observable from the resulting measurement locations.

The surprising behavior of DSs is illustrated by the two photographs in Figures 1 and 2 due to Crandall and Wittig [37]. The first natural frequency of the thin plate in Figure 1 is at 13 Hz, and the plate is being excited by a speaker coil at 5000 Hz. The pattern of the sand on the surface indicates the eigenfunction nodal lines at this resonance frequency. In Figure 2 the same plate is being excited by broad band noise; the lines show the regions of maximum energy. The behavior of the plate is verifiable but totally unexpected. The significant thing about Figure 2 is that if sensors were placed at points other than along the lines shown, a major part of the vibrational energy would be missed.

As indicated by Figures 1 and 2, observability questions for the DSPIP are complicated by several factors. If based on initial state recovery given the BCs and the PDE, observability criteria are neither necessary nor sufficient for the general DSPIP. The BCs may be unknown or nonlinear, and initial state recovery is

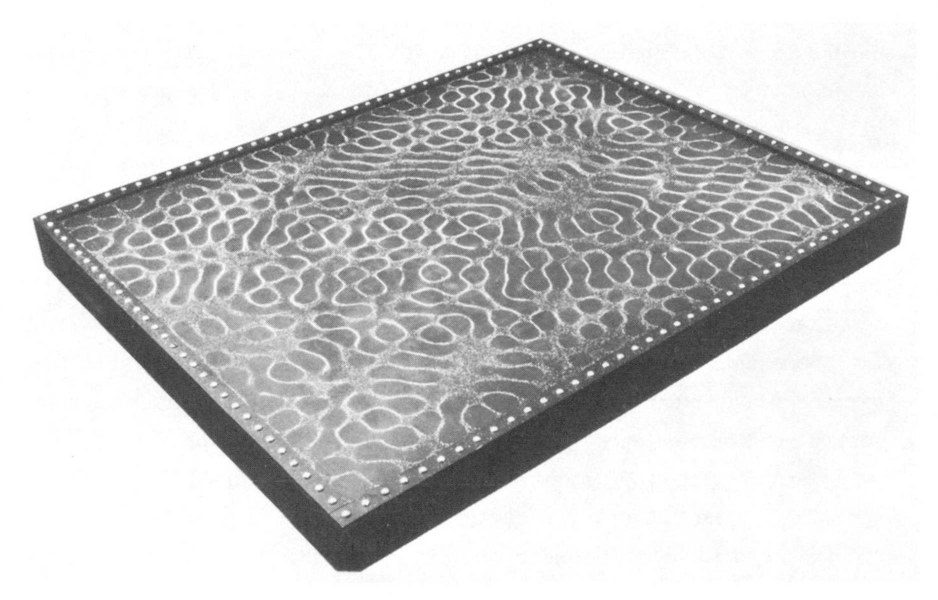

FIG. 1. Eigenfunction nodal lines for a thin plate being excited at 5000 Hz.

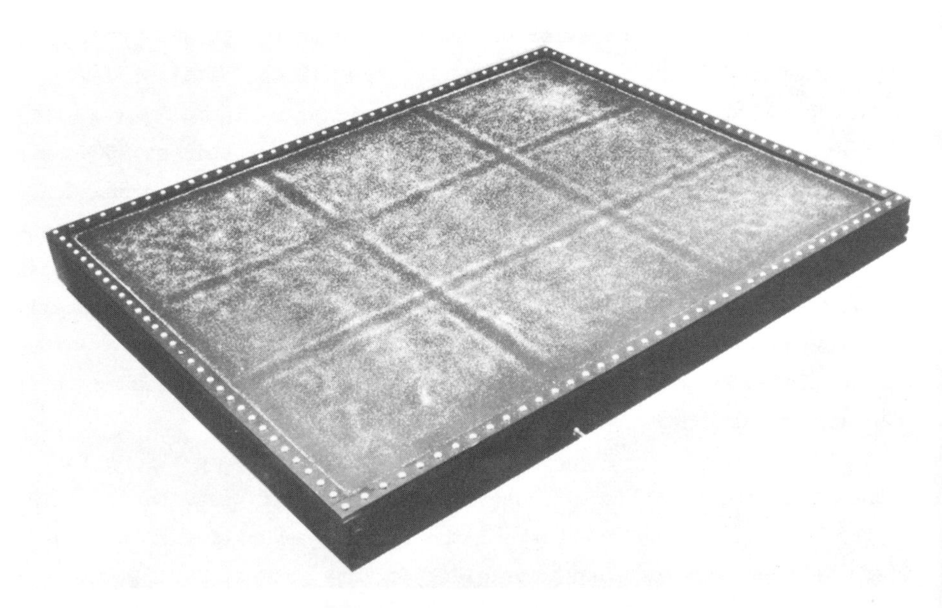

FIG. 2. Thin plate being excited by broadband noise.

impossible in some PDEs due to total information loss and difficult in other PDEs due to error propagation problems. In addition, variables other than elements of the dynamical system "state" may be measured such as the flux in diffusion processes and stress or moment in addition to strain in structural problems. For the DSPIP, observability by the Wang or Goodson-Klein definitions is sufficient in most cases but not necessary to determine the BC and PDE parameters. Parameter identifiability, then, is a less stringent requirement than observability. Thus, choosing measurement locations to maximize the sensitivity of the data to the parameters to be determined is generally acceptable. Such measurement locations may not be sufficient for observability, however. The basic question for measurement location determination for the DSPIP is whether the resulting data provide sufficient information to determine uniquely the unknowns among the BC and PDE parameters and those elements of the system state desired in a specified domain.

There are three basic techniques available to aid in measurement location determination for DSs. They are:

1. Avoidance of zeros of eigenfunctions for processes described by linear PDEs with solutions expressible by

$$v_i(\underline{x},t) = \sum_{m=0}^{\infty} a_m^i g_m^i(t) \varphi_m^i(\underline{x},t) \qquad i = 1, \ldots, r \tag{13}$$

Measurement locations can be chosen based on the properties of the eigenfunctions $\varphi_m^i(\underline{x},t)$ so as to determine the coefficients a_m^i and parameters in the PDE. Reference 60 discusses this technique.

2. Intersection of the full manifold of characteristic lines in hyperbolic equations by the measurements and a sufficiently long period of measurement time to account for the longest time delays which are inherent in hyperbolic systems: Characteristic theory [36] provides an excellent tool for answering observability questions. The characteristic lines in the spatial domain provide an analysis procedure which is directly applicable to observability questions. The characteristic theory is also applicable to a class of nonlinear

PDEs and may be used to examine observability questions for such hyperbolic partial differential equations.

3. Maximization of the sensitivity of the measurements $\underline{z}(t)$ with respect to the parameters $\underline{a}$ and $\underline{c}$ by choice of measurement location: This technique is neither necessary nor sufficient for dependent variable observability but provides a straightforward procedure for picking measurement locations for the DSPIP. The only requirement for the procedure is that the solution, $\underline{v}(\underline{x},t)$, to the PDE be a unique function of the parameters and that the measurements $\underline{z}(t)$ be a unique function of the parameters.

For further reading see the references listed in Table 2 under observability and sensor location. Specific results are given in the next section.

Once the various steps in the DSPIP solution procedure discussed above have been treated, the DSPIP has been converted into a problem where any one of a number of well-documented optimization methods may be applied. Various different performance criteria and approximation techniques may more readily lend themselves to optimization by a particular method. Tables 1 and 2 show that previous work treating the DSPIP has primarily addressed this step of the solution procedure. The extremization of J of Eq. (12) is a well-researched area. The majority of optimization methods available for lumped systems have been extended to DSs. The major difficulties encountered in extending optimization methods to DSs lie with the infinite dimensional state space and the possibly infinite dimensional parameter space. Since these methods are well documented in Tables 1 and 2 and the associated references, this discussion is purposely abbreviated.

Many authors [25-27,31,105,116,118,119,138] use optimization schemes which are based on calculation of the gradients. The introduction of the adjoint equations [25-27,31,116,119] facilitates the calculation of the required gradients. Repeated solution of the system and adjoint equations yields estimates of the parameters.

Several authors [1,31,32,65,83,110,117,120,122,135] derive nonlinear filters for DSs. Although not all the authors have expressly considered the DSPIP [83,110,135], those who do, treat the parameters in the usual way by augmenting the "state" of the system. The computational requirements of these filters would seem to dictate either a low-order approximation to the equations or other simplifications. Much work has been done on the filtering problem for linear DSs and, although this work is referenced in Table 2, it is, as noted previously, of little utility for the DSPIP.

The authors who use stochastic approximation use finite differences [146-148], modal approximation [114], or characteristics [23] to transform the distributed equations into lumped representations. Stochastic approximation is then used to estimate the parameters. A difficulty here is that the estimates are generally biased and considerable effort is required to calculate and eliminate these biases.

The "matrix inverse" method [2,24,35,63] usually requires sampling at a large number of spatial locations and accurate data. The parameter vector is found from

$$\underline{a} = R^{-1}\underline{b} \tag{14}$$

where the elements of R and $\underline{b}$ are generated from measurements of the dependent variable $\underline{v}$ and known forcing functions $\underline{u}$. The method is conceptually simple; however, errors in the measurements can cause the matrix R to be ill-conditioned.

A number of authors [1,2,32,34,85,147] have extended methods based on probabilistic decision theory to the identification of parameters in DSs. Bayesian estimation and maximum likelihood techniques are used by the majority of these authors.

Several other techniques [11,19,44,63,91,118] such as nonlinear programming [63] have also been used to optimize in the DSPIP. The reader is referred to the original works for further information.

III. SPECIFIC RESULTS FOR DISTRIBUTED SYSTEM IDENTIFICATION

In Section II, a comprehensive survey of literature relevant to the DSPIP was offered. In this section, specific results of the authors and their coworkers are given. The applications discussed are vibrations, the process industries, and heat transfer processes. Some results on observability are also presented. The examples are summarized from published work but presented together offer a wide range of techniques.

A. Identification of Parameters by the Method of Characteristics

This approach was published in reference 23. It puts the classical technique of identifying time delays, gains, and time constants in processes on a more solid theoretical foundation. The technique used is to formulate a general process industry transfer function problem as a set of simultaneous PDEs with unknown parameters. Noisy measurements are assumed and stochastic approximation used to determine the parameters after a gradient is computed.

A scalar first-order PDE including time delays and gains will be the model assumed to develop the basic results. The extension of these results to an important class of systems of first-order PDEs, as well as the incorporation of diffusion effects, is considered. Particular attention is paid to the practical requirement of a limited number of noisy measurements.

The estimation algorithms presented converge for all cases. However, more efficient algorithms with less rigorous convergence properties can be developed for specific cases. As is the case for most identification schemes, the proposed methods require a significant change in the input to yield good results; pulses, for example.

A unique feature of the proposed scheme is that the method of characteristics is used to reduce the PDE to a set of ODEs. The method of characteristics has two distinct advantages when compared to other techniques for reducing PDEs, such as the method of finite differences. First of all, a smaller number of ODEs must be solved and, secondly, the characteristic differential equations constitute an exact representation of the PDE.

1. Problem Statement

The identification algorithms developed in this section are addressed to the following general model. Consider a distributed system described by

$$\frac{\partial \underline{v}}{\partial t} + A(\underline{v},x,t) \frac{\partial \underline{v}}{\partial x} + \underline{b}(\underline{v},x,t) = 0 \tag{15}$$

where $\underline{v}(x,t)$ is the n x 1 vector of dependent variables, x is the spatial coordinate, and t is time. Assume that one or both the functions $A(\underline{v},x,t)$ and $\underline{b}(\underline{v},x,t)$ are unknown. However, both the functions are restricted to class C^1 in $\underline{v}$ and are piecewise continuous in x and t.

Assume that two noisy spatial measurements of v(x,t) are available

$$\begin{aligned} \underline{z}_1(t) &= \underline{v}(x_1,t) + \underline{\xi}_1(t) \\ \underline{z}_2(t) &= \underline{v}(x_2,t) + \underline{\xi}_2(t) \end{aligned} \tag{16}$$

where the additive noise has the following properties

$$\begin{aligned} E\{\underline{\xi}_1(t_i)\underline{\xi}_1^T(t_j)\} &= \begin{cases} \sigma^2_{i,j} & , i = j \\ 0 & , i \neq j \end{cases} \\ E\{\underline{\xi}_2(t_i)\underline{\xi}_2^T(t_j)\} &= \begin{cases} \sigma^2_{i,j} & , i = j \\ 0 & , i \neq j \end{cases} \end{aligned} \tag{17}$$

and

$$E\{\underline{\xi}_1(t_i)\underline{\xi}_2^T(t_j)\} = 0 \qquad , \qquad \text{for all } i,j$$

where E{ } denotes the expected value operator. These conditions are generally satisfied by transducer noise. Furthermore, assume that the boundary and initial conditions are known only through the measurements on $\underline{v}(x,t)$.

The basic problem is to identify the matrix function A and/or the vector $\underline{b}$ from $\underline{z}_1(t)$ and $\underline{z}_2(t)$. Also incorporated in this basic problem is the identification of industrial models which contain lag terms due to diffusion.

A common first-order model for an industrial process, v(x,t), with time delay T_d, gain K, and lag τ is given by the transfer function

$$G(s) = \frac{Ke^{-sT_d}}{\tau s + 1} \tag{18}$$

between $v(x_1,t)$ and $v(x_2,t)$. This process is partially described by a scalar version of Eq. (15), namely,

$$\frac{\partial v(x,t)}{\partial t} + a_1 \frac{\partial v(x,t)}{\partial x} + a_2 v(x,t) = 0 \tag{19}$$

where a_1 and a_2 are constants. The relationship between Eqs. (18) and (19) is

$$K = e^{-[a_2(x_2-x_1)]/a_1} \tag{20}$$

$$T_d = \frac{x_2 - x_1}{a_1} \tag{21}$$

The lag τ does not appear explicitly in the PDE formulation, but it is incorporated into the identification scheme by augmenting Eq. (19) with a first-order linear differential equation (see Case 2 below).

2. Derivation of Identification Algorithms

For the sake of exposition, the results are expressed in terms of several specific algorithms and then extended to more general conditions. Each successive algorithm represents a physically significant but theoretically more complex problem. The basic approach in all cases is to reduce the PDE to a set of ODEs using the method of characteristics and then to establish recursive relations which estimate unknown parameters in the differential equations by minimizing an appropriate error criterion along the characteristic curve. Each problem dictates one of two forms of the error criterion. The first error form is the expected value of the difference between the measured and predicted values; the second form is the integral over time of that error. Consequently, the two cases presented are categorized by the form of error criterion.

Case 1. Identification of a_2. Consider the scalar equation

$$\frac{\partial v}{\partial t} + a_1 \frac{\partial v}{\partial x} + a_2 v = 0 \tag{22}$$

where the constant a_1 is known and the constant a_2 is to be estimated from the noisy measurements, $z_1(t)$ and $z_2(t)$, defined earlier. The algorithm in this case is obtained by first guessing a value of

$a_2 = \hat{a}_2$ and then using the method of characteristics to extend $z_1(t_1)$ to the spatial location x_2 (Fig. 3). A measure of the error due to the guess $\hat{a}_2$ is obtained by comparing the extension of the measurement $z_1(t_1)$ to the actual measurement $z_2(t_2)$. This measure is then used to improve the initial estimate of a_2 via a recursive stochastic gradient scheme. The procedure is continued until further iteration results in no significant change in $\hat{a}_2$.

More specifically, the solution of the characteristic equations of (22) is

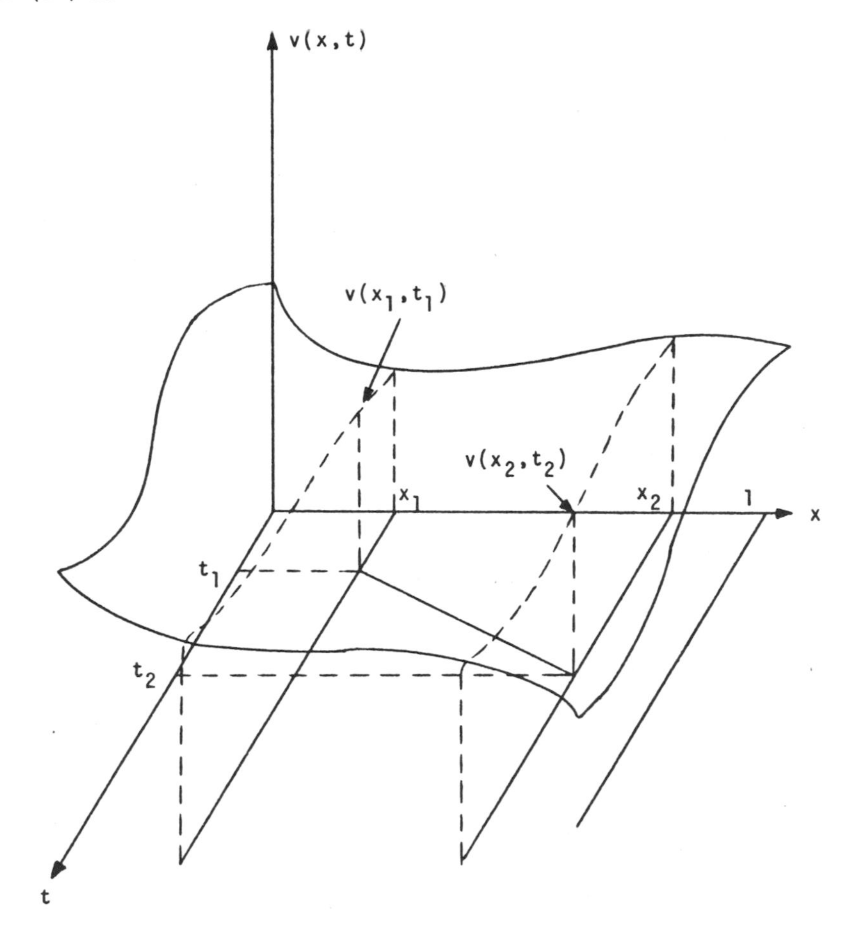

FIG. 3. Characteristic trajectory for scalar system.

$$x_2 - x_1 = a_1(t_2 - t_1) \tag{23}$$

$$v(x_2,t_2) = v(x_1,t_1)e^{-a_2(t_2 - t_1)} \tag{24}$$

where x_1,t_1 and $v(x_1,t_1)$ represent the initial values and $x_2,t_2,v(x_2,t_2)$ the terminal values (see Fig. 3). Since a_1 is known, then Eq. (23) specifies t_2. The predicted value of v at $x = x_2$, $t = t_2$ extended from the noisy measurement $z_1(t_1)$ is denoted by $y_p(t_2)$ and is

$$y_p(t_2) = z_1(t_1)e^{-a_2T_d} \tag{25}$$

where

$$T_d = t_2 - t_1 \tag{26}$$

The performance index, i.e., the average measure of error between the predicted value $y_p(t_2)$ and the measured value $z_2(t_2)$ at the spatial location $x = x_2$ is

$$J(a_2) = \frac{1}{2} E\{[z_2(t_2) - y_p(t_2)]^2\} \tag{27}$$

Since v(t) along a characteristic curve is an exponential, then $J(\hat{a}_2)$ is unimodal and the value of $\hat{a}_2 = a_2^*$ which minimizes Eq. (27) is the desired estimate of a_2.

A sequential algorithm of the Robbins-Munro type, which minimizes Eq. (27), is developed next. The instantaneous gradient error term of Eq. (27) is

$$e'(\hat{a}_2) = \frac{d}{d\hat{a}_2} \frac{1}{2} [z_2(t_2) - y_p(t_2)]^2 \tag{28}$$

To insure that the sequential estimation algorithm yields an unbiased estimate of a_2, the regression function

$$m(\hat{a}_2) = E[e'(\hat{a}_2)] \tag{29}$$

must satisfy $m(a_2^*) = 0$, where a_2^* is the true value of a_2. However, in this case

$$m(a_2^*) = -T_d e^{-2a_2^*T_d} \sigma_1^2 \tag{30}$$

Consequently, the instantaneous unbiased gradient error function becomes

$$e'(\hat{a}_2) = \left[z_2(t_2) - z_1(t_1)e^{-\hat{a}_2 T_d}\right]\left[T_d z_1(t_1)e^{-\hat{a}_2 T_d}\right] + T_d \sigma_1^2 e^{-2\hat{a}_2 T_d} \quad (31)$$

and the algorithm for the sequential estimation is

$$\hat{a}_2^{k+1} = \hat{a}_2^k - \rho_k e'(\hat{a}_2^k) \quad (32)$$

The conditions on the ρ_k sequence, as well as the convergence properties of this algorithm, are discussed in the original paper.

Generalization of Case 1. The previous method also is applicable to the case where the coefficient a_1 in Eq. (22) is a known function, $a_1 = a(x)$, since the characteristic curve, described by

$$\frac{dx}{dt} = a(x) \quad (33)$$

is still completely known. The extension to $a_1 = a(x,t)$ will depend on how well the algorithm can track the time changes.

The method can also be used to identify an unknown function, $a_2 = b(v)$, provided that the function can be adequately represented by a polynomial

$$b(v) = b_0 + b_1 v + b_2 v^2 + \cdots + b_k v^k \quad (34)$$

and that the coefficient a_1 is a constant. The most straightforward approach in this case is to use k transducers, identify straight-line segments between two successive transducers using a performance criterion of the form (27), and then fit the polynomial (34) through the segmented curve. It can be shown that two measurement points in x are sufficient to identify b(v).

The foregoing results can be extended directly to the following system of PDEs. Let Eq. (15) be of the form

$$\frac{\partial \underline{v}}{\partial t} + a_1 \frac{\partial \underline{v}}{\partial x} + B\underline{v} = 0 \quad (35)$$

where v is an n x 1 vector, a_1 is a known scalar function of v and x, and B is an unknown constant matrix to be determined. This equation represents, for example, certain multiple-pass heat exchangers and reactors where concentrations of the various components and temperatures are typical dependent variables. The characteristic theory for equations with the same principal parts results in characteristic equations

$$\frac{dx}{dt} = a_1 \qquad \frac{d\underline{v}}{dt} = -B\underline{v} \tag{36}$$

The performance index in this case is

$$J(B) = \frac{1}{2} E\{[z_2(t_2) - y_p(t_2)]^T [Q][z_2(t_2) - y_p(t_2)]\} \tag{37}$$

where the superscript T denotes transpose and Q is a constant weighting matrix. The stochastic algorithm for estimating the components $b_{i,j}$ of the B matrix is

$$\hat{b}^{k+1}_{i,j} = \hat{b}^k_{i,j} - \rho_k e'(\hat{b}^k_{i,j}) \tag{38}$$

where

$$e'(\hat{b}^k_{i,j}) = -\left(\frac{\partial \underline{y}_p(t_2)}{\partial b_{i,j}}\right)^T Q[\underline{z}_2(t_2) - \underline{y}_p(t_2)] + T_d \sigma^2_{i,j}\left[\left(e^{-\hat{B}T_d}\right)^T\right]_{i^{th}\,row} Q\left[e^{-\hat{B}T_d}\right]_{j^{th}\,col} \tag{39}$$

with

$$\frac{\partial \underline{y}_p(t_2)}{\partial b_{i,j}} = T_d\left[e^{-\hat{B}T_d}\right]_{i^{th}\,col}\left[z_1(t_1)\right]_{j^{th}\,row} \tag{40}$$

The last term in Eq. (39) serves to eliminate the bias from the estimator.

<u>Case 2. Simultaneous identification of a_1 and a_2.</u> The problem in this case is to simultaneously identify the constant coefficients a_1 and a_2 in

$$\frac{\partial v}{\partial t} + a_1 \frac{\partial v}{\partial x} + a_2 v = 0 \tag{41}$$

from two noise measurements, $z_1(t)$ and $z_2(t)$. The basic approach is to apply a pulse to the system and then vary a_1 and a_2 until the difference in area between the input pulse extended to the location x_2 and the measured pulse at x_2 is minimized. Thus, the required performance measure is

$$J(\hat{a}_1,\hat{a}_2) = \frac{1}{2} E\left\{ \int_{t_m}^{t_M} [z_2(\eta) - y_p(\eta)]^2 \, d\eta \right\} \tag{42}$$

where $(\hat{a},\hat{a}_2)$ are any estimates of the unknown constants, y_p is the data z_1 extended by characteristic theory to the location x_2, and the limits t_m,t_M bound the expected time interval in which the output pulse occurs. (Note that this case requires an integral type performance criterion as opposed to the terminal type used in Case 1.)

Carrying out the details, characteristic theory yields

$$y_p(\eta) = z_1(\eta - \hat{T}_d)e^{-\hat{a}_2\hat{T}_d} \tag{43}$$

where

$$\hat{T}_d = \frac{x_2 - x_1}{\hat{a}_1} \tag{44}$$

Consequently, Eq. (42) can be expressed as

$$J(\hat{a}_2,\hat{T}_d) = \frac{1}{2} E \int_{t_m}^{t_M} \left[y_2(\eta) - z_1(\eta - \hat{T}_d)e^{-\hat{a}_2\hat{T}_d}\right]^2 d\eta \tag{45}$$

The limits of integration, assuming an input pulse $u(x,t)$ of class C^1 and duration T units of time starting at t_1, are

$$\begin{aligned} t_m &= t_1 + T_{d\ min} \\ t_M &= t_1 + T + T_{d\ max} \end{aligned} \tag{46}$$

where $T_{d\ min}$, $T_{d\ max}$ are a priori estimates of the minimum and maximum values of T_d, respectively.

The bias terms are found from the regression functions for the instantaneous gradient of J with respect to $\hat{a}_2$,

$$m_1(\hat{a}_2, T_d) = E \int_{t_m}^{t_M} \left[z_2(\eta) - z_1(\eta - \hat{T}_d)e^{-\hat{a}_2\hat{T}_d}\right] \times \left[T_d z_1(\eta - \hat{T}_d)e^{-\hat{a}_2\hat{T}_d}\right] d\eta \tag{47}$$

and for the instantaneous gradient of J with respect to $\hat{T}_d$,

$$m_2(\hat{a}_2, \hat{T}_d) = E \left\{\int_{t_m}^{t_M} \left[z_2(\eta) - z_1(\eta - \hat{T}_d)e^{-\hat{a}_2\hat{T}_d}\right]\left[-\dot{z}_1(\eta - \hat{T}_d) - z_1(\eta - \hat{T}_d)\hat{a}_2\right]e^{-\hat{a}_2\hat{T}_d}\, d\eta\right\} \tag{48}$$

The bias term in Eq. (47) is

$$m_1(a_2^*, T_d^*) = \int_{t_m}^{t_M} \left[-z_1^2 T_d^* e^{-2a_2^* T_d^*}\right] d\eta \tag{49}$$

Consequently, the instantaneous gradient error term with respect to a_2 is

$$e_1'(\hat{a}_2, \hat{T}_d) = \int_{t_m}^{t_M} \left\{\left[z_2(\eta) - z_1(\eta - \hat{T}_d)e^{-\hat{a}_2\hat{T}_d}\right] \times \left[\hat{T}_d z_1(\eta - \hat{T}_d)e^{-\hat{a}_2\hat{T}_d}\right] - \sigma_1^2\hat{T}_d e^{-2\hat{a}_2\hat{T}_d}\right\} d\eta \tag{50}$$

The bias term in Eq. (48) is

$$m_2(a_2^*, T_d^*) = \int_{t_m}^{t_M} \left\{E\left[\xi_1(\alpha)\frac{d\xi_1(\alpha)}{d\alpha}\right] + a_2^*\sigma_1^2\right\}e^{-2a_2^* T_d^*}\, d\eta \tag{51}$$

where $\alpha = \eta - \hat{T}_d$. One must be careful in evaluating the term

$$E\left[\xi_1(\eta)\frac{d\xi_1(\eta)}{d\eta}\right] = -\frac{d}{d\eta}R_{\xi_1\xi_1}(\eta)\Big|_{\eta=0} \tag{52}$$

This term is zero if the derivative exists or if $\xi_1(\eta)$ and $(d/d\eta)\xi_1(\eta)$ are independent Gaussian processes. In the more general case where the derivative may not exist, approximation techniques must be used. In any case, the instantaneous gradient error term with respect to $\hat{T}_d$ is

$$e_2'(\hat{a}_2,\hat{T}_d) - \int_{t_m}^{t_M} d\eta \left\{\left[z_2(\eta) - z_1(\eta - \hat{T}_d)e^{-\hat{a}_2\hat{T}_d}\right]\left[-\dot{z}_1(\eta - \hat{T}_d) - z_1(\eta - \hat{T}_d)\hat{a}_2\right]e^{-\hat{a}_2\hat{T}_d} + \frac{d}{d\tau}R_{\xi_1\xi_1}(\tau)\Big|_{\tau=0} + \hat{a}_2\sigma_1^2 e^{-2\hat{a}\hat{T}_d}\right\} \tag{53}$$

The sequential algorithms for estimating b and T_d are

$$\hat{a}_2^{k+1} = \hat{a}_2^k - \rho_k e_1'(\hat{a}_2^k,\hat{T}_d^k) \tag{54}$$

$$\hat{T}_d^{k+1} = \hat{T}_d^k - \rho_k e_2'(\hat{a}_2^k,\hat{T}_d^k) \tag{55}$$

If the noise processes, $\xi_1(t)$, $\xi_2(t)$, are independent, the ρ_k may be chosen to satisfy the stochastic approximation theorem, e.g.,

$$\rho_k = \frac{1}{1+k} \tag{56}$$

If no knowledge of the noise is known, a least squares procedure for generating the ρ_k is possible.

This estimation procedure is an off-line technique, since the integrations in Eqs. (50) and (53) to obtain the error terms must be performed for each k. To indicate the procedure, consider the pulse to be the half sinusoid

$$u(x_1,t) = \begin{cases}\sin\frac{\pi t}{T}, & 0 \le t \le T\\ 0 & \text{otherwise}\end{cases} \tag{57}$$

Then, for the zero noise case,

$$e_1' = -\frac{1}{2} e^{-a_2 T_d} \cos \frac{\pi(T_d - \hat{T}_d)}{T} + \frac{1}{2} e^{-\hat{a}_2 \hat{T}_d} \tag{58}$$

$$e_2' = \frac{1}{2} e^{-a_2 T_d} \sin \frac{\pi(T_d - \hat{T}_d)}{T} \tag{59}$$

For $|T_d - \hat{T}_d| < 1$, a unique solution to Eqs. (58) and (59) results, namely

$$a_2 = \hat{a}_2, \qquad T_d = \hat{T}_d \tag{60}$$

Generalization of Case 2. The time constant τ in Eq. (18) approximates the effect of diffusion which tends to distort the shape of the input pulse as it propagates through the distributed system. Although diffusion effects are not accounted for in the model of Eq. (15), they can be incorporated by augmenting the model with an ODE. This is done by processing the predicted pulse through a filter with dynamics

$$G(s) = \frac{1}{\hat{\tau} s + 1} \tag{61}$$

and then comparing the resulting distorted pulse to the measured pulse at x_2. Consequently, the desired value of τ is the one which minimizes the performance criterion (for the deterministic case)

$$J(\tau) = \frac{1}{2} \int_{t_m}^{t_M} \left\{ \left[\frac{1}{\tau} e^{-t/\tau} * u_p(x_2, t) \right] - u_2(x_2, t) \right\}^2 dt \tag{62}$$

where * represents the convolution operator, and the algorithm is

$$\hat{\tau}^{k+1} = \hat{\tau}^k - \rho \left. \frac{\partial J(\tau)}{\partial \tau} \right|_{\tau = \hat{\tau}^k} \tag{63}$$

This algorithm can be extended to the case of noise measurements by the use of a stochastic approximation gain sequence ρ_k and the inclusion of bias correction terms.

3. Examples

The following examples were programmed on a digital computer and illustrate the use of algorithms developed in the previous section. The systems to be identified were simulated on a computer from given initial and boundary conditions. Measurements were made on the solutions and noise was added to simulate measurement errors. The noise was generated by a digital pseudo-random number generator program on the digital computer. The algorithms then processed the noisy measurements and extracted the estimates of the unknown parameters.

The following technique was used in the examples to speed up the convergence of the stochastic approximation algorithms. The gain ρ_k was held constant for ten iterations and eventually reduced by 1/n. That is, for an initial gain of one, the gain is one for the first ten iterations, then it is one-half for the second ten, then one-third for the third ten, etc. The gain is allowed to change at every iteration after about 100 iterations, that is to say, 1/101, 1/102, 1/103, etc. Since the gain sequence satisfies the convergence theorems from this point on, the algorithms still converge. This technique reduces the number of iterations for convergence by a substantial amount. The amount of improvement was dependent upon the nature of the system, the initial gains, the variance of the noise, and the system inputs. The initial gain values were chosen to be as large as possible without undue oscillations in the parameter estimates. These values were determined experimentally.

Three examples are included to illustrate the convergence of the algorithms applied to the model

$$\frac{\partial v}{\partial t} + a_1 \frac{\partial v}{\partial x} + a_2 v = 0 \tag{64}$$

This equation represents a simplified version of a heat exchanger shown in Figure 4. The dependent variable, $v(z,t)$, represents temperature, a_1, fluid velocity, and a_2, the coefficient of heat trans-

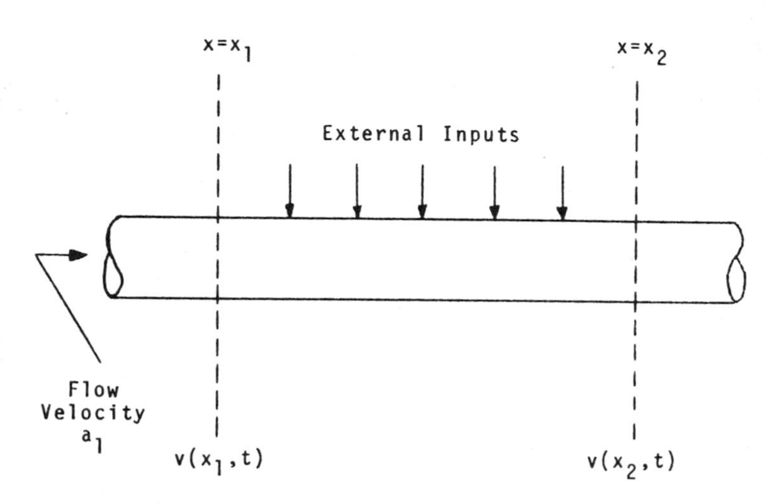

FIG. 4. Simplified version of heat exchanger.

fer (including some dimensional parameters). The space variable x ranges between 0 and 1, and the input boundary condition at x = 0 is known only through measurements made on the system. The input boundary condition used for simulation purposes in Examples 1 and 2 was a ramp with unity slope. The actual values of a_1 and a_2 used in Examples 1 and 3 were $a_1 = 1.0$ and $a_2 = 1.0$. The value of the initial guess of the parameter is shown on the graphs accompanying each case.

Example 1. Identification of a_2. Figure 5 illustrates the convergence properties of the algorithm for three values of noise variance with measurements at x = 0.25 and x = 0.75. The a priori estimate of a_2 was $\hat{a}_2^0 = 0.5$. Independent zero mean Gaussian noise was added to the measurements. The value of the initial gain used in this identification was $\rho_0 = 2.0$. The first sample was taken at t = 1.0 and x = 0.25, and the ramp input was repeated for every subsequent iteration.

Example 2. Identification of a function a(v). Equations with functions a = a(v) arise, for example, in certain ion-exchange processes. For the purposes of this example, $a(v) = v^2$ was used. The problem was run with four measurement locations (x = 0, x = 0.25,

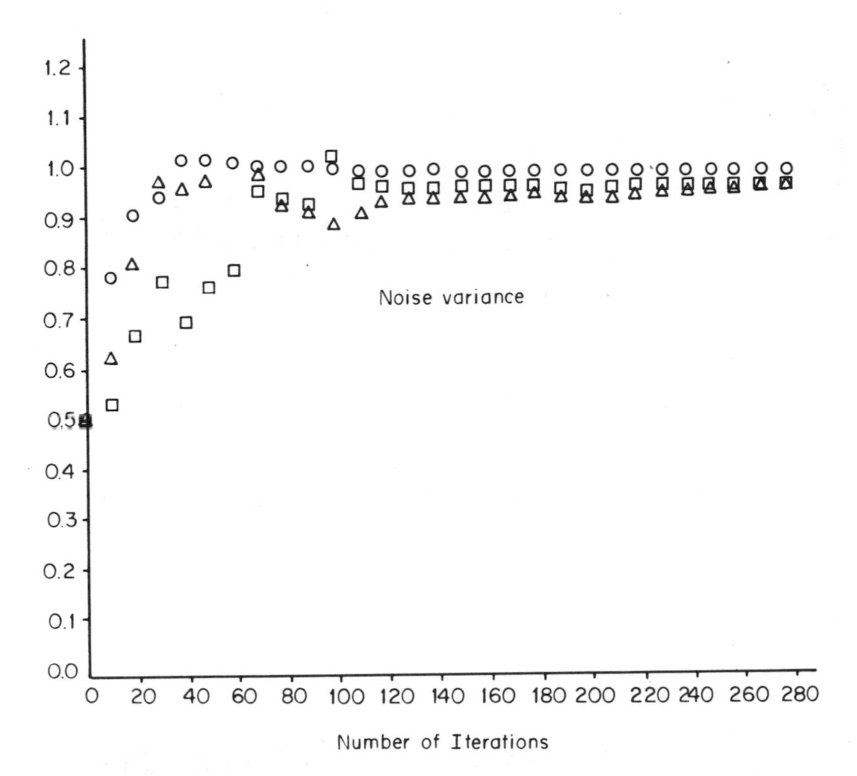

FIG. 5. Identification of a_2. Noise variance: ○, 0.01; △, 0.05; □, 0.10.

x = 0.75, x = 1.0) and two locations (x = 0, x = 1) using different schemes. The noisy measurements made at four transducer locations with independent, zero mean Gaussian noise and variance equal to 0.1, yielded the following polynomial coefficients after ten iterations:

$$a(v) = 0.07 + 0.11v + 1.05v^2 - 0.01v^3 \tag{65}$$

Four measurements were required at x = 0, x = 1 at each iteration for the two-transfer-location scheme. For the same noise conditions, the average coefficients after ten iterations are

$$a(v) = 0.002 + 0.003v + 0.999v^2 \tag{66}$$

where only a second-order polynomial was assumed.

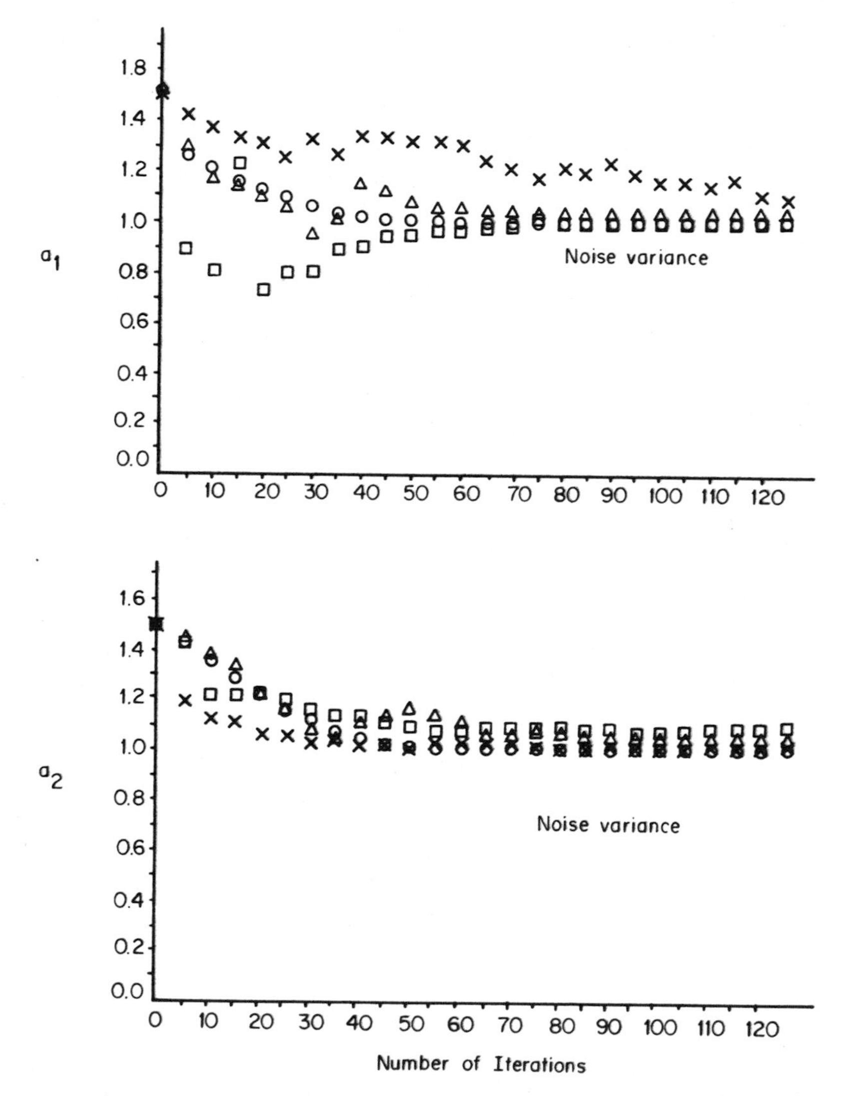

FIG. 6. Simultaneous identification of a_1 and a_2. Noise variance: ○, no noise; Δ, 0.01; □, 0.05; x, 0.10.

Example 3. Simultaneous identification of a_1 and a_2. Figure 6 shows the simultaneous identification of a_1 and a_2. The input for this experiment was a one-half cycle sinusoidal pulse $u(0,t) = \sin(t - 1)$ for $1 \le t \le \pi + 1$ and zero elsewhere. The integration limits,

t_m and t_M, were 1 and 10, respectively. A simple trapezoidal rule integration with a base of 0.01 was used. Measurements were at x = 0 and x = 1. The initial value of gain was $\rho_0 = 1$, and the a priori estimates of a and b were both 1.5.

4. Discussion

For systems with real characteristic trajectories, the method of characteristics is an effective means of identifying parameters. This method yields estimates of time delays and gain constants for mass transport systems. The estimates converge for noisy measurements. The largest variance chosen for experimental purposes (0.1) had a standard deviation on the order of the deterministic signal at the second measurement. Even with noise of this relative magnitude, the algorithms converge quite rapidly through careful selection of initial gains.

B. Use of Galerkin's Method to Facilitate Parameter Identification

Galerkin's method for approximating solutions to PDEs has been used successfully to facilitate the DSPIP. This technique has much previous research to aid in the approximate step. The examples which have appeared in detail in previous work by the authors [105,106] are given here to illustrate the use of Galerkin's criterion in the DSPIP. The first example uses simulated data while the second uses data obtained from a test beam.

Example 1. A diffusion process with a nonlinear term. A tubular reactor can be modeled by an energy equation of the form

$$\frac{\partial T}{\partial t}(x,t) + \bar{v}\,\frac{\partial T}{\partial x}(x,t) = -\frac{hA}{\rho V c_p}\,[T(x,t) - T_w(t)] + D\,\frac{\partial^2 T}{\partial x^2}(x,t) + \frac{k_0}{c_0}\,e^{-[E/RT(x,t)]} \qquad (67)$$

where D is the diffusivity, E is the activation energy, R is the thermodynamic constant, ρ is the mass density, $\bar{v}$ is the average velocity, c_p is the specific heat, ℓ is length, and T_0 is some initial value of the temperature $T(x,t)$. The control input is $T_w(t)$. The equation is normalized and values are assumed for the constant, thus

$$\frac{\partial v}{\partial t} + \frac{\partial v}{\partial x} = -10[v - u] + 0.50 \frac{\partial^2 v}{\partial x^2} + 2.86 \exp\left[\frac{-0.7}{1 + v}\right] \tag{68}$$

where $x \in [0,1]$ and

$$\begin{aligned} 1 + v &= T(x,t)/T_0 \qquad v = v(x,t) \\ 1 + u &= T_w(t)/T_0 \qquad u = u(x,t) \end{aligned} \tag{69}$$

Boundary and initial conditions are

$$\begin{aligned} v(0,t) &= 0.10 \\ \frac{\partial v(1,t)}{\partial x} &= 0 \end{aligned} \tag{70}$$

$$v(x,0) = 0 \tag{71}$$

Using finite differences the resulting equations in time are integrated on the digital computer with $\Delta t = 0.002$ and with the control $u = -0.1420$. The result of this simulation is used as "measured data" for the identification procedure. These data will be inexact due to the large step size $\Delta x = 0.10$ in the x direction.

Equation (68) is linearized by linearizing the exponential term

$$\begin{aligned} \exp\left[\frac{-0.70}{1 + v}\right] &\cong \exp[-0.7(1 - v)] \\ &\cong \exp(-0.7)\exp(0.7v) \\ &\cong 0.4966(1 + 0.70v) \end{aligned} \tag{72}$$

The linearized equation is

$$\frac{\partial v}{\partial t} + \frac{\partial v}{\partial x} = 0.5 \frac{\partial^2 v}{\partial x^2} + 9.0v + 10u + 1.420 \tag{73}$$

Letting the assumed parameter vector be $a^* = (a_1^*, a_2^*) = (0.50, 9.00)$ and choosing $u = -0.1420$, Eq. (73) is written below in terms of the parameters to be identified.

$$\frac{\partial v}{\partial t} + \frac{\partial v}{\partial x} = a_1 \frac{\partial^2 v}{\partial x^2} - a_2 v \qquad x \in [0,1] \tag{74}$$

The approximate solution is chosen as

$$v(x,t) = 0.10 + \sum_{j=1}^{N} F_j(x)\varphi_j(t)$$

The first four approximating functions are:

$$F_1(x) = 12x^2 - 24x$$

$$F_2(x) = 8x^3 - \frac{61}{4}x^2 + \frac{13}{2}x$$

$$F_3(x) = \frac{17290}{2717}x^4 - \frac{39368}{2717}x^3 + \frac{1431}{143}x^2 - 2x$$

$$F_4(x) = \frac{576}{11}x^5 - \frac{17190}{121}x^4 + \frac{14952}{121}x^3 - \frac{6051}{121}x^2 + 6x$$

These functions are chosen to satisfy the boundary conditions and were orthogonalized using a Gram-Schmidt process.

The equations resulting from application of Galerkin's criterion are

$$\dot{\varphi}(t) = A^{-1}K\varphi(t) + a_1A^{-1}B\varphi(t) + a_2[-0.1A^{-1}d - I\varphi(t)] \tag{75}$$

$$\varphi(0) = -0.10A^{-1}d \tag{76}$$

where I is the identity matrix and the other matrices and vectors are:

$$A^{-1}B = \begin{bmatrix} -2.5000 & 0.05208 & -0.0043706 & 0.005165 & -0.001235 \\ 12.9231 & -23.5308 & 1.03475 & -1.8942 & 0.3446 \\ -54.0000 & 51.5250 & -68.4692 & 35.10000 & -15.7891 \\ 22.0000 & -32.5155 & 12.1000 & -142.3000 & 13.36739 \\ -51.9998 & 58.4691 & -53.8000 & 132.1272 & -251.3522 \end{bmatrix}$$

$$A^{-1}K = \begin{bmatrix} -0.9375 & -0.18359 & -0.014897 & -0.040566 & -0.010898 \\ 16.4769 & -0.90865 & -0.70697 & -0.29037 & -0.28612 \\ -45.4500 & 20.8594 & -1.1368 & -12.185 & -0.64183 \\ 24.2786 & -4.2969 & 2.7296 & -1.3807 & -2.9654 \\ -48.1928 & 16.8594 & -2.8358 & 19.8827 & -1.6282 \end{bmatrix}$$

$$A^{-1}d = \begin{bmatrix} -0.10417 \\ 0.53846 \\ -2.25000 \\ 0.91667 \\ -2.16666 \end{bmatrix}$$

The matrices for lower-order approximations can be deduced directly from the above. The linearized problem is up to 10 mode G-K observable at $x^* = 0.50$ even if the transducer averages over ±2% of the normalized spatial domain. A search technique is used to give parameter estimates. For this example $t_0 = 0.0$, $t_f = 1.5$, and the initial parameter estimate is chosen arbitrarily to be $\hat{a}^0 = (1.0, 1.0)$. The results are shown in Table 3. From Table 3 it can be seen that for the four-term approximation the parameter estimates seem to be slightly better, but the performance index is greater than for either the three- or five-term approximation. The parameters are estimated from measurements taken on the nonlinear system. Thus, it is not expected that the parameter estimates will agree exactly with the assumed values. The estimates for three-, four-, and five-term approximations are all within 5% of the assumed values; furthermore, the estimates for the three- and five-term approximations agree to

TABLE 3

Summary of Results for the Diffusion Process with a Nonlinear Term

Number of terms in the approximation	2	3	4	5
$\hat{a}_1$	0.5855	0.52239	0.51747	0.52239
$\hat{a}_2$	9.150	9.1745	9.1291	9.1745
$J(\hat{a})$	7.56×10^{-6}	1.751×10^{-7}	1.6077×10^{-6}	1.593×10^{-7}
% error $\hat{a}_1$	17.1	4.48	3.49	4.48
% error $\hat{a}_2$	1.67	1.94	1.43	1.94

five significant figures. It is thus concluded that the three-term approximation is sufficient to give accurate parameter estimates.

Example 2. A thin uniform cantilevered test beam. The test beam apparatus and experimental procedure are described in detail in reference 106. The dynamic equations for the thin uniform beam can be expressed as:

$$v_2 = \frac{\partial v_1}{\partial t}(x,t)$$

$$v_3 = \frac{\partial v_1}{\partial x}(x,t)$$

$$v_4 = \frac{\partial v_3}{\partial x}(x,t) \tag{77}$$

$$v_5 = \frac{\partial v_4}{\partial x}(x,t)$$

$$\frac{\partial v_2}{\partial t} + a_1 v_2 + a_2 \frac{\partial v_5}{\partial x} = 0$$

The unknown parameter vector $a = (a_1, a_2) = [(D/M), (EI/ML^4)]$, where M = mass per unit length [lbm/(in.-386)], D = the internal damping factor term [lbf/(in./sec - in.)], E = the modulus of elasticity (psi), I = the moment of inertia (in.^{4}), L = the length (in.), and $v_1(x,t)$ is the displacement. In the above, x has been normalized so that $0 \leq x \leq 1$. The equations assume that the internal damping can be approximated by a constant term and that shear and rotation inertia effects are negligible.

The beam is started at rest, so the initial conditions are

$$v_1(x,0) = 0, \qquad v_2(x,0) = 0 \tag{78}$$

A shaker table was used to apply the forcing u(t) at the boundary x = 0, thus

$$v_1(0,t) = u(t) \tag{79}$$

The remaining BCs are those for a cantilever beam:

$$v_3(0,t) = 0, \qquad v_4(1,t) = 0, \qquad v_5(1,t) = 0 \tag{80}$$

Measurements are made of the normalized moment at the measurement location x_m, thus

$$z(t) \triangleq z(x_m,t) \equiv v_4(x_m,t) \tag{81}$$

The data used for the identification procedure were the sampled shaker table motion, u(t), and the normalized moment, z(t), from a strain gage.

An approximate solution, $\tilde{v}_1(x,t)$, is assumed of the form:

$$\tilde{v}_1(x,t) = u(t) + \sum_{j=1}^{N} F_j(x)\varphi_j(t) \tag{82}$$

The first three orthogonal approximating functions satisfying the boundary condition are:

$$F_1(x) = 1/6x^4 - 2/3x^3 + x^2$$

$$F_2(x) = 1.1166x^5 - 4.0552x^4 + 5.0552x^3 - 2.0x^2$$

$$F_3(x) = 153.04x^6 - 584.07x^5 + 815.72x^4 - 483.12x^3 + 100x^2$$

Using Galerkin's criterion the resulting initial value ODEs become:

$$\begin{aligned} \ddot{\underline{\varphi}}(t) + a_1 I\dot{\underline{\varphi}}(t) + a_2 S\underline{\varphi}(t) &= \underline{\gamma}[\ddot{u}(t) + a_1\dot{u}(t)] \\ \underline{\varphi}(0) &= -\underline{\gamma}u(0) \\ \dot{\underline{\varphi}}(0) &= -\underline{\gamma}\dot{u}(0) \end{aligned} \tag{84}$$

With the following definitions

$$A = \{a_{i,j}\} = \left\{\int_0^1 F_j(x)F_i(x)\,dx\right\}$$

$$B = \{b_{i,j}\} = \left\{\int_0^1 F_j''''(x)F_j(x)\,dx\right\}$$

$$\underline{d} = \mathrm{col}[d_i] = \mathrm{col}\left[\int_0^1 F_i(x)\,dx\right]$$

$$\underline{\varphi}(t) = \mathrm{col}[\varphi_i(t)]$$

$$i,j = 1, \ldots, N$$

the matrices and vectors of Eq. (84) are defined as $S = \{s_{i,j}\} \triangleq A^{-1}B$, $\underline{\gamma} = \mathrm{col}[\gamma_j] \triangleq A^{-1}d$, and I = identity. Equation (84) may be written in state variable form using the following definitions:

$$\dot{y}_k(t) = -a_1 y_k(t) + y_{k+1}(t)$$

$$\dot{y}_{k+1}(t) = -a_2 \sum_{j=1}^{N} s_{(k+1)/2,j}\varphi_j(t) \qquad k = 1,3,5, \ldots, (2N-1) \tag{85}$$

$$y_k(t) = \varphi_{(k+1)/2}(t) + \gamma_{(k+1)/2}u(t)$$

For a two-term approximation, $\underline{y}(t)$ is a 4 x 1 vector and the state equation is

$$\dot{\underline{y}}(t) = \begin{bmatrix} -a_1 & 1 & 0 & 0 \\ -a_2 s_{11} & 0 & -a_2 s_{12} & 0 \\ 0 & 0 & -a_1 & 1 \\ -a_2 s_{21} & 0 & -a_2 s_{22} & 0 \end{bmatrix} \underline{y}(t) + a_2 \begin{bmatrix} 0 \\ s_{11}\gamma_1 + s_{12}\gamma_2 \\ 0 \\ s_{21}\gamma_1 + s_{22}\gamma_2 \end{bmatrix} u(t)$$

$$\underline{y}(0) = \underline{0} \tag{86}$$

The strain gage was located at $x_m = 0.115$ and the beam is observable up to the fifth mode at x_m, even if the gage averages the measurements over ±1.5% of the normalized spatial domain.

A least squares measurement error criterion

$$J(a) = \int_{t_0}^{t_f} [z(t) - v_4(x_m,t)]^2\, dt$$

$$= \int_{t_0}^{t_f} [z(t) - \sum_{k=1}^{N} [y_{2k-1}(t) - \gamma_k u(t)]F''(x_m)]^2\, dt \tag{87}$$

was minimized using a Hooke-Jeeves search technique to yield estimates of the parameter vector $\underline{a}$. For this problem $t_0 = 0.00$, $t_f = 0.250$, and the initial parameter estimate was chosen arbitrarily to be $\underline{a}^0 = (10.0, 1000.)$. The stopping criterion was that the step size be reduced to 0.1% of the range of the parameters where $-100 \le a_1 \le 500$, and $500 \le a_2 < 2000$. The results for a one-, two-, three-, and four-term approximation are shown in Table 4.

TABLE 4

Results for the Beam Equation with Assumed Values
$a_1^* = D/M = 2.154$ and $a_2^* = EI/mL^4 = 889.8$

Number of terms in the approximation	1	2	3	4
a_1	2.910	1.933	2.037	2.639
a_2	861.0	872.9	874.3	878.7
J(a)	1.336×10^{-4}	4.232×10^{-5}	4.577×10^{-5}	1.536×10^{-4}
% error a_1	35.1	10.3	5.4	22.5
% error a_2	3.24	1.90	1.74	1.25

For comparison purposes, classical techniques were used to measure the "true" field parameter values which were found to be $\underline{a}^* = (2.154, 889.8)$.

The beam model given in Eq. (77) assumes $a_1 = D/m =$ constant. This is equivalent to assuming the product $\delta_n \omega_n$ constant for all modes (δ_n and ω_n are the damping ratio and natural frequency of the nth mode, respectively). However, the damping for the higher modes is in reality greater than for the first mode. The relatively large error in a_1 and the observation that the four-term approximation yields poorer results than the two- and three-term approximations is accounted for by the model's assumption of constant $\delta_n \omega_n$. The estimate of a_1 is actually some compromise and takes into account all modes present in the output. Even though the model is not completely accurate, it does give an adequate simple representation of the behavior of the beam since the first mode is dominant.

A good test of the identified parameters is to use the values and to compare the resulting estimate of the normalized moment with the actual measurements. Figure 7 shows the results of this comparison. From the results it is concluded that a two-term approximation (a computation time $\doteq$ 58 sec) gives adequate results even though the three-term approximation is somewhat more accurate.

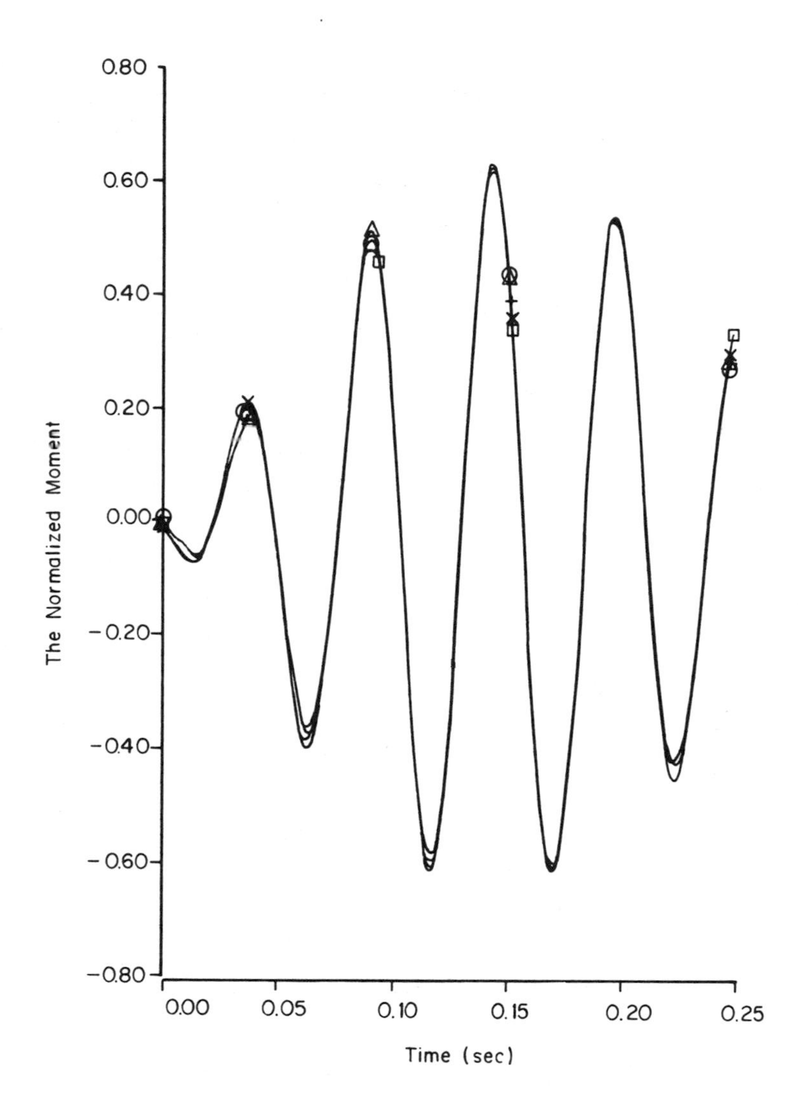

FIG. 7. The measured and estimated normalized moment at $x_m = 0.115$.

C. Identification of Nonlinear Boundary Conditions

Systems considered in this section are those which can be adequately represented by known one-dimensional, linear, stationary field dynamics but have boundary conditions which are nonlinear algebraic equations in the field energy variables. Heat conduction in a continuum with a radiation boundary condition is an example of such a system. The method presented in this section identifies the unknown parameters in these nonlinear algebraic relations between the energy variables at the boundaries.

1. Development of the Identification Procedure

The method presented in this section, as shown in Figure 8, takes advantage of several properties of distributed parameter systems. In simulating partial differential equations one can separately simulate the field dynamics and the boundary conditions and then combine them to obtain the interactive response [142]. Furthermore, since internal spatial data appropriately located are adequate along with the field model to describe response throughout the entire field, one can use information at a sensor located within the field to determine the energy variable response at the unknown boundary. Thus, the appropriate energy variable can be simulated at the unknown boundary using only as many appropriately located sensors as there are unknown boundary conditions. Having recovered the boundary energy variable by such simulation, the boundary forcing can then be expressed as a function of only the unknown boundary parameters. If the unknown boundary condition is appropriately modeled, this boundary forcing will be a linear function of the unknown parameters even though nonlinear in the field variables. A linear function in the unknown parameters is necessary for the full power of the proposed method to be utilized.

After forming the boundary forcing at the unknown boundary as a linear function of the unknown parameters, an error is defined at the original measurement location between the actual measured response and the simulated response using the unknown boundary condition as forcing. An ISE performance index is then minimized, yielding a set of linear algebraic equations which are solved digitally for the unknown parameters. Since a digital computer is used, a steep descent technique is not necessary to solve the optimization problem. The method is best explained by example.

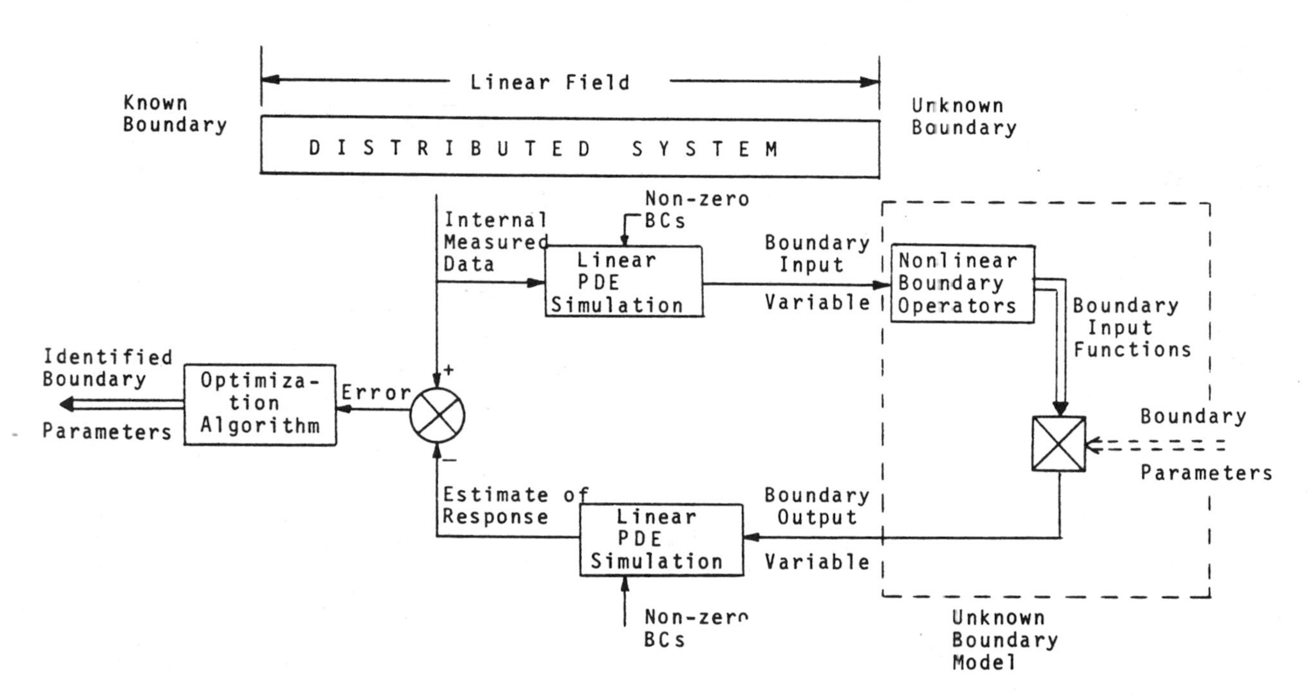

FIG. 8. Block diagram of the basic identification method.

2. Application of the Indentification Method

Example 1. Heat conduction with a radiation and forced convection boundary condition. The problem considered in this first example is one-dimensional heat conduction in a homogeneous slab with boundary radiation and forced convection. Constant diffusivity, environment temperature, and the initial uniform constant temperature are known. The field dynamics are

$$\frac{\partial v_1}{\partial t} = -\frac{\partial v_2}{\partial x}$$
$$v_2 = -\frac{\partial v_1}{\partial x} \tag{88}$$

The symbols are standard for the diffusion equation. Since the heat conduction is assumed symmetrical, only half of the slab is simulated and the boundary at the center, $x = 0$, is equivalent to an insulated boundary. At the unknown boundary condition, $x = 1$, heat transfer occurs by combined radiation and forced convection. Thus the boundary conditions are

$$v_2(0,t) = 0 \tag{89}$$

$$v_2(1,t) = c_1\{[v_1(1,t) + 1]^4 - \theta_c^4\} + c_2\{[v_1(1,t) + 1] - \theta_c\} \tag{90}$$

Data were available in the literature for this problem with $c_1 = 1.0$, $c_2 = 0$, and for $c_1 = 1.0$, $c_2 = 1.0$ [142]. These data in the literature were generated at $x = 1$ by an iterative numerical scheme given the c_1 and c_2 values. By further simulation, values of $v_1(0.5,t)$ were generated as the measurement data for the identification method. Figure 9 shows the necessary steps in applying the identification method to this example.

By applying the Laplace transform to the PDE, an ODE is obtained in terms of $v_1(x,s)$ and x where s is the Laplace variable. If the actual measured internal data at $x = x_m$ is $z(t)$, then the spatial conditions used to recover $v_1(1,t)$ are Eq. (77) and

$$v_1(x_m,t) = z(t) \tag{91}$$

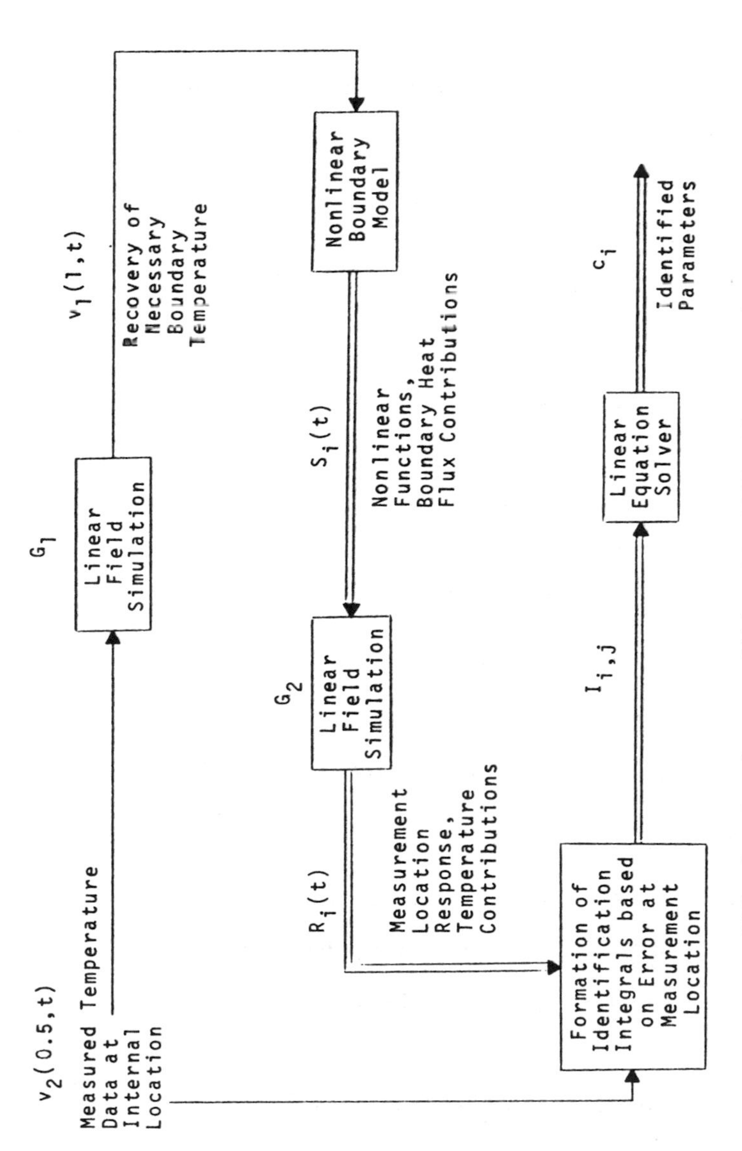

FIG. 9. Boundary condition identification procedure.

Using these spatial conditions, a transcendental transfer function, G_1, is obtained relating $v_1(1,t)$ to $z(t)$.

$$\hat{v}_1(1,s) = G_1[\hat{z}(s)] \tag{92}$$

To simulate this transfer function, defined in the Appendix, truncated infinite product expansions were used as proposed by Goodson [57].

Having recovered boundary temperature, $v_1(1,t)$, the unknown boundary heat flux is formed as

$$v_2(1,t) = S_1(t) + S_2(t) \tag{93}$$

where

$$S_1(t) = \{[v_1(1,t) + 1]^4 - \theta_c^4\} \tag{94}$$

and

$$S_2(t) = \{[v_1(1,t) + 1] - \theta_c\} \tag{95}$$

The response at the measurement location can be estimated as a function of the unknown vector $\underline{c} = (c_1,c_2)$ using $v_2(t)$ as a spatial condition. The spatial conditions for this simulation are Eq. (89) and $v_2(1,t)$ in Eq. (93). This yields

$$\hat{v}_1(x_m,s) = G_2[\hat{v}_2(1,s)] \tag{96}$$

Since $v_1(1,t)$ itself is not known, but $S_1(t)$ and $S_2(t)$ are known, the estimate of $v_1(0.5,t)$ is expressed as a function of the unknown parameters c_1 and c_2.

$$\hat{v}_1(0.5,s) = c_1G_2[\hat{S}_1(s)] + c_2G_2[\hat{S}_2(s)] \tag{97}$$

To identify c_1 and c_2, an error, $\epsilon(t)$, is defined as

$$\epsilon(t) = R_3(t) - [c_1R_1(t) + c_2R_2(t)] \tag{98}$$

where

$$\hat{R}_3(s) = \hat{z}(s), \quad \text{and} \quad \hat{R}_i(s) = G_2[\hat{S}_i(s)] \qquad i = 1,2 \tag{99}$$

An ISE performance index is minimized to find the "best" c_1 and c_2.

$$IP = \frac{1}{2}\int_{t_a}^{t_b} \epsilon^2(t)\, dt \tag{100}$$

Minimization of the performance index results in a set of linear algebraic equations

$$\begin{aligned} c_1 I_{11} + c_2 I_{12} &= I_{13} \\ c_1 I_{21} + c_2 I_{22} &= I_{23} \end{aligned} \tag{101}$$

where

$$I_{i,j} = \int_{t_a}^{t_b} R_i(t) R_j(t)\, dt \tag{102}$$

Since a digital computer was used in this method, this set of linear algebraic equations was solved directly using a Gaussian elimination routine.

Using data at $x_m = 0.500$ from Crosbie and Viskanta for radiation alone at the boundary ($c_2 = 0$, $c_1 = 1.0$), the simulated results in Figure 10 were obtained and $c_1 = 0.974$ was identified. Thus the radiation coefficient was identified to within 3% of the correct value. Using this identified value of c_1 the agreement with the actual measured data is quite good.

The results are shown in Figure 11 using data at $x_m = 0.500$ for combined radiation and forced convection at the boundary ($c_1 = 1.0$, $c_2 = 1.0$). In this case the sensitivity to the radiation contribution is much less than the convection contribution. Thus, the radiation coefficient, c_1, was identified only to within 20% of the correct value, but the convection coefficient, c_2, was correct to within 1%. More importantly, the estimated $v_1(0.5,t)$ resulting from these identified parameters virtually coincides with the measured data. It was also observed that a higher-order model for G_1 would allow better identification of c_1 (within 5-10%) because of inclusion of less significant eigenvalue terms.

Engineering judgment must be used to select the order of the simulation model. A more complete and therefore more costly simulation model results in the more accurate quantitative parameter identification. However, a much simpler field model, while resulting in less accurate parameter identification, may still yield equally accurate state estimation results. In actual practice, the agree-

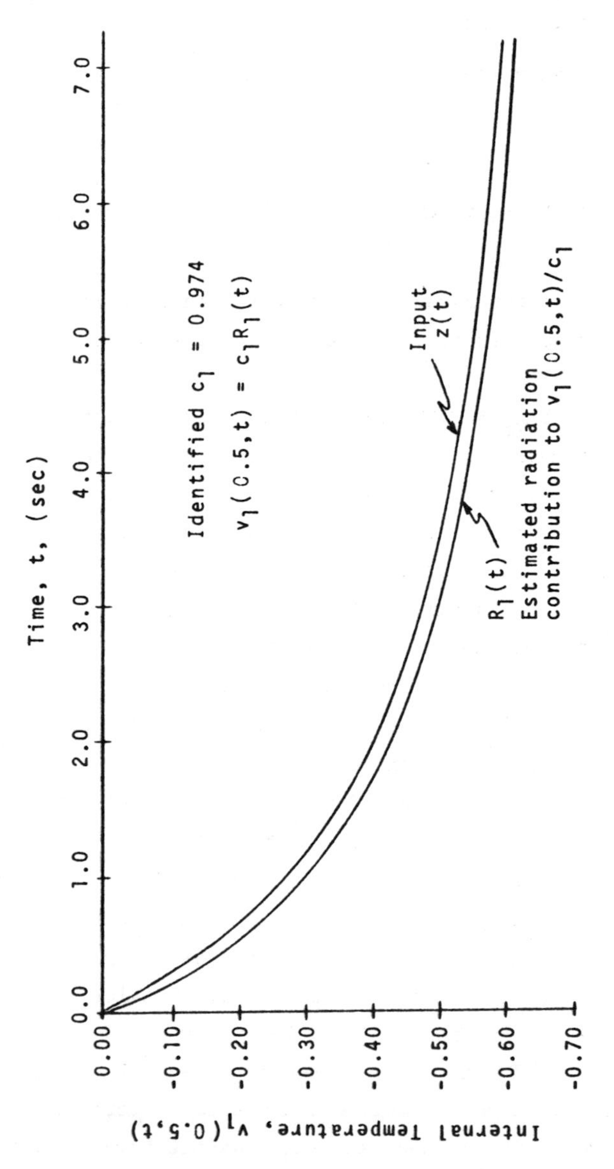

FIG. 10. Simulated $R_1(t)$ compared with internal input temperature, $z(t)$, $x_m = 0.5000$.

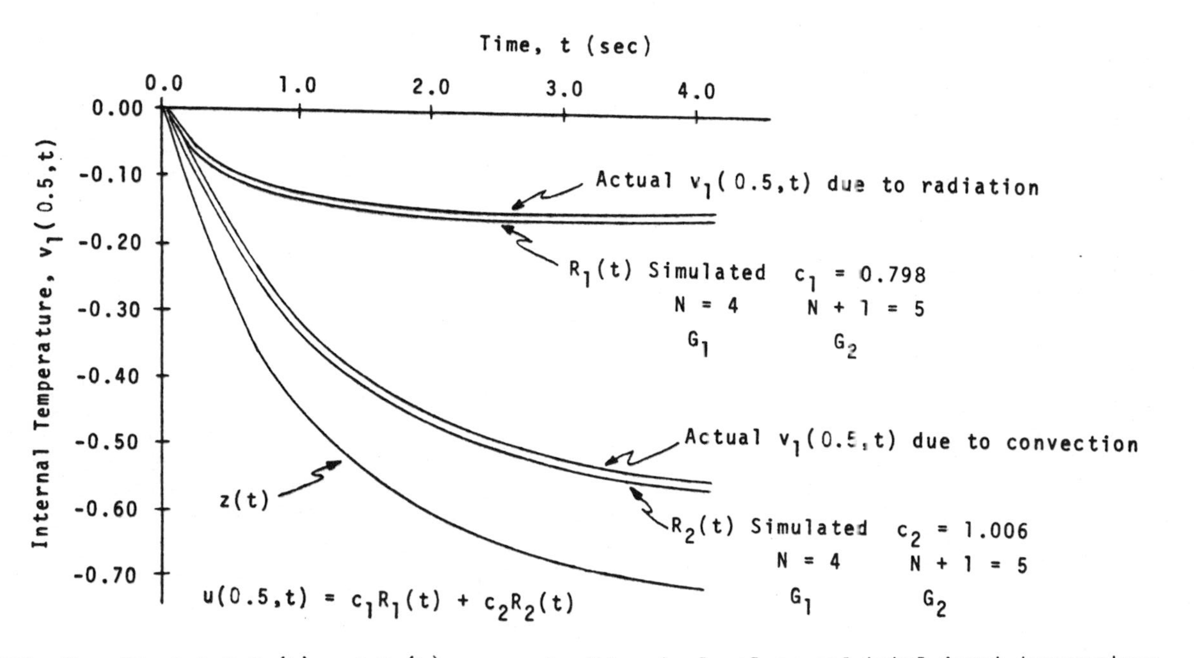

FIG. 11. Simulated $R_1(t)$ and $R_2(t)$ compared with actual values and total input temperature, $z(t)$, at $x_m = 0.500$.

ment between measured and estimated responses would generally be the final measure of a successful identification. Since the field dynamics are known, well-known linear system techniques such as frequency response can be used to determine which **eigenvalues** are significant and therefore where the model should be truncated. A more complete consideration of field model selection is found in reference 142.

Example 2. Transverse beam vibration with a nonlinear support. In the second example, a long slender beam, sketched in Figure 12, was excited in transverse motion by an electrodynamic shaker table clamped to the right end of the beam as shown in Figures 13 and 14. The boundary condition at $x = 0$ was the unknown nonlinear one which was identified. Furthermore, the energy variables at this boundary could not be measured directly due to the interference of the knife edges. The input motion of the shaker table, $z(t)$, was measured with a reluctance distance detector, and the measurement internal to the field was obtained from a set of strain gages mounted to provide a temperature compensated measurement of bending moment.

A PDE for this thin, uniform beam is

$$
\begin{aligned}
v_2(x,t) &= \frac{\partial v_1}{\partial t}(x,t) \\
v_3(x,t) &= \frac{\partial v_1}{\partial x}(x,t) \\
v_4(x,t) &= \frac{\partial v_3}{\partial x}(x,t) \qquad (103) \\
v_5(x,t) &= \frac{\partial v_4}{\partial x}(x,t) \\
\frac{ML^4}{EI}\frac{\partial v_2}{\partial t} &+ \frac{DL^4}{EI} v_2 + \frac{\partial v_5}{\partial x} = 0
\end{aligned}
$$

as in Eq. (77), $v_1(x,t)$ is the displacement, M is the mass per unit length, L is the length, D is the internal damping, E is the modulus of elasticity, and I is the area moment.

$$v_1(1,t) = u(t) \qquad (104)$$

and

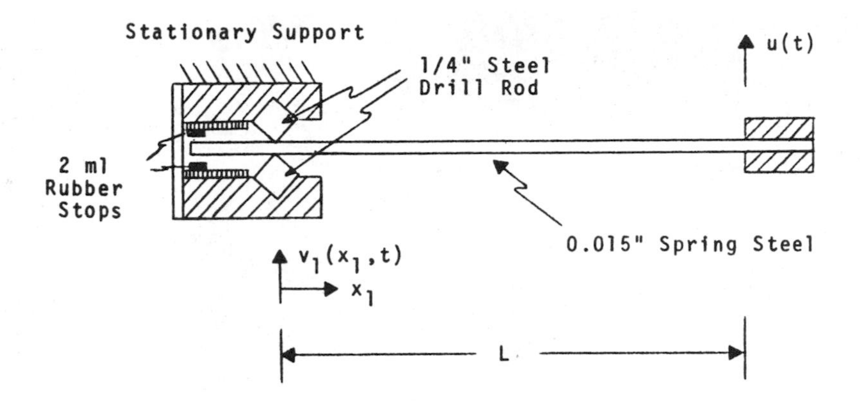

FIG. 12. Test beam mounting configuration.

$$v_3(1,t) = 0 \tag{105}$$

At x = 0, the knife edges restricted vertical motion such that

$$v_1(0,t) = 0 \tag{106}$$

The problem is to formulate and identify the fourth remaining boundary condition.

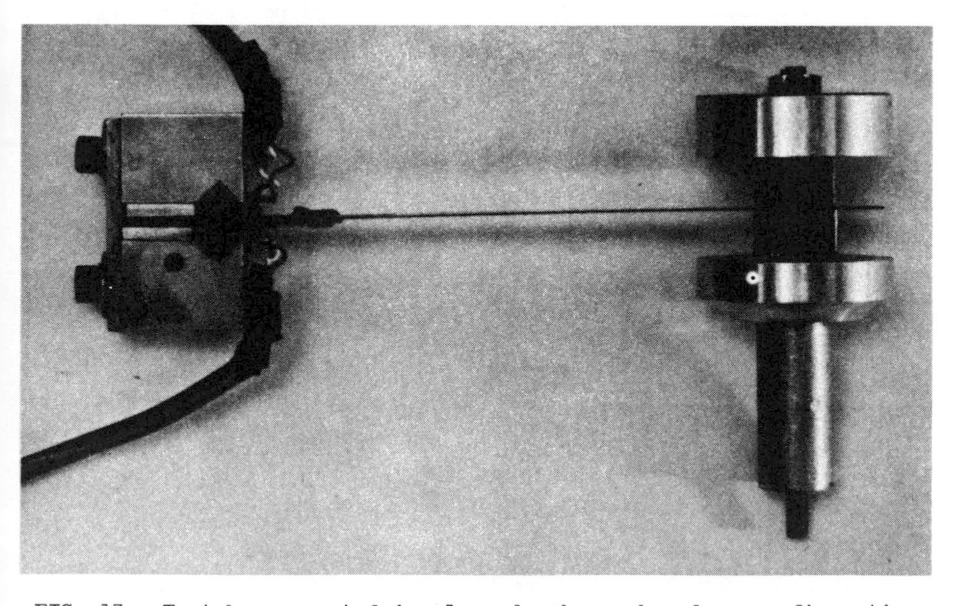

FIG. 13. Test beam mounted in clamped-unknown boundary configuration.

FIG. 14. Beam test stand.

Since the section of the beam inside the boundary at $x_1 = 0$ was only 10% of the total beam length, the displacement of the tip of the beam was assumed to be proportional to the slope at $x_1 = 0, v_3(0,t)$. Because of the eventual contact with the rubber stops, the force exerted on the tip of the beam would be similar to a hardening spring force. Furthermore, the moment at $x_1 = 0, EIv_4(0,t)$ would be proportional to the force on the beam tip. Thus a cubic, hardening spring relationship was assumed for this unknown boundary condition model:

$$v_4(0,t) = c_1S_1(t) + c_2S_2(t) + c_3S_3(t) \tag{107}$$

where

$$S_1(t) = v_3(0,t) \tag{108}$$

$$S_2(t) = [v_3(0,t)]^2 \tag{109}$$

$$S_3(t) = [v_3(0,t)]^3 \tag{110}$$

The c_2 term allows for a nonsymmetrical boundary condition. This boundary model is able to fit a dead band reasonably well.

Having completed the important modeling step, the application of the identification method proceeds in a fashion similar to Example 1. Laplace transforms are applied to the PDE in (103) above. The three known spatial conditions and the internal measured bending moment data, $z(t)$, at $x = x_m$ were used to simulate the boundary slope at $x = 0$, yielding

$$v_3(0,s) = G_1[\hat{u}(s)] + G_2[\hat{z}(s)] \tag{111}$$

where ($\hat{}$) denotes Laplace transformed variables and G_1 and G_2 are defined in the Appendix.

Having recovered the boundary slope using Eq. (111), the unknown boundary condition, $v_4(0,t)$, is formed as a function of the unknown parameters using Eq. (107). Using $v_4(0,t)$ and the three known boundary conditions, an estimate of the bending moment at $x = x_m, \mu(x_m,t)$ is simulated from

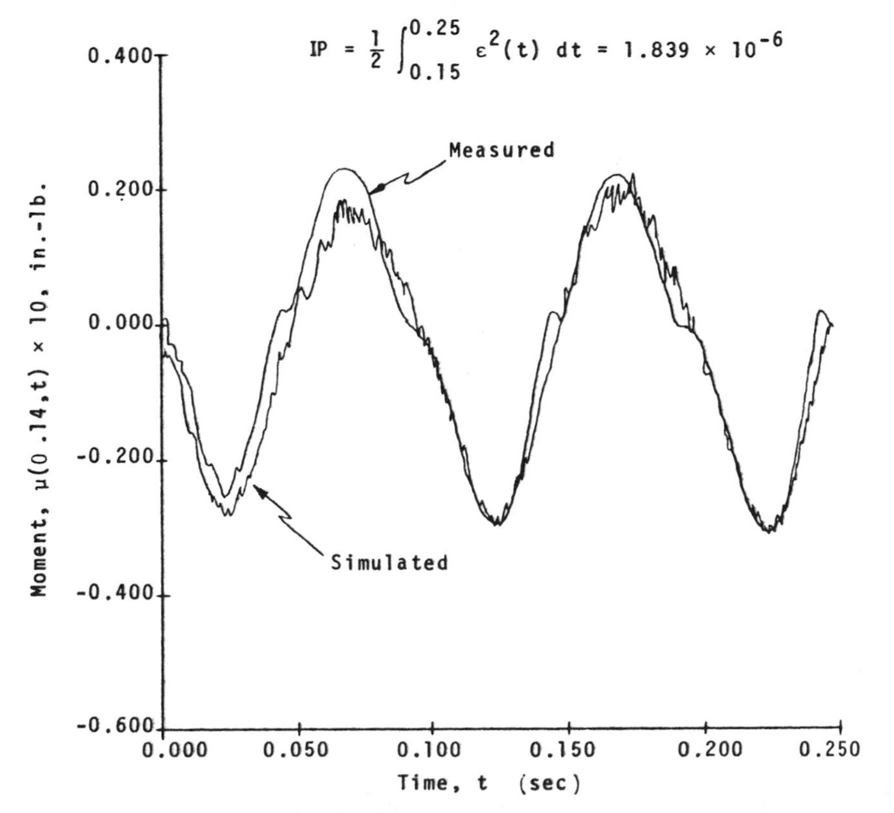

FIG. 15. Simulated $\mu(0.14,t)$. Given c's for $N = 4$, $\zeta_2 = 4\zeta_1$ identification model, $\zeta_2 = 4\zeta_1$ damping model.

$$\hat{\mu}(x_m,s) = G_3[\hat{u}(s)] + G_4[\hat{z}(s)] \tag{112}$$

Now the error at $x = x_m$ is

$$\hat{\varepsilon}(s) = \hat{z}_m(s) - \{c_1G_4[\hat{S}_1(s)] + c_2G_4[\hat{S}_2(s)] + c_3G_4[\hat{S}_3(s)] + G_3[\hat{u}(s)]\} \tag{113}$$

Linear regression analysis using an ISE performance index again yields a set of linear algebraic equations which are solved for the desired c's as in Example 1.

After the boundary parameters were identified, a separate closed loop simulation was used to check the results against the actual measured data. These results are presented in Figure 15 for the identified boundary condition which is plotted in Figure 16. Although ringing still exists in the simulation model due to an imprecise damping model, it is significant to note that the identification

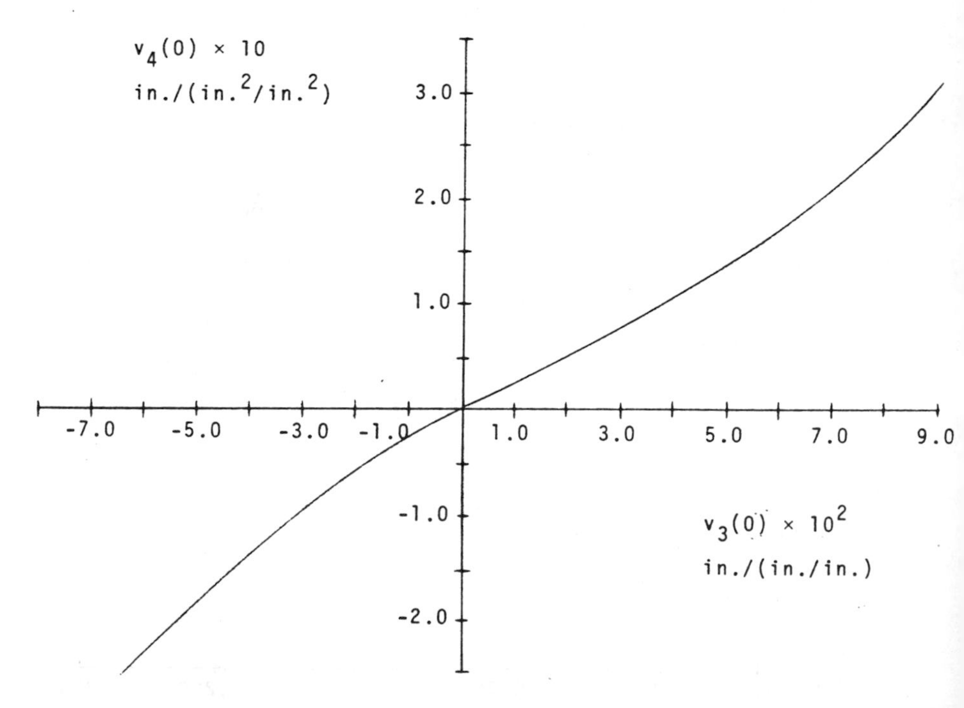

FIG. 16. Plot of identified boundary, $v_4(0)$ vs $v_3(0)$, for best identified c's.

method was nevertheless able to successfully match the peak amplitude after the first two peaks. These peak amplitudes are important since they aid in predicting failure of a part. Distortion of the first two peaks was caused by FM tape recording of the measured data.

IV. DISCUSSION

In summary, to implement the method presented in these examples, the user:

1. Develops the model for the linear field with the known boundary conditions
2. Formulates the unknown nonlinear boundary condition as an algebraic relationship linear in the unknown parameters
3. Solves, by simulation, for the boundary variable necessary as input in (2) using internal measured data for the missing spatial condition in the model in (1)
4. Generates the unknown boundary output variable as a function of the unknown parameters using the boundary model in (2) and the results of (3)
5. Solves, by simulation, for the response at the original measurement location as a function of the unknown parameters using the results of (4) for the missing spatial condition in the model in (1)
6. Identifies the unknown parameters using regression analysis applied to an error between the actual measured data and the estimate in (5)

A particularly important result is the indication that a single measurement not at the boundary includes enough information for identification of parameters in the boundary given some knowledge of the field dynamics.

APPENDIX: TRANSFER FUNCTIONS AND INFINITE PRODUCT APPROXIMATIONS

<u>Example 1. Heat conduction.</u>

$$G_1 = \frac{\cosh \sqrt{s}}{\cosh \sqrt{s}(x_m)} \simeq \frac{\prod_{m=1}^{N-1} \left[1 + \frac{s}{\left(\frac{2m-1}{s}\right)^2 \pi^2}\right]}{\prod_{n=1}^{N} \left[\frac{s}{\left(\frac{2n-1}{s}\right)^2 \frac{\pi^2}{x_m^2}}\right]}$$

For $x_m = 0.500$ and a fourth-order model ($N = 4$),

$$G_1 \simeq \frac{\left(\frac{s}{2.47}+1\right)\left(\frac{s}{22.2}+1\right)\left(\frac{s}{61.7}+1\right)}{\left(\frac{s}{9.87}+1\right)\left(\frac{s}{88.8}+1\right)\left(\frac{s}{246.7}+1\right)\left(\frac{s}{483.6}+1\right)}$$

$$G_2 = \frac{\cosh\sqrt{s}(x_m)}{\sqrt{s}\sinh\sqrt{s}} \simeq \frac{\prod_{m=1}^{N-1}\left[1+\frac{s}{\left(\frac{2m-1}{2}\right)^2\frac{\pi^2}{x_m^2}}\right]}{s\prod_{n=1}^{N}\left[1+\frac{s}{n^2\pi^2}\right]}$$

For $x_m = 0.500$ and a fifth-order model ($N = 4$),

$$G_2 \simeq \frac{\left(\frac{s}{9.87}+1\right)\left(\frac{s}{88.8}+1\right)\left(\frac{s}{246.7}+1\right)}{s\left(\frac{s}{9.87}+1\right)\left(\frac{s}{39.4}+1\right)\left(\frac{s}{88.8}+1\right)\left(\frac{s}{157.8}+1\right)}$$

Example 2. Beam equation.

$$G_1 = \frac{\gamma[(\sinh\gamma+\sin\gamma)(\sinh\gamma x_m+\sin\gamma x_m)]}{\text{Denom.}}$$

$$\frac{\gamma[(\cosh\gamma-\cos\gamma)(\cosh\gamma x_m+\cos\gamma x_m)]}{\text{Denom.}}$$

where

$$\text{Denom.} = [\cosh\gamma(1-x_m)+\cos\gamma(1-x_m)](\sinh\gamma-\sin\gamma)$$
$$+ [\sinh\gamma(1-x_m)+\sin\gamma(1-x_m)](\cos\gamma-\cosh\gamma)$$

and

$$\gamma^4 = -\left(\frac{mL^4}{EI}s^2+\frac{DL^4}{EI}s\right)$$

For $x_m = 0.14$ and a fourth-order model, infinite products yield

$$G_1 = \frac{1.367\left(1+\frac{\gamma^4}{1116.04}\right)}{\left(1-\frac{\gamma^4}{348.08}\right)\left(1-\frac{\gamma^4}{3947.53}\right)}$$

$$G_2 = \frac{-0.316 \frac{L^2}{EI}\left(1 - \frac{\gamma^4}{500.56}\right)}{\left(1 - \frac{\gamma^4}{348.08}\right)\left(1 - \frac{\gamma^4}{3947.53}\right)}$$

$$G_3 = \frac{\gamma^2(\cos\gamma \sinh\gamma x + \cosh\gamma \sin\gamma x)}{(\sinh\gamma \cos\gamma - \cosh\gamma \sin\gamma)}$$

$$G_4 = \frac{(\cos\gamma - \cosh\gamma)[\sinh\gamma(1-x) + \sin\gamma(1-x)]}{2(\sinh\gamma \cos\gamma - \cosh\gamma \sin\gamma)} + \frac{(\sinh\gamma - \sin\gamma)[\cosh\gamma(1-x) + \cos\gamma(1-x)]}{2(\sinh\gamma \cos\gamma - \cosh\gamma \sin\gamma)}$$

For $x_m = 0.14$ and a fourth-order model, infinite products yield

$$G_3 = \frac{-0.42 \frac{EI}{L^2}\left(1 + \frac{\gamma^4}{25.36}\right)}{\left(1 - \frac{\gamma^4}{237.72}\right)\left(1 - \frac{\gamma^4}{2496.49}\right)}$$

$$G_4 = \frac{0.790 \frac{EI}{L^2}\left(1 - \frac{\gamma^4}{348.08}\right)}{\left(1 - \frac{\gamma^4}{237.72}\right)\left(1 - \frac{\gamma^4}{2496.49}\right)}$$

Examples of Observability Results

The observability criterion of uniqueness as established by the PDE, BCs, and measurements allows the selection of measurement locations for several types of equations; some new and some previously published results are included here.

Example 1. The one-dimensional diffusion equation. The normalized diffusion equation is given by

$$\begin{bmatrix} 1 & 0 \\ 0 & 0 \end{bmatrix} \frac{\partial \underline{v}}{\partial t}(x,t) + \begin{bmatrix} 0 & -1 \\ 1 & 0 \end{bmatrix} \frac{\partial \underline{v}}{\partial x}(x,t) = \begin{bmatrix} 0 & 0 \\ 0 & 1 \end{bmatrix} \underline{v}(x,t) \qquad (114)$$

$$x \in \Omega = [0,1] \qquad t \geq 0$$

with BCs

$$\begin{aligned} c_{11}v_1(0,t) - c_{12}v_2(0,t) &= 0 \\ c_{21}v_1(1,t) + c_{22}v_2(1,t) &= 0 \end{aligned} \qquad (115)$$

and IC

$$v_1(x,0) = v^o(x) \tag{116}$$

For point measurements $z(x_i,t)$ of the dynamical state $v_1(x,t)$, $D(x,t)$ in Eq. (11) would become

$$d(x,t) = \begin{bmatrix} \delta(x - x_i) & 0 \\ 0 & 0 \end{bmatrix} \tag{117}$$

where $\delta(\cdot)$ is a delta function. For observability defined by the sufficiency of the BCs and data to provide a unique solution $v(x,t_s)$ over the domain $\Omega_D = (0 \le x \le 1,\ t_0 \le t_s \le t_1)$, the following measurement location results may be derived. The technique for determining uniqueness is to investigate the zeros of the infinite set of eigenfunctions for the solution [60].

(a) BC: No BC information, $c_{i,j}$ arbitrary. IC: $v^o(x)$ arbitrary. Measurement: $\underline{z} = [v_1(x_1,t), v_2(x_1,t)]$. This is equivalent to measurement of the temperature and heat flux simultaneously at a point.

Result: The system is observable for all $x_1, 0 \le x_1 \le 1$.

(b) BC: $c_{12} = c_{22} = 0$ (constant boundaries). IC: $v^o(x)$ arbitrary. Measurement: $z(t) = v_1(x_i,t)$, $i = 1,2, \ldots, S$.

Result: For all x_i rational, the system is unobservable for any S.

(c) BC: $c_{12} = c_{21} = 0$ (mixed BC). IC: $v^o(x)$ arbitrary. Measurement: $z(t) = v_1(1/2,t)$, i.e., a measurement transducer located at the center of the spatial domain.

Result: The system is observable.

(d) BC: $c_{12} \ne 0$. IC: $v^o(x)$ arbitrary. Measurement: $z(t) = v_1(0,t)$.

Result: The system is observable.

(e) BC: No BC information, $c_{i,j}$ arbitrary. IC: $v^o(x)$ arbitrary. Measurement:

$$D(x,t) = \begin{bmatrix} h(x) & 0 \\ 0 & 0 \end{bmatrix} \tag{118}$$

for $h(x) > 0$ in either of the ranges

$$0 < x < \frac{1}{2N - 1}, \qquad \frac{2N - 2}{2N - 1} < x < 1 \tag{119}$$

and zero elsewhere (spatial averaging of the point measurements).
Result: The system is N mode observable. This result provides a sufficiency condition for the observability of the first N modes for arbitrary BCs. The result follows from the fact that there are no zeros of the eigenfunctions

$$\varphi_n(x) = \sin \gamma_n x + \frac{c_{12}}{c_{11}} \gamma_n \cos \gamma_n x \tag{120}$$

for

$$\left(1 - \frac{c_{22}}{c_{21}} \frac{c_{12}}{c_{11}} \gamma_n^2\right) \sin \gamma_n = - \left(\frac{c_{12}}{c_{11}} + \frac{c_{22}}{c_{21}}\right) \gamma_n \cos \gamma_n \qquad n = 1,2, \ldots ,N \tag{121}$$

in the ranges of x specified in Eq. (21).

Example 2. A first-order, quasilinear PDE. For the first-order scalar equation

$$\frac{\partial v}{\partial t}(x,t) + a(x) \frac{\partial v}{\partial x}(x,t) = f(x,v,t) \tag{122}$$

$a(x) > 0$, $\Omega = (0 \leq x \leq 1)$, $t \geq 0$, and for BCs and ICs given by

$$v(0,t) = h(t), \qquad v(x,0) = v^0(x) \tag{123}$$

the following results apply for the observability which is defined by the existence of a uniqueness proof for $v(x,t_s)$ as given by the BC and measurement.

(a) BC: $h(t)$ arbitrary. IC: $v^0(x)$ arbitrary. Measurement: $z(t) = v(x_1,t)$. $0 \leq x_1 \leq 1$, $t_0 \leq t \leq t_1$.
Result: The system defined by Eqs. (24) and (25) is equivalent to the system of first-order ODs given by

$$\frac{dv}{dt} = f \qquad \frac{dx}{dt} = a \tag{124}$$

Where an existence proof exists for the solution $v(x(t),t)$, e.g., for f linear in v and time invariant, time invariant and power series in v, then $v(x,t_s)$ is uniquely determined in the domain

$$\Omega_D(t_s) = (0 \le x \le 1,\ t_0 + T_1 - T_m \le t_s \le t_1 - T_m) \tag{125}$$

where

$$T_1 = \int_0^1 \frac{1}{a(x)}\,dx, \qquad T_m = \int_0^{x_1} \frac{1}{a(x)}\,dx \tag{126}$$

Other examples can be cited for second-order hyperbolic equations, linear diffusion equations in more than one space dimension, and systems of first-order PDEs.

ACKNOWLEDGMENTS

We would like to thank the many authors who kindly sent us reprints of their articles and lists of references. We especially thank Professor A. G. Butkovskiy for some of the references to the literature published in the U.S.S.R., and Professor S. H. Crandall for permission to use the two photographs in Figures 1 and 2. This work was supported in part by a National Research Council of Canada Grant A-8670, and under Office of Naval Research Contract No. N-00014-67-A-0226-0012.

REFERENCES

1. J. Aguilar Martin, G. Alengrin, and C. Hernandez, Estimation of the diffusion coefficient of a fluidized bed catalytic reactor using non linear filtering techniques, Identification and System Parameter Estimation (P. Eykhoff, ed.), Proc. 3rd IFAC Symp., The Hague/Delft, the Netherlands, June 12-15, 1973, pp. 255-264.
2. J. Aguilar Martin, G. Continente, G. Alengrin, and C. Hernandez, Méthodes recurrentes d'estimation de modèles dynamiques de processus chimiques industriels. Applications, Fédération Européenne de Génie Chimique, Congrès International, Paris, April 24-28, 1973.
3. E. Angel and A. K. Jain, A dimensionality reducing model for distributed filtering, IEEE Trans. Auto. Contr., vol AC-18, no. 1: 59-62 (1973). See also A nearest neighbors approach to multidimensional filtering, Proceedings of the 1972 IEEE Decision and Control Conference, Dec. 13-15, New Orleans, pp. 84-88.
4. K. J. Astrom and P. Eykhoff, System identification-- A survey, Automatica, 7: 123-162 (1971).

5. M. Athans, Toward a practical theory for distributed parameter systems, IEEE Trans. Auto. Contr., vol. AC-15, no. 2, 245-247 (1970).

6. S. R. Atre, A note on Kalman-Bucy filtering for linear distributed parameter systems, IEEE Trans. Auto. Contr., vol. AC-17, no. 5, 712-713 (1972).

7. S. R. Atre and S. S. Lamba, Boundary filter for linear distributed parameter systems, Ninth Annual Allerton Conference on Systems and Circuit Theory, 1971, pp. 1213-1221.

8. S. R. Atre and S. S. Lamba, Filtering for linear distributed parameter systems via boundary measurements, Proc. IEE (London), 119: 757-759 (1972).

9. S. R. Atre and S. S. Lamba, Optimal estimation in distributed processes using the innovations approach, Preprints IFAC Symposium on the Control of Distributed Parameter Systems, Banff, Alberta, Canada, June 1971, paper no. 13-8. See also IEEE Trans. Auto. Contr., vol. AC-17, no. 5, 710-712 (1972).

10. S. R. Atre and S. S. Lamba, Derivation of an optimal estimator for distributed parameter systems via maximum principle, IEEE Trans. Auto. Contr., AC-17: 388-390 (1972). See also Optimal estimator for linear distributed parameter systems, Proc. IEEE Symposium on Adaptive Processes, Decision and Control Conference, Austin, Texas, Dec. 1970.

11. A. V. Balakrishnan, Identification and stochastic control of non-dynamic systems, Identification and System Parameter Estimation (P. Eykhoff, ed.), Proc. 3rd IFAC Symp., The Hague/Delft, The Netherlands, June 12-15, 1973, pp. 679-686.

12. A. V. Balakrishnan and J. L. Lions, State estimation for infinite-dimensional systems, J. Compu. Syst. Sci., 1, 391-403 (1967).

13. J. V. Beck, Analytical determination of optimum transient experiments for measurements of thermal properties, Amer. Institute of Chemical Engineers, Proceedings of the Third International Heat Transfer Conference, vol. IV, Aug. 1966, pp. 74-80.

14. J. V. Beck, Calculation of thermal diffusivity from temperature measurements, Trans. ASME J. Heat Transf., Series C, 85: 181-182 (1963).

15. J. V. Beck, "The optimum analytical design of transient experiments for simultaneous determinations of thermal conductivity and specific heat," Ph.D. thesis, Michigan State Univ., Dept. of Mechanical Engineering, 1954.

16. R. J. Bennett, Identification of the minimal representation of linear distributed parameter systems, in Identification of Parameters in Distributed Systems (R. E. Goodson and M. P. Polis, eds.), ASME Monograph, 1974, pp. 91-97.

17. A. Bensoussan, Filtrage Optimale des Systèmes Linéaires, Dunod, Paris, 1971.

18. A. Bensoussan, Optimization on sensor's location in a distributed filtering problem, in Stability of Stochastic Dynamical Systems, Lecture Notes in Mathematics, no. 294, Springer-Verlag, Berlin, 1972.

19. B. K. Bhagavan and L. R. Nardizzi, Parameter identification in linear distributed parameter systems, IEEE Trans. Auto. Contr., vol. AC-18, no. 6, 667-679 (1973).

20. A. G. Butkovskiy, Control in distributed systems (a review), Proceedings of the IFAC Symposium on the Control of Distributed Parameter Systems, vol. II, Banff, Alberta, Canada, June 1971, pp. 1-9.

21. J. R. Cannon and R. E. Klein, Optimal selection of measurement locations in a conductor for approximate determination of temperature distributions, 1970 JACC, Atlanta, Georgia, June 22-26, 1970, pp. 750-756.

22. W. T. Carpenter, "The identification of distributed parameter systems," Ph.D. thesis, Purdue Univ., Lafayette, Indiana, August 1969.

23. W. T. Carpenter, M. J. Wozny, and R. E. Goodson, Distributed parameter identification using the method of characteristics, Trans. ASME J. Dynam. Syst. Meas. Contr., Series G, vol. 93, no. 2, 73-78 (1971).

24. S. P. Chaudhuri, A class of distributed parameter system identification via differential approximation and quasi linearization, 5th Hawaii International Conference on System Sciences, Univ. Hawaii, Jan. 8-12, 1973.

25. G. Chavent, "Analyse fonctionnelle et identification de coefficients répartis dans les équations aux dérivées partielles," thesis, Univ. of Paris, Paris, 1971.

26. G. Chavent, Identification of distributed parameters, Identification and System Parameter Estimation (P. Eykhoff, ed.), Proc. 3rd IFAC Symp., The Hague/Delft, The Netherlands, June 12-15, 1973, pp. 649-660.

27. G. Chavent, Identification of functional parameters in partial differential equations, in Identification of Parameters in Distributed Systems (R. E. Goodson and M. P. Polis, eds.), ASME Monograph, 1974, pp. 31-48.

28. G. Chavent, Une méthode de résolution de problème inverse dans les équations aux dérivées partielles, Bulletin de l'académie polonnaise des Sciences, Série des sciences techniques, vol. XVIII, no. 8, 99-105 (1970).

29. G. Chavent and G.Ciligot-Travain, L'identification des systèmes, IRIA Bull. (1970).

30. G. Chavent, M. Dupuy, and P. Lemonnier, History matching by use of optimal control theory, Fall Meeting SPE, Las Vegas, 1973, paper no. SPE 4627.

31. W. H. Chen and J. H. Seinfeld, Estimation of spatially varying parameters in partial differential equations, Int. J. Contr., vol. 15, no. 3, 487-195 (1972).

32. G. Cilligot-Travain, "Estimation bayésienne et identification des systèmes gouvernés par des équations aux dérivées partielles," thesis Docteur Ingenieur, Univ. of Paris, Paris, 1970.

33. R. Cohen, Identification in vibratory systems: A survey, in Identification of Parameters in Distributed Systems (R. E. Goodson and M. P. Polis, eds.), ASME Monograph, 1974, pp. 49-67.

34. J. D. Collins, J. P. Young, and L. Kiefling, Methods and application of system identification in shock and vibration, in System Identification of Vibrating Structures, Mathematical Models from Test Data (W. O. Pilkey and R. Cohen, eds.), ASME, 1972, Chap. 3.

35. P. L. Collins and H. C. Khatri, Identification of distributed parameter systems using finite differences, Trans. ASME J. Basic Eng., Series D, 91: 239-245 (1969).

36. R. Courant and D. Hilbert, Methods of Mathematical Physics, Vol. II, Interscience, New York, 1962.

37. S. H. Crandall and L. E. Wittig, Chladni's patterns for random vibration of a plate, in Dynamic Response of Structures, Proc. Symp., Stanford Univ., June 28-29, 1971 (G. Hermann and N. Perrone, eds.), Pergamon Press, New York, 1972.

38. B. O. Dale, "Multiparameter identification and optimization methods for linear continuous vibratory systems," Ph.D. thesis, Purdue Univ., Lafayette, Indiana, Jan. 1970.

39. B. O. Dale and R. Cohen, Multiparameter identification in linear continuous vibratory systems, Trans. ASME J. Dynam. Syst. Meas. Contr., Series G, vol. 93, no. 1, 45-52 (1971).

40. M. C. Delfour and S. K. Mitter, Controllability and observability for infinite-dimensional systems, SIAM J. Contr., vol. 10, no. 2, 329-333 (1972).

41. M. C. Delfour and S. K. Mitter, Controllability, observability and optimal feedback control of affine hereditary differential systems, SIAM J. Contr., vol 19, no. 2, 298-328 (1972).

42. D. Detchmendy and R. Sridhar, Sequential estimation of states and parameters in noisy nonlinear dynamical systems, Trans. ASME J. Basic Eng., Series D, 88: 362-368 (1966).

43. S. E. Diamessis, On the simultaneous identification of distributed system parameters and of boundary conditions in the form of power series and Chebychev polynomials. Application to the wave equation of electromagnetics and transmission line theory,

Preprints, IFAC Symposium on the Control of Distributed Parameter Systems, Banff, Alberta, Canada, June 1971, paper no. 13-4.

44. G. Di Pillo and L. Grippo, Application of the ε-technique to distributed parameter systems identification, J. Optim. Theory and Appl., vol. 11, no. 1, 84-99 (1973).

45. G. Di Pillo, L. Grippo, and M. Lucertini, Identification of distributed parameter systems using the epsilon method: A new algorithm, in Identification and System Parameter Estimation (P. Eykhoff, ed.), Proc. 3rd IFAC Symp., The Hague/Delft, The Netherlands, June 12-15, 1973, pp. 687-690.

46. J. Douglass, Jr. and B. F. Jones, Jr., The determination of a coefficient in a parabolic differential equation. Part II, Numerical approximation, J. Math. Mech., vol. II, no. 6, 919-926 (1962).

47. G. E. Dreifke and J. O. Hougen, Experimental determination of system dynamics by pulse methods, Preprints JACC, June 19-21, 1963, pp. 452-458.

48. W. R. Elistratov, Determination of dynamic characteristics on nonstationary systems during their normal operation, Automa. Remote Contr., vol. 24, no. 7, 852-862 (1963).

49. D. J. Ewing and T. J. Higgins, A sensor location algorithm for distributed parameter systems, Ninth Annual Allerton Conference on Circuit and System Theory, 1971, pp. 1203-1212.

50. F. W. Fairman and D. W. C. Shen, Parameter identification for a class of distributed systems, Int. J. Contr., vol. 11, no. 6, 929-940 (1970).

51. P. L. Falb, Infinite dimensional filtering: the Kalman Bucy filter in Hilbert space, Information Contr., no. 11, 102-137 (1967).

52. B. A. Finlayson and L. G. Scriven, The method of weighted residuals: A review, Appl. Mech. Rev., 19: (1966).

53. G. E. Forsythe and W. R. Wasow, Finite-Difference Methods for Partial Differential Equations, Wiley, New York, 1967.

54. C. Gilath, Determination of the dilution of a gas discharged into a ventilated atmosphere, Int. J. Appl. Radia. Isotop., 22: 671-675 (1971).

55. C. Gilath, S. Blit, Y. Yoshpe-Purer, and H. I. Shuval, Radioisotope tracer techniques in the investigation of dispersion of sewage and disappearance rate of enteric organisms in coastal waters, Proceedings of the IAEA Symposium on the Use of Nuclear Techniques in the Measurement and Control of Environmental Pollution, Vienna, 1971, pp. 673-689.

56. C. Gilath and A. Stuhl, Concentration dynamics in a lake with a water current flowing through it, Proceedings of the IAEA Symposium on the Use of Nuclear Techniques in the Measurement and Control of Environmental Pollution, Vienna, 1971, pp. 483-496.

57. R. E. Goodson, Distributed system simulation using infinite product expansions, Simulation, vol. 15, no. 6, 255-273 (1970).

58. R. E. Goodson, "Identification in distributed systems," paper presented to the Japanese Society of Information and Control Engineers, Automatic Control Centre Report AC-72-5, School of Mechanical Engineering, Purdue Univ., Lafayette, Indiana.

59. R. E. Goodson, A. F. D'Souza, R. H. Kohr, and T. J. Williams, Distributed Systems Modeling and Control, Short Course Lecture Notes, Purdue Univ., Lafayette, Indiana, July 1967.

60. R. E. Goodson and R. E. Klein, A definition and some results for distributed system observability, IEEE Trans. Auto. Contr., AC-15, no. 2, 165-174 (1970).

61. R. E. Goodson and M. P. Polis, Parameter identification in distributed systems: A synthesising overview, in Identification of Parameters in Distributed Systems (R. E. Goodson and M. P. Polis, eds.), ASME Monograph, 1974, pp. 1-30.

62. J. Hadamard, Lectures on Cauchy's Problem in Linear Partial Differential Equations, Dover, New York, 1952.

63. M. H. Hamza and M. A. Sheirah, On-line identification of distributed parameter systems, Automatica, 9: 689-698 (1973); see also Preprints, IFAC Symposium on the Control of Distributed Parameter Systems, Banff, Alberta, Canada, June 1971, paper no. 13-6.

64. M. A. Hassan, M. A. R. Ghonaimi, and M. A. Adbel Shaheed, A computer algorithm for optimal discrete time state estimation of linear distributed parameter systems, Preprints, IFAC Symposium on the Control of Distributed Parameter Systems, Banff, Alberta, Canada, June 1971, paper no. 13-7.

65. M. Hwang, J. H. Seinfeld, and G. R. Gavalas, Optimal least square filtering and interpolation in distributed parameter systems, J. Math. Anal. Appl., vol. 39, no. 1, 49-74 (1972).

66. E. L. Itskovich, Determination of optimal distance between pickups in control of space distributed fields, Autom. Remote Contr., vol. 24, no. 2, 217-223 (1963).

67. A. K. Jain, Synthesis of distributed systems by network matching, IEEE Trans. Auto. Contr., vol. AC-19, no. 4, 388-390 (1974).

68. B. F. Jones, Jr., The determination of a coefficient in a parabolic differential equation, Part I: Existence and uniqueness, J. Math. Mech., vol. 11, no. 6, 907-918 (1962).

69. K. W. Kaiser, A method of determining the heater-sensor configuration in temperature distribution controllers, Preprints, IFAC Symposium on the Control of Distributed Parameter Systems, Banff, Alberta, Canada, June 1971, paper no. 1-3.

70. R. E. Kalman, Contributions to the theory of optimal control, International Symposium on Ordinary Differential Equations, Mexican Mathematical Society, 1961, pp. 102-119.

71. S. S. Kaniovsky et al., Some problems of substantiation of the mathematical models of thermal processes in systems with distributed parameters, Preprints, IFAC Symposium on the Control of Distributed Parameter Systems, Banff, Alberta, Canada, June 1971, paper no. 1-5.

72. L. V. Kantorovich and I. Krylov, Approximate Methods in Higher Analysis, Interscience, New York, 1964.

73. R. E. Klein, Non well posed problems in the sense of Hadamard and their relationship to distributed parameter state reconstructor problems, Preprints, IFAC Symposium on the Control of Distributed Parameter Systems, Banff, Alberta, Canada, June 1971, paper no. 7-2.

74. R. E. Klein, Observability: Its role in control and estimation theory, Proceedings of the Seventh Annual Allerton Conference on Circuit and System Theory, Allerton House, Univ. of Illinois, Urbana-Champaign, October 1969, pp. 291-297.

75. R. E. Klein, L. D. Metz, and G. L. Hogg, Identification via orthogonality for a class of hyperbolic distributed parameter systems, Proceedings 1972 IEEE Decision and Control Conference, paper no. WA 6-5, p. 101.

76. H. N. Koivo and A. J. Koivo, Optimal estimation of polluted stream variables, in Identification and System Parameter Estimation (P. Eykhoff, ed.), Proc. 3rd IFAC Symp., The Hague/Delft, The Netherlands, June 12-15, 1973, pp. 327-329.

77. A. J. Koivo and G. R. Phillips, Identification of mathematical models for DO and BOD concentrations in polluted streams from noise corrupted measurements, Water Resources Research, vol. 7, no. 4, 853-862 (1971).

78. O. S. Kozhinsky and N. S. Rajbman, Identification of distributed plants, Preprints, IFAC Symposium on the Control of Distributed Parameter Systems, Banff, Alberta, Canada, June 1971, paper no. 13-5.

79. G. R. V. Kumar and A. P. Sage, On error analysis in optimum and suboptimum filtering in distributed parameter systems, Preprints, IFAC Symposium on the Control of Distributed Parameter Systems, Banff, Alberta, Canada, June 1971, paper no. 8-3.

80. H. J. Kushner, Filtering for linear distributed parameter systems, SIAM J. Contr., 8: 346-359 (1970).

81. G. B. Lamont, Estimation of the states in systems described by partial differential equations, Proceedings of the 13th Midwest Symposium on Circuit Theory, Minneapolis, Minn., May 7-8, 1970.

82. G. B. Lamont and K. S. P. Kumar, Sequential interpolating estimator for nonlinear partial differential equation systems, Preprints, IFAC Symposium on the Control of Distributed Parameter Systems, Banff, Alberta, Canada, June 1971, paper no. 8-5.

83. G. Lamont and K. S. P. Kumar, State estimation in distributed parameter systems via least squares and invariant imbedding, J. Math. Anal. Appl., 38:588-606 (1972).

84. C. Lanezos, Linear Differential Operators, Van Nostrand, New York, 1961.

85. B. Leden, "Identification of dynamics of a one dimensional heat diffusion process," Report 7121, Lund Institute of Technology, Division of Automatic Control, Nov. 1971.

86. B. Leden, M. H. Hamza, and M. A. Sheirah, Different methods for estimation of thermal diffusivity of a heat diffusion process, in Identification and System Parameter Estimation (P. Eykhoff, ed.), Proc. 3rd IFAC Symp., The Hague/Delft, The Netherlands, June 12-15, 1973, pp. 639-648.

87. P. Lemonnier, "Identification dans une équation non-linéaire," Dept. d'Informatique Numérique, INF/7114-71023, Sept. 15, 1971.

88. J. L. Lions, Contrôle Optimal des Systèmes Gouvernés par des Equations aux Derivées Partielles, Dunod, Paris, 1968 (English translation: S. K. Mitter, Springer-Verlag, 1971).

89. J. L. Lions, Optimal control of deterministic distributed parameter systems, Proceedings of the IFAC Symposium on the Control of Distributed Parameter Systems, vol. 2, Banff, Alberta, Canada, pp. 10-24.

90. J. L. Lions and G. Stampacchia, Variational inequalities, Commun. Pure Appl. Math., vol XX, no. 3, 493-519 (1967).

91. D. L. Luckinbill and B. Childs, Parameter identification by Newton-Raphson expansions for partial differential equations, Proceedings of the IFAC Symposium on the Control of Distributed Parameter Systems, Banff, Alberta, Canada, June 1971, paper no. 13-3.

92. S. N. Malpani and J. K. Donnelly, Identification of a packed absorption column, Can. J. Chem. Eng., 51: 479-488 (1973).

93. S. N. Malpani and J. K. Donnelly, Identification of one phase flow processes, Can. J. Chem. Eng., vol. 50, no. 6, 791-795 (1972).

94. C. McGreavy and A. Vago, Estimation of spatially distributed decay parameters in a chemical reactor, Identification and System Parameter Estimation (P. Eykhoff, ed.), Proc. 3rd IFAC Symp., The Hague/Delft, The Netherlands, June 12-15, 1973, pp. 307-316.

95. R. K. Mehra, Time-domain synthesis of optimal inputs for system identification, Proceedings of the 1974 IEEE Conference on Decision and Control, Phoenix, Arizona, pp. 480-487.

96. J. S. Meditch, Least squares filtering and smoothing for linear distributed parameter systems, IFAC Symposium on Identification

and Process Parameter Estimation, Prague, Czechoslovakia, June 15-20, 1970.

97. J. S. Meditch, On state estimation for distributed parameter systems, J. Franklin Inst., vol. 290, no. 1, 49-59 (1970).

98. V. J. Mizel and T. I. Seidman, Observation and prediction for the heat equation, J. Math. Anal. Appl., 28: 303-312 (1969).

99. V. J. Mizel and T. I. Seidman, Observation and prediction for the heat equation, II, J. Math. Anal. Appl., 38:149-166 (1972).

100. F. J. Perdreauville and R. E. Goodson, Identification of systems described by partial differential equations, Trans. ASME J. Basic Eng., 463-468 (1966).

101. G. A. Phillipson, Identification of Distributed Systems, Elsevier, New York, 1971.

102. G. A. Phillipson and S. K. Mitter, State identification of a class of linear distributed systems, Preprints, IFAC Symposium, Warsaw, Poland, 1969, pp. 34-56.

103. W. O. Pilkey and R. Cohen, eds., System Identification of Vibrating Structures, Mathematical Models from Test Data, ASME, 1972.

104. M. P. Polis, "On problems of parameter identification for distributed systems with results using Galerkin's criterion," Ph.D. thesis, Purdue Univ., Lafayette, Indiana, Jan. 1972.

105. M. P. Polis, R. E. Goodson, and M. J. Wozny, On parameter identification for distributed systems using Galerkin's criterion, Automatica, vol. 9, no. 1, Jan. 1973, pp. 53-64.

106. M. P. Polis, R. E. Goodson, and M. J. Wozny, Parameter identification for the beam equation using Galerkin's criterion, Proceedings of the 1972 IEEE Decision and Control Conference, paper no. TP4-3, pp. 391-395.

107. M. P. Polis, M. J. Wozny, and R. E. Goodson, Identification of parameters in distributed systems using Galerkin's method, Preprints, IFAC Symposium on the Control of Distributed Parameter Systems, Banff, Alberta, Canada, June 1971, paper no. 13-2.

108. N. S. Rajbman, The application of identification methods (survey), in Identification and System Parameter Estimation (P. Eykhoff, ed.), Proc. 3rd IFAC Symposium, The Hague/Delft, The Netherlands, June 12-15, 1973, pp. 1-48.

109. A. I. Ruban, Identification of distributed dynamic objects on the basis of a sensitivity algorithm, Eng. Cyb., vol. 9, no. 6, 1137-1142 (1971).

110. R. Sahgal and R. P. Webb, Nonlinear distributed parameter estimation, Proceedings of the 1972 IEEE Decision and Control Conference, New Orleans, Dec. 13-15, 1972, pp. 89-93.

111. Y. Sakawa, Optimal filtering for linear distributed parameter systems, Preprints, IFAC Symposium on the Control of Distributed Parameter Systems, Banff, Alberta, Canada, June 1971, paper no. 8-2.

112. Yu. I. Samoilenko, Spectral properties of controlling media having periodic parameters, Autom. Remote Contr., no. 3, 371-383 (1970).

113. G. N. Saridis and P. C. Badavas, A performance adaptive self-organizing control of a class of distributed systems, Preprints, 1970 JACC, June 22-26, 1970, pp. 221-226.

114. G. N. Saridis and P. C. Badavas, Identifying solutions of distributed parameter systems by stochastic approximation, IEEE Trans. Auto. Contr., vol. AC-15, no. 3, 393-395 (1970).

115. T. I. Seidman, The observability and prediction problem for diffusion processes, Eleventh Annual Allerton Conference on Circuit and System Theory, Oct. 3-5, 1973, pp. 123-130.

116. J. H. Seinfeld, Identification of parameters in partial differential equations, Chem. Eng. Sci., 24:65-74 (1969).

117. J. H. Seinfeld, Nonlinear estimation for partial differential equations, Chem. Eng. Sci., 24: 75-83 (1969).

118. J. H. Seinfeld and W. H. Chen, Estimation of parameters in distributed systems, in Identification of Parameters in Distributed Systems (R. E. Goodson and M. P. Polis, eds.), ASME Monograph, 1974, pp. 69-89.

119. J. H. Seinfeld and W. H. Chen, Estimation of parameters in distributed systems from noisy experimental data, Chem. Eng. Sci., 26: 753-766 (1971).

120. J. H. Seinfeld, G. R. Gavalas, and M. J. Hwang, Nonlinear filtering in distributed parameter systems, Trans. ASME J. Dynam. Syst. Meas. Contr., Series G, 93:157-163 (1971); also Preprints JACC, 573-583 (1970).

121. Y. Sevely, G. Alengrin, B. Delval, A. Dignan, and D. Pinglot, Etudes sur unité pilote d'acetate de vinyle en lit fluide, Premier colloque Franco-Soviétique, Simulation et Modélisation de Processus et de Réacteurs Catalytiques, Nancy, France, May 10-12, 1973.

122. H. Sherry and D. W. C. Shen, Combined state and parameter estimation for distributed parameter systems using discrete observations, in Identification and System Parameter Estimation (P. Eykhoff, ed.), Proc. 3rd IFAC Symposium, The Hague/Delft, The Netherlands, June 12-15, 1973, pp. 671-677.

123. E. K. Shigin, Classification of dynamic models of chemical production plants III, Auto. Remote Contr., 1703-1713 (1968).

124. M. Shridhar and N. A. Balatoni, Application of cubic splines to system identification, in Identification and System Parameter Estimation (P. Eykhoff, ed.), Proc. 3rd IFAC Symp., The Hague/Delft, The Netherlands, June 12-15, 1973, pp. 787-791.

125. P. H. Sidles and G. C. Danielson, Thermal diffusivity of metals at high temperatures, J. Appl. Phys., 25: 58-66 (1954).

126. F. E. Thau, On optimum filtering for a class of linear distributed parameter systems, ASME Trans., J. Basic Eng., 91: 173-178 (1969).

127. A. Thowsen and W. R. Perkins, Observability conditions for two general classes of linear flow processes, IEEE Trans. Auto. Contr., vol. AC-19, no. 5, 603-604 (1974).

128. E. C. Titchmarsh, Eigenfunction Expansions, Pt. 1, London, Oxford, 1962.

129. S. G. Tzafestas, Bayesian approach to distributed parameter filtering and smoothing, Int. J. Contr., 15: 273-296 (1972).

130. S. G. Tzafestas, Boundary and volume filtering of linear distributed parameter systems, Electron. Lett., vol. 5, no. 9, 199-200 (1969).

131. S. G. Tzafestas, Identification of stochastic distributed parameter systems, Int. J. Contr., vol. 11, no. 4, 619-624 (1970).

132. S. G. Tzafestas, On optimum distributed-parameter filtering and fixed-interval smoothing for colored noise, IEEE Trans. Auto. Contr., AC-17, no. 2, 448-458 (1972).

133. S. G. Tzafestas, On the identification of hybrid distributed parameter systems, Int. J. Contr., vol. 13, no. 1, 145-154 (1971).

134. S. G. Tzafestas and J. M. Nightingale, Concerning optimal filtering theory for distributed parameter systems, Proc. IEE, vol. 115, no. 11, 1737-42 (1968).

135. S. G. Tzafestas and J. M. Nightingale, Maximum-likelihood approach to optimal filtering of distributed-parameter systems, Proc. IEE, vol. 116, no. 6, 1085-1093 (1969).

136. S. G. Tzafestas and J. M. Nightingale, Optimal filtering smoothing and prediction in linear distributed-parameter systems, Proc. IEE, vol. 115, no. 8, 1207-1212 (1968).

137. U.S. Dept. of Transportation, Final Report, Climatic Impact Assessment Program, Jan. 1975.

138. V. Vemuri and W. J. Karplus, Identification of non-linear parameters of ground water basins by hybrid computation, Water Resourc. Res., vol. 5, no. 1, 172-185 (1969).

139. M. Vidyasagar, R. E. Goodson, and R. E. Klein, Comments and author closure on a definition and some results for distributed system observability, IEEE Trans. Auto. Contr., AC-16, no. 1, 106 (1971).

140. P. K. C. Wang, Control of distributed parameter systems, in Advances in Control Systems, Vol. I (C. T. Leondes, ed.), Academic, New York, 1964, pp. 75-172.

141. P. K. C. Wang and F. Tung, Optimum control of distributed parameter systems, Trans. ASME J. Basic Eng., March 1964, pp. 67-79.

142. E. D. Ward, "Identification of parameters in nonlinear boundary conditions of distributed systems with linear fields," Ph.D. thesis, Purdue Univ., Lafayette, Ind., August 1971.

143. E. D. Ward and R. E. Goodson, Identification of nonlinear boundary conditions in distributed systems with linear fields, Trans. ASME J. Dynam. Syst. Meas. Contr., Series G, 95: 390-395 (1973).

144. M. J. Wozny, W. T. Carpenter, and G. Stein, Identification of Green's function for distributed parameter systems, IEEE Trans. Auto. Contr., vol. AC-15, no. 1, 155-157 (1970).

145. T. K. Yu and J. H. Seinfeld, Observability of a class of hyperbolic distributed parameter systems, IEEE Trans. Auto. Contr., vol. AC-16, no. 5, 495-497 (1971).

146. V. P. Zhivoglyadov and V. Kh. Kaipov, Application of the method of stochastic approximations in the problem of identification, Automatica i Telemetkanika, vol. 27, no. 10, 54-58 (1966).

147. V. P. Zhivoglyadov and V. Kh. Kaipov, Identification of distributed parameter plants in the presence of noises, Preprints, IFAC Symposium, Prague, Czechoslovakia, June 12-17, 1967, paper no. 3-5.

148. V. P. Zhivoglyadov, V. Kh. Kaipov, and I. M. Tsikunova, Stochastic algorithms of identification and adaptive control of distributed parameter systems, Preprints, IFAC Symposium on the Control of Distributed Parameter Systems, Banff, Alberta, Canada, June 1971, paper no. 13-1.

Chapter 3

DISTRIBUTED PARAMETER STATE ESTIMATION

Spyros G. Tzafestas[†]

Control Systems Laboratory
Electrical Engineering Department
University of Patras
Patras, Greece

[†]Also with the Department of Reactors, NRC Democritos, Aghia Paraskevi Attikis, Athens, Greece.

I. INTRODUCTION

The practical application of feedback control to a given physical system requires knowledge of the state functions which must be fed back into the controller input. More often than not, only some of these state functions are accessible to direct measurement, and hence the problem arises to estimate and compute the nonmeasured states by utilizing the measured outputs of the system at hand. Also, in almost all practical situations the measurements are influenced by errors, due to the measurement instruments and methods or to other reasons, and the systems are corrupted by internal and external disturbances of a random nature. These facts have naturally led to the evolution of the stochastic state estimation or state indentification theory of dynamic and control systems. Of course in many situations the mathematical model of the system under consideration contains unknown structural parameters which must also be estimated by using the same measured quantities. This requirement has led to the parameter estimation (or identification) problem which has been examined either on its own or as a subproblem of state estimation.

There are two major avenues which have been followed in control theory depending on whether the system under study is assumed to be concentrated at a single spatial point (lumped parameter system, LPS) or it is assumed to occupy a certain spatial domain (distributed parameter system, DPS). Actually, all real physical systems are DPS. LPS are modeled by ordinary differential equations (ODE), whereas DPS are modeled by partial differential equations (PDEs). Examples of distributed parameter (DP) physical systems are nuclear reactors, heat exchangers, diffusion processes, chemical reactors, fluid systems, vibration systems, atmospheric phenomena, drying systems, steel, petroleum, papermaking, and glass processes, magnetohydrodynamic systems, etc.

Our purpose in this chapter is to give a comprehensive survey of the DP state estimation theory developed thus far. The chapter will naturally include works dealing with physical systems composed by LP and DP subsystems interacting with each other, and also works which treat generalized classes of systems described by functional equations, containing time delays or having general infinite dimensional states. A survey of some parameter estimation methods for DPSs available in 1972 has been made [1], and a more recent survey was provided in [2-3]. The reader is referred to [4] and the IEEE special issue on the linear-quadratic-Gaussian problem for the parameter and state estimation of LPS [142]. A survey of the optimal control theory for DPS has been provided in [5], which also contains a section on parameter and state identification of DPS.

II. STOCHASTIC DISTRIBUTED PARAMETER MODELS

Actually, there is not a general and unique model covering all DPS. This is particularly true in the nonlinear case. For linear DPS the model considered by most authors has the form [6-25]:

$$X_t(x,t) = A_x X(x,t) + B(x,t)W(x,t), \qquad x \in D \tag{1a}$$

$$\beta_x X(x,t) = 0, \qquad x \in \partial D \qquad (t > 0) \tag{1b}$$

$$X(x,0) = X_0(x), \qquad x \in D \tag{1c}$$

$$Z(x,t) = M_x(x,t) + V(x,t), \qquad x \in D \tag{1d}$$

where $X_t(x,t) = \partial X(x,t)/\partial t$, D is an open spatial n-dimensional domain with boundary ∂D; $X(x,t)$, $x \in D$ is the state vector function of the system; $X_0(x)$ is the initial state function, assumed to be a zero-mean Gaussian variable, independent of $W(x,t)$ and $V(x,t)$, with covariance matrix $P_0(x,y), x,y \in D$; A_x and β_x are well-posed linear spatial differential or integrodifferential operators; M_x is a linear spatial measurement operator; $Z(x,t)$ is the measured output function; and $W(x,t)$, $V(x,t)$ are zero-mean Gaussian and white (in time) distributed processes with covariance matrices.

$$\begin{aligned} E\{W(x,t)W^T(y,\tau)\} &= Q(x,y,t)\delta(t - \tau) \\ E\{V(x,t)V^T(y,\tau)\} &= R(x,y,t)\delta(t - \tau) \end{aligned} \qquad x,y \in D \tag{2}$$

and crosscovariance matrix zero, where $E\{\cdot\}$ and $\delta(\cdot)$ are the expectation operator and the unit impulse function, respectively. Here $Q(x,y,t)$ is assumed to be semipositive definite in the sense that $\int_D\int_D X^T(x,t)Q(x,y,t)X(y,t)\,dx\,dy \geq 0$ for all $X(x,t) x \in D$. Similarly, $R(x,y,t)$ is assumed to be positive definite. In most references, all white processes are assumed to be defined as formal derivatives of DP Wiener processes interpreted in the Ito sense [26], and hence their approach is called the "formal approach" in contrast to the "rigorous approach" in which the DP stochastic processes are defined in appropriate rigorous ways. This will be reviewed in a following section [27-35].

Concerning the nonlinear DPS, there appear two main models [36-46], namely the model

$$X_t(x,t) = N_d(X,x,t) + B(x,t)W(x,t), \qquad x \in D \tag{3a}$$

$$N_b(X,x,t) = 0, x \in D, \qquad X(x,0) = X_0(x), x \in D \qquad (t > 0) \tag{3b}$$

$$Z(x,t) = M_x X(x,t) + V(x,t), \qquad x \in D \tag{3c}$$

and the model

$$X_t(x,t) = F(x,t,X,X_x,X_{xx},U(x,t)) + W(x,t), \qquad x \in (0,1)$$

$$X(x,0) = X_0(x), \qquad x \in (0,1)$$

$$G_0(t,x,X_x,\Omega_0(t)) + V_{b0}(t) = 0, \qquad x = 0 \qquad (t > 0) \tag{4}$$

$$G_1(t,x,X_x,\Omega_1(t)) + V_{b1}(t) = 0, \qquad x = 1$$

$$Z(x,t) = M(x,t,X(x,t)) + V(x,t), \qquad x \in (0,1)$$

where N_d, N_b, and M_x are nonlinear spatial differential or integro-differential operators, and F, G_0, G_1, and M are conventional functions of their arguments sufficiently smooth, with $U(x,t)$, $\Omega_0(t)$, and $\Omega_1(t)$ being known inputs. All stochastic processes are zero-mean white (in time) DP processes. Knowledge of their statistical properties in the model (4) is not required.

The preceding models constitute the core of DP state estimation theory but do not cover all cases. For example, in [11,12] $W(x,t)$ is colored; in [17,19,47] the boundary equation is not homogeneous; in [37-44] and [48-49] the system is not a pure DPS; in [50-52] the system involves time delays, etc. All these cases will be treated in the appropriate sections of the chapter.

III. A BRIEF OUTLINE OF THE PRESENT SURVEY

The procedure which must be followed when estimating the states and/or structural parameters of any DPS involves the following steps:

1. Select the mathematical model of the physical system under consideration
2. Select a solution (or simulation) method appropriate for this model
3. Select the form of measurements which are convenient to be made, e.g., dicrete-point or scanning form, etc.
4. Select the estimation method which is to be used and derive the appropriate state estimator

5. Investigate the properties (stability, sensitivity, etc.) of this estimator

Step 1 requires knowledge of the physical laws governing the process at hand and the utilization of justifiable assumptions which lead to a simple but realistic model. Step 2 requires knowledge of the available methods for solving ODEs and PDEs together with their numerical computational algorithms. Steps 1 and 2 are common to all engineering design problems concerning DPS. These steps will not be studied here. For brief discussions see, e.g. [2,53].

Step 3 is of fundamental importance and affects the success of the optimal state estimator. In most cases the measurement model used is of the discrete spatial point type. In this case one is faced with the problem of choosing the number and location of the measurement points where the transducers must be placed. This step will be studied in Section VIII.

Step 4 will be the central theme of the present survey chapter. The basic linear distributed parameter state filtering results are reviewed, i.e., the orthogonal projection, the least squares, the conditional characteristic functional, the Bayesian, the maximum likelihood, the innovation, and the maximum principle methods. Section V is devoted to smoothing and prediction problems, and Section VI gives an exposition of the extensions of the basic DP filtering theory available until the present time, which cover the cases of (i) discrete-point measurements, (ii) filtering with boundary noise and measurements, (iii) filtering for colored measurement noise, and (iv) filtering for time-delay systems. In Section VII we review the three primary state estimation methods applied to nonlinear DPS, namely the maximum likelihood, the least squares, and the Fokker-Planck equation technique, and outline the solutions of the filtering problem for nonlinear systems with time delays and moving boundaries.

Section VIII presents the available observability results for DPS, reviews the existing solutions to the optimal selection of spatial measurement points, and outlines some results concerning the spatial and time sampling of DPS. In Section IX the rigorous infinite-dimensional filtering which covers the DPS as special case is exposed, and in Section X a collection of seven application examples are given which support the largest part of the theoretical results. Due to limitations of space, the problems and their solutions are simply stated and not derived because the derivations are usually lengthy. In most cases, however, we have tried to provide fast descriptions of the solution methods.

Step 5 actually has not been investigated very much for DPS and is an area requiring further research work. Some questions concerning the observability and stability properties have been answered in [9,23,54] and [55-59].

IV. BASIC LINEAR FILTERING THEORY OF DPS

The state estimation theory of infinite dimensional systems dates back to 1967 [16,27,60]. In 1968 and 1969 there appeared the works of [6,7,13,18,47] and in later years one finds [8,9,11,12,14,15,17, 19-22,29-34], etc. The estimation methods principally used are: (1) orthogonal projection [6,13,19,47]; least squares [9,14-16,18, 37-44]; (3) conditional characteristic function [7,11,12]; (4) Bayesian procedure [8]; (5) maximum likelihood [36]; (6) innovations approach [9,10,21]; and (7) maximum principle [22].

The basic filtering (i.e., estimation of current state) problem is: Given a DPS of type (1) and a record of measured data $Z(x,t)$, $x \in D$ in the time interval $[t_0,t]$, find a best estimate $\hat{X}(x,t)$, $x \in D$ of $X(x,t)$, $x \in D$ for $t \geq t_0$.

A. Orthogonal Projection Approach

The key point in this approach is the assumption that the best estimate $\hat{X}(x,t)$ has the linear form

$$\hat{X}(x,t) = \int_0^t\int_D H(x,t;y,t')Z(y,t')\, dy\, dt' \tag{5}$$

where $H(x,t;y,t')$ is a kernel matrix to be determined. Using the orthogonality principle in a way similar to the LP case [61], one concludes that $H(x,t;y,t')$ satisfies the following DP Wiener-Hopf equation for $y \in D$, $t_0 \leq t_1 < t$:

$$E\{X(x,t)Z^T(y,t_1)\} = \int_0^t \int_D H(x,t;s,\tau)E\{Z(s,\tau)Z^T(y,t_1)\}\, ds\, d\tau \qquad (6)$$

Defining $\tilde{X}(x,t) = X(x,t) - \hat{X}(x,t)$ as the filtering error, its covariance matrix is $P(x,y,t) = E\{\tilde{X}(x,t)\tilde{X}^T(y,t)\}$. The equations for $\hat{X}(x,t)$ and $P(x,y,t)$ are found by appropriately manipulating the Wiener-Hopf equation (6), and are

$$\hat{X}_t(x,t) = A_x\hat{X}(x,t) + \int_D H(x,t;s,\tau)\tilde{Z}(s,\tau)\, ds, \qquad x \in D$$

$$H(x,t;s,\tau) = \int_D P(x,\zeta;t)M_\zeta^T R^+(\zeta,s,t)\, d\zeta$$

$$\begin{aligned} P_t(x,y,t) &= A_x P(x,y,t) + P(x,y,t)A_y^T + B(x,t)Q(x,y,t) \\ &\quad \times B^T(y,t) - \int_D\int_D P(x,s,t)M_s^T R^+(s,\zeta,t) \\ &\quad \times M_\zeta P(\zeta,y,t)\, ds\, d\zeta, \qquad x \in D \end{aligned} \qquad (7)$$

$$\beta_x\hat{X}(x,t) = 0,\ x \in \partial D; \qquad \beta_x P(x,y,t) = 0,\ x \in \partial D,\ y \in D$$

$$\hat{X}(x,0) = 0, \qquad P(x,y,0) = P_0(x,y), \qquad x \in D,\ y \in D$$

where $\tilde{Z}(x,t) = Z(x,t) - M_x\hat{X}(x,t)$, and $R^+(s,\zeta,t)$ is the inverse of $R(x,s,t)$ defined by

$$\int_D R(x,s,t)R^+(s,\zeta,t)\, ds = I\delta(x - \zeta), \qquad x,\zeta \in D \qquad (8)$$

The filter equations derived in [13] are a special case of (7) since they have been derived only for the stochastic heating system subject to LP noise disturbances.

B. Least Squares Approach

The key point of the least squares (or functional minimization) approach is the definition of an appropriate error criterion [14,15]:

$$J = \frac{1}{2}\int_D X_0^T(x)A_{x,y}^0 X_0(y)\, dx + \int_0^{t_f} \tilde{Z}^T(t)R^{-1}(t)\tilde{Z}(t)\, dt$$

$$+ \int_0^{t_f} \int_D [X_t(x,t) - A_x X(x,t)]^T B^o_{x,y} [X_t(y,t) - A_y X(y,t)] \, dx \, dt \tag{9}$$

where

$$A^o_{x,y}[\cdot] = [\int_D P_0(x,y)[\cdot] \, dy]^{-1}$$
$$B^o_{x,y}[\cdot] = [\int_D Q(x,y,t)[\cdot] \, dy]^{-1} \tag{10}$$

In [14] the measurement process is assumed to be of the averaging (or scanning type, i.c., $Z(t) = \int_D M(x,t)X(x,t)\,dx + V(t)$. The matrices $P_0(x,y)$, $Q(x,y,t)$, and $R(t)$ are bounded, continuous, symmetric, and positive definite.

The solution is found by selecting $X(x,t)$ and $W(x,t)$ so as to minimize J subject to the constraints (1a,b). In [14] the minimization is performed by the calculus of variations method, i.e., by minimizing the augmented functional

$$\bar{J} = J + \int_0^{t_f} \int_D \lambda^T(x,t)[X_t - A_x X - B(x,t)W(x,t)] \, dx \, dt \tag{11}$$

without any constraint, where $\lambda(x,t)$, $x \in D$ is the Lagrange multiplier function. The result is the following pair of DP Euler-Lagrange equations:

$$X_t(x,t) = A_x X(x,t) + B(x,t) \int_D Q(x,y,t)\lambda(y,t) \, dy$$
$$\lambda_t(x,t) = -A^*_x \lambda(x,t) - M^T R^{-1}\{Z(t) - \int_D MX(s,t) \, ds\} \tag{12}$$

together with the transversality conditions

$$\lambda(x,0) = A^o_{x,y}[X(y,0) - X_0(y)], \qquad \lambda(x,t_f) = 0 \tag{13}$$

Equations (12) and (13) constitute a DP two-point boundary value problem (TPBVP) which is solved by the sweep method [62], i.e., by assuming that $X(x,t)$ has the form

$$X(x,t) = \int_D P(x,y,t)\lambda(y,t) \, dy + \sigma(x,t) \tag{14}$$

where the symmetric matrix $P(x,y,t)$ and the vector $\sigma(x,t)$ have to be determined. The estimator equations derived are similar to (7), i.e.,

$$\sigma_t(x,t) = A_x\sigma(x,t) + K(x,t)\{Z(t) - \int_D M(s,t)\sigma(s,t)\,ds\}$$

$$P_t(x,y,t) = A_xP + (A_yP^T) + B(x,t)Q(x,y,t)B^T(y,t)$$

$$- \iint_{DD} PM^T(s,t)R^{-1}(t)M(\varsigma,t)P\,ds\,d\varsigma \qquad (15)$$

$$K(x,t) = \int_D P(x,s,t)M^T(s,t)R^{-1}(t)\,ds$$

$$\sigma(x,0) = X_0(x), \qquad P(x,y,0) = P_0(x,y), \qquad x,y \in D$$

Equations (13) and (14) imply that $X(x,t_f) = \sigma(x,t_f)$ is the required state estimate at time t_f.

The same approach is followed in [15] for the discrete-point measurement case $Z(x_i,t) = M(x_i,t)X(x_i,t) + V(x_i,t)$ $(i = 1,2, \ldots ,M)$ with $B(x,t) = 0$, and also in [16-18]. It is useful to note that this approach is a deterministic one that converts a stochastic optimal filtering problem to a deterministic optimal control problem, thus bypassing the requirement of a probabilistic treatment. To the contrary, a probabilistic treatment of this problem is given in [29-33]. This approach will be reviewed in Section IX.

C. Conditional Characteristic Function Approach

This approach is based on the definition of the characteristic functional of a DP generalized random variable $X(D) = \{X(x),\ x \in D\}$ as

$$C_X\{j\lambda(x),\ x \in D\} = E[\exp\{j \int_D \lambda^T(x)X(x)\,dx]$$

which for the case of a Gaussian variable $X(D)$ becomes

$$C_X\{j\lambda(x),\ x \in D\} = \exp[j \int_D \lambda^T(x)\overline{X}(x)\,dx - \frac{1}{2}\iint_{DD} \lambda^T(x)$$

$$\times\ P_x(x,y)\lambda(y)\,dx\,dy]$$

where $\overline{X}(x) = E[X(x)]$, and $P_X(x,y) = E\{\tilde{X}(x)\tilde{X}^T(y)\}$ with $\tilde{X}(x) = X(x) - \overline{X}(x)$.

The optimal filter equations are derived in discrete-time form by using a DP learning theorem which states "Given two Gaussian DP variables $X_1(D)$ and $X_2(D)$, then the conditional variable $X(D) = \{X_1(D)$ given $X_2(D)\}$ is also Gaussian with mean value $\overline{X}(x)$ and covariance matrix $P(x,y)$ given by

$$\overline{X}(x) = \overline{X}_1(x) + \iint_{DD} P_{12}(x,s)P_2^+(s,\zeta)\{X_2(\zeta) - \overline{X}_2(\zeta)\}\, ds\, d\zeta \tag{16}$$

$$P(x,y) = P_1(x,y) - \iint_{DD} P_{12}(x,s)P_2^+(s,\zeta)P_{12}^T(\zeta,y)\, ds\, d\zeta\text{"}$$

The system equations are discretized in time as

$$X(x, k+1) = A_x^1 X(x,k) + B^1(x,k)W(x,k)$$

$$Z(x,k) = M(x,k)X(x,k) + V(x,k)$$

where Δt is the sampling period, $X(x,k) = X(x, k\,\Delta t)$, etc., and

$$A_x^1 = I + \Delta t A_x, \qquad B^1(x,k) = \Delta t B(x,k)$$

Let $X(x, k+1 \mid k-1)$ be the conditional variable "$\hat{X}(x, k+1)$ given $Z(x,0), \ldots ,Z(x, k-1)$," with mean value $\hat{X}(x, k+1 \mid k-1)$, and let $\hat{Z}(x, k \mid k-1)$ be the mean value of the variable "$Z(x,k)$ given $Z(x,0), \ldots ,Z(x, k-1)$." Using the preceding learning theorem one then finds that the mean value $\hat{X}(x, k+1 \mid k)$ and covariance matrix $P(x, y, k+1)$ of the updated random variable "$X(x, k+1)$ given $Z(x,0), \ldots ,Z(x, k-1)$ given $Z(x,k)$ given $Z(x,0), \ldots ,Z(x, k-1)$" satisfy the equations

$$\frac{[\hat{X}(x, k+1 \mid k) - \hat{X}(x, k \mid k-1)]}{\Delta t} = A_x\hat{X}(x, k \mid k-1)$$

$$+ \iint_{DD} [P(x, s, k \mid k-1)M^T + \Delta t A_x P(x, s, k \mid k-1)M^T]$$

$$[\Delta t MP(s, \zeta, k \mid k-1)M^T + R]^+ [Z - M\hat{X}(\zeta, k \mid k-1)]\, ds\, d\zeta \tag{17a}$$

$$\frac{[P(x, y, k+1 \mid k) - P(x, y, k \mid k-1)]}{\Delta t} = A_x P(x, y, k \mid k-1)$$

$$+ P(x, y, k \mid k-1)A_y^T + B(x,k)Q(x,y,k)B^T(y,k)$$

$$+ \Delta t A_x P(x, y, k \mid k-1)A_y^T - \iint_{DD} [P(x, s, k \mid k-1)M^T$$

$$+ \Delta t A_x P(x, s, k \mid k-1)M^T][\Delta t MP(s, \zeta, k \mid k-1)M^T + R(s,\zeta,k)]^+$$

$$[P(\zeta, y, k \mid k-1)M^T + \Delta t A_y P(\zeta, y, k \mid k-1)M^T] \, ds \, d\zeta \tag{17b}$$

which in the limit as $\Delta t \to 0$ reduce to the optimal filter (7).

D. Bayesian Approach

This approach consists in maximizing the marginal probability density functional of $X_k(x) = X(x, k\,\Delta t)$ conditional upon the data $\{Z_0, Z_1, \ldots, Z_N\}$ where $Z_k(x) = Z(x,k)$. Defining $X_k(D) = \{X_k(x), x \in D\}$ and $Z_k(D) = \{Z_k(x), x \in D\}$, and using Baye's theorem one obtains

$$p(X_k(D) \mid Z_0(D), \ldots, Z_k(D)) = \text{const. } p(Z_k(D) \mid X_k(D))$$
$$p(X_k(D) \mid Z_0(D), \ldots, Z_{k-1}(D)) \tag{18}$$

Denote the estimate of $X_k(x)$ based on data up to time k-1 and its covariance matrix by $\hat{X}_{k|k-1}(x)$ and $P(x, y, k \mid k-1)$, and the estimate and covariance matrix of $X_{k-1}(x)$ based on data up to time k-1 by $\hat{X}_{k-1|k-1}(x)$, $P(x, y, k-1 \mid k-1)$. Then, the one step predicted estimate $\hat{X}_{k+1|k}(x)$ and its covariance matrix $P(x, y, k+1 \mid k)$ are equal to

$$\hat{X}_{k+1|k}(x) = A_x^1 X_{k|k}(x)$$

$$P(x, y, k+1 \mid k) = A_x^1 P(x, y, k \mid k)A_y^{1T} + B^1(x,k)Q(x,y,k) \times B^{1T}(y,k) \tag{19}$$

where A_x^1 and $B^1(x,k)$ have already been defined.

Also

$$E[Z_k \mid X_k] = M_x X_k(x), \qquad \text{cov}[Z_k \mid X_k] = R(x,y,k) \tag{20}$$

Equations (18) to (20) imply that

$$p(X_k(D) \mid Z_0(D), \ldots, Z_k(D)) = \text{const.} \exp\{-\frac{1}{2}\iint_{DD} (Z_k - M_x X_k)^T R^+(x,y,k)(Z_k - M_y X_k)\, dx\, dy$$

$$-\frac{1}{2}\iint_{DD} (X_k - \hat{X}_{k|k-1})^T P^+(x, y, k \mid k-1)(X_k - \hat{X}_{k|k-1})\, dx\, dy \quad (21)$$

Maximizing (21), i.e., minimizing its negative exponent and applying two DP matrix inversion lemmas [8], yields

$$\hat{X}_{k|k}(x) = A_x^1 X_{k-1|k-1}(x) + \int_D K(x,y,k)\tilde{Z}_{k|k-1}(y)\, dy \quad (22)$$

where $\tilde{Z}_{k|k-1} = Z_k(y) - M_y\hat{X}_{k|k-1}(y)$, and

$$K(x,y,k) = \int_D P(x, s, k \mid k-1)M_s^T[M_s P(s, y, k \mid k-1)M_y^T + R(s,y,k)]^+\, ds$$

Using (18) and (22) one can now find the equation for the error $\tilde{X}_{k|k}(x) = X_k(x) - \hat{X}_{k|k}(x)$ by which the equation for the covariance matrix $P(x, y, k \mid k) = E\{\tilde{X}_{k|k}(y)\tilde{X}^T_{k|k}(x)\}$ may be determined. The resulting discrete-time optimal filter can be converted in continuous-time form as before by Kalman's limiting procedure. The result is again the DP filter (7).

E. Maximum Likelihood Approach

The maximum likelihood approach [36] is essentially the same as the least squares approach outlined in Section IV.B. One difference is that in the (deterministic) least squares approach the matrices Q(x,y,t) and R(t) need not be interpreted as the covariance matrices of W and V [14,15], but are considered as arbitrary bounded, continuous, symmetric, and positive definite weighting matrices as originally was done in [63] for LPS. This has the advantage of being able to work with any stochastic disturbances not necessarily Gaussian. Of course the results are correct to second-order statistics. A second difference between [14,15] and [36] is that in [36] the maximization of the functional is performed by the dynamic programming technique and not by the calculus of variations. Since the maximum likelihood method is primarily concerned with nonlinear DPS it will be described in Section VII together with the similar method of [37-44].

F. Innovations Approach

The innovations approach [64,65] is actually a modified form of the least squares or the orthogonal projection approach. The key point is the definition and utilization of the innovation process $\hat{V}(x, t \mid t) = Z(x,t) - M_x\hat{X}(x, t \mid t)$ which carries all the information contained in the new observation $Z(x,t)$. The innovation process $\hat{V}(x, t \mid t)$ is a white (in time) process having the same covariance matrix as $V(x,t)$, and is equivalent to $Z(x,t)$ with respect to linear operations. This property implies that the estimate $\hat{X}(x, t \mid t)$ can be written in the form

$$\hat{X}(x, t \mid t) = \int_0^{t_f}\int_D K(x,t;s,\tau)\hat{V}(s, \tau \mid \tau)\, ds\, d\tau, \qquad x \in D \tag{23}$$

where $K(x,t;s,\tau)$ must be selected such that $\tilde{X}(x, t \mid t) = X(x,t) - \hat{X}(x, t \mid t)$ is orthogonal to $Z(x,\tau)$, $x \in D$, $\tau \in \mid [0, t)$, or equivalently to $\hat{V}(x, \tau \mid \tau)$, $x \in D$, $\tau \mid [0, t)$, i.e.,

$$\begin{aligned} E\{X(x,t)\hat{V}^T(y, t_1 \mid t_1)\} &= \int_0^t\int_D K(x,t;s,\tau)E\{\hat{V}(s, \tau \mid \tau) \\ &\quad \times \hat{V}^T(y, t_1)\}\, ds\, d\tau \\ &= \int_D K(x,t;s,t_1)R(s,y,t_1)\, ds, \\ &\qquad y \in D,\ t_1 \in [0,t] \end{aligned} \tag{24}$$

Solving (24) for K and introducing the result into (23) yields:

$$\begin{aligned} \hat{X}(x, t \mid t) = \int_0^t\int_D\int_D E\{X(x,t)\hat{V}^T(s, t_1 \mid t_1)\}R^+(s,\zeta,t_1) \\ \times \hat{V}(\zeta, t_1 \mid t_1)\, ds\, d\zeta\, dt_1 \end{aligned} \tag{25}$$

The difference between (24) and the Wiener-Hopf equation (6) is that (24) involves $\hat{V}(x, t \mid t)$ in place of $Z(x,t)$, which leads to the solution more directly.

The filter equations are now derived in the usual way, i.e., (25) is differentiated with respect to t under the assumption that $E\{W(x,t)\hat{V}^T(s, \tau \mid \tau)\} = 0$, $x, s \in D$, $\tau < t$. The resulting equations are identical to (7).

G. Maximum Principle Approach

The maximum principle approach was first used for LPS in [66] and extended to DPS in [22]. The starting point is again the assumption that $\hat{X}(x, t \mid t)$ has the form (5), which implies that $\hat{X}(x, t \mid t)$ may be viewed as the state of a DPS of the type

$$\hat{X}_t(x, t \mid t) = L_x\hat{X}(x, t \mid t) + \int_D H(x,t;s,t)Z(s,t)\, ds \tag{26}$$

The problem then reduces to that of determining the spatial operator L_x and the kernel H such that $\hat{X}(x, t \mid t)$ is an unbiased minimum variance estimate of $X(x,t)$. The estimate $\hat{X}(x, t \mid t)$ is unbiased if $E\{\tilde{X}(x, t \mid t)\} = 0$, i.e., if

$$(A_x - L_x)E\{X(x,t)\} - \int_D H(x,t;s,t)M(s,t)E\{X(s,t)\}\, ds = 0$$

which implies that

$$L_x[\cdot] = A_x[\cdot] - \int_D H(x,t;s,t)M(s,t)[\cdot]\, ds \tag{27}$$

since in general $E\{X(x,t)\} \neq 0$.

The matrix kernel $H(x,t;s,t)$ is determined from the minimum variance condition. To this end, the covariance equation is derived in terms of $H(x,t,s,t)$, i.e.,

$$P_t(x,y,t) = A_xP + PA_y^T + B(x,t)Q(x,y,t)B^T(y,t)$$

$$- \int_D P(x,s,t)M^T(s,t)H(s,t;y,t)\, ds$$

$$- \int_D H(x,t;s,t)M(s,t)P(s,y,t)\, ds$$

$$+ \int\int_{DD} H(x,t;s,t)R(s,\zeta,t)H^T(\zeta,t;y,t)\ ds\ d\zeta \tag{28}$$

and $H(x,t;s,t)$ is chosen so as to minimize the error functional $J = \int_D\int_D \mathrm{tr}\{N(x,s,t_f)P(x,s,t_f)\}\ dx\ ds$, where $N(x,s,t_f)$ is symmetric and positive definite, subject to the dynamic constraint (28). This minimization is performed by using the minimum principle developed in [67], the result being

$$H(x,t;y,t) = \int_D P(x,s,t)M^T(s,t)R^+(s,y,t)\ ds \tag{29}$$

Introducing (27) into (26), and (29) into (27), (28) directly yields the desired filter which is identical to (7).

V. SMOOTHING AND PREDICTION

The smoothing or interpolation is the problem of estimating past states on the basis of data up to the present time. Solutions of various forms of the smoothing problem, as well as of the prediction problem (i.e., estimation of future state), are given in [6-9, 12-16, 18,21].

The smoothing problem in general is to find a best estimate $\hat{X}(x, t \mid t_f)$, $x \in D$ of $X(x,t)$, $x \in D$ for $t \in [0,t_f]$ on the basis of data $Z(x,\tau)$, $x \in D$, $0 \leq \tau \leq t_f$, where $X(x,t)$ and $Z(x,t)$ are given by (1). The smoothing problem is classified in three cases, namely, (1) fixed-interval smoothing (the time interval $[0,t_f]$ is fixed); (2) fixed-point smoothing (t fixed, say T, and t_f increasing); and (3) fixed-lag smoothing ($t_f = t + \Theta$, where Θ is fixed positive constant).

The derivation in [6] follows the LP method of [60], the derivation in [8] follows the discrete-time Bayesian procedure, the solution in [12] is found via the DP theorem concerning the combination of two independent estimates of the state (see [36,68]), and finally the solution in [14,15] is determined by reverse-time integration of the Euler-Lagrange equations (12) with terminal conditions $X(x,t_f) = \sigma(x,t_f)$ and $\lambda(x,t_f) = 0$ [69].

An outline will be given here of the smoothing filters derivation by the innovation technique [9,21]. In this technique the smoothing estimate is written as

$$\hat{X}(x, t \mid t_f) = \int_0^{t_f}\int_D H_1(x,t;s,\tau)\hat{V}(s, \tau \mid \tau)\, ds\, d\tau$$

$$= \hat{X}(x, t \mid t) + \int_D P(x, s, t \mid t)\lambda(s,t)\, ds, \qquad x \in D \quad (30)$$

where the adjoint variable $\lambda(s,t)$, $s \in D$ is given by

$$\lambda(s,t) = \int_0^{t_f}\int_D \Phi^T(s,\tau;\zeta,t)\hat{V}_m(\zeta, \tau \mid \tau)\, d\zeta\, d\tau$$

$$\hat{V}_m(\zeta, \tau \mid \tau) = \int_D M_\zeta^T R^+(\zeta,\xi,\tau)\hat{V}(\xi, \tau \mid \tau)\, d\xi$$

where $\Phi(s,\tau;\zeta,t)$ is the Green's function of the error equation. Equation (30) implies that the smoothed estimate $\hat{X}(x, t \mid t_f)$ is expressed in terms of the filtered estimate $\hat{X}(x, t \mid t)$, its covariance matrix $P(x, s, t \mid t)$, and the innovation process $\hat{V}(\xi, \tau \mid \tau)$ in the interval $[t,t_f]$. Utilizing (30), the following smoothing filters are obtained.

A. Fixed-interval Smoothing

$$\hat{X}_t(x, t \mid t_f) = A_x\hat{X}(x, t \mid t_f) + \iint_{DD} Q_0P^+(s, \zeta, t \mid t)[\hat{X}(\zeta, t \mid t_f) - \hat{X}(\zeta, t \mid t)]\, ds\, d\zeta,$$

$$P_t(x, y, t \mid t_f) = A_xP + PA_y^T - Q_0(x,y,t) + \iint_{DD} Q_0P^+(s, \zeta, t \mid t)P(\zeta, y, t \mid t_f)\, ds\, d\zeta + \iint_{DD} P(x, s, t \mid t_f)P^+(s, \zeta, t \mid t)Q_0\, ds\, d\zeta,$$

$$\beta_x \hat{X}(x, t \mid t_f) = 0, \quad \beta_x P(x, y, t \mid t_f) = 0 \quad x \in \partial D, y \in D,$$

$$\hat{X}(x, t \mid t_f)\Big]_{t=t_f} = \hat{X}(x, t \mid t)\Big]_{t=t_f},$$

$$P(x, y, t \mid t_f)\Big]_{t=t_f} = P(x, y, t \mid t)\Big]_{t=t_f} \tag{31}$$

where $Q_O(x,y,t) = B(x,t)Q(x,y,t)B^T(y,t)$.

B. Fixed-point Smoothing

$$\hat{X}_t(x, T \mid t) = \iint_{DD} P(x,T;s,t)M_s^T R^+(s,\zeta,t)\{Z(\zeta,t) - M_\zeta \hat{X}(\zeta, t \mid t)\}\, ds\, d\zeta,$$

$$P_t(x, y, T \mid t) = -\iiiint_{DDDD} \Phi(x,T;s,t)P(s, \zeta, t \mid t)M^T R^+(\zeta,\zeta',t) \times M_{\zeta'} P(\zeta', s, t \mid t)\Phi(s',T;y,t)\, ds\, d\zeta\, ds'\, d\zeta',$$

$$\beta_x \hat{X}(x, T \mid t) = 0, \quad \beta_x P(x, y, T \mid t) = 0 \quad x \in \partial D, y \in D$$

$$\hat{X}(x, T \mid t)\Big]_{t=T} = \hat{X}(x, t \mid t)\Big]_{t=T},$$

$$P(x, y, T \mid t)\Big]_{t=T} = P(x, y, t \mid t)\Big]_{t=T} \tag{32}$$

where $P(x,T,s,t) = \int_D \Phi(x,T;\zeta,t)P(\zeta, s, t \mid t)\, d\zeta$.

C. Fixed-lag Smoothing

$$\hat{X}_t(x, t \mid t+\Theta) = A_x \hat{X}(x, t \mid t+\Theta) + \iint_{DD} Q_O(x,s,t)P^+(s, \zeta, t \mid t)\tilde{X}(\zeta, t \mid t+\Theta)\, ds\, d\zeta + \iiint_{DDD} \Phi(x,t; s, t+\Theta)P(s, \zeta, t \mid t+\Theta)M_\zeta^T \times R^+(\zeta, \zeta', t+\Theta)\tilde{Z}(\zeta', t+\Theta)\, ds\, d\zeta\, d\zeta',$$

$$P_t(x, y, t \mid t+\Theta) = A_x P + PA_y^T - \{Q(x,y,t) + \Sigma(x,y,t)\}$$

$$+ \iint_{DD} Q_o(x,s,t)P^+(s, \zeta, t \mid t)P(\zeta, y, t \mid t+\Theta)\, ds\, d\zeta$$

$$+ \iint_{DD} P(x, s, t \mid t+\Theta)P^+(s, \zeta, t \mid t)Q_o(\zeta,y,t)\, ds\, d\zeta,$$

$$\beta_x \hat{X}(x, t \mid t+\Theta) = 0, \qquad \beta_x P(x, y, t \mid t+\Theta) = 0,$$

$$x \in \partial D,\ y \in D \tag{33}$$

where

$$\tilde{X}(\zeta, t \mid t+\Theta) = \hat{X}(\zeta, t \mid t+\Theta) - \hat{X}(\zeta, t \mid t)$$

$$\tilde{Z}(\zeta, t+\Theta) = Z(\zeta, t+\Theta) - M_\zeta \hat{X}(\zeta, t+\Theta \mid t+\Theta)$$

and

$$\Sigma(x,y,t) = \iiiint_{DDDD} \Psi^T(x, s', s, t, t+\Theta)R^+(s,\zeta,t)$$

$$\times\ \Psi(\zeta, \zeta', y, t, t+\Theta)\, ds\, ds'\, d\zeta\, d\zeta'$$

where

$$\Psi(x, s', s, t, t+\Theta) = M_s P(s, s', t+\Theta \mid t+\Theta)\Phi(x, t+\Theta, s', t)$$

The initial conditions $\hat{X}(x, t_o \mid t_o+\Theta)$ and $P(x, y, t_o \mid t_o+\Theta)$ are obtained by computing the fixed-point filter equations in the interval $[t_o, t_o+\Theta]$ with fixed point $T = t_o$.

D. State Prediction

The optimal state prediction (or extrapolation) problem is to find an estimate $\hat{X}(x, t_1 \mid t)$ of $X(x,t_1)$ for $t_1 \geq t$ based on data $Z(x,\tau)$, $x \in D$ for $0 \leq \tau \leq t$.

In general, the solution of (1) for $X(x,t_1)$ with initial time t is given by

$$X(x,t_1) = \int_D G(x,t_1;s,t)X(s,t)\, ds$$

$$+ \int_t^{t_1} \int_D G(x,t_1;s,\tau)B(s,\tau)W(s,\tau)\, ds\, d\tau$$

from which one directly obtains

$$\hat{X}(x, t_1 \mid t) = \int_D G(x,t_1;s,t)\hat{X}(s, t \mid t)\, ds \tag{34}$$

where of course the Green's function $G(x,t_1;s,t)$ of the state equation must be known.

Equation (34) implies that the PDE which describes the predicted estimate $\hat{X}(x, t_1 \mid t)$ is

$$\hat{X}_t(x, t_1 \mid t) = A_x\hat{X}(x, t_1 \mid t)$$

with initial condition $\hat{X}(x, t_1 \mid t)\big\}_{t_1=t} = \hat{X}(x, t \mid t)$.

VI. FURTHER LINEAR FILTERING RESULTS

The basic linear DP state estimation results of Sections IV and V have been extended to cover other cases such as: (1) discrete-point measurements [11-14,17]; (2) filtering with boundary noise [17,19,47]; (3) filtering with colored noise [11,12]; and (4) filtering for time-delay DPS [6,50,51,60,70].

A. Discrete-point Measurements

To handle the situations where $Z(x,t)$ is measured at m distinct spatial points $x = s_i$, $i = 1,2, \ldots ,m$, one puts

$$Z(x,t) = \sum_{i=1}^{m} Z(s_i,t)\delta(x - s_i), \qquad M(x,t) = \sum_{i=1}^{m} M(s_i,t)\delta(x - s_i)$$

or

$$R^+(x,y,t) = \sum_{i,j=1}^{m} R^+(s_i,s_j,t)\delta(x - s_i)\delta(y - s_j)$$

The resulting discrete-measurement-point filter is

$$\hat{X}_t(x, t \mid t) = A_x\hat{X}(x, t \mid t) + \sum_{i,j=1}^{m} P(x,s_i,t)M^T(s_i,t) \times R^+(s_i,s_j,t)\tilde{Z}(s_j,t) \tag{35a}$$

$$P_t(x, y, t \mid t) = A_x P + P A_y^T + B(x,t)Q(x,y,t)B^T(y,t)$$

$$- \sum_{i,j=1}^{m} P(x, s_i, t \mid t)M^T(s_i,t)$$

$$\times R^+(s_i,s_j,t)M(s_j, y, t \mid t) \tag{35b}$$

Similar equations are derived for the discrete-measurement-point smoothing filters.

B. Boundary Noise and Measurements

The case of boundary noise was first considered in [47] by the orthogonality principle and later in [17,19] by using the extended operator concept [71,72]. The case of boundary measurements was considered in [20] by the innovation process.

Consider the case of system (1) with the exception that the boundary condition is

$$\beta_x X(x,t) = B_b(x,t)W_b(x,t)$$

where $W_b(x,t)$ is a zero-mean white noise independent of $W(x,t)$ and $V(x,t)$, with covariance matrix $Q_b(x,y,t)\delta(t - \tau)$. This boundary condition may be replaced by the homogeneous one $\beta_x X(x,t) = 0$ if the noise term in (1a) is replaced by $B(x,t)W(x,t) + F_{x_b} \delta(x-x_b)W_b(x,t)$, where $x_b \in D$, $\delta(x - x_b)$ is the Dirac function and F_{x_b} is a scalar spatial PD operator being determined as shown in [71].

The equations for the filtered state estimate are identical to (7), but the covariance equations become

$$P_t(x,y,t) = A_x P + P A_y^T + BQB + B_b Q_b B_b F_{x_b} \delta(x - x_b)$$

$$- \iint_{DD} P(x,s,t)M^T(s,t)R^+(s,\zeta,t)M(\zeta,t)P(\zeta,y,t)\, ds\, d\zeta$$

$$\beta_{x_b} P(x_b,y,t) = 0, \qquad x_b \in \partial D,\ y \in D$$

which are reduced to

$$P_t(x,y,t) = A_x P + PA_y^T + B(x,t)Q(x,y,t)B^T(y,t)$$

$$- \iint_{DD} P(x,s,t)M^T(s,t)R^+(s,\zeta,t)M(\zeta,t)P(\zeta,y,t)\, ds\, d\zeta$$

$$\beta_x P(x,y,t) = \int_{\partial D} B_b(x,t)Q_b(x,\zeta,t)B_b^T(\zeta,t)\Lambda_b^T(y,t;\zeta,t)\, d\zeta$$

where $\Lambda_b(y,t;\zeta,t)$ is the Green's function on the boundary for the error equation.

Now consider the case where $B(x,t) = 0$ and, in addition to having boundary noise, the measurements are taken only on the boundary ∂D. In this case the filter equations are [20]:

$$\hat{X}_t(x,t) = A_x \hat{X}(x,t) + \iint_{\partial DD} P(x,s,t)M^T(s,t)R^+(s,\zeta,t)\tilde{Z}(\zeta,t)\, d\zeta\, ds$$

$$P_t(x,y,t) = A_x P + PA_y - \int_{\partial D}\int_{\partial D} P(x,s,t)M^T R^+ MP(\zeta,y,t)\, ds\, d\zeta$$

$$+ \int_{\partial D}\int_{\partial D} \Lambda_b(x,t;s,t)B_b Q_b B_b^T \Lambda_b^T(y,t;\zeta,t)\, ds\, d\zeta$$

$$\beta_x \hat{X}(x,t) = 0, \qquad \beta_x P(x,y,t) = 0, \qquad x \in \partial D,\ y \in D$$

C. Filtering for Colored Measurement Noise

When the observation noise is colored (correlated) of the type

$$Z(x,t) = M(x,t)X(x,t) + N(x,t)\Omega(x,t), \qquad x \in D$$

$$\Omega_t(x,t) = B_x \Omega(x,t) + D(x,t)V(x,t), \qquad x \in D$$

$$\omega_x \Omega(x,t) = 0, \qquad x \in \partial D$$

one must transform the equations for $Z(x,t)$ and $V(x,t)$ to a single equation involving white noise. This was done in [11,12], where the equivalent single measurement equation is found to be

$$Z_t(x,t) = \Lambda(x,t)Z(x,t) + M_O(x,t)X(x,t) + V_O(x,t)$$

where $\Lambda(x,t)$ and $M_O(x,t)$ are uniquely defined and $V_O(x,t)$ is a new white noise. With this transformation the filter derivation may be accomplished by any one of the methods discussed in Section IV.

D. Filtering for Time-delay Systems

The filtering problem for time-delay LPS was first solved in [60]. This problem for time-delay DPS has been treated in [50,70,73,74]. Here a brief review is provided of the work in [50]. The system model is

$$X_t(x,t) = \sum_{i-0}^{k} A_{x,i}X(x,\ t-\sigma_i) + W(x,t)$$

$$Z(x,t) = \sum_{i=0}^{k} M_i(x,t)X(x,\ t-\sigma_i) + V(x,t)$$

with homogeneous spatial boundary conditions.

The solution is derived by the innovation approach combined with the results of [60]. The resulting filter equations are

$$\hat{X}_t(x,t,\delta) + X_\delta(x,t,\delta) = \iint_{DD} K(x,s,\zeta,t,\delta,t)\gamma(s,t)\ ds\ d\zeta, \qquad \delta > 0$$

$$\hat{X}_t(x,t,0) = \sum_{i=0}^{k} A_{x,i}\hat{X}(x,t,\sigma_i) + \iint_{DD} K(x,s,\zeta,t,0,t)\gamma(s,t)\ ds\ d\zeta$$

$$K(x,s,\zeta,t,\delta,t) = \sum_{i=0}^{k} P(x,\zeta,t,\delta,\sigma_i)M_i^T(\zeta,t)R^+(\zeta,s,t), \qquad \delta > 0$$

$$P_t(x,y,t,\delta_1,\delta_2) + P_{\delta_1}(x,y,t,\delta_1,\delta_2) + \Gamma_{\delta_2}(x,y,t,\delta_1,\delta_2)$$

$$= -\iiiint_{DDDD} K(x,s,s',t,\delta_1,t)R(s,\zeta,t)K^T(y,\zeta,\zeta',t,\delta_2,t)\ ds\ ds'\ d\zeta\ d\zeta'$$

$$P_t(x,y,t,0,\delta_2) + P_{\delta_2}(x,y,t,0,\delta_2) = \sum_{i=0}^{k} A_{x,i}P(x,y,t,\sigma_i,\delta_2)$$

$$- \iiiint_{DDDD} K(x,s,s',t,0,t)R(s,\zeta,t)K^T(y,\zeta,\zeta',t,\delta_2)\ ds\ ds'\ d\zeta\ d\zeta'$$

$$P_t(x,y,t,\delta_1,0) + P_{\delta_1}(x,y,t,\delta_1,0) = \sum_{i=0}^{k} P(x,y,t,\delta_1,\sigma_i)A^T_{x,i}$$

$$- \iiiint\limits_{DDDD} K(x,s,s',t,\delta_1,t)R(s,\zeta,t)K^T(y,\zeta,\zeta',t,0,t)\, ds\, ds'\, d\zeta\, d\zeta'$$

$$P_t(x,y,t,0,0) = \sum_{i=0}^{k} \{A_{xi}P(x,y,t,\sigma_i,0) + P(x,y,t,0,\sigma_i)A^T_{yi}\} + Q(x,y,t)$$

$$-\iiiint\limits_{DDDD} K(x,s,s',t,0,t)R(s,\zeta,t)K^T(y,\zeta,\zeta',t,0,t)\, ds\, ds'\, d\zeta\, d\zeta'$$

A generalized filtering theory of time-delay systems was developed in [51]. Since this work concerns nonlinear systems, it will be described in the next section together with other nonlinear filtering results.

VII. NONLINEAR DISTRIBUTED PARAMETER STATE ESTIMATION

The state estimation of nonlinear DPS has been treated by the maximum likelihood approach [36,75], the least squares approach [37,45], and the Fokker-Planck-Kolmogorov approach [46,49]. A short description of all methods will be given.

A. Maximum Likelihood Technique

The problem is to find a maximum likelihood estimate $\hat{X}(x, t_1 \mid t)$ of $X(x,t_1)$ in the model (3) based on data $Z(x,\tau)$, $x \in D$, $0 \le \tau \le t$. Clearly we have the filtering problem for $t_1 = t$, the prediction problem for $t_1 > t$, and the smoothing problem for $t_1 < t$.

First consider the filtering problem. The maximum likelihood estimate $\hat{X}(x, t \mid t)$, $x \in D$ is found by minimizing the negative exponent

$$J(0,t) = \frac{1}{2} \iint\limits_{DD} \langle \tilde{X}_0(x),\ P_0^+(x,s)\tilde{X}_0(s)\rangle\, dx\, ds$$

$$+ \frac{1}{2} \iiint\limits_{0DD}^{t} \langle Z - M_x,\ R^+(x,s,\tau)(Z - M_s)\rangle\, dx\, ds\, d\tau$$

$$+ \frac{1}{2} \int_0^t\iint_{DD} \langle W(x,\tau), Q^+(x,s,\tau)W(s,\tau)\rangle \, dx \, ds \, d\tau \tag{36}$$

of the likelihood functional with respect to both $W(x,t)$ and $X(x,t)$ $x \in D$ under the constraint (3a-b).

Applying the differential dynamic programming technique [76], [77] yields the approximate nonlinear DP filter

$$\begin{aligned}
\hat{X}_t(x, t \mid t) &= N_d(\hat{X},x,t) + \iint_{DD} P(M_s)_{\hat{X}} R^+(s,\zeta,t)\tilde{Z}(\zeta,t) \, ds \, d\zeta \\
P_t(x, y, t \mid t) &= (N_d)_{\hat{X}}P + P(N_d)^T_{\hat{X}} + B(x,t)Q(x,y,t)B^T(y,t) \\
&\quad - \iint_{DD} P(x, s, t \mid t)(M_s)^T_{\hat{X}}R^+(M_\zeta)_{\hat{X}} \\
&\quad \times P(\zeta, y, t \mid t) \, ds \, d\zeta \\
N_b(\hat{X},x,t) &= 0, \qquad (N_b)_{\hat{X}}P(x, y, t \mid t) = 0, \qquad x \in \partial D, \ y \in D \\
\hat{X}(x, 0 \mid 0) &= E\{X_0(x)\}, \qquad P(x, y, 0 \mid 0) = P_0(x,y), \\
&\qquad x, y \in D
\end{aligned} \tag{37}$$

For linear systems, i.e., $N_d(X,x,t) = A_x X(x,t)$, $N_b(X,x,t) = A_x X(x,t)$, and $M_x(X,x,t) = M(x,t)X(x,t)$, the filter (37) reduces to the linear filter (7).

To solve the prediction problem one has to minimize the cost functional

$$J' = J(0,t) + \int_t^{t_f} \iint_{DD} \langle W(x,\tau), R^+(x,y,\tau)W(y,\tau)\rangle \, dx \, dy \, d\tau$$

where $J(0,t)$ is given by (36). To this end, one first minimizes $J(0,t)$, i.e., finds the filter (37), and then for $t_1 \geq t$ sets

$$\begin{aligned}
\hat{X}_t(x, t_1 \mid t) &= N_d(\hat{X}(x, t_1 \mid t),x,t_1), \qquad t_1 > t \\
N_b(\hat{X}(x, t_1 \mid t),x,t_1) &= 0, \qquad x \in \partial D \\
\hat{X}(x, t_1 \mid t)\Big\}_{t_1=t} &= \hat{X}(x, t \mid t) \quad \text{initial condition}
\end{aligned} \tag{38}$$

To solve the smoothing problem, i.e., to find an estimate $\hat{X}(x, t_1 \mid t_f)$ of $X(x,t_1)$ on the basis of a data record in the time interval $[t_0,t_f]$, the functional under minimization is decomposed in two parts $J(0,t_1)$ and $J(t_1,t_f)$. The first part is minimized in forward time leading to the forward filter (37), and the second part is minimized in backward time leading to a similar backward filter. The smoothed estimate is then determined by combining the estimates of $X(x,t_1)$ resulting from the two independent filters. This combination is performed as in Section V [12,36,68,78].

The same approach was also used in [75], but the solution was determined by the algorithm of [79] which contains some additional second-order terms and leads to more accurate filter equations at the expense of more computing time.

B. Least Squares Approach

This approach originally used for LPS in [63] has been applied to DPS in [37-45]. The model considered in [40] is that described by Eq. (4), which is a single-dimensional model involving noises in the interior and on the boundary of the spatial domain. Here the work of [42] will be reviewed where the DP system is composed by interacting distributed- and lumped-parameter subsystems and also involves noise on the boundary. This model is

$$X_t(x,t) = F(x,t,X,X_x,X_{xx},a(t)) + W_1(x,t), \qquad x \in (0,1)$$

$$G_0(t,X,X_x) + W_3(t) = 0 \quad (x = 0), \qquad G_1(t,X,X_x)\beta(t)) + W_4(t) = 0 \quad (x = 1) \tag{39}$$

$$\dot{a} = A(t,a(t)) + W_2(t), \qquad \dot{\beta} = B(t,\beta(t)) + W_5(t)$$

$$Z(x,t) = M(x,t,X(x,t)) + V(x,t), \qquad x \in (0,1)$$

where $X_x = \partial X/\partial x$, etc., and the statistical characteristics of $W_i (i = 1,2,3,4,5)$ and $V(x,t)$ are unknown.

Here together with $X(x,t)$ one has to find estimates of the structural parameters $a(t)$ and $\beta(t)$ at time t_1 on the basis of a data record in the interval $[0,t_f]$.

The solution is determined by minimizing the squared error functional

$$J = \int_0^{t_f} \left\{ \int_0^1\int_0^1 [\tilde{Z}^T(x,t)S(x,s,t)\tilde{Z}(s,t) + \tilde{X}_t^T(x,t)T_1(x,s,t)\tilde{X}_t(s,t)]\, dx\, ds \right.$$

$$+ \tilde{Z}^T(0,t)S(0,0,t)\tilde{Z}(0,t) + G_0^T T_3(t)G_0 + G_1^T T_4(t)G_1$$

$$+ \tilde{\dot{a}}^T(t)T_2(t)\tilde{\dot{a}}(t) + \tilde{\dot{\beta}}^T(t)T_5(t)\tilde{\dot{\beta}}(t) \tag{40}$$

where

$$\tilde{Z}(x,t) = Z(x,t) - M(x,t,X), \qquad \tilde{X}_t = X_t - F(x,t,X,X_x,X_{xx},a)$$

$$\tilde{\dot{a}}(t) = \dot{a}(t) - A(t,a), \qquad \tilde{\dot{\beta}}(t) = \dot{\beta}(t) - B(t,\beta)$$

The weighting matrices $S(x,s,t)$ and $T(x,s,t)$ are assumed symmetric in s,t, and all matrices $T_i(i=1, \ldots ,5)$ are assumed positive definite. If the structural parameter functions a and β are constant, then one simply puts $\dot{a} = 0$ and $\dot{\beta} = 0$.

The minimization method is the calculus of variations as in [14,15] for the linear case. The resulting TPBVP is approximately solved by using first-order linearizations about the optimal estimates, and yields the result:

$$\hat{X}_t(x,t) = \hat{F} + \int_0^1\int_0^1 P^{(\upsilon\upsilon)}(x,s,t)(\hat{M}_X)^T S(s,\zeta,t)\tilde{Z}(\zeta,t)\, ds\, d\zeta$$

$$+ P^{(\upsilon\upsilon)}(x,0,t)M_X^T(0,t,\hat{X})S(0,0,t)\tilde{Z}(0,t) \tag{41a}$$

$$\dot{\hat{a}}(t) = A(t,\hat{a}) + \int_0^1\int_0^1 P^{(a\upsilon)}(s,t)(\hat{M}_X)^T S(s,\zeta,t)\tilde{Z}(\zeta,t)\, ds\, d\zeta$$

$$+ P^{(a\upsilon)}(0,t)M_X^T(0,t,\hat{X})S(0,0,t)\tilde{Z}(0,t) \tag{41b}$$

$$\dot{\hat{\beta}}(t) = B(t,\beta) + \int_0^1\int_0^1 P^{(\beta\upsilon)}(s,t)(\hat{M}_X)^T S(s,\zeta,t)\tilde{Z}(\zeta,t)\, ds\, d\zeta$$

$$+ P^{(\beta\upsilon)}(0,t)M_X^T(0,t,\hat{X})S(0,0,t)\tilde{Z}(0,t) \tag{41c}$$

$$P_t^{(\upsilon\upsilon)}(x,y,t) = \hat{F}_X(x)P^{(\upsilon\upsilon)} + P^{(\upsilon\upsilon)}\hat{F}_X^T(y) + \hat{F}_{X_x}(x)P^{(\upsilon\upsilon)} + P_y^{(\upsilon\upsilon)}\hat{F}_{X_y}^T(s)$$

$$+ \hat{F}_{X_{xx}}(x)P_{xx}^{(\upsilon\upsilon)} + P_{yy}^{(\upsilon\upsilon)}\hat{F}_{X_{yy}}^T(y) + \hat{F}_a(x)P^{(\upsilon a)}(s,t)^T$$

$$+ P^{(\upsilon a)}(x,t)\hat{F}_a^T(y) + T_1^{-1}(x,y,t)$$

$$+ P^{(\upsilon\upsilon)}(x,0,t)\Lambda(0,0,t)P^{(\upsilon\upsilon)}(0,y,t)$$

$$+ \int_0^1\int_0^1 P^{(\upsilon\upsilon)}(x,\zeta,t)\Lambda(\zeta,s,t)P^{(\upsilon\upsilon)}(s,y,t)\, d\zeta\, ds \qquad (41d)$$

$$P_t^{(\upsilon a)}(x,t) = \hat{F}_X P^{(\upsilon a)} + \hat{F}_{X_x} P_x^{(\upsilon a)} + \hat{F}_{X_{xx}}^T P_{xx}^{(\upsilon a)} + \hat{F}_a(x)P^{(aa)} + P^{(\upsilon a)}\hat{A}_a^T$$

$$+ \int_0^1\int_0^1 P^{(\upsilon\upsilon)}(x,s,t)\Lambda(s,\zeta,t)P^{(\upsilon a)}(\omega,t)\, ds\, d\zeta$$

$$+ P^{(\upsilon\upsilon)}(x,0,t)\Lambda(0,0,t)P^{(\upsilon a)}(\omega,t) \qquad (41e)$$

$$P_t^{(\upsilon\beta)}(x,t) = \hat{F}_X P^{(\upsilon\beta)} + \hat{F}_{X_x} P_x^{(\upsilon\beta)} + \hat{F}_{X_{xx}} P_{xx}^{(\upsilon\beta)} + \hat{F}_a(x)P^{(a\beta)} + P^{(\upsilon\beta)}\hat{B}_\beta^T$$

$$+ \int_0^1\int_0^1 P^{(\upsilon\upsilon)}(x,s,t)\Lambda(s,\zeta,t)P^{(\upsilon\beta)}(\omega,t)\, ds\, d\zeta$$

$$+ P^{(\upsilon\upsilon)}(x,0,t)\Lambda(0,0,t)P^{(\upsilon\beta)}(\omega,t) \qquad (41f)$$

$$\dot{P}^{(aa)}(t) = \hat{A}_a P^{(aa)} + P^{(aa)}\hat{A}_a^T + P^{(a\upsilon)}(0,t)\Lambda(0,0,t)P^{(\upsilon a)}(\omega,t)$$

$$+ \int_0^1\int_0^1 P^{(a\upsilon)}(s,t)\Lambda(s,\zeta,t)P^{(\upsilon a)}(\omega,t)\, ds\, d\zeta + T_2^{-1}(t) \qquad (41g)$$

$$\dot{P}^{(a\beta)}(t) = \hat{A}_a P^{(a\beta)} + P^{(a\beta)}\hat{B}_\beta^T + P^{(a\upsilon)}(0,t)\Lambda(0,0,t)P^{(\upsilon\beta)}(\omega,t)$$

$$+ \int_0^1\int_0^1 P^{(a\upsilon)}(s,t)\Lambda(s,\zeta,t)P^{(\upsilon\beta)}(\omega,t)\, ds\, d\zeta \qquad (41h)$$

$$\dot{P}^{(\beta\beta)}(t) = \hat{B}_\beta P^{(\beta\beta)} + P^{(\beta\beta)}\hat{B}_\beta^T + \int_0^1\int_0^1 P^{(\upsilon\beta)^T}\Lambda(s,\zeta,t)^{(\upsilon\beta)}(\omega,t)\, ds\, d\zeta$$

$$+ T_5^{-1}(t) \tag{41i}$$

where

$$\Lambda(s,\zeta,t) = \int_0^1 \{[Z(\zeta,t) - M(\zeta,t,\hat{X})]^T S(\zeta,s,t) M_{XX}(s,t,\hat{X})\delta(\omega - s)$$

$$\times M_X^T(s,t,\hat{X})S(s,\zeta,t)M_X(\zeta,t,\hat{X})\delta(\omega - \zeta)\}\, d\omega \tag{42}$$

The initial conditions are assumed to be given, and the spatial boundary conditions are

$$\{G_0(t,\hat{X},\hat{X}_x)\}_{x=0} = 0, \qquad \{G_1(t,\hat{X},\hat{X}_x,\beta)\}_{x=1} = 0$$

$$[\hat{G}_{0x}P^{(\upsilon\upsilon)}(x,y,t) + \hat{G}_{0X_x}P_x^{(\upsilon\upsilon)}(x,y,t) + T_3^{-1}\hat{G}_{0X_x}\hat{F}_{X_{xx}}^T\delta(s)]_{x=0} = 0$$

$$[\hat{G}_{1x}P^{(\upsilon\upsilon)}(x,y,t) + \hat{G}_{1X_x}P_x^{(\upsilon\upsilon)} + \hat{G}_{1\beta}P^{(\upsilon\beta)^T}(x,t) - T_4^{-1}\hat{G}_{1X_x}\hat{F}_{X_{xx}}$$

$$\times \delta(s - 1)]_{x=1} = 0$$

$$[\hat{G}_{0X}P^{(\upsilon a)}(x,t) + \hat{G}_{0X_x}P_x^{(\upsilon a)}(x,t)]_{x=0} = 0$$

$$[\hat{G}_{0X}P^{(\upsilon\beta)}(x,t) + \hat{G}_{0X_x}P_x^{(\upsilon\beta)}(x,t)]_{x=0} = 0$$

$$[\hat{G}_{1X}P^{(\upsilon a)}(x,t) + \hat{G}_{1X_x}P_x^{(\upsilon a)}(x,t) + \hat{G}_{1\beta}P^{(a\beta)}(t)]_{x=1} = 0$$

$$[\hat{G}_{1X}P^{(\upsilon\beta)}(x,t) + \hat{G}_{1X_x}P_x^{(\upsilon\beta)}(x,t) + \hat{G}_{1\beta}P^{(\beta\beta)}(t)]_{x=1} = 0$$

In the preceding equations $\hat{F}$, $\hat{F}_X$, and $\hat{F}_{XX}$, etc., denote the values of F, F_X, and F_{XX} at $X = \hat{X}$. In [42] the same approach is applied for solving the fixed-time smoothing problem as originally done in [80] for LPS.

C. Fokker-Planck Equation Approach

The model treated by this approach [46,49] is more general than the model (3), i.e.,

$$\partial X(x,t) = N_d(X,x,t)\ dt + B_d(X,x,t)\partial W(x,t), \qquad x \in D$$

$$\partial Z(x,t) = M_d(X,Z,x,t)\ dt + C(Z,x,t)\partial V(x,t), \qquad x \in D \qquad (43)$$

$$N_b(X,x,t) = 0,\ x \in \partial D, \qquad M_b(Z,x,t) = 0,\ x \in \partial D$$

with $E\{W(x,t)\} = 0$, $E\{V(x,t)\} = 0$, and

$$E\{W(x,t)W^T(s,\tau)\} = Q_W(x,s,t)\ \min(t,\tau), \qquad x,\ s \in D$$

$$E\{V(x,t)V^T(s,\tau)\} = Q_V(x,s,t)\ \min(t,\tau), \qquad x,\ s \in D$$

The estimate selected is the minimum variance unbiased estimate which is derived by utilizing the forward DP Fokker-Planck equation

$$\frac{\partial p(S,t;X,t')}{\partial t'} = A^*_{(X,t')}P(S,t;X,t'), \qquad t' > t \qquad (44)$$

where

$$A^*_{(X,t')}[\cdot] = \frac{1}{2}\sum_{i,j=1}^{m}\iint_{DD}\frac{\delta^2\{B_d(X,x,t)Q_W(x,s,t)B_d^T(X,s,t)[\cdot]\}}{\delta X_i(x)\delta X_j(s)}\,dx\,ds$$

$$-\sum_{i=1}^{m}\int_D \frac{\delta\{N_{d,i}(X,x,t)[\cdot]\}}{\delta X_i(x)}\,dx \qquad (45)$$

is the forward DP stochastic diffusion operator.

This equation, together with the backward DP Fokker-Planck equation, was derived by means of the characteristic functional concept defined in Section IV.C.

Using (44) one derives the forward Fokker-Planck equation

$$\partial P(X,t) = [dtA^*_{(X,t)} + \iint_{DD}\Gamma^T(X,Z,x,s,t)\partial\tilde{Z}(s,t)\ dx\ ds]P(X,t) \qquad (46)$$

for the conditional probability function $P(X,t)$ of $X(x,t)$, where $\partial\tilde{Z}(s,t) = \partial Z(s,t) - dt\hat{M}_d(Z,s,t)$.

Equation (46) is employed to derive the equations

$$M_k(\{x_k\},t) = \int_{-\infty}^{\infty} \left\{ \prod_{i=1}^{k} [X - \hat{X}(x_i,t)] \right\} P(X, dt)\, dX$$

of $\hat{X}(x,t)$ where $\{x_k\} = \{x_1, x_2, \ldots, x_k\}$. The final result is $(k = 1,2,3, \ldots)$:

$$\begin{aligned}
\partial M_k(\{x_k\},t) = {} & [A_k(\{x_k\},t) - \sum_{i=1}^{k} A_1(x_i,t) M_{k-1}(\{x_k\}^i,t) \\
& + \int_D \sum_{i=1}^{k} M_{k-1}(\{x_k\}^i,t)\Gamma_1(Z;x_i,x',t)\hat{M}_d(Z,x',t)\, dx' \\
& - \iint_{DD} \Gamma_k(Z,\{x_k\},x,x',t)\hat{M}_d(Z,x',t)\, dx'\, dx \\
& + \sum_{\substack{i,s=1 \\ i \neq s}}^{k} \Delta(Z,x_i,x_s,t) M_{k-2}(\{x_k\}^{i,s},t) \\
& - \iint_{DD} \sum_{i=1}^{k} \Gamma_1(Z;x_i,x')R(x',x'',t)\Gamma_{k-1}(Z;\{x_k\}^i,x'') \\
& \qquad dx'\, dx''\,]\, dt \\
& + \int_D \Big[\int_D \Gamma_k(Z;\{x_k\},x,x',t)\, dx \\
& - \sum_{i=1}^{k} M_{k-1}(\{x_k\}^i,t)\Gamma_1(Z,x_i,x',t) \Big]\, \partial Z(x',t)\, dx' \qquad (47)
\end{aligned}$$

where

$$\{x_k\}^i = \{x_j, j = 1,2, \ldots, k,\ j \neq i\}$$

$$\{x_k\}^{i,s} = \{x_j,\ j = 1,2, \ldots, k,\ j \neq i,\ j = s\}$$

and

$$A_k(\{x_k\},t) = \int_{-\infty}^{\infty} [\prod_{i=1}^{k} (X - \hat{X}(x_i,t))] A^*_{(X,t)} P(X,t)\, dX$$

$$\Gamma_k(Z;\{x_k\}, x, x',t) = \int_{-\infty}^{\infty} \left[\prod_{i=1}^{k} (X - \hat{X}(x_i,t)) \right] \Gamma(X,Z,x,x',t)P(X,t)\, dX$$

$$\Gamma(X,Z,x,x',t) = R^{+}(Z,x,x',t)[M_d(X,Z,x,t) - \hat{M}_d(Z,x,t)]$$

$$\hat{M}_d(Z,x,t) = \int_{-\infty}^{\infty} M_d(S,Y,x,t)P(S,t)\, dS$$

$$\Delta(Z,x_i,x_s,t) = \iint_{DD} \Gamma_1(Z,x_i,x,t)R(x,y,t)\Gamma_1(Z,y,x_s,t)\, dx\, dy$$

$$R(Z,x,x',t) = C(Z,x,t)Q_v(x,x',t)C^T(Z,x',t)$$

Retaining only the first two moments (i.e, the state estimate and covariance matrix) Eq. (47) yields

$$\partial\hat{X}(x,t) = dtA_1(x,t) + \iint_{DD} \Gamma_1(Y,x,s,t)\partial\tilde{Z}(s,t)\, dx\, ds \tag{48a}$$

$$\begin{aligned}\partial M_2(x,y,t) = \{&A_2(x,y,t) + \Delta(Z,x,y,t) + \Delta(Z,y,x,t)\\ &- \iint_{DD} \Gamma_2(Z,x,y,s,\zeta,t)\hat{M}_d(Z,\zeta,t)\, ds\, d\zeta\}\, dt\\ &+ \iint_{DD} \Gamma_2(Z,x,y,s,\zeta,t)\partial Z(\zeta,t)\, ds\, d\zeta\end{aligned} \tag{48b}$$

which, by using approximations of the type $\Gamma_X(\hat{X},Z,x,t) = R^{+}(Z,x,s,t)(M_d)_X(\hat{X},Z,x,t)$, reduce to (37).

In [49] the generalized composite DP and LP model

$$\partial X(x,t) = N_d(X,X_d(t),x,t)\, dt + B_d(X,X_d,x,t)\partial W(x,t), \qquad x \in D$$

$$dX_d(t) = A_d(X_d,t)\, dt + C_d(X_d,t)\, dW_d(t)$$

$$dX_b(t) = A_b(X_b,t)\, dt + C_b(X_b,t)\, dW_b(t)$$

$$\partial Z(x,t) = M_d(X,X_d,X_b,Z,x,t) + C(Z,x,t)\partial V(x,t), \qquad x \in D$$

$$N_b(X,X_b,x,t) = 0, \qquad M_b(Z,x,t) = 0, \qquad x \in \partial D$$

is treated by using an increased state vector $X' = [X(x,t),X_d(t),$

$X_b(t)]^T$ involving DP and LP components, and applying the filter (48a,b). The resulting filter is similar to that derived in [42].

It is noted here that the backward DP Fokker-Planck equation derived in [46] is more general than that derived in [81] which concerns only a linear stochastic heat-conductionlike system.

D. Nonlinear Systems with Time Delays

The purpose here is to review the work described in [51]. The general model considered is

$$\dot{X}(t) - F(X,Y(x_1,t), \ldots ,Y(x_\beta,t),t) + \int_0^1 K(Y,x,t)\,dx + W(t)$$

$$Y_t(x,t) = -H(x,t)Y_x(x,t) + G(Y,x,t) + W_d(x,t), \qquad x \in [0,1]$$

$$Z(t) = M(X,Y(x_1^*,t), \ldots ,Y(x_\gamma^*,t),t) + \int_0^1 N(Y,x,t)\,dx + V(t) \tag{49}$$

$$X(0) = X_0, Y(x,0) = Y_0(x), \qquad x \in [0,1]$$

$$Y(0,t) = b(X(t)), \qquad x = 0$$

where $0 < x_1 < \cdots < x_\beta \leq 1$ and $0 < x_1^* < \cdots < x_\gamma^* \leq 1$.

This model contains as special cases (1) nonlinear LPS with multiple constant time delays; (2) nonlinear LPS with functional time delays; and (3) mixed nonlinear LP and hyperbolic DPS. The last model is

$$\begin{aligned}
\dot{X}(t) &= F(X,Y(1,t),t) + W(t)\\
Y_t(x,t) &= -H(x,t)Y_x(x,t) + G(Y,x,t) + W_d(x,t)\\
Z(t) &= M(X,Y(x_1^*,t), \ldots ,Y(x_\gamma^*,t),t) + V(x,t)\\
X(0) &= X_0, \quad Y(x,0) = Y_0(x), \quad Y(0,t) = b(X(t))
\end{aligned} \tag{50}$$

This model is obtained from (49) by setting $K = N = 0$, $\beta = 1$, and $x_1 = 1$. The approach followed is that of minimizing the estimation error criterion

$$J = \int_0^{t_f} \langle \dot{X} - F, Q_0(\dot{X} - F)\rangle\, dt + \int_0^{t_f} \langle Z - M, R(Z - M)\rangle\, dt$$

$$+ \int_0^{t_f} \left[\int_0^1\int_0^1 \langle Y_t + HY_x - G, Q_1(x,s,t)(Y_t + HY_x - G)\, dx\, ds \right] dt$$

where the weighting matrices Q_0, Q_1, and R are symmetric positive definite. The problem is again formulated as an optimal control problem and is solved by the calculus of variations method. The associated TPBVP is

$$\hat{X}_t(t \mid t_f) = \hat{F} - \frac{1}{2} Q_0^{-1}(t)\hat{\lambda}(t \mid t_f)$$

$$\hat{Y}_t(x, t \mid t_f) = -H\hat{Y}_x + \hat{G} - \frac{1}{2}\int_0^1 Q_1^{+}(x,s,t)\hat{\sigma}(s, t \mid t_f)\, ds$$

$$\hat{\lambda}(t \mid t_f) = 2\hat{M}_X^T R(t)(Z - \hat{M}) - \hat{F}_X^T\hat{\lambda}(t \mid t_f)$$

$$- \tilde{b}_X^T H^T(0,t)\hat{\sigma}(0, t \mid t_f) \tag{51}$$

$$\hat{\sigma}_t(x, t \mid t_f) = 2\sum_{i=1}^{\gamma} \hat{M}_Y^T\hat{\lambda}(t \mid t_f)\delta(x - x_i) - \hat{G}_Y^T\hat{\sigma}(x, t \mid t_f)$$

$$- [H^T(x,t)\hat{\sigma}(x, t \mid t_f)]_x$$

$$\hat{\lambda}(0 \mid t_f) = \hat{\lambda}(t_f \mid t_f) = 0, \qquad \hat{Y}(0, t \mid t_f) = b(\hat{X}(t \mid t_f))$$

$$\hat{\sigma}(x, 0 \mid t_f) = \hat{\sigma}(x, t_f \mid t_f) = 0, \qquad \hat{\sigma}(1, t \mid t_f) = 0$$

where $\hat{F}$ symbolizes $F(\hat{X}(t \mid t_f), \hat{Y}(x_1, t \mid t_f), \ldots, t)$, etc.

Solving these canonical Hamilton equations yields the required least squares estimate (smoothed). The required covariance matrices are represented as the following sensitivity matrices:

$$P^{XX}(t \mid t_f) = -2\,\frac{\partial\hat{X}(t \mid t_f)}{\partial\hat{\lambda}(t \mid t_f)}, \qquad P^{XY}(x, t \mid t_f) = -2\,\frac{\delta\hat{X}(t \mid t_f)}{\delta\hat{\sigma}(x, t \mid t_f)}$$

$$P^{YX}(x, t \mid t_f) = -2 \frac{\partial \hat{Y}(x, t \mid t_f)}{\partial \hat{\lambda}(t \mid t_f)}, \qquad P^{YY}(x, s, t \mid t_f) = -2 \frac{\delta \hat{Y}(x, t \mid t_f)}{\delta \hat{\sigma}(s, t \mid t_f)}$$

When the kernels K and N in (49) are not zero, then M and F in J must be replaced by $M + \int_0^1 N\,dx$ and $F + \int_0^1 K\,dx$, and the method works in exactly the same way as for the model (50). The reader is referred to [51] for the details.

E. Filtering of DPS with Moving Boundaries

Many practical systems contain boundaries which move owing to phase change, e.g., heat and mass transfer, chemical reaction, melting or solidification, etc. These systems are usually described by mixed PDE and ODE, the ODE describing the movement of the boundary. Such systems have been considered in [54] for control purposes. Our aim here is to give a short account of the work in [82] concerning the filtering of such systems.

The model considered in a single spatial dimensional of the type

$$\begin{aligned} \frac{\partial X_i(x,t)}{\partial t} &= g_i(t,x,X_i,X_{ix},X_{ixx}) + W_i(x,t), \qquad t > 0 \\ & b_i - 1 < x < b_i, \qquad i = 1,2, \dots ,n_2 + 1 \\ \frac{db_i(t)}{dt} &= f_i(t,b_i,X_i,X_{i+1},X_{ix},X_{i+1,x}) + W_{bi}(t) \\ & i = 1,2, \dots ,n_2 \end{aligned} \tag{52}$$

where each n_1-dimensional state vector $X_i(x,t)$, $i = 1,2, \dots ,n_2 + 1$ of the system is assumed to be piecewise continuous, $b_i(t)$ is the location function of each moving boundary, $X_{ix} = \partial X_i/\partial x$, and $b_0 = 0$, $b_{n_2+1} = 1$. The boundary equations for the states X_i are

$$\beta_0(t,X_1,X_{1x}) + \omega_0 = 0, \qquad x = b_0 = 0$$

$$\left.\begin{aligned} \beta_i(t,x,X_i,X_{i+1},X_{ix},X_{i+1,x}) + \omega_i &= 0 \\ B_i(t,x,X_i,X_{i+1},X_{ix},X_{i+1,x}) + \Omega_i &= 0 \end{aligned}\right\} x = b_i \quad (i = 1, \ldots ,n_2) \tag{53}$$

$$\beta_{n_2+1}(t,X_{n_2+1},X_{n_2+1,x}) + \omega_{n_2+1} = 0, \qquad x = b_{n_2+1} = 1$$

The m-dimensional vectorial measurement process equations are

$$Z(t) = M(b,X(x_j^*,t)) + V(t), \quad j = 1,2, \ldots ,a \tag{54}$$

where the measurement points are ordered as $0 \le x_1^* \le x_2^* \le \ldots \le x_a^* \le 1$. Here $W_i(x,t)$, $W_{bi}(t)$, $\omega_0,\omega_1,\Omega_i$, and ω_{n_2+1} are zero-mean stochastic processes with unknown statistical properties. The initial conditions $X_i(x,0)$ and $b_i(0)$ may be unknown. The general description (54) of the measurement process includes the practical situation where it is not known which region is actually at a measurement location each time. Indeed, if we have n_2+1 regions and the measurements consist of all state variables, then (54) takes the specific form

$$Z(t) = \sum_{i=1}^{n_2+1} \begin{bmatrix} X_i(t,x_1^*)[U(b_i - x_1^*) - U(b_{i-1} - x_1^*)] \\ \vdots \\ X_i(t,x_a^*)[U(b_i - x_a^*) - U(b_{i-1} - x_a^*)] \end{bmatrix} + V(t)$$

where $U(x)$ is the unit step function, i.e., $U(x) = 1$ for $x > 0$ and $U(x) = 0$ for $x < 0$.

The filter derivation is made through the nonlinear least squares filtering approach described in Section VII.B [41-43,48], i.e., through the calculus of variations minimization of an error functional of the type (40). The equations for the filtered estimates $X_i(x,t)$ and $b_i(t)$ are

$$\frac{\partial \hat{X}_i(x,t)}{\partial t} = g_i(t,x,\hat{X}_i,\hat{X}_{ix},\hat{X}_{ixx}) + \sum_{k=1}^{n_2} P^{ib_k}(x,t)\hat{M}_{b_k}^T R^{-1}(t)(Z - \hat{M})$$
$$+ \sum_{j=1}^{n_2+1} \sum_{k=1}^{a} P^{i,j}(x,x_k^*,t)\hat{M}_{X_j}^T (x_k^*,t)R^{-1}(t)(Z - \hat{M}) \tag{55a}$$
$$(i = 1,2, \ldots ,n_2 + 1)$$

$$\frac{db_i(t)}{dt} = \{f_i(t,x,\hat{X}_i,\hat{X}_{i+1},\hat{X}_{ix},\hat{X}_{i+1,x})\}_{x=\hat{b}_i}$$

$$+ \sum_{k=1}^{n_2} P^{b_i b_k}(t)\hat{M}^T_{b_k} R^{-1}(t)(Z - \hat{M})$$

$$+ \sum_{j=1}^{n_2+1} \sum_{k=1}^{a} P^{jb_i}(x^*_k,t)\hat{M}^T_{X_j}(x^*_k,t)R^{-1}(t)(Z - \hat{M})$$

$$(i = 1,2, \ldots ,n_2) \qquad (55b)$$

$$\beta_0(t,\hat{X}_1,\hat{X}_{1x}) = 0, \qquad x = 0$$

$$\left.\begin{aligned}\beta_i(t,x,\hat{X}_i,\hat{X}_{i+1},\hat{X}_{ix},\hat{X}_{i+1,x}) &= 0\\ B_i(t,x,\hat{X}_i,\hat{X}_{i+1},\hat{X}_{ix},\hat{X}_{i+1,x}) &= 0\end{aligned}\right\} x = \hat{b}_i (i = 1,2, \ldots ,n_2) \qquad (55c)$$

$$\beta_{n_2+1}(t,\hat{X}_{n_2+1},\hat{X}_{n_2+1,x}) = 0, \qquad x = 1$$

where $\hat{M}_{b_k} = \partial M(\hat{b},\hat{X}(x^*_j,t))/\partial b_k$, and the quantities $P^{i,j}$ and $P^{ib_k}(t)$, which have the covariance interpretation, are described by a set of equations similar to (41d-i).

In the same paper [82], some important applications of the filter (55a-c) are described and studied, e.g., solidification and melting problems, detection of degree of conversion in metal oxide reduction, hydrodynamic problems, estimation of concentration, and temperature profiles in packed bed reactors. One of them will be discussed in Section X.E.

VIII. OBSERVABILITY AND OPTIMAL LOCATION OF MEASUREMENT POINTS

A. Distributed Parameter Observability

Observability was originally defined in [83,61] for linear LPS as the ability of fully reconstructing the value of the system's state at a past time on the basis of an output record over a certain period of time. Observability of DPS was studied in [54] as the dual of DP controllability; in [9] within the frames of DP linear unbiased estimation; in [55] for a class of linear hyperbolic systems; in [56] as the ability of establishing the uniqueness of a system solution;

in [23] by using the modal representation of DPS, also in [57,59]; in [84] for systems in Banach space; and in [85] for a heating system in high-dimensional space with general boundary conditions.

State observability is closely related to the stability of the optimal filter associated with the system under consideration as stated in the following theorem of [9]: "If the DPS (1) is uniformly completely controllable and uniformly completely observable, then the optimal filter (7) is uniformly asymptotically stable."

Of particular interest is the observability definition provided in [56] which states that: "A DPS of the type

$$C_0(X,x,t;\gamma)X_t(x,t) + \sum_{i=1}^{n} C_i(X,x,t;\gamma)X_{x_i}(x,t)$$

$$= B(X,x,t;\gamma) + W(x,t), \ x \in D \qquad (56)$$

$$A(X,x,t;a) + W_b = 0, \ x \in \partial D, \qquad X(x,0) = X^0(x)$$

$$Z(t) = \int_D M(X,x,t)\,dx + V(x,t)$$

is said to be observable in the spatial domain D if a unique solution X(x,t) is established in D by the boundary condition and the measured output Z(t)."

With the help of this definition the basic question of where should measurement points be located and when does noise-free measurement data provide sufficient information to ensure a unique solution to (56), in the absence of initial conditions and possibly of boundary conditions, can be answered.

For the situation where the system solution can be expressed in the form of infinite sums of certain orthogonal functions (e.g., eigenfunctions), one can use the so-called N-mode observability definition which states:

> Let the system (56) with stationary boundary and linear boundary conditions. Then for all admissible solutions of the type $X_i(x,t) = \Sigma_{m=0}^{\infty} C_m^i \Phi_m^i(t) \Psi_m^i(x,t)$, $i = 1,2, \ldots ,n$, with $\Phi_m^i(t)$ and $\Psi_m^i(x,t)$ known, the system is N-mode observable in D if and only if the uniqueness of the coefficients

C_m^i ($m \leq N$) is established for each i by the measured output $Z(t)$.

In general, uniqueness of the system solution or of the expansion coefficients can be established by the classical technique of analytic continuation or by the method of characteristics or by eigenfunction expansion methods.

In [23] state observability of a DPS is defined through its modal form as follows:

The DPS (1) is completely observable and detectable when its modal representation:

$$\dot{\zeta}(t) = (A + F)\zeta(t) + w(t), \qquad y(t) = M\zeta(t) + v(t) \quad (57)$$

where A+F is an mN x mN matrix and M is a d x mN matrix, is completely observable and detectable.

It is known that the system (57) is completely observable if [61]:

$$\text{rank}\{M^T, (A^T + F^T)M^T, \ldots, (A^T + F^T)^{mN-1}M^T\} = mN$$

The system (57) is completely detectable if and only if there exists a real matrix C such that $(A^T + F^T) + M^TC$ has all eigenvalues negative [9,54].

The optimal filter for the LPS (57) is

$$\dot{\hat{\zeta}} = (A + F)\hat{\zeta} + P(t)M^TR^{-1}(y - M\hat{\zeta}) \quad (58a)$$

$$\dot{P} = (A + F)P + P(A + F)^T - PM^TR^{-1}MP + Q \quad (58b)$$

The filter (58) is said to be convergent if (58a) is asymptotically stable and there exists a unique matrix $\overline{P}(t)$ such that $P(t) \to \overline{P}(t)$ as $t \to \infty$, for every symmetric nonnegative initial condition P(0). If the LPS (57) is completely observable, then the filter is convergent. It is easy to verify [6,7] that the filter (58) is actually the modal representation of the DP filter (7) with the distributed state estimate being given by

$$\hat{X}(x,t) = \sum_{i=1}^{N} \hat{\zeta}_i(t)\Psi_i(x) \quad (59)$$

B. Optimal Location of Measurement Points

Clearly, in practice it is not always possible to measure the system outputs or states over the entire spatial domain $D \times \partial D$ but only at a finite number of spatial points. Hence one is faced with the problem of optimally locating the measurement points. This problem was first studied in [86] by regarding it as an optimal control problem of a dynamic Riccati-type system with control variables the measurement locations.

The results were motivated by the work in [87] concerning LPS, and were derived by a pure algebraic argument. Other works on the optimum measurement design of LPS include [88] and [89]. Further works concerning the optimum measurement location design of DPS are [23,24,90-93,85]. A brief exposition will be given of [23] and [24].

In [23] use is made of the modal representation (57) for determining the effect of measurement locations on observability and filter convergence. The two principal questions are: (a) Is it always possible to determine a finite number of measurement points in $D \times \partial D$ which guarantees system observability? (b) Is it possible to select a minimum number of measurement points which ensures system observability? The answers are given in the following two theorems [23]:

1. It is always possible to find measurement points $x_{i,j}$, $i = 1,2, \ldots ,m$, $j = 1,2, \ldots ,N$ for which the system is observable.
2. Under certain conditions (see [23]) complete observability is guaranteed if each component $X_i(x,t)$ of $X(x,t)$ is measured at only one point x_i^*, $i = 1,2, \ldots ,m$ of $D \times \partial D$.

The optimal selection of the measurement locations is meaningful only in the stochastic case, since if the system is deterministic a measurement set either determines $X(x,t)$ uniquely or not. A reasonable criterion function is $J = \text{trace } \bar{P}$. The minimization of J can be done by an exhaustive search over a large grid of points in $D \times \partial D$. But this is time consuming and hence the following suboptimal procedure was suggested in [23].

1. Select that single (measurement) point x_1 for which J is minimized.

2. Keep x_1 fixed and select a second point x_2 for which J is minimized.
3. Repeat steps (1) and (2) by adding and selecting further points until the required number of points have been selected.

In [24] the problem is treated directly in its DP form by using the discrete-data filter of Section VI.A. The performance function is

$$J(t_f) = \iint_{D} \text{trace } P(x,y,t_f) \, dx \, dy \tag{60a}$$

or

$$J = \int_0^{t_f} \iint_{D} \text{trace } P(x,y,t) \, dx \, dy \, dt \tag{60b}$$

To exclude the possibility that several points might be clustered in a small region, the best m points are sought from among an a priori set of points $x_i, i = 1,2, \ldots, m'$ $(m' > m)$. In this case one must replace m in (35a,b) by m'. Now since only m of the m' possible points are to be used, one introduces the parameters

$$u_i = \begin{cases} 1, & \text{if measurement location is at point } i \\ 0, & \text{otherwise} \end{cases}$$

for $i = 1,2, \ldots, m'$, where $u_1 + u_2 + \cdots + u_{m'} = m$. The problem then reduces to that of selecting the parameters $u_1, u_2, \ldots, u_{m'}$ so as to minimize J subject to the constraints imposed by the discrete-data filter (35a,b) with m replaced by m'. The first requirement in solving the problem is to ensure observability. In [24] a special DP matrix minimum principle is utilized which is analogous to the one developed in [94] for LPS. A Hamiltonian H is defined as

$$H = \text{trace } P(x,y,t) + \text{trace}\{\Lambda^T P_t\} \tag{61}$$

where the adjoint matrix $\Lambda(x,y,t)$ associated with the state matrix $P(x,y,t)$ is governed by the adjoint equation of the Riccati equation, i.e., by

$$\Lambda_t = -I - A_x^T \Lambda - \Lambda A_y$$

$$+ \sum_{i,j=1}^{m'} \Lambda(x,s,t)P^T(s,\sigma_j,t)M^T(\sigma_j,t)u_jR^+(\sigma_j,\sigma_i,t) \times u_iM(\sigma_i,t)\delta(x - \sigma_i)\, ds$$

$$+ \sum_{i,j=1}^{m'} M^T(\sigma_j,t)u_jR^+(\sigma_j,\sigma_i,t)u_iM(\sigma_i,t) \times P^T(s,\sigma_i,t)\Lambda(s,y,t)\delta(y - \sigma_j)\, ds \tag{62a}$$

$$\beta_x\Lambda(x,y,t) = 0,\ x \in \partial D, \qquad \Lambda(x,y,t_f) = 0,\ x,\ y \in D \tag{62b}$$

The DP matrix minimum principle is: "Let $\{u^o,P^o\}$ be the optimal pair satisfying (35a-b). Then there exists an optimal adjoint matrix function Λ^o satisfying (62a-b) such that

$$\int_0^{t_f}\iint_{DD} H(P^o,\Lambda^o,u^o)\, dx\, dy\, dt \le \int_0^{t_f}\iint_{DD} H(P^o,\Lambda^o,u)\, dx\, dy\, dt \tag{63}$$

for any admissible u." Clearly this provides a necessary condition for optimality.

Further, consider the case where $R^+(x_i,y_j,t) = R^{-1}(x_i,t)\delta(x_i - y_j)$. Since $u_i^2 = u_i$ the Hamiltonian (61) is a linear function of u_i, and since u_i may be 0 or 1, the problem is of the bang-bang (on-off) type in space. The switching function, i.e., the part of H not depending on u_i, is

$$\Sigma_i = \int_0^{t_f}\iint_{DD} P(x,\sigma_i,t)M^T(\sigma_i,t)R^{-1}(\sigma_i,t)M(\sigma_i,t) \times P(\sigma_i,y,t)\Lambda^T(x,y,t)\, dx\, dy\, dt \qquad (i = 1,2,\ \ldots\ ,m) \tag{64}$$

Using (63), the following algorithm for selecting the optimal measurement points can be employed [24].

1. Choose a starting parameter set u_i, $i = 1,2, \ldots ,m'$ with only m values equal to 1 (the remaining equal to 0).
2. Integrate the system state equations (i.e., the variance equations) in forward time.

3. Integrate the adjoint equations (62a-b) in reverse time.
4. Compute the m′ switching functions Σ_i using (64).
5. Choose the m u_i values corresponding to the m largest Σ_i as equal to 1, the rest as equal to zero.
6. If step (5) does not lead to a change in the u_i values, terminate the algorithm. Otherwise go to step (2).

C. Space-Time Sampling in DPS

The purpose here is to review the work of [95] in which equivalence is established between the scanned sampling and the continuous time-spatial sampling (or the continuous spatial-time sampling). In general, the measurement of a DP variable can be made by one of the following methods: (1) Spatial sampling-continuous time; (2) time sampling over the entire continuum (spatial domain); (3) continuous-time scanning; and (4) spatial sampling-time sampling.

The basic questions arising are:

1. Is there a way by which a variable of the system sampled by means of one method can be obtained from the same variable sampled by the other methods?
2. Is it possible to find a common method for studying and relating the various sampling methods?
3. Is it possible to use a more practical or convenient sampling method in place of another method in a given situation?

Obviously complete and general answers are difficult. Actually, in [95] the following results are provided.

1. Transformation of Scanned Sampling into a Single Fixed Spatial Point Measurement

Let x,t be the independent variables of the scanning plane (x,t), x being single-dimensional, and x′,t′ the same variables in the fixed spatial point plane (x′,t′). Then

$$x' = x - (L/T)t + L \sum_{k=1}^{\infty} u(t - kT), \qquad t' = t \tag{65}$$

where $u(\cdot)$ is the unit step function, L is the length of the spatial domain, and T is the sampling time period.

2. Transformation of Scanned Sampling into Continuous Space-Time Sampling

If (x,t) is the scanned sampling plane and (x',t') the continuous space-time sampling plane, then

$$x' = x, \qquad t' = t - xT/L \tag{66}$$

Special care must be taken when applying these transformations to ensure that measurements and disturbances propagate forward in time. Also, the causality requirement restricts the application and utilization of the method.

The transformations (65) and (66) were derived by constructing the trajectories of the new sampled variables from the corresponding ones of the original sampled variables. In [95] an elegant example is also provided, where the transformation of scanned sampling into continuous space is used to stabilize an unstable stretched string by means of scanned point feedback.

IX. RIGOROUS GENERALIZED STATE ESTIMATION THEORY

The rigorous generalized state estimation theory was originated in [27] and further or independently explored in [28-34,73,96,97-99, 74,100,101]. The work in [27] is based on a generalization of the Ito integral in Hilbert space by means of which one can use an infinite-dimensional stochastic equation of the type

$$dX(t) = A(t)X(t)\,dt + B(t)\,dW(t)$$

$$dZ(t) = M(t)X(t)\,dt + dV(t)$$

where $W(t)$ and $V(t)$ are independent Wiener processes in Hilbert space. In [27], $A(t)$ is assumed to be a bounded operator (thus excluding differential operators), but in [28] this restriction was removed, thus enlarging the applicability of the theory to systems described by ODE, PDE, and mixed ODE-PDE.

A generalized state estimation theory in Banach space was given in [96], where the system model is described in terms of the equations

$$dX(t) = i_1 a(t)X(t)\,dt + dB_1(t)$$
$$dZ(t) = i_2 \gamma(t)X(t)\,dt + dB_2(t) \tag{67}$$

with initial conditions X_0 and $Z_0 = 0$, where X_0 is a Gaussian random variable, independent of the Brownian motions B_1, B_2 having mean value zero and covariance operator $\Gamma_0 \in L(H_1,H_1)$. Here, $L(H_1,H_1)$ is the family of linear continuous mappings from H_1 to H_1, where H_1 is a separable Hilbert space with norm $|\cdot|_1$. It is assumed that the linear mappings a and γ are continuous in the uniform operator topology, and

$$a(t)\colon \mathbb{B}_1 \to H_1, \qquad \gamma(t)\colon \mathbb{B}_1 \to H_2$$
$$i_1\colon H_1 \to \mathbb{B}_1, \qquad i_2\colon H_2 \to \mathbb{B}_2$$

where H_2 is a separable Hilbert space which is measurable with respect to the family of canonical normal distributions on H_2, and $\mathbb{B}_1, \mathbb{B}_2$ are the completions of H_1, H_2 with respect to the norms $|\cdot|_1, |\cdot|_2$ correspondingly.

The main result is the following theorem.

Let the model (67). Then the conditional mean $\hat{X}(t) = E\{X(t) \mid Z(u), u \le t\}$ and its covariance matrix $P(t)$ exist as an $\mathbb{B}_1$-valued process and a self-adjoint linear operator on H_1 respectively, and satisfy the following equations

$$d\langle \ell,\hat{X}\rangle = \langle \ell, i_1 a\hat{X}\rangle\,dt + \langle \gamma i_1 P\ell, dZ - \gamma\hat{X}\,dt\rangle, \quad \hat{X}_0 = 0$$

$$\frac{d}{dt}\langle \ell_1, P\ell_2\rangle = \langle \ell_1, a i_1 P\ell_2\rangle + (\ell_1, P j_1 a^* \ell_2)$$
$$- \langle \ell_1, P j_1 \gamma^* \gamma i_1 P\ell_2\rangle + \langle \ell_1, I\ell_2\rangle, \qquad P_0 = \Gamma_0$$

where $\ell, \ell_1, \ell_2 \in H_1, i_1\colon H_1 \to \mathbb{B}_1$, and $j_1 = i_1^*$.

Clearly, the operators involved are bounded, so this theorem is applicable only to integral systems.

A complete treatment along these lines is involved in [29-33, 73,97-99,74]. In summary, the model considered is of the type

$$\dot{X}(t) + A(t)X(t) = f(t) + B(t)W(t), \qquad t_0 \le t \le t_f \tag{68a}$$

$$Z(t) = M(t)X(t) + V(t), \quad X(t_0) = X_0 + \xi \tag{68b}$$

where the state function $X(t)$ is assumed to belong to a Hilbert space H, the measured variable to a Hilbert space F, $A(t)$ is assumed to be a linear and closed operator (i.e., a differential operator), and $B(t)$, $M(t)$ are assumed to be linear bounded operators.

The stochastic processes $\{\xi, W(\cdot), V(\cdot)\}$ are modeled as a linear random functional (LRF) on $\Phi = H \times L^2(0,T;H) \times L^2(0,T;F)$, which implies that to any test function $\{\zeta, w(\cdot), v(\cdot)\} = \Psi$ there corresponds a generalized random variable $\tilde{\mu}_\varphi(\omega)$ with: (1) $\Psi \to \tilde{\mu}_\psi(\omega)$ is a.s. linear; (2) $E\tilde{\mu}_\psi = 0$ for every Ψ, and

$$(3)\ \operatorname{cov}(\tilde{\mu}_{\psi_1}, \tilde{\mu}_{\psi_2}) = \langle P_0\zeta_1, \zeta_2\rangle + \int_0^{t_f} \langle Q(t)w_1(t), w_2(t)\rangle\, dt + \int_0^{t_f} \langle R(t)v_1(t), v_2(t)\rangle\, dt$$

where $\Psi_1 = \{\zeta_1, w_1(\cdot), v_1(\cdot)\}$ and $\Psi_2 = \{\zeta_2, w_2(\cdot), v_2(\cdot)\}$.

The estimation problem, i.e., the problem of determining the optimal estimate of $X(t)$, is reduced to that of minimizing the functional

$$J(\xi, W) = \langle P_0^{-1}\xi, \xi\rangle + \int_0^{t_f} \langle Q^{-1}(t)W(t), W(t)\rangle\, dt + \int_0^{t_f} \langle R^{-1}(t)\{Z(t) - M(t)X(t)\}, Z(t) - M(t)X(t)\rangle\, dt \tag{69}$$

with respect to $(\xi, W(\cdot))$, where $P_0, Q(t)$ and $R(t)$, the covariance operators for the LRF associated with $\{\xi, W(\cdot), V(\cdot)\}$, are self-adjoint, positive, and invertible.

Thus the problem has again been converted to a deterministic control problem (as in the formal least squares approach described in Sections IV.B and VII.B) with control variable $(\xi, W(\cdot))$ and operator system dynamics described by (68a). This generalized optimal control problem was studied in [102]. The solution of the state filtering problem is $\hat{X}(t) = X(t;\hat{\xi},\hat{W}(\cdot))$ where $(\hat{\xi},\hat{W}(\cdot))$ is the optimal pair which minimizes J under the assumption that Z(t) is known in the interval $(0,t_f)$.

The result is, when t_f varies, the mapping $t_f \to \hat{X}(t_f) = r(t_f)$ can be obtained in a recursive manner by the generalized Kalman-Bucy filter

$$\begin{aligned} &\dot{r} + Ar + P(t)M^*R^{-1}Mr = f + P(t)M^*R^{-1}Z, \qquad r(0) = X_0 \\ &\dot{P} + PA^* + AP + PM^*R^{-1}MP = BQB^*, \qquad P(0) = P_0 \end{aligned} \tag{70}$$

Actually, this filter also determines the optimal LRF estimate for the more general situation where P_0 and Q(t) are not invertible operators.

To facilitate the solution of (70), an approximate filter is derived in [32] by using the following approximation:

$$M_m(t)h = \sum_{j=1}^{m} \langle h,h_j\rangle M(t)h_j = M(t)h^m \tag{71}$$

of M(t), where $h_1, \ldots ,h_m$ is an orthonormal basis of H, and h^m is the mth-order approximation

$$h_m = \sum_{j=1}^{m} \langle h,h_j\rangle h_j \tag{72}$$

of h. By using $M_m(t)$ in place of M(t) in (69), an approximation of $J(\xi,W)$ denoted by $J_m(\xi,W)$ is obtained. The importance of this approximation is based on the following theorem: "If $(\hat{\xi},\hat{W})$ is the optimal pair for $J(\xi,W)$ and $(\hat{\xi}_m,\hat{W}_m)$ the optimal pair for $J_m(\xi,W)$, then $\lim_{m\to\infty} (\hat{\xi}_m,\hat{W}_m) = (\hat{\xi},\hat{W})$ and hence $\lim_{m\to\infty} X(t_f;\hat{\xi}_m,\hat{W}_m) = X(t_f,\hat{\xi},\hat{W})$."

The resulting approximate filter, when $M_m(t)$ is used in place of M(t), is

$$\begin{aligned} &\dot{r}_m + Ar_m + P_mM_m^*R^{-1}M_mr_m = f + PM_m^*R^{-1}Z, \qquad r_m(0) = X_0 \\ &\dot{P}_m + P_mA^* + AP_m + P_mM_m^*R^{-1}M_mP_m = BQB^*, \qquad P_m(0) = P_0 \end{aligned} \tag{73}$$

Clearly, if for every $(0,t_f)$, Z(t) is given on $(0,t_f)$, the preceding theorem implies that

$$\lim_{m\to\infty} r_m(t_f) = r(t_f)$$

An example of applying the approximate generalized filter (73) is when the system is a DPS. In this case, the basis $h_1, h_2, \ldots,$ h_m is chosen such that $A^*h_j = \lambda_j h_j$ (i.e., it is the set of the first m eigenfunctions of A^*) and Eqs. (73) reduce to ODEs.

It is remarked that $X(t)$ in the model (68a,b) is not a well-defined stochastic process, since it is defined only by means of an LRF, and so (68a) does not represent a stochastic differential equation in the strict sense. However, following [27,28], this may be overcome by using the model (68a,b) in the form

$$X(t,\omega) + \int_0^t A(\delta,\omega)X(\delta,\omega)\, d\delta = X_0(\omega) + \int_0^t f(\delta)\, d\delta + \int_0^t B(\delta)\, dW(\delta,\omega) \tag{74}$$

$$Z(t,\omega) = \int_0^t M(\delta)X(\delta,\omega)\, d\delta + V(t,\omega)$$

where the integration is defined in the generalized Ito sense, $W(t,\omega)$ and $V(t,\omega)$ are independent Wiener processes in Hilbert space, and $X_0(\omega)$ is a Gaussian process (with zero mean and covariance matrix P_0) independent of $W(t,\omega)$ and $V(t,\omega)$.

In this case, the optimal estimate $\hat{X}(t_f,\omega)$ of $X(t_f,\omega)$ based on $\{Z(t,\omega),\ 0 \le t \le t_f\}$ is described by

$$\hat{X}(t_f,\omega) + \int_0^{t_f} A(\delta)\hat{X}(\delta,\omega)\, d\delta + \int_0^{t_f} P(\delta)M^*(\delta)R^{-1}(\delta)C(\theta)\hat{X}(\delta,\omega)\, d\delta$$

$$= \int_0^{t_f} f(\delta)\, d\delta + \int_0^{t_f} P(\delta)M^*(\theta)R^{-1}(\delta)\, dZ(\delta,\omega) + X_0(\omega) \tag{75}$$

where $P(t)$ is described by the second equation in (70).

In [73] and [74] the rigorous filtering problem is treated for infinite-dimensional hereditary systems by combining the preceding results with those presented in [102]. The model considered is

$$\dot{X}(t) = A_0(t)X(t) + \sum_{i=1}^{N} A_i(t)X(t + \sigma_i) + \int_{-\sigma}^{0} A_{01}(t,s)X(t + s)\, ds + B(t)W(t) + f(t) \tag{76}$$

$$X(0) = X_0(0) + \xi^0, \qquad X(s) = X_0(s) + \xi^1(s), \qquad -\sigma \le s \le 0$$

$$Z(t) = M(t)X(t) + V(t)$$

where $\sigma > 0$, $0 = \sigma_0 > \sigma_1 \ldots. > \sigma_N = -\sigma$ are real numbers, A_0 and A_i $(i = 1,2, \ldots ,N)$ belong to $L^\infty(0,t_f;\mathcal{L}(H))$, and A_{01} belongs to $L^\infty(0,t_f,-\sigma,0;\mathcal{L}(H))$ where H is a Hilbert space, B belongs to $L^\infty(0,t_f;\mathcal{L}(E,H))$, and C belongs to $L^\infty(0,t_f;L(H,F))$.

The solution of (76) belongs to $AC^2(0,t_f,H)$, i.e., to the space of absolutely continuous functions from $(0,t_f) \to H$. Again, $\{\xi^0,\xi^1, W(\cdot),V(\cdot)\}$ is modeled as a zero-mean Gaussian LRF having the covariance operator $A = \text{diag}\{P_0,P_1(s);Q(t),R(t)\}$.

The problem is formulated in terms of the functional (69) and is solved with the aid of the generalized control results presented in [91]. The optimal estimator is first derived in the form of a TPBVP and then in the standard estimator-Riccati form.

In [101] the filtering problem is treated for a system of the type

$$dX(t,\omega) = A(t)X(t,\omega)\, dt + B(t)\, dW(t,\omega),$$

$$dZ(t,\omega) = M(t)X(t,\omega)\, dt + D(t)\, dV(t,\omega)$$

where $A(t)$ is an unbounded operator, $B(t)$, $M(t)$, and $D(t)$ are bounded operators, $W(t,\omega)$ is a Hilbert space-valued Wiener process, and $V(t,\omega)$ is a finite-dimensional Wiener process. The assumptions imposed are slightly weaker than those of the previous works in [29-33].

This section closes with a brief description of the work in [34]. Two models are studied, namely a parabolic-type DPS (involving an integral operator in the state equation) with Dirichlet boundary conditions, and a parabolic DPS with mixed boundary conditions. The approach followed is an extension of the method presented for LPS in [100] by which the results are derived under very weak assumptions. Here we only indicate the results for the first model which has the form

$$dX(x,t) = [A_x X(x,t) + \int K(s,x,t)X(s,t)\, ds]\, dt + \sigma(x,t)\, dW$$

$$Z(\tau) = \int_0^\tau h_r\, dr + V_\tau \qquad \tau \leq t$$

where $A_x X(x,t) \to 0$ and $A_x X(x,0) \to 0$ as $x \to \partial D$, $h_r = \int M(x,r)X(x,r)\, dx$, $A_x = \Sigma a_{i,j}(x,t)D_i D_j + \Sigma b_i(x,t)D_i$ with $D_i = \partial/\partial x_i$, and $V_\tau = \int_0^\tau B(s)\, d\tilde{V}_s = \int_0^\tau \Sigma_s^{1/2}\, d\tilde{V}_s$, with $\tilde{V}_s$ being a normalized Wiener process independent of the process W_s.

Upon the assumption that $Z(\tau)$ is available at time $\tau = t$, the resulting filter is

$$d\hat{X}(x,t) = \{A_x \hat{X}(x,t) + \int K(s,x,t)\hat{X}(s,t)\, ds\}\, dt$$

$$+ \{dZ - \int M(s,t)\hat{X}(s,t)\, ds\}^T \Sigma_t^{-1} \{\int M(s,t)P(s,x,t)\, ds\}$$

$$P_t(x,y,t) = \{A_x + A_y\}P(x,y,t) + \sigma(x,t)\sigma(y,t)$$

$$+ \int K(y,s,t)P(x,s,t)\, ds + \int K(s,x,t)P(s,y,t)\, ds$$

$$- \{\int M(s,t)P(x,s,t)\, ds\}^T \Sigma_t^{-1} \{\int M(s,t)P(s,y,t)\, ds\}$$

where with probability 1, $\hat{X}(x,t)$, $A_x \hat{X}(x,t)$, $P(x,y,t)$, $A_x P(x,y,t)$, and $A_y P(x,y,t)$ tend to zero as $x \to \partial D$. Analogous results are obtained for the second model as well. In both cases, existence and uniqueness of the solution are established.

X. APPLICATION EXAMPLES

Most of the works previously described provide illustrative and numerical results. In particular, [48,82,104-107] contain some important application examples with certain numerical and experimental results. The purpose in this section is to give a small but representative subset of these examples in order to provide the reader with a feeling of how the theory presented thus far might be applied in particular cases.

A. Example with Negligible State Disturbance

This is a simple example in which the solution of the optimal filter can be found analytically [14,15]. The system is a heat conduction system of the type

$$X_t(x,t) = (1/\pi^2)\partial^2 X(x,t)/\partial x^2, \qquad 0 < x < 1 \tag{77a}$$

$$X(0,t) = X(1,t) = 0 \tag{77b}$$

$$Z(t) = \int_0^1 \delta(x - \upsilon_0 t)X(x,t)\, dt + V(t)$$

$$= X(\upsilon_0 t,t) + V(t) \tag{77c}$$

where $X(x,t)$ denotes temperature.

The measurement model (77c) implies that the temperature sensor moves from left to right on the x-axis with a constant scanning velocity υ_0 or, equivalently, that the measurement process is of the averaging-scanning type.

Setting $P_0(x,y) = P_0 \sin(\pi x) \sin(\pi y)$, $P_0 > 0$, and $R(t) = 1$, the filter equations (15) give

$$\sigma_t(x,t) = \frac{1}{\pi^2}\frac{\partial^2\sigma}{\partial x^2} + P(x,\upsilon_0 t,t)\{Z(t) - \sigma(\upsilon_0 t,t)\},$$

$$P_t(x,y,t) = \frac{1}{\eta^2}\left\{\frac{\partial^2 P}{\partial x^2} + \frac{\partial^2 P}{\partial y^2}\right\} - P(x,\upsilon_0 t,y,t)P(\upsilon_0 t,y,t)$$

$$\sigma(0,t) = \sigma(1,t) = 0, P(0,1,t) = P(1,0,t) = 0,$$

$$P(x,y,0) = P_0 \sin(\pi x) \sin(\pi y) \tag{78}$$

The solution for $P(x,y,t)$ is

$$P(x,y,t) = P_0 e^{-t} \sin(\pi x) \sin(\pi y)/\{e^t + P_0\gamma(t)\} \tag{79}$$

where

$$\gamma(t) = \frac{\pi\upsilon_0}{4(\pi^2\upsilon_0^2 + 1)}\left[2\pi\upsilon_0 \sinh t - e^{-t}\left(\frac{2}{\pi\upsilon_0} \sin^2 \pi\upsilon_0 t + \sin \ell\pi\upsilon_0 t\right)\right]$$

For sufficiently small scanning rate ($\pi^2\upsilon_0^2 \ll 1$), the solution (79) has the asymptotic property

$$P(x,y,t) \to 4P_0 e^{-2t} \sin(\pi x) \sin(\pi y)/\{P_0\pi^2\upsilon_0^2 + 4\}, \quad t \gg 0$$

Now consider the case where $Z(x,t) = X(x,t) + V(x,t)$, $0 < x < 1$, and take $P_0(x,y) = 6 \sin(\pi x) \sin(\pi y)$, $0 < x,y < 1$, and $R(x,t) = R(t) = \exp(-5t)$. Then the covariance equations become

$$P_t(x,y,t) = \frac{1}{\pi^2}\left[\frac{\partial^2 P}{\partial x^2} + \frac{\partial^2 P}{\partial y^2}\right] - e^{-5t}\int_0^1 P(x,s,t)P(s,y,t)\, ds$$

$$P(x,0,t) = P(0,x,t) = P(x,1,t) = P(1,x,t) = 0, \quad 0 < x < 1 \tag{80}$$

Equations (80), if solved by the separation method, give the solution:

$$P(x,y,t) = 6e^{-5t} \sin(\pi x) \sin(\pi y), \quad 0 < x,y < 1$$

The form of the exact, filtered, and smoothed state estimates and their covariances of this case with $X(x,0) = \sin \pi x$, $\hat{X}(x,0) = 1$, $P_0(x,y) = \sin(\pi x) \sin(\pi y)\delta(x-y)$, $Q(x,y,t) = R(x,y,t) = \delta(x-y)$ is shown in Figure 1 [21].

B. Example with Boundary Measurements

This is an example in which the measurements are taken on the boundary of the spatial domain. The system under consideration is [20]:

$$X_t = a\partial^2 X/\partial x^2, \quad 0 < x < 1$$

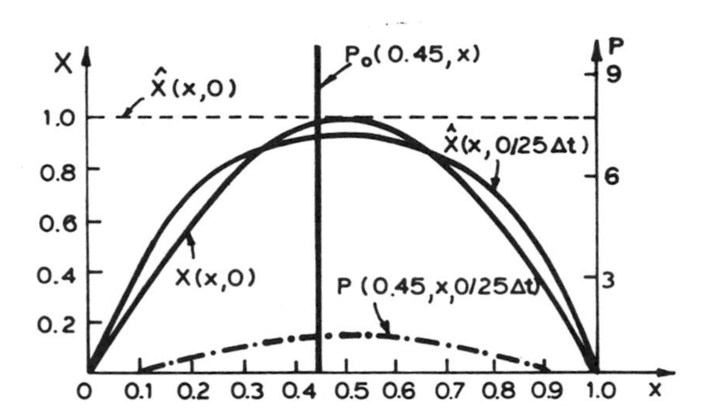

FIG. 1. Exact state, filtered, and smoothed state estimates and their covariances.

$$\{\partial X/\partial x\}_{x=0} = \{\partial X/\partial x\}_{x=1} = 0$$

$$Z(0,t) = X(0,t) + V(t)$$

The associated optimal filter is (see Section VI.B):

$$\hat{X}_t = a\partial^2\hat{X}/\partial x^2 + P(x,0,t)(1/R)\{Z(0,t) - \hat{X}(0,t)\}$$

$$\{\partial\hat{X}/\partial x\}_{x=0} = \{\partial\hat{X}/\partial x\}_{x=1} = 0$$

$$P_t(x,y,t) = a(\partial^2 P/\partial x^2 + \partial^2 P/\partial y^2) - P(x,0,t)(1/R)P(0,y,t)$$

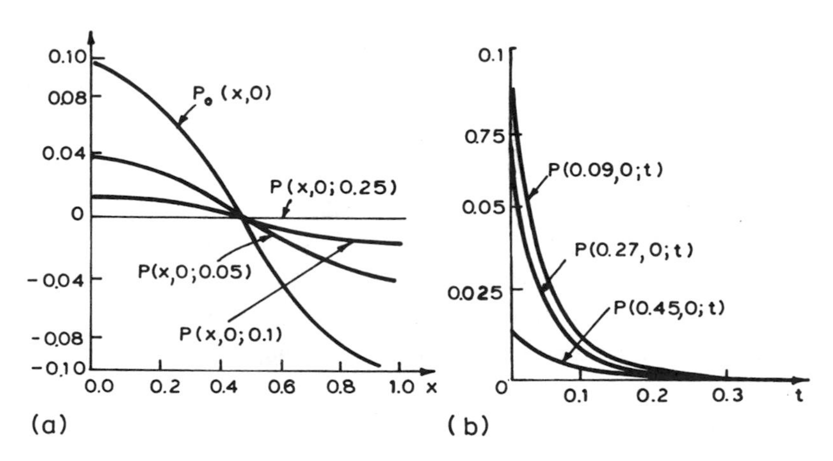

FIG. 2. Covariance function P(x,y,t) at y = 0, (a) with the time variable as parameter (t = 0,0.05,0.1,0.25; (b) with the spatial variable as parameter (x = 0.09,0.27,0.45).

$$\{\partial P/\partial x\}_{x=0} = \{\partial P/\partial y\}_{y=0} = 0$$

The solution of the covariance equation with $a = 1$ and initial condition $P(x,y,0) = 0.1 \cos(\pi x) \cos(\pi y)$ has the form of Figure 2 [20].

C. Example of Optimal Measurement Location

The purpose in this example is to illustrate the application of the optimal measurement location procedures described in Section VII.B. The system considered is [23]:

$$X_t = \alpha_1 \partial^2 X/\partial x^2 + \alpha_2 X + W(x,t), \qquad 0 < x < 1$$

$$\{\partial X/\partial x\}_{x=0} = \{\partial X/\partial x\}_{x=1} = 0$$

The first three eigenvalues and eigenfunctions are

$$\lambda_1 = \alpha_2, \qquad \lambda_2 = \alpha_2 - \pi^2\alpha_1, \qquad \lambda_3 = \alpha_2 - 4\pi^2\alpha_1$$

$$\Psi_1(x) = 1, \qquad \Psi_2(x) = \sqrt{2}\cos(\pi x), \qquad \Psi_3(x) = \sqrt{2}\cos(2\pi x)$$

This system is unobservable for $x = 0.25$, $x = 0.5$, and $x = 0.75$. It is desired to optimally locate the measurements at two points when $\alpha_1 = 0.1$ and $\alpha_2 = 0$. For one measurement point the system is detectable at any x. Figure 3 shows the value of trace $\overline{P}$ for a single measurement point and two measurement points with one point being at $x = 0$. In the one-measurement-point case trace $\overline{P}$ takes its minimum at $x = 0$ or $x = 1$. In the two-point case the minimum occurs at $x = 0$ and $x = 1$.

Now consider that $\alpha_1 = 0.5$ and $\alpha_2 = 0$, and try to optimally locate the measurement sensors by the DP matrix maximum principle of Section VIII.B. The problem is to select the m points x_i, $i = 1,2, \ldots, m$ such that the performance function (60b) is minimized [24]. In this case the covariance equation and its adjoint are

$$P_t(x,y,t) = 0.5\,(\partial^2 P/\partial x^2 + \partial^2 P/\partial y^2) + Q(x,y,t)$$

$$- \sum_{i=1}^{m} P(x,s_i,t)R^{-1}(s_i,t)P(s_i,t,y)$$

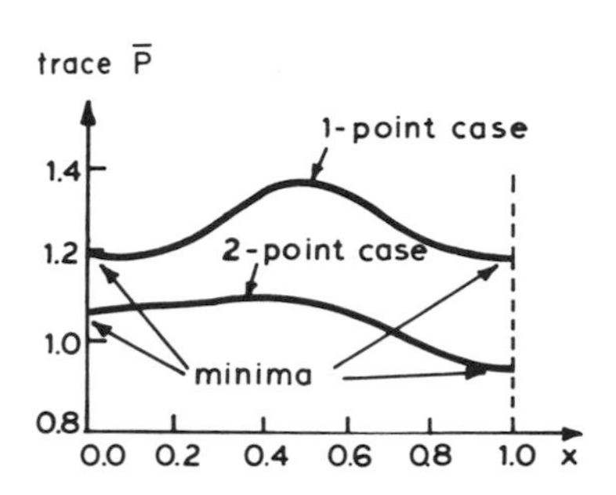

FIG. 3. Curves of trace $\overline{P}$ versus x for the 1-measurement point and 2-point cases.

$$\Lambda_t(x,y,t) = -1 - 0.5(\partial^2\Lambda/\partial x^2 + \partial^2\Lambda/\partial y^2)$$

$$+ \sum_{i=1}^{m} \int_0^1 \Lambda(x,s,t)P(s_i,s,t)R^{-1}(s_i,t)\delta(x-s_i)\, ds$$

$$+ \sum_{i=1}^{m} \int_0^1 R^{-1}(s_i,t)P(\sigma,s_i,t)\Lambda(\sigma,y,t)\delta(y-s_i)\, d\sigma$$

$$\{\partial P/\partial x\}_{x=0,1} = \{\partial P/\partial y\}_{x=0,1} = \{\partial\Lambda/\partial y\}_{x=0,1} = 0$$

$$P(x,y,0) = P_0(x,y)$$

under the assumption that $R^+(x_i,y_j,t) = R^{-1}(x_i,t)\delta(x_i-y_j)$.

Clearly, the parameters which serve as the control inputs are $P_0(x,y)$, $Q(x,y,t)$, and $R(x_i,t)$. In a series of numerical experiments described in [24] the effect of variations in $R(x_i,t)$ on the optimal measurement points was examined for $P_0(x,y) = 3$ and $Q(x,y,t) = 1.33$. A summarized form of the results is given in Table 1 where $t_f = 1$.

In the $m' = 5$ case the points x_i were taken as 0.1, 0.3, 0.5, 0.7, and 0.9, and in the $m' = 11$ case as 0, 0.1, 0.2, ... ,0.9, 1.0. It is seen that for the first form of $R(x_i,t)$ the level of observation error is minimum at x = 0.5, increasing as the observation points move toward the boundaries of the domain. For the second form of R the observation level error is minimum at the boundary points and maximum at x = 0.5. Hence one would naturally expect

TABLE 1

Optimum Location of Measurement Points

$R(x_i,t) = 0.5e^{2\|x_i-0.5\|}$				$R(x_i,t) = 0.5e^{-2\|x_i-0.5\|}$			
m	m′	Initial guess	$x_{i,opt}$	m	m′	Initial guess	$x_{i,opt}$
2	5	(0.1,0.3)	(0.5,0.7)	2	5	(0.3,0.5)	(0.1,0.9)
2	11	(0.1,0.3)	(0.5,0.6)				

that in the first case the optimal measurement points are close to the middle point x = 0.5, whereas in the second case they are close to the boundary points. In all cases the procedure has led to the optimal solution in one step. The minimum value of J obtained in the m′ = 5 case is 0.984 and in the m′ = 11 case one has more degrees of freedom in selecting the optimal points. For more details see [24].

D. Nonlinear Filtering Example

The problem is to estimate the state and the constant parameter of a plug-tubular chemical reactor described by the equations [40,42]:

$$X_t + \partial X/\partial x = -aX^2, \qquad \dot{a}(t) = 0, \qquad X(0,t) = 1$$

with unknown true parameter value a = 2, and unknown steady-state solution $X(x,0) = (1 + ax)^{-1}$. The measurements are

$$Z(x_i,t) = X(x_i,t)\{1 + 0.1G(0,1)\}, \qquad i = 1,2,3$$

where $G(0,\sigma)$ is a normally distributed random variable with zero mean and standard deviation σ. The measurement points are $x_1 = 0.25$, $x_2 = 0.5$, and $x_3 = 0.75$. The filter equations are

$$\hat{X}_t + \partial\hat{X}/\partial x = -\hat{a}\hat{X}^2 + \sum_{i=1}^{3} P^{(\upsilon\upsilon)}(x,x_i,t)[Z(x_i,t) - \hat{X}(x_i,t)]$$

$$\dot{\hat{a}} = \sum_{i=1}^{3} P^{(a\upsilon)}(x_i,t)[Z(x_i,t) - X(x_i,t)]$$

$$P_t^{(\upsilon\upsilon)}(x,y,t) = -2\hat{a}\hat{X}(x,t)P^{(\upsilon\upsilon)}(x,y,t) - 2\hat{a}P^{(\upsilon\upsilon)}(x,y,t)\hat{X}(y,t)$$

$$-\hat{X}^2(x,t)P^{(a\upsilon)}(y,t) - P^{(a\upsilon)}(x,t)\hat{X}^2(y,t)$$

$$-P_x^{(\upsilon\upsilon)}(x,y,t) - P_y^{(\upsilon\upsilon)}(x,y,t)$$

$$- \sum_{i=1}^{3} P^{(\upsilon\upsilon)}(x,x_i,t)P^{(\upsilon\upsilon)}(x_i,y,t)$$

$$P_t^{(a\upsilon)}(x,t) = -2a\hat{X}P^{(a\upsilon)} - P_x^{(a\upsilon)} - \hat{X}^2P^{(\upsilon a)}(t)$$

$$- \sum_{i=1}^{3} P^{(\upsilon\upsilon)}(x,x_i,t)P^{(\upsilon\upsilon)}(x_i,x,t)$$

$$\dot{P}^{(aa)}(t) = - \sum_{i=1}^{3} \{P^{(a\upsilon)}(x_i,t)\}^2, \qquad P^{(\upsilon\upsilon)}(0,y,t) = P^{(a\upsilon)}(0,t) = 0$$

$$\hat{X}(x,0) = 0, \qquad \hat{a}(0) = 1, \qquad P^{(a\upsilon)}(x,0) = 15 \sin(0.8\pi x)$$

$$P^{(\upsilon\upsilon)}(x,y,0) = 20 \sin(0.8\pi x) \sin(0.8\pi y), \qquad P^{(aa)}(0) = 20$$

The numerical results are shown in Figure 4 [42]. Another useful application concerning the on-line estimation of catalyst activity profiles in packed-bed reactors may be found in [104].

E. Example of Filtering with Moving Boundaries

The system treated in this example is a simplified solidification process taking place in a steel mill reheating furnace in which a partially solidified ingot is being externally heated in preparation for rolling. Specifically, it is assumed that a cylindrical solidifying ingot is exchanging heat by radiation with its surroundings at temperature T_w. The problem is to estimate the temperature profile and position of the phase boundary in real time on the basis

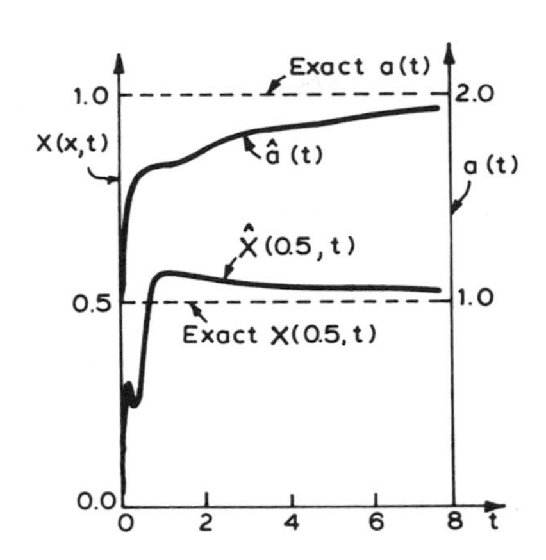

FIG. 4. Exact values and filtered estimates of X(0.5,t) and a(t).

of temperature measurements. For simplicity one pure component is assumed to be solidified and hence one phase boundary is considered.

The system model is [82]:

$$X_t^1 = a_1[\partial^2 X^1/\partial x^2 + (1/x)\partial X^1/\partial x] + W(x,t), \qquad 0 < x < b(t)$$

$$X_t^2 = a_2[\partial^2 X^2/\partial x^2 + (1/x)\partial X^2/\partial x)] + W(x,t), \qquad b(t) < x < 1$$

$$\{\partial X^1/\partial x\}_{x=0} + \omega_0 = 0$$

$$[\partial X^2/\partial x - \epsilon(T_W^4 - (X^2)^4)]_{x=1} + \omega_2 = 0$$

$$X^1(b,t) - T_m = 0, \qquad X^2(b,t) - T_m = 0, \qquad x = b(t)$$

$$\dot{b}(t) = -\beta_1\{\partial X^1/\partial x\}_{x=b} + \beta_2\{\partial X^2/\partial x\}_{x=b} + W_b(t)$$

$$Z(t) = X^2(1,t) + V(t)$$

where the state variables X^1 and X^2 are the temperatures of the liquid and solid, respectively, T_m is the melting temperature of the material, and all other variables and parameters have the meaning explained in (52 to (55).

The state estimate equations (55a-c) give in the present case

$$\hat{X}_t^1 = a_1[\partial^2\hat{X}^1/\partial x^2 + (1/x)\partial\hat{X}^1/\partial x] + P^{12}(x,1,t)R^{-1}(t)[Z - \hat{X}^2(1,t)]$$

$$0 < x < b(t)$$

$$\hat{X}^2_t = a_2[\partial^2\hat{X}^2/\partial x^2 + (1/x)\partial\hat{X}^2/\partial x] + P^{22}(x,1,t)R^{-1}[Z - \hat{X}^2(1,t)]$$

$$b(t) < x \le 1$$

$$\dot{\hat{b}}(t) = -\beta_1\{\partial\hat{X}^1/\partial x\}_{x=b} + \beta_2\{\partial\hat{X}^2/\partial x\}_{x=b} + P^{2b}(1,t)R^{-1}[Z - \hat{X}^2(1,t)]$$

$$\{\partial\hat{X}^1/\partial x\}_{x=0} = 0, \qquad [\partial\hat{X}^2/\partial x - \epsilon(T_W^4 - (\hat{X}^2)^4]_{x=1} = 0$$

$$\hat{X}^1(\hat{b},t) - T_m = 0, \qquad \hat{X}^2(\hat{b},t) - T_m = 0 \qquad \text{at } x = \hat{b}$$

The covariances (differential sensitivities) P^{12}, P^{22}, and P^{2b} are governed by the appropriate Riccati-type equations which, due to limitations of space, are omitted here (see [82] for details).

F. Nuclear Reactor Filtering Example

The problem is to estimate the space-time power distribution of a slab homogeneous nuclear reactor having one prompt step disturbance of power density at a point $x = x_0$ of the spatial domain $D = (0,1)$, and one detector at the point $x_m \neq x_0$. The system model has the form [106]:

$$X_t(x,t) = (D_u\partial^2/\partial x^2 + \Sigma_u)X + U_0\delta(x - x_0) + W(x,t)$$

$$X(0,x) = 0, \qquad X(t,0) = X(t,1) = 0, \qquad D_u = D\epsilon_1, \qquad \Sigma_u = \Sigma\epsilon_2 \tag{81}$$

$$Z(t) = \int_0^1 [S_m X(x,t) + V(x,t)]\delta(x - x_m)\, dx$$

where D and Σ are the diffusion constant and the neutron absorption cross-section, respectively, and ϵ_1 and ϵ_2 are proportionality coefficients.

The filtering approach considered in [106] is the least squares one as in [14,15] except that the differential system model is transformed in integral form prior to applying the filtering procedure, and that the optimal estimate $\hat{X}(x,t)$ is required to satisfy an inequality constraint of the form

$$\hat{X}(x,t) \ge 0, \qquad x \in D x \partial D, \qquad t \in [0,t_f]$$

The performance index to be minimized is

$$J = \int_0^{t_f} \int_0^1 [\hat{W}^2 + \hat{V}^2 \delta(x-x_m)]\, dx\, dt$$

$$+ \int_0^{t_f} \{Z(t) - \int_0^1 [S_m \hat{X}(x,t) + \hat{V}(x,t)] \delta(x-x_m)\, dx\}^2\, dt \qquad (82)$$

under the constraint $\hat{X}(x,t) + X_p(x) \geq 0$, where $X_p(x)$ is a known shape function of power.

Transforming the state equation of system (81) into integral form and applying the calculus of variations method, the optimal estimate $\hat{X}^o$ is found to satisfy the integral equation

$$\hat{X}^o(x,t) = \Xi(x,t) + \mu \int_0^t K(x,t,t';\hat{X}^o)\, dt \qquad (83)$$

where

$$\Xi(x,t) = \int_0^t G(x, x_0, t-t') X_0\, dt'$$

$$+ \frac{1}{2} S_m \int_0^1\int_0^1 G(x, x', t-t') \int_{t'}^{t_f} G(x_m, x', t'') Z(t'')\, dx'\, dt'\, dt''$$

$$\mu = -S_m^2 \qquad (84)$$

$$K(x,t,t';\hat{X}^o) = \int_{t'}^{t_f} \sum_{i=1}^{\infty} \exp[-(i^2\pi^2 D_u - \Sigma_u)(t - 2t' + t'')]$$

$$\times \sin(i\pi x_m) \sin(i\pi x) \hat{X}^o(x_m, t'')\, dt''$$

and the optimal estimates $\hat{W}^o$ and $\hat{V}^o$ are given by

$$\hat{V}^o(x_m,t) = \frac{1}{2} [Z_m(t) - S_m \hat{X}^o(x_m,t)]$$

$$\hat{W}^o(x,t) = S_m \int_t^{t_f} G(x, x_m, t'-t) \hat{V}^o(x_m,t')\, dt' \qquad (85)$$

$$\hat{V}^o(x,t) = 0 \qquad \text{for } x \neq x_m$$

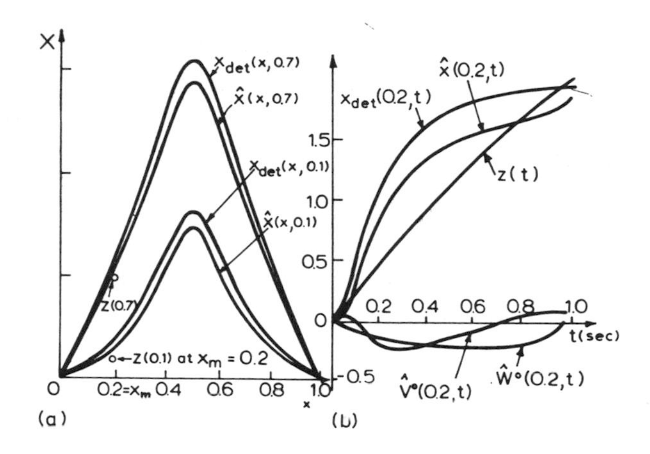

FIG. 5. Numerical results for $x_m = 0.2$. (a) Comparison of $X_{det}(x,t)$ and $\hat{X}(x,t)$ for $t = 0.1$, and $t = 0.7$ and $z(t)$ for $t = 0.1, 0.7$; (b) Comparison of $X_{det}(x,t)$ and $\hat{X}(x,t)$ for $x = x_m = 0.2$ and optimal estimates $\hat{W}^o(x_m,t)$, $\hat{V}^o(x_m,t)$.

The integral equation (83) was solved by the successive approximation method. Some numerical results are given in Figure 5 [105] for the case where $Z(t) = a \exp\{\beta(t)\} - a$, with a and β being constant parameters. For comparison, the deterministic solution $X_{det}(x,t)$ of the state equations in (81) with $W(x,t) = 0$ is also plotted.

G. Example of Optimum Detection in the Presence of Distributed Interference

In this example we illustrate the application of the optimal filter (7) to the problem of detecting the return from a slowly fluctuating point target in the presence of distributed interference [107-109]. This kind of detection problem usually arises in active sonar systems. The slowly fluctuating point target is located at a known delay τ_d and a known Doppler ω_d. The transmitted signal has an envelope $\sqrt{E_t}F(t)$, where E_t is the transmitted energy and F(t) is the normalized complex envelope, and as it is propagated through the sea it meets several objects and inhomogeneities which cause reflections. The return of this distributed interference is known as reverberation in the case of sonar systems and as clutter in the radar systems. This kind of reflection process is modeled as a spatial Poisson stochastic process or as a Gaussian stochastic process when the number of reflectors is very large. The Poisson model is satisfactory in many cases.

The complex envelope of the reverberation process is symbolized as $N_R(t)$ and is given by

$$N_R(t) = \sqrt{E_t} \int_{-\infty}^{\infty} F(t-s)b(s,t)\, ds$$

Aside from the reverberation return there exists a complex white noise $V(t)$ so that we obtain the following hypothesis testing problem

$$\begin{aligned} H_1:&\quad Z(t) = \sqrt{E_t}\, bF_d(t) + N_R(t) + V(t), \qquad t \in (-\infty,\infty) \\ H_0:&\quad Z(t) = N_R(t) + V(t), \qquad t \in (-\infty,\infty) \end{aligned} \tag{86}$$

where $F_d(t)$ is the complex envelope of the desired signal. The reverberation noise process $N_R(t)$ is Gaussian with zero-mean value and covariance functional

$$\begin{aligned} \operatorname{cov}\{N_R(t),N_R(\tau)\} &= E_t \iint_{-\infty}^{\infty} F(t-s)S\{f,s)F^*(\tau-s)e^{j2\pi f(t-\tau)}\, df\, ds \\ &= C(t,\tau) \end{aligned}$$

where $S(f,s)$ is the scattering function of reverberation.

The problem determined by (86) is a detection problem in colored noise [the colored noise is $N_R(t)$] in which we are interested to find whether or not the point target lies in a given range and velocity. The multiplier b is a complex Gaussian variable (with mean value $2\sigma_b^2$) which represents the effect of the target on the transmitted signal.

This detection problem can be reduced in the form of a DP optimum filtering problem if the reverberation noise $N_R(t)$ is modeled by its PDE model, which has the form

$$X_t(s,t) = A(s)X(s,t) + B(s)\, W(s,t), \qquad s \in \Omega_L,\ t > T_i$$

$$b(s,t) = M(s)X(s,t) \tag{87}$$

$$N_R(t) = \int_{-\infty}^{\infty} \sqrt{E_t}F(t-s)b(s,t)\, ds$$

where Ω_L is the range of s in which the reverberation scattering is nonzero, and the process W(s,t) is white in both s and t with $E\{W(s,t)W^T(s',\tau)\} = Q\delta(s-s')\delta(t-\tau)$.

Applying the optimal DP filter (7) in the present case yields the desired optimal receiver equations:

$$\hat{X}_t(s,t) = A(s)\hat{X}(s,t) + K(s,t)[Z(t) - \hat{N}_R(t)], \qquad s \in \Omega_L,\ t \geq T_i$$

$$K(s,t) = \frac{1}{N_0}\int_{\Omega_L} P(s,s',t)M^T(s')\sqrt{E_t}F^*(t-s')\, ds'$$

$$\hat{N}_R(t) = \int_{-\infty}^{\infty} \sqrt{E_t}F(t-s)M(s)\hat{X}(s,t)\, ds, \qquad \hat{X}(s,T_i) = 0, \qquad s \in \Omega_L$$

$$P_t(s,s',t) = A(s)P + PA^T(s') + B(s)QB^T(s)\delta(s-s')$$
$$- \frac{E_t}{N_0}\{\int_{\Omega_L} P(s,\zeta,t)M^T(\zeta)F^*(t-\zeta)\, d\zeta$$
$$\times \int_{\Omega_L} F(t-\zeta')M(\zeta')P(\zeta',s',t)\, d\zeta'\}, \qquad s,\ s' \in \Omega_L$$

$$P(s,s',T_i) = P_0(s,T_i)\delta(s-s')$$

A similar problem in encountered when a double spread target is being detected.

XI. CONCLUSIONS

In this chapter we have presented a survey of the DP state estimation results available in the current literature. Our goal was to include and illustrate the wide range of problems solved and the various estimation techniques that have been applied.

Actually, DP state estimation theory (formal and rigorous) is attracting more and more scientists and engineers, since with the evolution of large digital and hybrid computers it now seems possible to utilize more accurate models, such as DP models or functional differential models of the physical systems under investigation. Another but more specialized chapter on DP estimation theory has been provided in [110].

It is clear from the present survey that DP state estimation theory has now reached a very advanced level, and so, if combined with the available DP identification theory (concerning the estimation of structural parameters and functions other than states) and DP state variable control theory can be used in the design of large-scale physical and industrial systems.

Concerning the DP estimation techniques we conclude that we have two main categories of methods, namely control-like functional minimization estimation techniques and probabilistic estimation techniques, as in the LP case.

The first category involves all versions of least squares and the second category includes the conditional characteristic function, the Bayesian, the maximum likelihood, and the rigorous estimation techniques. Some of the methods in the second category (e.g., maximum likelihood) may be considered after a certain point as belonging to the first category.

A great deal remains to be done in studying various properties of DP estimators. Such properties are stability, observability, sensitivity, error analysis, etc. The computational problem is somewhat open and requires a transfer in the field of estimation and control of results from computational mathematics. Useful in this respect seems to be the hybrid computation approach which combines the benefits of analog and digital computation [111-115]. Also the study of stochastic PDE on their own is considerably open. This study will be useful in both the estimation and control of stochastic DPS [116-120].

Until now only a limited number of applications to real systems of the DP estimation theory were reported. The field of applications seems to be a promising area of research work of both theoretical and practical value. Some possible application fields of DP estimation theory are oceanography, meteorology, air and water pollution problems, medical image data filtering, etc.

For treating medical imaging data by distributed parameter theory it is necessary to extend some recent results [121-124] concerning doubly-stochastic (or conditionally) Poissonian counting processes. A preliminary study in this direction was made in [125].

ACKNOWLEDGMENT

The author gratefully acknowledges the receipt of various reprints and papers still in manuscript form from several research workers.

REFERENCES

1. S. G. Tzafestas, Parameter estimation in distributed parameter dynamic models, I. Chem. E. Symp. Ser., 35: 5-43 (1972).

2. R. E. Goodson and M. P. Polis, Parameter identification in distributed systems: A synthesizing overview, in Identification of Parameters in Distributed Systems (R. E. Goodson and M. P. Polis, eds.), ASME Monograph, 1974.

3. R. E. Goodson and M. P. Polis, A survey of parameter identification in distributed systems, Proceedings of the IFAC 1975 World Congress, Boston/Cambridge, August 24-30, 1975.

4. K. J. Astrom and P. Eykhoff, System Identification: A Survey, Automatica, 7: 123 (1971).

5. A. C. Robinson, A survey of optimal control of distributed-parameter systems, Automatica, 7: 371 (1971).

6. S. G. Tzafestas and J. M. Nightingale, Optimal filtering smoothing and prediction in linear distributed parameter systems, Proc. IEE, 115: 1207 (1968).

7. S. G. Tzafestas and J. M. Nightingale, Concerning optimal filtering theory of linear distributed-parameter systems, Proc. IEE, 115: 1737 (1968).

8. S. G. Tzafestas, Bayesian approach to distributed parameter filtering and smoothing, Int. J. Contr., 15: 273 (1972).

9. S. G. Tzafestas, On the distributed-parameter least-squares estimation theory, Int. J. Syst. Sci., 4: 833 (1973).

10. S. G. Tzafestas, Theory and practice of optimal filtering, Proceedings of the NATO Advanced Study Institute on Network and Signal Theory, Bournemouth, September 4-16, 1972, p. 504.

11. S. G. Tzafestas, On optimum distributed-parameter filter design for correlated measurement noise, Proceedings of the IEEE Symposium on Adaptive Processes. Decision and Control, paper no. V2, Univ. of Texas, Dec. 7-9, 1970.

12. S. G. Tzafestas, On optimum distributed-parameter filtering and fixed interval smoothing for colored noise, IEEE Trans. Auto., Contr., 17: 448 (1972).

13. F. E. Thau, On optimum filtering for a class of linear distributed-parameter systems, ASME Trans. J. Basic Eng., 91: 173 (1969); also in Proceedings of the 1968 JACC, Univ. of Michigan, June, 1968, p. 610.

14. J. S. Meditch, Least squares filtering and smoothing for linear distributed-parameter systems, Automatica, 7: 315 (1971).

15. J. S. Meditch, On state estimation for distributed-parameter systems, J. Franklin Inst., 290: 49 (1970).

16. A. V. Balakrishnan and J. L. Lions, State estimation for infinite dimensional systems, J. Comp. Syst. Sci., 1: 391 (1967).

17. Y. Sakawa, Optimal filtering in linear distributed-parameter systems, Int. J. Contr., 16: 115 (1972).

18. G. A. Phillipson and S. K. Mitter, State identification of a class of linear distributed systems, Proceedings of the Fourth IFAC Congress, Warsaw, Poland, June 1969.

19. S. R. Atre and S. S. Lamba, A note on Kalman-Bucy filtering for linear distributed-parameter systems, IEEE Trans. Auto. Contr., 17: 712(1972).

20. S. R. Atre and S. S. Lamba, Filtering for linear distributed-parameter systems via boundary measurements, Proc. IEE, 119: 757 (1972).

21. S. R. Atre and S. S. Lamba, Optimal estimation in distributed processes using the innovations approach, IEEE Trans. Auto. Contr., 17: 710 (1972).

22. S. R. Atre and S. S. Lamba, Derivation of an optimal estimator for distributed-parameter systems via maximum principle, IEEE Trans. Auto. Contr., 17: 358 (1972).

23. T. K. Yu and J. H. Seinfeld, Observability and optimal measurement location in linear distributed-parameter systems, Int. J. Contr., 18: 785 (1973).

24. W. H. Chen and J. H. Seinfeld, Optimal location of process measurements, Int. J. Contr., 21: 1003 (1975).

25. T. M. Pell, Jr. and R. Aris, Some problems in chemical reactor analysis with stochastic observations, Industr. Eng. Chem. Fundam., 9: 15 (1970).

26. R. S. Bucy, Optimal filtering for correlated noise, J. Math. Anal. Appl., 20: 1 (1967).

27. P. Falb, Infinite dimensional filtering, Inf. Contr., 11: 102 (1967).

28. R. Curtain and P. Falb, Stochastic differential equations in Hilbert space, J. Diff. Eqns., 10: 412 (1971).

29. A. Bensoussan, Filtrage optimal des systèmes lineaires, Dunod, Paris, 1971.

30. A. Bensoussan, Identification et filtrage, Cahiers de l'IRIA, no. 1, 1-253 (1969).

31. A. Bensoussan, Filtering theory: A comparison between lumped and distributed systems, Mem. Conf. Intern. Sobre Sistemas Redes y Computadores, Mexico, 1971, pp. 160-163.

32. A. Bensoussan, On the separation principle for distributed-parameter systems, IFAC Conf. DPS, Banff, Canada, 1971.

33. A. Bensoussan, A. Bossavit, and J. C. Nedelec, Approximation des problèmes de contrôle, Cahiers de l'IRIA, no. 11, 104-172 (1970).

34. H. J. Kushner, Filtering for linear distributed-parameter systems, SIAM J. Contr., 8: 346 (1970).

35. R. Curtain, Filtering in distributed-parameter systems: A survey introduction, Report 7301, Division of Automatic Control, Lund Institute of Technology, Jan. 1973.

36. S. G. Tzafestas and J. M. Nightingale, Maximum likelihood approach to the optimal filtering of distributed-parameter systems, Proc. IEE, 116: 1085 (1969).

37. J. H. Seinfeld, Nonlinear estimation for partial differential equations, Chem. Eng. Sci., 24: 75 (1969).

38. J. H. Seinfeld and M. Hwang, Some remarks on nonlinear filtering in distributed-parameter systems, Chem. Eng. Sci., 25: 741 (1970).

39. J. H. Seinfeld, Optimal stochastic control of nonlinear systems, AIChE J., 16: 1016 (1970).

40. J. H. Seinfeld, G. R. Gavalas, and M. Hwang, Nonlinear filtering in distributed-parameter systems, J. Dynam. Syst. Meas. Contr., 936: 157 (1971).

41. J. H. Seinfeld, G. R. Gavalas, and M. Hwang, Control of nonlinear stochastic systems, Industr. Eng. Chem. Fundam., 8:257 (1969).

42. M. Hwang, J. H. Seinfeld, and G. R. Gavalas, Optimal least square filtering and interpolation in distributed-parameter systems, J. Math. Anal. Appl., 39: 49 (1972).

43. T. K. Yu and J. H. Seinfeld, Suboptimal control of stochastic distributed-parameter systems, AIChE J., 19: 389 (1973).

44. T. K. Yu and J. H. Seinfeld, Control of stochastic distributed parameter systems, J. Optim. Theory Appl., 10: 362 (1972).

45. G. Lamont and K. S. P. Kumar, State estimation in distributed-parameter systems via least squares and invariant embedding, J. Math. Anal. Appl., 38: 588 (1972).

46. S. G. Tzafestas, Fokker-Planck equation approach to nonlinear distributed-parameter filtering, Proceedings of the IFAC Symposium on Stochastic Control, Budapest, Hungary, 1974.

47. S. G. Tzafestas, Boundary and volume filtering of linear distributed-parameter systems, Electron. Lett., 5: 199 (1969).

48. M. B. Ajinkya, W. H. Ray, T. K. Yu, and J. H. Seinfeld, The application of an approximate nonlinear filter to systems governed by coupled ordinary and partial differential equations, Int. J. Syst. Sci., 6: 313 (1975).

49. S. G. Tzafestas, Nonlinear distributed-parameter filtering using the Fokker-Planck equation approach, J. Franklin Inst. 301: 429 (1976).

50. V. Shukla and M. Srinath, Optimal filtering in linear distributed-parameter systems with multiple time delays, Int. J. Contr., 16: 673 (1972).

51. T. K. Yu, J. H. Seinfeld, and W. H. Ray, Filtering in nonlinear time delay systems, IEEE Trans. Auto. Contr., 19: 324 (1974).

52. H. Koivo, Least-squares estimator for hereditary systems with time-varying delay, IEEE Trans. Syst. Man. Cyb., 4: 276 (1974).

53. S. G. Tzafestas, "Optimal estimation and control of distributed-parameter systems," Ph.D. thesis, Electrical Engineering Department, Southampton Univ., 1968.

54. P. K. C. Wang, Control of distributed parameter systems, in Advances in Control Systems, vol. 1 (C. T. Leondes, ed.), Academic Press, New York, 1964, p. 75.

55. T. K. Yu and J. H. Seinfeld, Observability of a class of hyperbolic distributed-parameter systems, IEEE Trans. Auto. Contr., 16: 495 (1971).

56. R. E. Goodson and R. E. Klein, A definition and some results for distributed systems observability, IEEE Trans. Auto. Contr., 15: 15 (1970).

57. G. N. Saridis and P. C. Badavas, Identifying solutions of distributed-parameter systems by stochastic approximation, IEEE Trans. Auto. Contr., 15: 393 (1970).

58. E. Angel and A. Jain, A dimensionality reducing model for distributed filtering, IEEE Trans. Auto. Contr., 18: 59 (1973).

59. A. Thowsen and W. Perkins, Observability conditions for two general classes of linear flow processes, IEEE Trans. Auto. Contr., 19: 603 (1974).

60. H. Kwakernaak, Optimal filtering in linear systems with time delays, IEEE Trans. Auto. Contr., 12: 169 (1967).

61. R. Kalman and R. Bucy, New results in linear filtering and prediction theory, Trans. ASME J. Basic Eng., 83D: 95 (1961).

62. I. M. Gelfand and S. V. Fomin, Calculus of Variations, Prentice-Hall, Englewood Cliffs, N.J., 1963.

63. D. M. Detchmendy and R. Sridhar, Sequential estimation of states and parameters in noisy nonlinear dynamical systems, Trans. ASME, Ser. D, J. Basic Eng., 88:362 (1966).

64. T. Kailath, An innovations approach to least-squares estimation (Part I), IEEE Trans. Auto. Contr., 13: 646 (1968).

65. T. Kailath and P. Frost, An innovations approach to least-squares estimation (Part II), IEEE Trans. Auto. Contr., 13: 655 (1968).

66. M. Athans and E. Tse, A direct derivation of the optimal linear filter using the maximum principle, IEEE Trans. Auto. Contr., 12: 690 (1967).

67. S. G. Tzafestas, A minimum principle in Hilbert space, Int. J. Contr., 11: 917 (1970).

68. R. Deutsch, Estimation Theory, Prentice-Hall, Englewood Cliffs, N.J., 1965, p. 108.

69. A. E. Bryson, Jr. and M. Frazier, Smoothing for linear and nonlinear dynamic systems, TDR-63-119 Aero. Syst. Div., Wright-Patterson AFB, Ohio, 1963, p. 353.

70. A. J. Koivo, Optimal estimation for linear stochastic systems described by functional differential equations, Inf. Contr., 19: 232 (1972).

71. W. L. Brogan, Optimal control theory applied to systems described by partial differential equations, in Advances in Control Systems, vol. 5 (C. T. Leondes, ed.), Academic, New York, 1968, p. 221.

72. I. M. Gelfand and G. E. Shirov, Generalized Functions, vol. 1, Academic, New York, 1964.

73. S. K. Mitter and R. B. Vinter, Filtering for linear stochastic hereditary differential systems, Intern. Symp. Contr. Theory: Num. Methods and Computer Syst. Modelling, June 1974 (Springer-Verlag, Lecture Notes in Economics and Math Systems).

74. A. Bensoussan, Filtrage des Systèmes lineaires avec retard, IRIA Report Technique, no. 7118/71027, Oct. 1971.

75. S. G. Tzafestas, On the nonlinear distributed-parameter filtering, Proceedings of the International Conference on Systems and Control, PSG College, Coimbatore, India, Aug. 30-Sept. 1, 1973.

76. D. Q. Mayne, A second-order gradient method for determining optimal trajectories of nonlinear discrete-time systems, Int. J. Contr., 3: 85 (1966).

77. D. H. Jacobson and D. Q. Mayne, Differential Dynamic Programming, Elsevier, New York, 1970.

78. S. Fujita and T. Fukao, Optimal linear fixed-interval smoothing for colored noise, Inf. Contr., 17: 313 (1970).

79. D. H. Jacobson, "Differential dynamic programming," Ph.D. thesis, Imperial College, Univ. of London, 1967; also Int. J. Contr., 7: 175 (1968).

80. H. H. Kagiwada, R. E. Kalaba, A. Schumitzky, and R. Sridhar, Invariant imbedding and sequential interpolating filters for nonlinear processes, J. Basic Eng., 91: 195 (1969).

81. F. E. Thau, A backward equation for a randomly excited diffusion process, IEEE Trans. Auto. Contr., 13: 14 (1968).

82. W. H. Ray and J. H. Seinfeld, Filtering in distributed-parameter systems with moving boundaries, Automatica, 11: 509 (1975).

83. R. Kalman, Contributions to the theory of optimal control, International Symposium on Ordinary Differential Equations, Mexican Mathematical Society, 1961, p. 102.

84. R. Triggiani, Controllability and observability in Banach space with bounded operators, SIAM J. Contr., 13: 462 (1975).

85. Y. Sakawa, Observability and related problems for PDE of parabolic type, SIAM J. Contr., 13: 14 (1975).

86. A. Bensoussan, Optimization of sensor's location in a distributed filtering problem, in Stability of Stochastic Dynamical Systems, Proceedings of the International Symposium, Coventry, England, Springer-Verlag, Berlin, 1972, p. 62.

87. M. Athans, On the determination of optimal costly measurement strategies for linear stochastic systems, Automatica, 18: 397 (1972).

88. L. Meier, J. Peschon, and R. Dressler, Optimal control of measurement subsystems, IEEE Trans. Auto. Contr., 12: 528 (1967).

89. K. D. Herring and J. L. Melsa, Optimum measurements for estimation, IEEE Trans. Auto. Contr., 19: 264 (1974).

90. J. R. Cannon and R. E. Klein, Optimal selection of measurement locations in a conductor for approximate determination of temperature distributions, 1970 JACC, Atlanta Georgia, 1970, p. 750.

91. M. C. Delfour and S. K. Mitter, Controllability, observability, and optimal feedback control of affine hereditary differential systems, SIAM J. Contr., 10: 298 (1972).

92. K. W. Kaiser, A method of determining the heater-sensor controllers, Proceedings of the IFAC Symposium on Control of Distributed Parameter Systems, paper no. 1-3, Banff, Alberta, Canada, 1971.

93. D. J. Ewing and T. J. Higgins, A sensor location algorithm for distributed parameter systems, Proceedings of the Ninth Annual Allerton Conference on Circuits and System Theory, 1971, p. 1203.

94. M. Athans, The matrix minimum principle, Inf. Contr., 11: 592 (1968).

95. W. G. Heller, Some equivalences between types of sampling in distributed-parameter systems, Int. J. Contr., 18: 915 (1973).

96. T. L. Duncan, Some Banach-valued processes with application, Proceedings of the Conference on Stability of Stochastic Dynamical Systems, 1972, Lecture Notes in Mathematics, vol. 294, Springer-Verlag, Berlin, 1972.

97. A. Bensoussan, On the approximate Kalman-Bucy filter, Proceedings of the Conference on System Science, Hawaii, 1971, pp. 462-464.

98. A. Bensoussan, Statistical problems in Hilbert spaces: Applications to filtering theory, Prague Conference 1970, pp. 270-279 (also Cahiers de l'IRIA Inf/69020, Nov. 1969).

99. A. Bensoussan, Statistical estimation in Hilbert spaces: Application to Bucy's representation theorem, Information Theory Statistical Decision Function Random Processes, Trans. Sixth Prague Conference, Sept. 19-25, 1971, pp. 107-124.

100. M. Zakaï, On the optimal filtering of diffusion processes, Z. Wahrscheinlich Keitstheorie, verw. Geb., 11: 230 (1969).

101. R. Curtain, Infinite-dimensional filtering, SIAM J. Contr., 13: 89 (1975).

102. T. L. Lions, Contrôle Optimal de Systèms Gouverné par des Equations aux Derivées Partielles, Dunod, Paris, 1968.

103. M. C. Delfour and S. K. Mitter, Hereditary differential systems with constant delays (I) and (II), J. Diff. Eqs., 12: 213 (1972).

104. M. B. Ajinkya, W. H. Ray, and G. F. Froment, On-line estimation of catalyst activity profiles in packed-bed reactors having catalyst decay, I & EC Process Design & Develop., 13: 107 (1974).

105. M. B. Ajinkya, M. Köhne, H. F. Mäder, and W. H. Ray, The experimentation of a distributed-parameter filter, Automatica, 11: 571 (1975).

106. A. Maslowski, Optimal estimation for space-time reactor processes, Nucl. Sci. Eng., 52: 274 (1973).

107. H. L. Van Trees, Optimum signal design and processing for reverberation-limited environments, IEEE Trans. Milit. Electron., 9: 212 (1965).

108. H. L. Van Trees, Applications of state-variable techniques in detection theory, Proc. IEEE, 58: 653 (1970).

109. R. R. Kurth, "Distributed-parameter state variable techniques applied to communications over dispersive channels," Sc.D. thesis, M.I.T. Department of Electronics Engineering, June 1969.

110. R. C. Desai and C. S. Lalwani, Identification Techniques, Tata, McGraw-Hill, New York, 1972.

111. S. G. Tzafestas, Hybrid computation in optimal distributed-parameter control systems, Proceedings of the AICA-IFIP Conference on Hybrid Computation, paper 1c, Munich, 1970.

112. R. Vichnevetsky, Use of functional approximation methods in the computer solution of initial value partial differential equation problems, IEEE Trans. Compu., 18: 499 (1969).

113. D. Newman and J. Strauss, Hybrid assumed mode solution of nonlinear partial differential equations, Proc. AFIPS, 39: 575 (1968).

114. H. Hara and W. Karplus, Application of functional optimization techniques for the serial hybrid computer solution of partial differential equations, Proc. AFIPS, 39: 565 (1968).

115. S. K. Chan, Analog-digital methods of solving partial differential equations, M.I.T. Final Report ESL-FR-330, Oct. 1967.

117. S. G. Tzafestas, Moment equations for distributed-parameter processes with Poissonian noise, Int. J. Contr., 15: 665 (1972).

118. S. G. Tzafestas, Some nonlinear functional formulas for stochastic distributed-parameter processes, IEEE Trans. Auto. Contr., 20: 692 (1975).

119. W. E. Boyce, Random vibration of elastic strings and bars, Proceedings of the Fourth U.S. National Congress on Applied Mechanics, American Society of Mechanical Engineers, New York, 1962.

120. J. M. Richardson, The application of truncated hierarchy techniques in the solution of stochastic linear differential equations, Proceedings of the Symposium on Applied Mathematics, vol. 16, American Mathematical Society, Providence, Rhode Island, 1964.

121. D. L. Snyder, Filtering and detection for doubly stochastic Poisson processes, IEEE Trans. Inform. Theory, 18: 91 (1972).

122. D. L. Snyder, Information processing for observed jump processes, Inform. Contr., 22: 69 (1973).

123. A. I. Yashin, Filtering of jump processes, Automat. i. Telemek., 5: 52 (1970).

124. A. Segall, M. H. A. Davis, and T. Kailath, Nonlinear filtering with counting observations, IEEE Trans. Inform. Theory, 21: 143 (1975).

125. S. G. Tzafestas and J. M. Nightingale, Stochastic distributed filtering approach to gamma ray imaging, Intern. J. Syst. Sci., 7: 1249 (1976).

SUPPLEMENTARY BIBLIOGRAPHY

126. R. F. Curtain and P. L. Falb, Ito's lemma in infinite dimensions, J. Math. Anal. Appl., 31: 434 (1970).

127. R. F. Curtain, Stochastic parabolic equations of higher order in t, J. Math. Anal. Appl., 46: 93 (1974).

128. R. F. Curtain, The infinite-dimensional Riccati equation, J. Math. Anal. Appl., 47: 43 (1974).

129. R. F. Curtain and A. J. Pritchard, The infinite dimensional Riccati equation for systems defined by evolution operators, Report No. 36, Control Theory Centre, Univ. of Warwick, England, Aug. 1975.

130. R. F. Curtain, Infinite dimensional estimation theory for linear systems, Report No. 38, Control Theory Centre, Univ. of Warwick, England, Aug. 1975.

131. R. F. Curtain, Estimation theory for abstract evolution equations excited by general white noise processes, Report No. 40, Control Theory Centre, Univ. of Warwick, England, Aug. 1975.

132. R. F. Curtain and A. Ichikawa, The separation principle for stochastic evolution equations excited by general white noise processes, Report No. 42, Control Theory Centre, Univ. of Warwick, England, Nov. 1975.

133. R. F. Curtain, A Kalman-Bucy filtering theory for affine hereditary differential equations, Report No. 25, Control Theory Centre, Univ. of Warwick, England (also International Symposium on Control Theory, Numerical Methods and Computer System Modeling, June 1974, Springer-Verlag, Berlin, Lecture Notes).

134. R. F. Curtain, The infinite dimensional Riccati equation with application to affine hereditary differential systems, Report No. 24, Control Theory Centre, Univ. of Warwick, England, 1974.

135. R. F. Curtain, Stochastic evolution equations with general white noise disturbance, Report No. 41, Control Theory Centre, Univ. of Warwick, England, 1975.

136. R. F. Curtain, A survey of infinite dimensional filtering, SIAM Rev., 17 (1975).

137. R. F. Curtain, Infinite dimensional estimation theory applied to a water pollution problem, Proceedings of the Seventh IFIP Conference, Nice, France, 1975.

138. T. Kato, Abstract evolution equation of parabolic type in Banach and Hilbert spaces, Nagoya Math. J., 19: 93 (1961).

139. T. Kato and H. Tanabe, On the abstract evolution equation, Osaka Math. J., 14: 107 (1962).

140. R. B. Vinter, Some results covering perturbed evolution operators with applications to delay equation, Report, Department of Computing and Control, Imperial College, London, 1975.

141. R. B. Vinter, On the evolution of the state of linear differential delay equations in M^2: Properties of the generator, Report ESL-R-541, Electronics Systems Laboratory, M.I.T., Boston.

142. Special issue on the Linear-quadratic-Gaussian problem, IEEE Trans. Auto. Contr., 16 (1971).

Chapter 4

CONTROL OF STOCHASTIC PARTIAL DIFFERENTIAL EQUATIONS

Alain Bensoussan

Paris-IX University
Paris, France
and
IRIA-LABORIA
Rocquencourt, France

I. INTRODUCTION

In this chapter we present some of the most important results in the theory of the control of stochastic distributed parameter systems. The reader is referred to Chapter 1 for the control of _deterministic_ distributed parameter systems and to Chapter 3 for filtering theory.

Needless to say, the field is extensive and many problems are still unsolved. This justifies the fact that some of the algorithms presented here are partially formal; a complete treatment is of course impossible. We will emphasize the intrinsic difficulties arising from the infinite dimensionality of the system.

II. A REVIEW OF STOCHASTIC PDEs

A. Some Examples

Let $\mathcal{O}$ be a bounded open subset of R^n, with a smooth boundary Γ. We consider a physical system, whose evolution is described by

$$\begin{aligned} &\frac{\partial y}{\partial t} - \Delta y = f + \zeta \quad \text{in } \mathcal{O} \\ &y \mid_{\Gamma} = 0 \\ &y(x,0) = y_0(x) + \zeta \end{aligned} \tag{1}$$

The state of the system at time t is the function $y(x,t)$; f and y_0 are given deterministic inputs, whereas $\xi(x,t)$ and $\zeta(x)$ are _perturbations_ of random type. In applications, it is often assumed that $\zeta(x,t)$ is a _white noise_ process in time and depends smoothly on the space variable x, i.e. to fix the ideas

$$E\zeta = 0$$

$$E\zeta(x_1,t_1)\zeta(x_2,t_2) = \delta(t_1-t_2)q(t_1,x_1,x_2)^{\dagger}$$

---

†E denotes the mathematical expectation.

where δ is the Dirac function and q is a given symmetric positive function. Of course (1) must be precise from the mathematical viewpoint. Other practical situations involve the case when the noise appears on the boundary (Neumann or Dirichlet conditions).

We may also encounter the case when the system is governed by a stationary (elliptic) equation. The noise corresponds in general to unknown parameters (modeled as random variables) (see Kernevez, Quadrat, and Viot [1] for examples in biochemistry).

B. Stochastic Calculus in Hilbert Spaces

As emphasized in Chapter 1, a correct formulation of PDEs is achieved by considering y(t), the state of the system at time t, as an element of a Hilbert space $H = L^2(\mathcal{O})$, or $V = H_0^1(\mathcal{O})$.† When there are noises, we are naturally led to consider stochastic processes with values in Hilbert spaces. Since we want to model noisy inputs of the white noise type, we need a stochastic differential calculus in Hilbert spaces.

Let $(\Omega,\mathcal{A},P)$ be a probability space, and $\mathfrak{F}^t$ an increasing family of sub σ-algebras of $\mathcal{A}$ ($\mathcal{A} = \mathfrak{F}^\infty$). A Wiener process with values in a separable Hilbert space E is a stochastic process $t \to \xi(t)$, which is adapted to $\mathfrak{F}^t$ and such that for any $e \in E$, $(\xi(t),e)$ is a real Wiener process and an $\mathfrak{F}^t$-martingale, with the correlation function

$$E(\xi(t_1),e_1)(\xi(t_2),e_2) = (Qe_1,e_2)\min(t_1,t_2) \tag{2}$$

where Q is a positive self adjoint nuclear operator‡ on E (Q is called the covariance operator).

Let $\sigma(t)$ be an adapted process with values in $\mathcal{L}(E;H)$, where H is also a separable Hilbert space, such that $t,\omega \to \sigma$ is measurable and

---

†See Chapter 1 for the notation of Sobolev spaces. The spaces mentioned are the ones which are useful for (1).

‡That is, $\text{tr } Q = \Sigma\ (Qe_n,e_n) < +\infty$ for any othonormal basis of E. Note that $E|\ \xi(t)\ |^2 = \text{tr } Q$.

$$E \int_0^T \|\sigma(t)\|^2_{\mathcal{L}(E;H)}\, dt < +\infty, \qquad \forall T \text{ finite}$$

We can define the stochastic integral

$$I = \int_0^T \sigma(t)\, d\xi(t)$$

as an element of $L^2(\Omega,\mathcal{A},P;H)$, in a way similar to what is done in the finite-dimensional case. In particular we have

$$EI = 0, \qquad E\|I\|^2_H = E\int_0^T \operatorname{tr} \sigma(t)Q\sigma^*(t)\, dt \tag{3}$$

Setting

$$I(t) = \int_0^t \sigma(s)\, d\xi(s)$$

one defines a stochastic process with values in H, and it can be shown that $I(t)$ is a continuous process [approximating $\xi(t)$ by $\xi_n(t) = \Sigma_{i=1}^n \sigma_i(\xi(t),e_i)$, where e_i is an orthonormal basis of E, and using the continuity property on $I_n(t) = \int_0^t \sigma(s)\, d\xi_n(s)$ and the uniform convergence of $I_n(t) \to I(t)$ in $L^2(\Omega,\mathcal{A},P;H)$].

Next we state Ito's formula in a Hilbert space. Let $a(t)$ be a stochastic process with values in H, which is adapted and satisfies the condition

$$\text{a.s.} \int_0^T \|a(t)\|_H\, dt < +\infty, \qquad \forall T$$

We define a continuous stochastic process in H by setting

$$z(t) = z(0) + \int_0^t a(s)\, ds + \int_0^t \sigma(s)\, d\xi(s) \tag{4}$$

where $z(0)$ is supposed to be a random variable in H which is $\mathcal{F}^0$ measurable.

Let $\Phi(z,t)$ be a functional on $H \times [0,T]$, which is twice continuously (Frechet) differentiable in H and once continuously differentiable in t. We assume that $(\partial\Phi/\partial z),(\partial^2\Phi/\partial z^2)$ are bounded on bounded sets of H.

The continuity property of the derivatives that we require is more precisely the following

$$z,t \to \frac{\partial\Phi}{\partial z}(z,t) \text{ is continuous from } H \times [0,T] \to H \times [0,T]$$

$$z,t \to \frac{\partial^2\Phi}{\partial z^2}(z,t) \text{ is continuous from } H \times [0,T] \to \sigma(\mathcal{L}(H), \mathcal{L}^1(H)) \times [0,T]$$

where $\sigma(\mathcal{L}(H),\mathcal{L}^1(H))$ denotes the space $\mathcal{L}(H)$ equipped with the weak * topology induced by the duality between $\mathcal{L}^1(H)$ (space of nuclear operators in H) and $\mathcal{L}(H)$

$$U \in \mathcal{L}^1(H), \qquad V \in \mathcal{L}(H) \to \text{tr } UV$$

One has the following formula (Ito's formula in Hilbert spaces)

$$\Phi(z(t),t) = \Phi(z(0),0) + \int_0^t \left(\frac{\partial\Phi}{\partial z}, a\right) ds + \int_0^t \left(\frac{\partial\Phi}{\partial z}, \sigma\, d\xi(s)\right)$$

$$+ \frac{1}{2}\int_0^t \text{tr } \sigma^* \frac{\partial^2\Phi}{\partial z^2}\sigma Q\, ds + \int_0^t \frac{\partial\Phi}{\partial t}\, ds \tag{5}$$

Formula (5) can be deduced from the usual Ito's formula.

Let $e_1, \ldots, e_m, \ldots$ be an orthonormal basis of E and $h_1, h_2, \ldots, h_m, \ldots$ be an orthonormal basis of H. We denote the projector on $[e_1, \ldots, e_m]$ by P_m and the projector on $[h_1, \ldots, h_m]$ by Π_m. We set

$$z_m(t) = \Pi_m\left[z(0) + \int_0^t a\, ds + \int_0^t \sigma\, d\xi_m(s)\right]$$

where

$$\xi_m(t) = P_m\xi(t)$$

We may take as $e_1, \ldots, e_m, \ldots$ the basis of eigenvectors of $Q(Qe_i = \lambda_i e_i)$. We have

$$z_m(t) = \sum_{i=1}^{m} z_m^i(t) h_i$$

where

$$z_m^i(t) = (z(0),h_i) + \int_0^t (h_i,a)\, ds + \sum_{j=1}^{m} \int_0^t (h_i,\sigma e_j) \sqrt{\lambda_j}\ db_j(t)$$

where the $b_j(t)$ are independent standard real-valued Wiener processes. We can then apply the standard Ito's formula, which gives, as is easily seen,

$$\Phi(z_m(t),t) = \Phi(\Pi_m z(0),0) + \int_0^t \left[\frac{\partial\Phi}{\partial t} + \sum_i \left(\frac{\partial\Phi}{\partial z}, h_i \right) dz_m^i(t) \right]$$

$$+ \frac{1}{2} \sum_{i,j,\ell=1}^{m} \int_0^t \left(\frac{\partial^2\Phi}{\partial z^2} h_i, h_j \right) \lambda_\ell (h_i,\sigma e_1)(h_j, \partial e_\ell)\, ds$$

or also

$$\Phi(z_m(t),t) = \Phi(\Pi_m z(0),0) + \int_0^t ds \left[\frac{\partial\Phi}{\partial t} + \left(\frac{\partial\Phi}{\partial t}, \Pi_m a \right) + \frac{1}{2} \operatorname{tr} P_m \sigma^* \Pi_m \right.$$

$$\left. \times \frac{\partial^2\Phi}{dz^2} \Pi_m \sigma P_m Q \right] + \int_0^t \left(\frac{\partial\Phi}{\partial z}, \Pi_m \sigma P_m\, d\xi \right) \qquad (6)$$

One can then go to the limit in (6) (details are omitted), hence (5).

Remark 1. As is usual in the finite-dimensional case, the process $z(t)$ in (4) is said to have a stochastic differential

$$dz = a(t)\, dt + \sigma(t)\, d\xi(t)$$

Remark 2. Stochastic differential calculus in Hilbert spaces has been considered by several authors [2-10]. Formulas such as (5) have been generalized in many directions. In Pardoux [11] one will find a good review of the existing results.

C. Stochastic Linear PDEs

We now give a precise statement for (1). Suppose $H = L^2(\mathcal{O})$, $V = H_0^1(\mathcal{O})$, and $V' = H^{-1}(\mathcal{O})$ = dual of V (note that H is identified with its dual, which prohibits identifying V and its dual). The operator

$A = -\Delta \in \mathcal{L}(V;V')$. Denoting the scalar product in V by ((,)) and the scalar product in H by (,) we have the following coercivity property:

$$\langle Av,v\rangle \geq \alpha \, \|v\|^2 \qquad \forall v \in V,\ \alpha > 0$$

where $\langle\ ,\ \rangle$ denotes the duality between V and V'.

Now let E be another Hilbert space and $\xi(t)$ be a Wiener process in E whose covariance operator is Q, as above. Let B(t) be an adapted process with values in $\mathcal{L}(E;H)$ such that $E\int_0^T \|B(t)\|^2\,dt < \infty$ and $y_0 \in H$, $\zeta \in L^2(\Omega,\mathcal{A},P;H)$. We consider the equation

$$\begin{aligned} &dy + Ay\,dt = B(t)\,d\xi \\ &y(0) = y_0 + \zeta \end{aligned} \tag{7}$$

$$\begin{aligned} &y \in L^2(\Omega \times (0,T),\ dP \otimes dt;\ V), \qquad y \text{ a.s.} \in C^0(0,T;H) \\ &y \in C^0(0,T;L^2(\Omega,H)), \qquad y(t) \text{ is adapted (with values in H)} \end{aligned} \tag{8}$$

Then there exists one and only one solution of (7), (8) (Bensoussan [12]†). One will interpret (7) as (1) provided B = identity (hence E = H), f = 0, and $q(t_1,x_1,x_2) = q(x_1,x_2)$ is the kernel of Q, i.e.,

$$(Qh_1,h_2) = \iint q(x_1,x_2)h_1(x_2)h_2(x_1)\,dx_1\,dx_2$$

Let us briefly sketch the proof of (7), (8). Let $h_1, \ldots, h_m, \ldots$ be an orthonormal basis of H made up with elements of V (which is possible since V is dense in H). Let $V_m = [h_1 \cdots h_m]$ and $A_m \in \mathcal{L}(V_m;V_m)$ be defined by

$$A_m h_i = \sum_{j=1}^{m} \langle Ah_i,h_j\rangle h_j \qquad i = 1 \cdots m$$

$$B_m(t)e = \sum_{j=1}^{m} (B(t)e,h_j)h_j \qquad \text{for } e \in E$$

$$y_{om} = \sum_{j=1}^{m} (y_0,h_j)h_j, \qquad \zeta_m = \sum_{j=1}^{m} (\zeta,h_j)h_j$$

and let the next y(t) be defined by the ordinary Ito equation

†The a.s. continuity has been proved by Pardoux [10].

$$dy_m(t) + A_m y_m(t)\,dt = B_m(t)\,d\xi(t)$$
$$y_m(0) = y_{om} + \zeta_m \tag{9}$$

We apply (5) with $z(t) = y_m(t)$ and $\Phi(z,t) = \frac{1}{2} \mid z \mid_H^2$. We obtain

$$\frac{1}{2} \mid y_m(t) \mid^2 = \frac{1}{2} \mid y_0 + \zeta \mid^2 - \int_0^t (y_m(s), A_m y_m(s)\,ds$$
$$+ \int_0^t (y_m(s),\ B_m\,d\xi(s)) + \frac{1}{2}\int_0^t \mathrm{tr}\ B_m Q B_m^*\,ds$$

or also

$$\frac{1}{2} \mid y_m(t) \mid^2 + \int_0^t \langle A y_m(s), y_m(s)\rangle\,ds = \frac{1}{2} \mid y_0 + \zeta \mid^2$$
$$+ \int_0^t (y_m(s), B\,d\xi(s)) + \frac{1}{2}\int_0^t \mathrm{tr}\ B_m Q B_m^*\,ds \tag{10}$$

From (10) we deduce the following a priori estimates:

$$E \mid y_m(t) \mid^2 \le C \qquad \forall t \in [0,T]$$

$$E \int_0^T \| y_m(t) \|^2\,dt \le C$$

We extract a subsequence $y_m \to y$ in $L^2(\Omega \times 0, T; V)$ weakly and $L^\infty(0,T;L^2(\Omega,H))$ weak star. Let h_i be an arbitrary element of the basis, but fixed. For $m \ge i$, we have

$$(A_m y_m, h_i) = \langle A y_m, h_i\rangle$$

hence from (9) it follows that

$$(y_m(t), h_i) - (y_0 + \zeta, h_i) + \int_0^t \langle A y_m(s), h_i\rangle\,ds = \left(\int_0^t B\,d\xi, h_i \right)$$

and passing to the limit we obtain

$$(y(t), h_i) - (y_0 + \zeta,\ h_i) + \int_0^t \langle A y(s), h_i\rangle\,ds = \left(\int_0^t B\,d\xi,\ h_i \right)$$

from which it follows that there is a solution of (7) which belongs to the class

$$L^2((0,T) \times \Omega; V) \cap L^\infty(0,T;L^2(\Omega;H))$$

Since the solution is obviously unique in that class, we obtain

$y_m \to y$ in $L^2((0,T) \times \Omega; V)$ weakly and $L^\infty(0,T;L^2(\Omega;H))$ weak star.

Furthermore, for any t

$$y_m(t) \to y(t) \quad \text{in } L^2(\Omega;H) \text{ weakly}$$

and since $y_m(t)$ is $\mathcal{F}^t$ measurable, it follows that $y(t)$ is $\mathcal{F}^t$ measurable. We also have

$$\text{a.e.t,} \quad y(t) \text{ is } \mathcal{F}^t \text{ measurable with values in } V$$

Furthermore, from (10) and weak convergence of $y_m \to y$, we obtain the energy inequality

$$\frac{1}{2} E \mid y(t) \mid^2 + E \int_0^t \langle Ay(s),y(s)\rangle\, ds \le \frac{1}{2} E \mid y_0 + \zeta \mid^2 + \frac{1}{2} E \int_0^t \text{tr } BQB^* \, ds \tag{11}$$

Let us prove the final step, i.e., the regularity of $y(t)$. We need a stochastic energy equality. We shall proceed as follows: let $h_1 \cdots h_m \cdots$ now be an orthonormal basis of H made up with elements of D(A) (the domain of A, considered as an unbounded operator in H).

We define $B_m(t)$ as before and from the above results there exists $y_m(t)$ solution of [note that $y_m(t)$ is different from the Galerkin approximation defined above]:

$$\begin{aligned} dy_m(t) + Ay_m(t)\, dt &= B_m \, d\xi \\ y_m(0) &= y_0 + \zeta \end{aligned} \tag{12}$$

Defining

$$\eta_m(t) = \int_0^t B_m \, d\xi(s)$$

which is a continuous process with values in $D(A)$, it is easy to see that

$$y_m(t) = z_m(t) + \eta_m(t)$$

where $z_m(t)$ is the solution of

$$\frac{dz_m}{dt} + Az_m(t) = -A\eta_m(t)$$

$$z_m(0) = y_0 + \zeta$$

Therefore $z_m(t)$ is also a continuous process with values in $D(A)$ [since $A\eta_m(t)$ is continuous with values in H], hence $y_m(t)$ has a stochastic differential which satisfies the conditions of validity of (5).

Therefore, we obtain, noticing that $(Ay_m, y_m) = \langle Ay_m, y_m \rangle$,

$$\frac{1}{2} | y_m(t) |^2 + \int_0^t \langle Ay_m(s), y_m(s) \rangle \, ds = \frac{1}{2} | y_0 + \zeta |^2$$

$$+ \int_0^t (y_m(s), B \, d\xi(s)) + \frac{1}{2} \int_0^t \operatorname{tr} B_m Q B_m^* \, ds \tag{13}$$

But applying (11) to $y - y_m$, we obtain

$$\frac{1}{2} E | y(t) - y_m(t) |^2 + E \int_0^t \langle A[y(s) - y_m(s)], y(s) - y_m(s) \rangle \, ds$$

$$\leq \frac{1}{2} E \int_0^t \operatorname{tr}(B - B_m) Q (B - B_m^*) \, ds$$

hence

$$y_m \to y \qquad \text{in } L^2((0,T) \times \Omega; V) \text{ strongly}$$

$$y_m(t) \to y(t) \qquad \text{in } C(0,T; L^2(\Omega; H)) \text{ strongly}$$

Let us fix t, and choose a subsequence such that

$$\text{a.s.} \quad y_m(\cdot) \to y(\cdot) \qquad \text{in } L^2(0,T;V)$$

$$y_m(t) \to y(t) \qquad \text{in } H$$

Passing to the limit in (13), we obtain the stochastic energy equality,

$$\frac{1}{2} | y(t) |^2 + \int_0^t \langle Ay(s), y(s) \rangle \, ds = \frac{1}{2} | y_0 + \zeta |^2 + \int_0^t (y(s), B \, d\xi(s)) + \frac{1}{2} \int_0^t \operatorname{tr} BQB^* \, ds, \qquad \text{a.s.} \tag{14}$$

From Eqs. (7) and (14) one first proves that $y(t) \in C(0,T;L^2(\Omega,H))$. Hence, up to an equivalence, there exists a separable version of $y(t)$ (in H). Using the separability and again (14) it follows that $y(t)$ is a continuous process in H. Equation (8) is thus completely proved.

We have additional information, namely

$$y \in L^2(\Omega;C(0,T;H)) \tag{15}$$

Indeed, denoting the orthonormal basis of eigenvectors of Q by $e_1, \ldots, e_n, \ldots$ we have

$$E \sup_{0\le t\le T} \left| \int_0^t (y(s), B(s) \, d\xi(s)) \right| = E \sup_t \left| \int_0^t \sum_i (y(s), Be_i)(e_i, d\xi) \right|$$

$$\le E \sum_i \sup_t \left| \int_0^t (y(s), Be_i)(e_i, d\xi(s)) \right|$$

and by the martingale inequality (see for instance Meyer [13])

$$\le \sum_i E \sqrt{\int_0^T (B^*(t)y(t), e_i)^2 \lambda_i}$$

$$\le E \sqrt{\Sigma \lambda_i} \sqrt{\int_0^T | B^*(t)y(t) |^2 \, dt}$$

$$\le CE \sup_t | y(t) | \sqrt{\int_0^T \| B^*(t) \|^2 \, dt}$$

Using this estimate in (14), after taking the sup in t and the mathematical expectation, one easily obtains

$$E \sup_{0\le t\le T} |\, y(t)\, |^2 < +\infty$$

hence (15).

Remark 3. Considering the semigroup $G(t)$ on H generated by A, one obtains a representation formula for the solution of (7), namely

$$y(t) = G(t)(y_0 + \zeta) + \int_0^t G(t - \tau)B(\tau)\, d\xi(\tau) \tag{16}$$

Formulas like (16) have been used to define a priori the solution of (7) (see Curtain and Falb [14]).

D. Nonlinear Stochastic PDEs

One may generalize (7) in two directions: Take A to be a nonlinear operator, or (and) let B depend on y.

Let us first consider the second extension. By analogy with the finite-dimensional case, we will now consider that B is a Lipschitz function of y, i.e.,

$$\|B(y) - B(z)\|_{\mathcal{L}(E;H)} \le C\, |\, y - z\, |_E \tag{17}$$

We consider the following problem: find $y(t)$ such that

$$\begin{aligned} & y \in L^2(\Omega; C(0,T;H)) \cap L^2(\Omega \times (0,T); V) \\ & y(t) \text{ is adapted to } \mathcal{F}^t \text{ (as a process with values in } H) \\ & \text{a.e.t } \ y(t) \text{ is } \mathcal{F}^t \text{ measurable with values in } V \end{aligned} \tag{18}$$

$$\begin{aligned} & dy(t) + Ay(t)\, dt = B(y)\, d\zeta \\ & y(0) = y_0 + \zeta \end{aligned} \tag{19}$$

Then there exists one and only one solution of (18), (19).

To prove (18), (19) we proceed by analogy with the finite-dimensional case. We define an iterative process: start with

$$y^0(t) = y_0 + \zeta$$

and knowing $y^n(t)$, define $y^{n+1}(t)$ by solving

$$dy^{n+1}(t) + Ay^{n+1}(t)\,dt = B(y^n)\,d\xi$$
$$y^{n+1}(0) = y_0 + \zeta \tag{20}$$

which uniquely defines y^{n+1} according to the linear theory.

Applying the energy equality (14) to the process $y^{n+1}(t) - y^n(t)$ yields

$$\frac{1}{2}\,|\,y^{n+1}(t) - y^n(t)\,|^2 + \int_0^t \langle A(y^{n+1} - y^n)(s),(y^{n+1} - y^n)(s)\rangle\,ds$$

$$= \int_0^t [(y^{n+1} - y^n)(s),[B(y^n) - B(y^{n-1})]\,d\xi(s)]$$

$$+ \frac{1}{2}\int_0^t \mathrm{tr}\,[B(y^n) - B(y^{n-1})]Q[B(y^n) - B(y^{n-1})]\,ds$$

from which it follows

$$E \sup_{0\le\theta\le t} |\,y^{n+1}(\theta) - y^n(\theta)\,|^2 \le CE\int_0^t \|B(y^n) - B(y^{n-1})\|^2\,ds$$

and from (17)

$$E \sup_{0\le\theta\le t} |\,y^{n+1}(\theta) - y^n(\theta)\,|^2 \le CE\int_0^t \|y^n(s) - y^{n-1}(s)\|^2\,ds$$

Setting

$$\varphi^n(t) = E \sup_{0\le\theta\le t} |\,y^{n+1}(\theta) - y^n(\theta)\,|^2$$

we see that

$$\varphi^n(t) \le C\int_0^t \varphi^{n-1}(s)\,ds$$

Iterating, we finally obtain that

y^n is a Cauchy sequence in $L^2(\Omega;C(0,T;H))$

It is also easy to prove that y^n remains in a bounded subset of $L^2(\Omega \times (0,T); V)$, hence

$$y^n \to y \quad \text{in } L^2(\Omega; C(0,T;H)) \text{ and in } L^2(\Omega \times (0,T); V) \text{ weakly}$$

Passing to the limit in (20) we obtain (18), (19).

Remark 4. The same method can be applied to the case of a Lipschitz perturbation of the operator A, changing A into

$$A + F$$

where $F(y) \in \mathcal{L}(H;H)$ and

$$\|F(y) - F(z)\|_{\mathcal{L}(H;H)} \leq C\| y - z\|$$

Operators of the type

$$Ay = -\Delta y + y^3$$

belong to the class of monotone operators.

One can prove (see Bensoussan and Temam [15]) the existence and uniqueness of the solution of

$$dy(t) - \Delta y(t)\, dt + y^3\, dt = B(t)\, d\xi(t)$$

$$y \Big|_{\Gamma} = 0$$

$$y(x,0) = y_0(x)$$

in the functional space $L^2(\Omega \times (0,T); H_0^1(\mathcal{O})) \cap L^4(\Omega \times (0,T); L^4(\mathcal{O}))$, and $y \in C(0,T); L^2(\Omega, L^2(\mathcal{O}))$. Pardoux [10] has extended this result to the case when B depends on y and obtained regularity results. Note that operators such as those of Remark 4 can also enter into the framework of monotone operators.

Thus far, all the equations have been considered in a strong sense. By that we mean, as in Ito equations, that the Wiener process is given a priori. It is well known that this concept is too narrow in the finite-dimensional case since it requires a regularity assumption on the coefficients. Aside from the strong theory, a theory of weak equations has been extensively developed, namely by Stroock and Varadhan [16], and Yamada and Watanabe [17].

A "weak sense equation" means that a system made with a probability space, a measure, a Wiener process, and a process solution of a stochastic differential equation corresponding to this Wiener process are looked for. The objective in the finite-dimensional case is to weaken the assumptions on the coefficients which guarantee the existence and uniqueness of the solution (in an appropriate sense). It turns out that this concept of weak solutions is also extremely useful for stochastic PDEs. A very simple reason is the following: Unlike the monotonicity method which extends from the deterministic case to the stochastic case, the powerful compactness method extensively used for studying deterministic PDEs (see Lions [18] for many examples) fails to extend to the stochastic case, if one keeps the concept of a strong solution--indeed, even if the injection of $H_0^1(\mathcal{O})$ into $L^2(\mathcal{O})$ is compact (for bounded $\mathcal{O}$), the injection of $L^2(\Omega;H_0^1(\mathcal{O}))$ into $L^2(\Omega;L^2(\mathcal{O}))$ is no more compact. However, if one works with the concept of weak solution, then the compactness method can be recovered. Along these lines, Viot [19,20] has studied the following stochastic PDE proposed by Fleming [21]:

$$dy - \Delta y \, dt + F(y) \, dt = \sqrt{y(1-y)_+} \, d\xi(t)$$

$$y \Big|_{\Gamma} = 0$$

$$y(x,0) = y_0(x)$$

where F is a bounded continuous function from $R \rightarrow R$, and $y(1 - y)_+$ is the positive part of $y(1 - y)$.

Finally, let us mention the following situation which may be useful in practice: One considers $\omega \in \Omega$ as a parameter such that for any of its values a <u>deterministic</u> problem can be solved. One then obtains a family of solutions depending on the parameter ω. This family is useful only if some <u>measurability</u> properties with respect to ω are satisfied. This approach has been used for stochastic Navier-Stokes equations by Bensoussan and Temam [22] (see also an example arising in biochemistry in Kernevez, Quadrat, and Viot [1]).

When the perturbations are white noise in both the time and the space variables, then the situation is much more complicated. Linear equations can be handled by using formulas similar to (16) (see Bensoussan [23] and also Kree [24]).

Remark 5. All the results stated above for operators A independent of time extend to the case when $A = A(t)$ depends on time (with some natural assumptions on the dependence with respect to time).

Remark 6. Problems with noise on the boundary have been considered by Balakrishnan [25]. He has also considered stochastic bilinear PDEs [26].

III. OPTIMAL CONTROL OF STOCHASTIC PDEs: CASE WHEN THE INFORMATION DOES NOT DEPEND ON THE CONTROL

A. Setting of the Problem

We consider a system whose evolution is governed by (6), with some changes. Firstly, $B(t)$ is no longer random. Secondly, we introduce a control term. Let

$$\begin{aligned} & v \in L^2(\Omega \times (0,T), H), \text{ a.e. } t \; v(t) \text{ is } \mathcal{F}^t \text{ is measurable (with values in H)} \\ & v(t) \in \mathcal{U}_{ad} \text{ (convex closed subset of H) a.e. } t, \text{ a.s. } \omega \end{aligned} \tag{22}$$

We will say that v satisfying (22) is an admissible control and denote the set of admissible controls by $\mathcal{U}$. We change (6) into

$$\begin{aligned} & dz + Az\, dt = B(t)\, d\xi + v(t)\, dt \\ & z(0) = y_0 + \zeta \end{aligned} \tag{23}$$

which uniquely defines y with the regularity properties (7) and (15).

Remark 7. $\mathcal{F}^t$ can be taken to be the σ-algebra generated by $(\zeta, \xi(s),\ s \le t)$, and represents all the available information at time t, if one is allowed to observe ζ and $\xi(s)$ until time t. Since an admissible control can depend only on the information which is available, we implicitly assume in (22) that we can observe ζ and $\xi(s)$. We can weaken this restriction by taking $v(t)$ $\mathcal{B}^t$ measurable where $\mathcal{B}^t \subset \mathcal{F}^t$. All results extend to this case. What is important here is that the information available at time t, which is summarized in $\mathcal{B}^t$ or $\mathcal{F}^t$ does not depend on the past controls. It will be the case if, for instance, we observe the state of the system instead of directly observing the noise on the system. This situation will be considered in the next paragraph.

We now define the payoff function. Let $\ell(z,v,t)$ and $\Lambda(z)$ be two functionals on $H \times H \times [0,T]$ and H, respectively, satisfying the assumptions

ℓ is continuous in z,v and measurable in t

$$| \ell | \le C_1(| z |^2 + | v |^2 + 1) \tag{24}$$

$z,v \to \ell$ is convex and Gateaux differentiable

$$\left| \frac{\partial \ell}{\partial z} \right| \le C_2(| z | + | v | + 1)$$

$$\left| \frac{\partial \ell}{\partial v} \right| \le C_3(| z | + | v | + 1)$$

$$| \Lambda | \le C_4(| z |^2 + 1) \tag{25}$$

Λ is convex and Gateaux differentiable;

$$\left| \frac{d\Lambda}{dz} \right| \le C_5(| z |^2 + 1)$$

For $v \in \mathcal{U}$ we write

$$J(v) = E\left[\int_0^T \ell(z(t),v(t),t)\, dt + \Lambda(z(T)) \right] \tag{26}$$

We notice that (as is easily verified; for details see Bensoussan and Viot [27]) the mapping

$$z(\cdot),\ v(\cdot) \to E\left[\int_0^T \ell(z(t),v(t),t)\, dt + \Lambda(z(T)) \right]$$

from $C(0,T;L^2(\Omega;H)) \times L^2(\Omega \times (0,T);\ H) \to R$ is continuous convex and Gateaux differentiable. This implies that

$$v \to J(v)$$

is convex and Gateaux differentiable on $L^2(\Omega \times (0,T); H)$. Moreover, considering the space $L^2(\Omega \times (0,T); H)$ (subspace of $v(\cdot)$ such that $v(t)$ is $\mathfrak{F}^t$ measurable a.e.) which is a closed subspace of $L^2(\Omega \times (0.T); H)$, we see that $\mathcal{U}$ is a closed convex subset of the Hilbert space $L^2_{\mathfrak{F}}(\Omega \times (0,T); H)$.

Our problem, therefore, can be formulated as the minimization of a convex Gateaux differentiable functional on a closed convex subset of a Hilbert space, which is a standard problem of calculus of variations. We will state the necessary and sufficient conditions of optimality and will interpret them from the probabilistic viewpoint.

B. Necessary and Sufficient Conditions of Optimality: The Stochastic Maximum Principle

Let u,v be two admissible controls. We have the formula

$$
\begin{aligned}
(J'(u),v)_{\mathfrak{L}^2_{\mathfrak{F}}} &= E\int_0^T \left(\frac{\partial \ell}{\partial z}(y(t),u(t),t),\tilde{z}(t)\right)_H dt \\
&\quad + E(\Lambda'(y(T)),\tilde{z}(T))_H \\
&\quad + E\int_0^T \left(\frac{\partial \ell}{\partial v}(y(t),u(t),t),v(t)\right)_H dt
\end{aligned} \tag{27}
$$

where $y(t)$ is the solution of (23), corresponding to the control u, $\tilde{z}(t)$ the homogeneous solution of (23), i.e., the solution of

$$
\begin{aligned}
&d\tilde{z} + A\tilde{z}\, dt = v(t)\, dt \\
&\tilde{z}(0) = 0
\end{aligned} \tag{28}
$$

By standard theory of calculus of variations for $\hat{u}$ to be a solution of

$$
J(\hat{u}) = \inf_{v\in\mathcal{U}} J(v) \tag{29}
$$

it is necessary and sufficient that

$$
[J'(\hat{u}),\ v - \hat{u}] \geq 0 \qquad \forall v \in \mathcal{U},\ \hat{u} \in \mathcal{U} \tag{30}
$$

We can make (30) explicit by virtue of (27). We obtain

$$E \int_0^T \left(\frac{\partial \ell}{\partial z}[(\hat{y},\hat{u},t),\hat{z}(v - \hat{u})] \, dt + E \int_0^T \left[\frac{\partial \ell}{\partial v}(\hat{y},\hat{u},t), \; v - \hat{u} \right] dt \right.$$

$$\left. + E(\Lambda'(\hat{y}(T)),\hat{z}(T; \; v - \hat{u})) \right) \geq 0 \tag{31}$$

Next we introduce the adjoint system

$$\begin{aligned} -\frac{d\hat{p}}{dt} + A^*\hat{p} &= -\frac{\partial \ell}{\partial z}(\hat{y},\hat{u},t) \\ \hat{p}(T) &= -\Lambda'(\hat{y}(T)) \end{aligned} \tag{32}$$

Notice that (31) is not a stochastic PDE in the sense of (23). It is for ω <u>fixed</u> (ω comes in through y,u) an ordinary PDE. By this assumption it is easy to check that

$$\hat{p} \in L^2(\Omega \times (0,T); V), \qquad \frac{d\hat{p}}{dt} \in L^2(\Omega \times (0,T); V')$$

However, $\hat{p}$ is clearly not an adapted process, since (31) is a backward equation. Nevertheless, using (32) in (31) yields

$$E \int_0^T [-\hat{p}(t) + \frac{\partial \ell}{\partial v}(\hat{y},\hat{u},t), \; v(t) - \hat{u}(t)] \, dt \geq 0 \qquad \forall v \in \mathcal{U} \tag{33}$$

One has to be careful in manipulating (33) because v and $\hat{u}$ are adapted, whereas $\hat{p}$ is not. But we may rewrite (33) as

$$E \int_0^T [-E^{\mathcal{F}^t}\hat{p}(t) + \frac{\partial \ell}{\partial v}(\hat{y},\hat{u},t), \; v(t) - \hat{u}(t)] \, dt \geq 0$$

from which, one can deduce (see Bensoussan and Viot [27]) that

$$\begin{aligned} &[-E^{\mathcal{F}^t}\hat{p}(t) + \frac{\partial \ell}{\partial v}(\hat{y}(t),\hat{u}(t),t), \; y - \hat{u}(t)]_H \geq 0 \\ &\qquad\qquad\qquad\qquad \text{a.e.t, a.s.}\omega, \quad \forall v \in \mathcal{U}_{ad} \end{aligned} \tag{34}$$

Hence (32), (34) together with the state relationship between $\hat{y},\hat{u}$ form a set of necessary and sufficient conditions of optimality for $\hat{u}$.

IV. OPTIMAL CONTROL OF STOCHASTIC PDEs: CASE WHEN THE INFORMATION DEPENDS ON THE CONTROL

A. Setting of the Problem

The dynamics of the system is still described by (23). We will now assume that we can observe the process

$$\varphi(t) = \int_0^t C(\tau)z(\tau)\, d\tau + \eta(t) \tag{35}$$

where F is a Hilbert space and $C \equiv C(t) \in L^\infty(0,T;\, \mathcal{L}(H;F))$

$$\eta \in L^2(\Omega \times (0,T);\, F) \cap \text{meas}\, (\Omega;\, C^0(0,T;H)) \tag{36}$$

Notice that $\varphi(t)$ also satisfies (36).

We must now precise the set of admissible controls. We first notice that for any $v \in L^2(\Omega \times (0,T);\, H)$, z and φ are well defined, and in particular φ belongs to the space (36).

Let $\alpha(t)$ and $\beta(t)$ be the processes (independent of v) given by

$$\begin{aligned} & d\alpha(t) + A\alpha(t)\, dt = B(t)\, d\xi(t) \\ & \alpha(0) = y_0 + \zeta \end{aligned} \tag{37}$$

$$\beta(t) = \int_0^t C(\tau)\alpha(\tau)\, d\tau + \eta(t) \tag{38}$$

We define

$$\mathcal{F}^t = \sigma\text{-algebra generated by } \beta(s), \qquad s \le t \tag{39}$$

and let

$$\begin{aligned} & W = L^2_{\mathcal{F}}(\Omega \times (0,T);\, H) \\ & \mathcal{U} = \{v \in W \mid v(t) \in \mathcal{U}_{ad} \text{ a.e.t, a.s}\} \end{aligned} \tag{40}$$

Now for any v (for instance $v \in \mathcal{U}$), the process φ is, as we said, well defined. Then let

$$\Phi_v^t = \sigma\text{-algebra generated by } \varphi(s), \qquad 0 \le s \le t \tag{41}$$

(note that φ depends on v which justifies the notation)

The set of admissible controls is finally defined by

$$\tilde{\mathcal{U}} = \{v \in \mathcal{U} \mid v \in L^2_{\Phi_v}(\Omega \times (0,T); H)\} \tag{42}$$

We see that $\tilde{\mathcal{U}}$ is defined in an implicit way, but it is not empty since it contains at least the deterministic controls. Let us explain the choice of $\tilde{\mathcal{U}}$. We first notice that admissible controls must be defined implicitly in one way or another since the observation depends on the control and admissible controls must depend only on the past observation.

However, one may criticize the choice of W, since a more natural way to proceed would have been to take $L^2(\Omega \times (0,T); H)$ instead of W and let the rest remain unchanged. We will examine this point in the next paragraph and give some very useful properties of $\tilde{\mathcal{U}}$.

B. Some Properties of $\tilde{\mathcal{U}}$

One fundamental property is the following:

$$\text{if } v \in \tilde{\mathcal{U}} \Longleftrightarrow \Phi_v^t = \mathcal{F}^t \qquad \forall t \tag{43}$$

The left-hand implication is obvious (for $v \in \mathcal{U}$). Let us prove the right-hand implication. Indeed, defining z_1, φ_1 as follows

$$\frac{dz_1}{dt} + Az_1 = v$$
$$z_1(0) = 0 \tag{44}$$

$$\varphi_1(t) = \int_0^t C(\tau)z_1(\tau)\, d\tau \tag{45}$$

we clearly have

$$z(t) = z_1(t) + \alpha(t)$$
$$\varphi(t) = \varphi_1(t) + \beta(t) \tag{46}$$

Therefore, if $v \in \tilde{u}$, then by (44) and (45) it follows that $\varphi_1(t)$ is $\mathcal{F}^t$ and Φ_v^t measurable, which, with the second relation (46), implies that $\varphi(t)$ is $\mathcal{F}^t$ measurable and $\beta(t)$ is Φ_v^t measurable, hence (43).

Let $\epsilon > 0$. We define

$$\mathcal{F}_\epsilon^t = \begin{cases} \mathcal{F}^{t-\epsilon} & \text{if } t > \epsilon \\ (\Omega, \emptyset) & \text{if } 0 \le t \le \epsilon \end{cases}$$

then we have

$$\text{if } v \in L^2_{\mathcal{F}_\epsilon}(\Omega \times (0,T); H) \text{ and } v(t) \in \mathcal{U}_{ad} \quad \text{a.e.t. a.s.,} \tag{47}$$

then $v \in \tilde{\mathcal{U}}$

To prove (47), it is enough, by virtue of (43), to prove that

$$\Phi_v^t = \mathcal{F}^t \qquad \forall t \tag{48}$$

(notice that $v \in \mathcal{U}$). Clearly, $\Phi_v^t \subset \mathcal{F}^t$. But for $t \in [0,\epsilon]$, since v is deterministic, the reverse if true by virtue of the second relation (46). Therefore, we have

$$\Phi_v^t = \mathcal{F}^t \qquad \text{for } 0 \le t \le \epsilon \tag{49}$$

Let us then prove that we have

$$\Phi_v^t = \mathcal{F}^t \qquad \text{for } \epsilon \le t \le 2\epsilon \tag{50}$$

Indeed, for a.e. $t \in]\epsilon, 2\epsilon[$, $v(t)$ is $\mathcal{F}^{t-\epsilon}$ measurable, hence if $0 < t - \epsilon < \epsilon$, it follows from (49) that $v(t)$ is also $\Phi_v^{t-\epsilon}$ measurable and therefore also Φ_v^t measurable. But then $\varphi_1(t)$ is also Φ_v^t measurable, and from the second relation (46) we obtain that $\beta(t)$ is Φ_v^t measurable, which implies $\mathcal{F}^t \subset \Phi_v^t$. Thus, (50) is proved, as is (47).

From (47) it is easy to obtain the second fundamental property of $\tilde{\mathcal{U}}$, namely

$$\tilde{\mathcal{U}} \text{ is } \underline{\text{dense}} \text{ in } \mathcal{U} \text{ (for the topology of } L^2_{\mathcal{F}}(\Omega \times (0,T); H)) \tag{51}$$

Properties (43) and (51) will be used extensively to derive the stochastic maximum principle. Let us denote

$$W^0 = L^2(\Omega \times (0,T); H)$$

and $\mathcal{U}^{\rho}$, $\tilde{\mathcal{U}}^{\rho}$ are obtained from W^{o} in the same way as $\mathcal{U}$ and $\tilde{\mathcal{U}}$ are defined from W. Then properties (43) and (51) are not satisfied for $\mathcal{U}^{\rho}$, $\tilde{\mathcal{U}}^{\rho}$ (see a counter example in Bensoussan and Viot [27]). This explains why one restricts the admissible controls to belong to $\tilde{\mathcal{U}}$ instead of $\tilde{\mathcal{U}}^{\rho}$ (see previous paragraph).

The set $\tilde{\mathcal{U}}^{\rho}$ (which contains $\tilde{\mathcal{U}}$) is actually too big. Let us clarify this point. For any $v \in \mathcal{U}^{\rho}$, denote by $\Phi^{t}_{\epsilon,v}$ the σ-algebra obtained from Φ^{t}_{v} as $\mathcal{F}^{t}_{\epsilon}$ is obtained from $\mathcal{F}^{t}$, then denote by $\tilde{\mathcal{U}}^{\epsilon}$ the subset of $\mathcal{U}^{\rho}$ of the controls which belong $L^2_{\Phi_{\epsilon,v}}(\Omega \times (0,T); H)$. Clearly

$$\tilde{\mathcal{U}}^{\epsilon} \subset \tilde{\mathcal{U}}^{\rho}$$

Then we have

$$\tilde{\mathcal{U}}^{\epsilon} \subset \tilde{\mathcal{U}} \qquad \forall \epsilon > 0 \tag{52}$$

and

$$\tilde{\mathcal{U}}^{\rho} \cap \bar{\tilde{\mathcal{U}}} = \tilde{\mathcal{U}} \tag{53}$$

where $\bar{\tilde{\mathcal{U}}}^{\epsilon}$ denotes the closure of the sets $\tilde{\mathcal{U}}^{\epsilon}$ in $\mathcal{U}^{\rho}$.

The proof of (52) is similar to the proof of (47). Let us prove (53). If $v \in \tilde{\mathcal{U}}$, define

$$v_{\epsilon}(t) = \begin{cases} v(t - \epsilon) & \text{if } t > \epsilon \\ w \in \mathcal{U}_{ad} & \text{if } 0 < t \leq \epsilon \end{cases}$$

Then $v_{\epsilon} \to v$ in $\mathcal{U}^{\rho}$, and $v_{\epsilon} \in \tilde{\mathcal{U}}$ from (47), which implies that $\mathcal{F}^{t} = \Phi^{t}_{v_{\epsilon}}$, hence since $v_{\epsilon}(t)$ is $\mathcal{F}^{t-\epsilon}$ measurable, $(t \geq \epsilon) v_{\epsilon}(t)$ is $\Phi^{t-\epsilon}_{\epsilon,v_{\epsilon}}$ measurable $(t \geq \epsilon)$, i.e., Φ^{t} measurable for any t.

Thus, $v \in \bar{\tilde{\mathcal{U}}}^{\epsilon}$.

Conversely, suppose that $v \in \tilde{\mathcal{U}}^{\rho} \cap \bar{\tilde{\mathcal{U}}}^{\epsilon}$. Since $v \in \tilde{\mathcal{U}}^{\rho}$, we obtain [as in the proof of (43)]

$$\mathcal{F}^{t} \subset \Phi^{t}_{v} \tag{54}$$

Now since $v \in \bar{\tilde{\mathcal{U}}}^{\epsilon}$, there exists a sequence $v_{\epsilon} \to v$ in $\mathcal{U}^{\rho}$, such that $v_{\epsilon} \in \tilde{\mathcal{U}}^{\epsilon}$. From (52) it follows that $v_{\epsilon} \in \tilde{\mathcal{U}}$, hence $v_{\epsilon} \in \mathcal{U}$, which is closed in $\mathcal{U}^\rho$. Therefore, $v \in \mathcal{U}$, which implies [as in the proof (43)]

$$\Phi_v^t \subset \mathcal{F}^t$$

which with (54) and (43) completes the proof of (53).

Let us emphasize that $\tilde{\mathcal{U}}$ contains the set of Lipschitz feedbacks. More precisely, suppose that $v \in \mathcal{U}^\rho$ is such that

$$v(t) = \gamma(t,\varphi_v(\cdot))$$

where $\gamma(t,\varphi)$ is a continuous mapping from $[0,T] \times C(0,T;F) \to \mathcal{U}_{ad}$, which is

Nonanticipative, i.e., $\gamma(t,f) = \gamma(t,g)$ if $f(s) = g(s)\ \forall s \le t$.

There exists $k > 0$ such that for any $t \in [0,T]$

$$|\gamma(t,f) - \gamma(t,g)| \le k \sup_{0\le s\le t} |f(s) - g(s)|$$

Then $v \in \tilde{\mathcal{U}}$. (For another approach using feedbacks, see Bensoussan [28] and Kushner [29].)

C. Derivation of the Necessary and Sufficient Conditions

Our problem is to minimize $J(v)$ other $\tilde{\mathcal{U}}$, where $J(v)$ is still given by (26). It is easy to check that $J(v)$ is continuous in $\mathcal{U}$. This and the density property (51) imply

$$\inf_{v\in\mathcal{U}} J(v) = \inf_{v\in\tilde{\mathcal{U}}} \tag{55}$$

Therefore, if $\hat{u} \in \tilde{\mathcal{U}}$ solves the problem on the right-hand side of (55), then it obviously solves the problem on the left-hand side. But the problem on the left-hand side of (55) belongs to the class which has been considered in the previous paragraph. Therefore u must satisfy the necessary and sufficient conditions of optimality stated in (32), (33). Moreover, since $\hat{u} \in \tilde{\mathcal{U}}$, we know that $\mathcal{F}^t = \Phi_{\hat{u}}^t$, hence the necessary and sufficient can be rewritten as follows (stochastic Maximum Principle):

$$\left(-E^{\Phi_{\hat{u}}^t}\hat{p}(t) + \frac{\partial \ell}{\partial v}(\hat{y}(t),\hat{u}(t),t),\ v - \hat{u}(t)\right)_H \ge 0 \tag{56}$$

a.e.t, a.s.ω, $\forall v \in \mathcal{U}_{ad}$

where the adjoint variable $\hat{p}(t)$ is still given by (32).

Remark 8. The role of the linearity in both the state equation and the information is obviously fundamental to derive (55). The point here is that even if $\tilde{\mathcal{U}}$ can be defined in a quite natural fashion, without linearity properties, it has no structure and therefore it is impossible to derive necessary conditions for the problem on the right-hand side of (55). This difficulty has, of course, nothing to do with the fact that the system is infinite-dimensional.

Remark 9. In the case of perfect information, i.e.,

$$\varphi(t) = z(t) \qquad \forall t$$

then it would be very interesting to develop a dynamic programming approach of the problem, analogous to the finite-dimensional case. The drawback here is that little is known about the theory of Hamilton-Jacobi-Bellman equations in infinite-dimensional spaces. Some work in that direction has been done by Bensoussan and Lions [30].

We now turn to the quadratic case, and derive the extension of the separation principle for infinite-dimensional systems. We now are going to assume that

$$\ell(z,v,t) = \frac{1}{2}(L(t)z,z)_H + \frac{1}{2}(N(t)v,v)_H + (g(t),z)_H$$

$$\Lambda(z) = \frac{1}{2}(Mz,z) \tag{57}$$

$$\mathcal{U}_{ad} = H \text{ (no constraints)}$$

where $L \in L^\infty(0,T;\mathcal{L}(H;H))$, $L \geq 0$ self-adjoint, $N \in L^\infty(0,T;\mathcal{L}(H;H))$, N self-adjoint, $N \geq \nu I$, $\nu > 0$, $g \in L^\infty(0,T;H)$, $M \in \mathcal{L}(H;H)$, $M \geq 0$ self-adjoint. From the standard theory of calculus of variations, we know that the problem on the left-hand side of (55) has one and only one solution [which does not imply that the problem on the right-hand side of (55) has a solution]. Let us write the necessary and sufficient conditions of optimality. First, the adjoint system (32) becomes

$$- \frac{d\hat{p}}{dt} + A^*\hat{p} = -L(t)\hat{y}(t) - g(t)$$
$$\hat{p}(T) = -M\hat{y}(T) \tag{58}$$

Secondly, (34) becomes

$$[-E^{\mathcal{F}^t}\hat{p}(t) + N(t)\hat{u}(t), v - \hat{u}(t)] \geq 0$$

and since there are no constraints we obtain

$$\hat{u}(t) = +N^{-1}(t)E^{\mathcal{F}^t}\hat{p}(t) \tag{59}$$

Therefore, the pair $\hat{p}(t),\hat{y}(t)$ satisfy (58) and

$$d\hat{y} + A\hat{y}(t)\,dt = B(t)\,d\xi(t) + [N^{-1}(t)E^{\mathcal{F}^t}\hat{p}(t)]\,dt$$
$$\hat{y}(0) = y_0 + \zeta \tag{60}$$

To go further, define

$$\hat{y}_t(\tau) = E^{\mathcal{F}^t}\hat{y}(\tau), \qquad \tau \geq t$$
$$\hat{p}_t(\tau) = E^{\mathcal{F}^t}\hat{p}(\tau) \tag{61}$$

and we make the assumption

$$\eta(t) \text{ is independent from } \xi(t) \text{ and } \zeta \tag{62}$$

From (58) and (60), together with assumption (62), one can deduce (for details see Bensoussan and Viot [27]) that the pair $\hat{y}_t(\tau),\hat{p}_t(\tau)$ is the solution of the following system

$$\frac{d\hat{y}_t}{d\tau} + A\hat{y}_t(\tau) = N^{-1}(\tau)\hat{p}_t(\tau), \qquad \tau > t$$
$$\hat{y}_t(t) = \text{given} \tag{63}$$

$$- \frac{d\hat{p}_t(\tau)}{d\tau} + A^*\hat{p}_t(\tau) = -L(\tau)\hat{y}_t(\tau) - g(\tau), \qquad \tau > t$$
$$\hat{p}_t(T) = -M\hat{y}_t(T) \tag{64}$$

However, to (54), (64) we can apply the <u>decoupling</u> theory of Lions [18] (see also Chapter 1), obtaining

$$\hat{p}_t(\tau) = - [P(\tau)\hat{y}_t(\tau) + r(\tau)], \qquad \tau \geq t \tag{65}$$

where $P(\tau), r(\tau)$ are the solutions of, respectively,

$$\begin{aligned} &\frac{dP}{dt} - PA - A^*P - PN^{-1}P + L = 0 \\ &P(T) = M \end{aligned} \tag{66}$$

$$\begin{aligned} &\frac{dr}{dt} - A^*r - PN^{-1}r + g = 0 \\ &r(T) = 0 \end{aligned} \tag{67}$$

We notice that (66) is the Riccati equation arising in the <u>deterministic</u> problem; (67) also arises in the deterministic problem.

Writing (65) for $\tau = t$, and remembering that

$$\hat{p}_t(t) = E^{\mathfrak{F}^t}\hat{p}(t)$$

we see that (59) can be rewritten as

$$\hat{u}(t) = -N^{-1}(t)[P(t)E^{\mathfrak{F}^t}\hat{y}(t) + r(t)] \tag{68}$$

Let us again emphasize that (68) solves the problem on the left-hand side of (55). Concerning the problem on the right-hand side of (55), we can only say that if it has a solution, then it coincides with (68), more precisely with

$$\hat{u}(t) = -N^{-1}(t)[P(t)E^{\Phi^t_{\hat{u}}}\hat{y}(t) + r(t)] \tag{69}$$

which is equal to (68).

We can, however, say that the problem on the right-hand side of (55) has a <u>minimizing sequence</u> given by

$$u_n(t) = -N^{-1}(t)[P(t)E^{\mathfrak{F}^{t-1/n}}y_n(t) + r(t)] \tag{70}$$

Indeed, if we take in the state equation (23), v to be given by the right-hand side of (70) with z instead of y_n, then we obtain a functional equation in z, and since the operator $E^{\mathfrak{F}^{t-1/n}}$ is a contraction, it is easy to see that this equation has one and only one solution, called y_n, to fix the ideas. But from (47), with $\epsilon = 1/n$, it follows that u_n belongs to $\tilde{\mathcal{U}}$. Since $u_n \to \hat{u}$ in $\mathcal{U}$, it follows from the equality of the two infima (55) that u_n is a minimizing sequence for the problem on the right-hand side of (55).

In order to prove that the problem on the right-hand side of (55) indeed has a solution, we need to make an additional assumption on the noise $\eta(t)$, which thus far was quite general (independent of ξ,ζ).

We now assume that

$$\zeta \text{ is Gaussian [i.e., } (\zeta,h) \text{ is Gaussian for any } h \in H] \text{ and}$$

$$E\zeta = 0, \qquad E(\zeta,h_1)(\zeta,h_2) = (P_0h_1,h_2) \tag{71}$$

where P_0 is a <u>nuclear</u> operator on H

F is <u>finite</u>-dimensional and $\eta(t)$ is a Wiener process with values in F independent of $\xi(t)$. Furthermore, its covariance matrix is denoted by R and assumed to be invertible. (72)

<u>Remark 10.</u> The fact that F must be finite-dimensional is a consequence of the fact that R is at the same time invertible and nuclear (for a more general discussion see Bensoussan [12]).

If (71), (72) are now satisfied, we can state an <u>existence</u> result for the problem on the right-hand side of (55). This will be a consequence of filtering theory for infinite-dimensional systems (the reader is referred to Chapter 3, and Bensoussan [12]).

We consider a new Riccati equation (the Riccati equation of the filter)

$$\frac{d\Pi}{dt} + A\Pi + \Pi A^* + \Pi C^* R^{-1} C\Pi = BQB^*$$
$$\Pi(0) = P_0 \tag{73}$$

and the following system of infinite-dimensional Ito equations (whose solution is the triplet $\hat{y},\hat{s},\hat{\varphi}$)

$$d\hat{y} + A\hat{y}\,dt = B(t)\,d\xi(t) - N^{-1}(t)P(t)\hat{s}(t)\,dt - N^{-1}(t)r(t)\,dt$$
$$\hat{y}(0) = y_0 + \zeta \tag{74}$$

$$d\hat{s} + A\hat{s}\,dt + \Pi(t)C^*(t)R^{-1}C(t)\hat{s}\,dt = -N^{-1}(t)P(t)\hat{s}\,dt$$
$$- N^{-1}(t)r(t)\,dt + \Pi(t)C^*(t)R^{-1}C(t)\hat{y}\,dt \tag{75}$$
$$+ \Pi(t)C^*(t)R^{-1}\,d\eta(t)$$
$$\hat{s}(0) = y_0$$

$$d\hat{\varphi} = C(t)\hat{y}(t)\,dt + d\eta(t) \tag{76}$$
$$\hat{\varphi}(0) = 0$$

We can apply the results of Section II.C concerning stochastic linear PDEs. They insure the existence and uniqueness of $\hat{y},\hat{s},\hat{\varphi}$. The <u>additional</u> result is that

$$\hat{s}(t) = E^{\hat{\Phi}^t}\hat{y}(t) \tag{77}$$

where by $\hat{\Phi}^t$ we mean the σ-algebra generated by $\hat{\varphi}(s)$, $s \le t$ (for details see Bensoussan [28]). Let us then define a control $\hat{u}(t)$ by

$$\hat{u}(t) = -N^{-1}(t)[P(t)\hat{s}(t) + r(t)] \tag{78}$$

After some manipulations of (74) to (76) and (37), (38), one can prove that $\hat{u} \in \mathcal{U}$. To that $\hat{u}$ corresponds an observation given obviously by $\hat{\varphi}(t)$. Therefore,

$$\hat{\Phi}^t = \hat{\Phi}^t_{\hat{u}}$$

and formula (77) implies that $\hat{u} \in \tilde{\mathcal{U}}$. Since $\hat{u}$ is a solution of the problem on the left-hand side of (55), we have

$$J(\hat{u}) = \inf_{v\in\tilde{u}} J(v)$$

This proves that $\hat{u}$ is optimal for the problem on the right-hand side of (55). Formula (78) expresses the so-called <u>separation principle</u>. This principle, when true, states that the optimal solution of the stochastic control problem is obtained by the <u>same feedback</u> as the deterministic problem, with the only transformation that the state $\hat{y}(t)$ is changed into its best estimate with respect to the observations. Hence, as in the finite-dimensional case, we have seen that the separation principle holds true for the linear quadratic Gaussian problem.

Remark 11. For the finite-dimensional case, Wonham [31] has proved a separation principle under weaker assumptions, namely he allows the payoff to be nonquadratic. We do not know whether his results can be extended to the infinite-dimensional case.

Let us make the optimal stochastic control explicit in the following example: We take $\zeta = 0$, $E = R$, hence $\xi(t)$ is a standard one-dimensional Wiener process and

$$B(t) = \sigma(x,t)$$

Thus the evolution of the system is governed by

$$\begin{aligned} &dz(x,t) - \Delta z(x,t)\, dt = v(x,t)\, dt + \sigma(x,t)\, d\xi(t) \\ &z(x,t) = 0 \qquad \text{on } \Gamma = \partial\mathcal{O} \\ &z(x,0) = y_0(x) \end{aligned} \tag{79}$$

Next we define the observation process. Take $F = R$,

$$C(t)z = \int_{\mathcal{O}_0} z(x)\, dx \tag{80}$$

where $\mathcal{O}_0 \subset \mathcal{O}$, and $\eta(t)$ to be a real Wiener process with covariance σ_1^2. The payoff is defined by

$$J(v) = E \iint_{0\mathcal{O}}^{T} z(x,t)^2\, dx\, dt + NE \iint_{0\mathcal{O}}^{T} v(x,t)^2\, dx\, dt \tag{81}$$

The separation principle can be stated as follows:

$$\hat{u}(x,t) = \int_{\sigma} K(x,\xi,t)\hat{s}(\xi,t)\, d\xi \tag{82}$$

where K is solution of the following integrodifferential equation

$$-\frac{\partial K}{\partial t} - (\Delta_x + \Delta_\xi)K + \frac{1}{N}\int_{\mathcal{O}} K(x,\xi_1,t)K(\xi_1,\xi,t)\, d\xi_1 = \delta(x - \xi) \tag{83}$$

$$\begin{aligned} &K(x,\xi,t) = K(\xi,x,t) \\ &K(x,\xi,t) = 0 \qquad \text{if } x \in \Gamma,\ \xi \in \mathcal{O} \\ &K(x,\xi,T) = 0 \end{aligned}$$

$\hat{s}$ is obtained by the generalized Kalman filter

$$d\hat{s}(x,t) - \Delta\hat{s}(x,t)\,dt = \left(\int_{\mathcal{O}} K(x,\xi,t)\hat{s}(\xi,t)\,d\xi\right)dt$$

$$+ \frac{1}{\sigma_1^2}\left[\int_{\mathcal{O}_0} \Pi(x,\xi,t)\,d\xi(d\varphi - dt\int_{\mathcal{O}_0} \hat{s}(\xi,t)\,d\xi)\right] \tag{84}$$

$$\hat{s}(x,t) = 0 \qquad \text{on } \Gamma$$

$$\hat{s}(x,0) = y_0(x)$$

and Π is solution of another integrodifferential equation

$$\frac{\partial\Pi}{\partial t} - (\Delta_x + \Delta_\xi)\Pi + \frac{1}{\sigma_1^2}\left(\int_{\mathcal{O}_0} \Pi(x,\eta,t)\,d\eta\right)\left(\int_{\mathcal{O}_0} \Pi(\eta,\xi,t)\,d\eta\right)$$

$$= \sigma(x,t)\sigma(\xi,x,t) \tag{85}$$

$$\Pi(x,\xi,t) = \Pi(\xi,x,t)$$

$$\Pi(x,\xi,t) = 0 \qquad \text{if } x \in \Gamma,\ \xi \in 0$$

$$\Pi(x,\xi,0) = 0$$

Remark 12. In practice, formulas (82) to (85) can be handled (if the dimension is not too high) generalizing the numerical method of Nedelec [32] (see also Chapter 1, and Bensoussan [12]).

V. CONTROL BY A PRIORI FEEDBACK

A. General Comments

As we have seen in the previous paragraphs, obtaining an optimal stochastic control is generally out of reach. Therefore, it is quite natural to work with some special classes of control; since we are considering a stochastic problem, we clearly cannot be satisfied with open loop controls. Thus, it is very natural to search for an optimal control among a class of a priori feedbacks of a certain type (linear, nonlinear Lipschitz, etc.). Along these lines, we present some results and methods of Robin [33].

B. Optimization Among Lipschitz Controls

We return to the model (23), where we will choose v in terms of feedback, namely

$$v = v(z)$$

The set $\mathcal{U}$ of admissible controls is now

$$\mathcal{U} = \{v: R \to R, \text{ uniformly Lipschitz with constant } K, \quad | v(0) | \leq M; \ x \to v(z(x)) \in \mathcal{U}_{ad}, \text{ for } z \in M\} \tag{86}$$

Then $\mathcal{U}$ is a compact subset of $C(R)$, equipped with the topology of uniform convergence on compact sets of R (remember that $\mathcal{U}_{ad}$ is closed in H).

The payoff will be quadratic. More specifically,

$$J(v) = E\left[\int_0^T | z - z_d |_H^2 \, dt + N \int_0^T | v(t) |_H^2 \, dt \right] \tag{87}$$

We notice (Remark 4) that for fixed v the equation

$$\begin{aligned} dz + Az \, dt &= B(t) \, d\xi + v(z) \, dt \\ z(0) &= y_0 + \zeta \end{aligned} \tag{88}$$

has one and only one solution satisfying the regularity properties (17), and the payoff $J(v)$ is well defined. It is then possible to give an existence result for the problem $\inf_{v \in \mathcal{U}} J(v)$. This results from the fact that the mapping $v \to J(v)$ is continuous on $\mathcal{U}$. Indeed, take

$$v_n \to v \text{ in } C(R), \qquad v_n \in \mathcal{U} \text{ (thus } v \in \mathcal{U})$$

Noting z_n, the solution of (88) for $v = v_n$, we obtain

$$d(z_n - z) + A(z_n - z) \, dt = [v_n(z_n) - v(z)] \, dt$$

from which it is easy to deduce, using the uniform Lipschitz condition on v_n, v (for details, see Robin [33]) that

$$z_n \to z \text{ in } L^2(\Omega \times (0,T); V) \text{ and } L^\infty(0,T; L^2(\Omega;H))$$

strongly. Hence $J(v_n) \to J(v)$. From the continuity of J and compactness of $\mathcal{U}$, the existence of an optimal control follows. This approach can be used also to solve stochastic control problems, where the control is a boundary control (see Robin [33]). Furthermore, Robin has given the following algorithm: Approximate the problem by a discrete-time problem. The discrete form of (88) is

$$\frac{z^{n+1} - z^{n}}{\Delta t} Az^{n+1} = \eta^{n+1} + v(z^{n})$$
$$z^{0} = y_{0} + \zeta \tag{89}$$

where η^{n+1} is a Gaussian random variable with values in H. For each n, one generates M realizations of ζ and η^{n}, denoted by ζ_i, η_i^n, $i = 1 \cdots M$. Then, instead of (89), one writes M equations, whose solutions are denoted by z_i^n. One defines an approximate cost as follows:

$$J_M(v) = \frac{1}{M} \sum_{i=1}^{M} \left[\sum_{n} | z_i^n - z_d^n|_H^2 \, \Delta t + N \sum_{n} |v(z_i^n)|_H^2 \, \Delta t \right] \tag{90}$$

To minimize $J_M(v)$, there are two possibilities: first, no a priori knowledge on the form of v is assumed. The set of values of z is discretized and v(z) is computed as a piecewise linear function. This approach relies on some previous work by Chavent [34] in identification problems. The second possibility is to define v as a linear combination (with unknown weights) of given functions.

C. Case of Imperfect Observation

The method of a priori feedback can also be adapted to the case of imperfect observation, with serious difficulties for nonlinear systems, however. We shall rely on work by Robin [35] (see also Hwang, Seinfeld, and Gavalas [36] and Yu and Seinfeld [37]). We assume that the system is described by

$$dz + Az\,dt + F(z)\,dt = B(t)\,d\xi(t) + v(t)\,dt$$
$$z(0) = y_0 + \zeta \tag{91}$$

with payoff (87). The operator F is assumed to be Lipschitz from $H \to H$. The set of admissible controls will be made precise later.

The observation on the system is defined by the relation (see Section IV.A.)

$$d\varphi = C(t)z(t)\,dt + d\eta(t) \qquad (92)$$
$$\varphi(0) = 0$$

By analogy with (75) (filter equation) we consider the equation

$$d\hat{s} + A\hat{s}\,dt + \Pi(t)C^*(t)R^{-1}(t)C(t)\hat{s}\,dt + F(\hat{s})\,dt$$
$$= \Pi(t)C^*(t)R^{-1}(t)C(t)\hat{y}\,dt + \Pi(t)C^*(t)R^{-1}\,d\eta(t) + \hat{u}(t)\,dt \qquad (93)$$
$$\hat{s}(0) = y_0$$

where $\Pi(t)$ is the solution of the Riccati equation

$$\frac{d\Pi}{dt} + (A + F_z' I)\Pi + H(A^* + F_z'^* I) + \Pi C^* R^{-1} C\Pi = BQB^* \qquad (94)$$
$$\Pi(0) = P_0$$

In (94), F_z' (which is unknown) is <u>evaluated on some chosen trajectory</u>. In (93), $\Pi(t)$ is thus a given operator (assumed to be computed off-line). In (93) we choose $\hat{u}(t)$ in a class of a priori linear feedbacks with respect to the $\hat{s}$ [still by analogy with (78)], i.e.,

$$\hat{u}(t) = u_1 + u_2[\hat{s}(t) - z_d(t)] \qquad (95)$$

where u_1, u_2 are two real parameters to be chosen. To (93), (95) one must add the state equation, namely

$$d\hat{y} + A\hat{y}\,dt + F(\hat{y})\,dt = B(t)\,d\xi(t) + \hat{u}(t)\,dt \qquad (96)$$
$$\hat{y}(0) = y_0 + \zeta$$

Thus (93), (96) [where $\hat{u}$ is given by (95)] define a system in $\hat{y}$, $\hat{s}$ (remember that $\Pi(t)$ is computed off-line), depending only on the two parameters u_1, u_2. One chooses u_1, u_2 in order to minimize the payoff (87). The problem is solved numerically using the simulation approach described in the previous paragraph.

<u>Remark 13.</u> In the finite-dimensional case, the simulation method has been justified by Quadrat and Viot [38].

<u>Remark 14.</u> It is clear from the described algorithm that the filtering part is poorly treated. Little is known on the theory of

nonlinear filtering for distributed systems, making the development of algorithms a very difficult task. Let us mention that Eqs. (93), (94) can be obtained by optimizing a least square functional associated with (91). However, unlike the linear case, it is well known that this approach does not lead to the best filter.

Remark 15. Another example of the method of a priori feedback can be found in Kernevez, Quadrat, and Viot [1] for a problem arising in biochemistry.

VI. MISCELLANEOUS QUESTIONS

Let us mention a few other problems.

A. Stochastic Differential Games for Systems Governed by PDEs

The linear quadratic theory for finite-dimensional systems can be extended to infinite-dimensional systems, provided the players have the *same* observation on the system (particularly in the case of full information). For results along these lines, see Bensoussan [39].

B. Optimization of Sensors' Location in Distributed Systems Filtering

Consider the state equation

$$\begin{aligned} dz + Az\,dt &= B(t)\,d\xi(t) \\ z(0) &= y_0 + \zeta \end{aligned} \tag{97}$$

with the observation process

$$\begin{aligned} d\varphi &= C(t;v)z\,dt + d\eta(t) \\ \varphi(0) &= 0 \end{aligned} \tag{98}$$

In (98), the operator $C(t;v)$ depends on a *control* v. Such a situation arises when there are some degrees of freedom in the choice of sensors (as far as the number, quality, and location are concerned). The decision variable v (which may be a function of time) is *deterministic*, however, since no information whatsoever is available before choosing the value of the control. For each value of v a

Riccati equation corresponds to the filtering problem (97), (98), namely

$$\frac{d\Pi}{dt} + \Lambda\Pi + \Pi A^* + \Pi C^*(v)R^{-1}C(v)\Pi = BQB^*$$
$$\Pi(0) = P_0 \tag{99}$$

Since $\Pi(T;v)$ is the covariance operator of the estimation error process (see Chapter 3, or Bensoussan [12]), it is natural to choose v in order to minimize

$$J(v) = N(v) + \text{tr } P(T;v) \tag{100}$$

where tr is the trace operator and $N(v)$ is a given cost of the control. In Bensoussan [40], the problem of minimizing the functional $J(v)$ has been considered. Necessary conditions and existence results are given.

C. Optimal Stopping Time for Distributed Systems

The system is governed by (97). The control is a stopping time τ, to which the payoff corresponds:

$$J(\tau) = E\left[\int_0^\tau \ell(z(t),t)\ dt + \Lambda(z(\tau),\tau \right]$$

The problem of existence and characterization of a stopping time $\hat{\tau}$ such that

$$J(\hat{\tau}) = \inf J(\tau)$$

has been considered by Bensoussan and Lions [30].

D. Open Problems

Many results which are known for finite-dimensional systems can be extended to infinite-dimensional systems. In many cases, the difficulties which are met are already existing for finite-dimensional systems.

However, one serious difficulty arising from the infinite-dimensionality concerns the Hamilton-Jacobi theory, which is very powerful in the finite-dimensional case. The reason is because in the finite-dimensional case one can rely on the theory of elliptic or parabolic PDEs (see Fleming and Rishel [41]), whereas such a theory is almost nonexistent in the infinite-dimensional case. For similar reasons, the theory of nonlinear filtering for stochastic PDEs has still not been formulated.

Numerical algorithms which can save computer time are also desperately needed. The method of a priori feedbacks, coupled with the simulation approach developed above, is a partial attempt in that direction. Aside from this method, little else is known.

REFERENCES

1. J. P. Kernevez, J. P. Quadrat, and M. Viot, Control of a non-linear stochastic boundary value problem, Proceedings of the I.F.I.P. Congress, Rome, 1973.

2. R. Curtain, On the Ito's stochastic integral in a Hilbert space, Technical Report, no. 20, Control Theory Centre, Warwick University, 1970.

3. H. Kuo, Stochastic integrals in abstract Wiener spaces, Pac. J. Math., vol. 41, no. 2 (1972).

4. J. Neveu, "Intégrales stochastiques et applications," third cycle course, Paris-VI Univ., 1971-1972.

5. M. Yor, "Les intégrales stochastiques hilbertiennes," third cycle thesis, Paris IV Univ., 1974.

6. B. Gaveau, Intégrale stochastique radonifiante, C.R.A.S., Paris, 1973.

7. D. Lepingle-Jouvrard, Martingales browniennes hilbertiennes, I.R.M.A., Grenoble, France, 1973.

8. M. Métivier, Martingales à valeurs vectorielles. Application à la dérivation des mesures vectorielles, Annales Institut Fourier (1967).

9. M. Métivier, Intégrales stochastiques par rapport à des processus à valeur dans un espace de Banach réflexif, Theory Prob. Appl., T.19 (1972).

10. E. Pardoux, thesis, Paris IV Univ., 1975.

11. E. Pardoux, Intégrales stochastiques hilbertiennes, to be pub-Technical Report, IRIA, 1975.

12. A. Bensoussan, Filtrage optimal des systèmes linéaires, Dunod, Paris, 1971.

13. P. A. Meyer, Probabilités et Potentiel, Hermann, Paris, 1966.

14. R. Curtain and P. L. Falb, Stochastic differential equations in Hilbert space, J. Diff. Equs., vol. 10, no. 3 (1971).

15. A. Bensoussan and R. Temam, Equations aux dérivées partielles stochastiques non linéaires I, Isr. J. Math., 11: 95-129 (1972).

16. D. Stroock and S. R. S. Varadhan, Diffusion processes with continuous coefficients, I, II, Comm. Pure Appl. Math., 12: 345-400, 470-530 (1969).

17. T. Yamada and S. Watanabe, On the uniqueness of solutions of stochastic differential equations, T. Math. Kyoto Univ., 11.1: 155-167 (1971).

18. J. L. Lions, Contrôle Optimal de Systèmes Distribués, Dunod, Paris, 1968. (English translation: S. K. Mitter, Springer-Verlag, Berlin.)

19. M. Viot, thesis, Paris, 1975, to appear.

20. M. Viot, A stochastic partial differential equation arising in population genetics theory, Brown University Reports (1975).

21. W. H. Fleming, Distributed parameter stochastic systems in population biology, International Symposium, IRIA, June 1974, in Lecture Notes in Economics and Mathematical Systems, vol. 107, Springer, New York, 1974, pp. 179-191.

22. A. Bensoussan and R. Temam, Equations de Navier Stokes stochastiques, J. Funct. Anal., vol. 13, no. 2 (1973).

23. A. Bensoussan, Contrôle optimal stochastique de systèmes gouvernés par des équations aux dérivées partielles, Rendi Conti di Matematica, Rome, 1969.

24. P. Krée, Images de probabilités cylindriques par certaines applications non linéaires, accouplement de processus linéaires, C.R.A.S., vol. 274, no. 4 (1972).

25. A. V. Balakrishnan, Identification and stochastic control of a class of distributed systems with boundary noise, IRIA Conference on Control Theory, Paris, June 1974, in Lecture Notes in Economics and Mathematical Systems, vol. 107, Springer, New York, 1974, pp. 163-178.

26. A. V. Balakrishnan, Stochastic bilinear partial differential equations, Proceedings of the Conference on Variable Structure Systems, Oregon, 1974.

27. A. Bensoussan and M. Viot, Optimal stochastic control for linear distributed parameter systems, SIAM J. Contr. (1975).

28. A. Bensoussan, On the separation principle for distributed parameter systems, Proceedings of the IFAC Conference on Control of Distributed Parameter Systems, Banff, Canada, 1970.

29. H. J. Kushner, On the optimal control of a system governed by a linear parabolic equation with white noise inputs, SIAM J. Contr., 6: 596-614 (1968).

30. A. Bensoussan and J. L. Lions, Temps d'Arrêt Optimal et Contrôle Impulsionnel, Dunod, Paris, to be published.

31. W. M. Wonham, On the separation principle for stochastic systems, SIAM J. Contr., vol. 6, no. 2, 1968.

32. J. C. Nedelec, thesis, Paris, 1971.

33. M. Robin, Feedback control for a class of stochastic distributed parameter systems, Proceedings of the IFAC Symposium on Stochastic Control, Budapest, Hungary, 1974.

34. G. Chavent, "Identification," thesis, Paris, 1972.

35. M. Robin, Contrôle par feedback d'un système stochastique distribué, Colloque IRIA, June 1974, in Lecture Notes in Economics and Mathematical Systems, vol. 107, Springer, New York, 1974.

36. M. Hwang, I. H. Seinfeld, and G. R. Gavalas, Optimal filtering and interpolation for distributed parameter systems, J. Math. Anal. Appl., 39 (1972).

37. Yu and I. H. Seinfeld, Control of stochastic distributed parameter systems, J. Optim. Theory Appl., vol. 10, no. 6, 1976.

38. J. P. Quadrat and M. Viot, Méthodes de simulation en programmation dynamique stochastique, RAIRO, 1 (1973).

39. A. Bensoussan, Game theory for systems governed by partial differential equations, Proceedings of the First International Conference on Differential Games, Amherst, Massachusetts, 1969.

40. A. Bensoussan, Optimization of sensors location in a distributed filtering problem, Proceedings of the Warwick Conference on Stochastic Stability, Springer Lecture Notes, 1972.

41. W. H. Fleming and R. Rishel, Control Theory, Springer-Verlag, New York, 1976.

Chapter 5

CONTROL AND ESTIMATION OF SYSTEMS WITH TIME DELAYS

H. N. Koivo[†]

Department of Electrical Engineering
Tampere University of Technology
Tampere, Finland

A. J. Koivo

Department of Electrical Engineering
Purdue University
West Lafayette, Indiana

[†]Part of the chapter was written while the author was with the Department of Electrical Engineering, University of Toronto, Toronto, Canada.

I. INTRODUCTION

The systems considered in this chapter are those modeled by the following delay differential equations:

$$\dot{x}(t) = f(x(t), x(t-\tau), u(t), u(t-\nu), t), \qquad t > t_0$$

$$y(t) = h(x(t), x(t-\tau), t), \qquad t > t_0$$

$$x(t) = g(t), \qquad t \in [t_0-\tau, t_0]$$

$$u(t) = v(t), \qquad t \in [t_0-\tau, t_0)$$

The linear time-invariant differential-difference equation

$$\dot{x}(t) = A_0x(t) + A_1x(t-\tau) + B_0u(t) + B_1u(t-\nu)$$

will be discussed in detail, since the linear-quadratic theory is quite complete.

The design of a controller for systems of the above type is quite difficult because of the sluggishness of the response. In the engineering literature, the scalar case, usually with a single control delay, has received the most attention. The reason for this is that many chemical processes can be modeled approximately by a single delay and a lumped first- or second-order system. The control design is commonly accomplished by a classical (analog) PID-controller or by first discretizing the system equation and then applying a three-term controller (a discrete PID-controller) [1,2] or by a Smith-predictor [3].

However, when the overall <u>multivariable</u> plant has to be controlled, scalar models are inadequate most of the time. In fact, because of couplings and transport delays, they often lead to erroneous, unstable designs. Although the multivariable design techniques for time-delay systems have not reached the same level of maturity as systems without delays, we feel that there are a number of useful multivariable design techniques for delay systems which can help and provide insight in practical design tasks.

We will attempt to summarize the most recent research results for the design of time-delay systems. The linear-quadratic theory, in particular, has reached the point where the theory is fairly complete and most of the results for systems without delay are also known for delay systems. In addition, the estimation theory for delay systems has, at least formally, been well developed, so that a comprehensive coverage of modern design techniques for linear systems can be provided. The state-space approach is used throughout, but sometimes it is possible to achieve good results using frequency domain. Such procedures are discussed by Rosenbrock [4].

In Section II we will pose the optimization problem for delay systems and state the maximum principle. We will then study the linear-quadratic problem, and determine the optimal feedback control and the equations specifying the optimal feedback gains. We will also discuss computational considerations, which play an important role in the design of (optimal) controls for delay systems. We then present an observer theory for time-delay systems using spectral decomposition.

Section III is devoted to the state estimation in time-delay systems. We will first give the theory for linear time-delay systems and observe the relationship between the constructed state estimator and observer. We will then formally present two specific approaches to nonlinear state estimation with examples. Section IV concludes the chapter with a discussion on the stochastic control problem.

A. Notation and Terminology

The k-dimensional vector space over the field $\underline{R}$ of real numbers (or the field $\underline{C}$ of complex numbers) will be denoted by $\underline{E}^k$. $\underline{C}^-$ (resp. $\underline{C}^+$) is the open left-half (resp. closed right-half) complex plane. $\mathcal{C}([-\tau,0],\underline{E}^n)$ is the Banach space of continuous functions mapping the interval $[-\tau,0]$ into $\underline{E}^n$ with the topology of uniform convergence. For convenience we will write $\mathcal{C} = \mathcal{C}([-\tau,0],\underline{E}^n)$. $\mathcal{U}$ and $\mathcal{Y}$ are real vector spaces of dimension m and p, respectively. Roman capital letters denote linear transformations (maps) as well as the matrices corresponding to the maps. The identity operator is denoted by I. The image and the kernel of a map are denoted as usual by $\mathrm{Im}(\cdot)$ and $\mathrm{Ker}(\cdot)$, respectively. The transpose of a matrix A is denoted by A'.

B. Preliminaries

Some fundamental theory of the linear time-invariant, differential-difference equations is needed later. The results presented here are derived in Hale [5] for linear functional differential equations and are briefly summarized below.

Consider the system

$$\dot{x}(t) = A_0 x(t) + A_1 x(t-\tau) + Bu(t) \tag{1}$$

with the initial condition

$$x(t) = g(t) \tag{2}$$

and the output relation

$$y(t) = Cx(t) \tag{3}$$

Here $x \in \underline{E}^n$, $y \in \mathcal{Y}$, $u \in \mathcal{U}$, A_0, A_1, B, and C are constant matrices of appropriate sizes, the initial function $g \in \mathcal{C}$, and the control $u \in L^1_{loc}([0,\infty],\underline{E}^m)$.

Equation (1) has the unique solution $x_t \in \mathcal{C}$ given by

$$x_t = T(t)g + \int_0^t T(t-s)\hat{Q}Bu(s)\, ds, \qquad t \geq 0 \tag{4}$$

where T(t) is a solution operator $T: \mathcal{C} \to \mathcal{C}$, $\hat{Q}$ is defined by

$$\hat{Q}(\xi) = \begin{matrix} 0, & \xi \in [-\tau,0) \\ I, & \xi = 0 \end{matrix}$$

and $x_t(\xi) = x(t+\xi)$, $\xi \in [-\tau,0]$. The state of the system at time t is x_t and, to be consistent, the output equation is then rewritten as

$$y(t) = C\hat{P}x_t \tag{5}$$

where $\hat{P}: \mathcal{C} \to \underline{E}^n$ is defined by

$$\hat{P}x_t = x_t(0) \tag{6}$$

Since the operator T(t) is strongly continuous, the infinitesimal generator $\mathcal{A}$ of T(t) is given by

$$\mathcal{A}\varphi(\xi) = \begin{cases} \dfrac{d\varphi(\xi)}{d\xi}, & \xi \in [-\tau, 0) \\ A_0\varphi(0) + A_1\varphi(-\tau), & \xi = 0 \end{cases} \tag{7}$$

with the domain $\mathcal{D}(\mathcal{A}) = \{\varphi\colon \varphi \in \mathcal{C}, \dot{\varphi} \in \mathcal{C}, \dot{\varphi}(0) = A_0\varphi(0) + A_1\varphi(-\tau)\}$. The spectrum $\sigma(\mathcal{A})$ of $\mathcal{A}$ is a point spectrum and is given by

$$\sigma(\mathcal{A}) = \{\lambda \in \underline{C}\colon \det(A_0 + A_1 e^{-\tau\lambda} - \lambda I) = 0\} \tag{8}$$

It is also known that each $\lambda \in \sigma(\mathcal{A})$ is of finite multiplicity $d(\lambda)$, and there are only a finite number of spectral points to the right of any vertical line in the complex plane.

Assuming that the initial function belongs to the domain of $\mathcal{A}$ and $u \equiv 0$, the system (1) is equivalent to

$$\dot{x}_t = \mathcal{A}x_t, \qquad x_0 = g \tag{9}$$

By defining a pseudo-adjoint operator $\mathcal{A}^*$ relative to the bilinear form

$$\langle\psi, \varphi\rangle = \psi(0)\varphi(0) + \int_{-\tau}^{0} \psi(\xi + \tau)A_1\varphi(\xi)\, d\xi \tag{10}$$

where $\varphi \in \mathcal{C}$, $\psi \in \mathcal{C}^* = \mathcal{C}([0,\tau], \underline{E}^{n*})$, $\underline{E}^{n*}$ being the n-dimensional space whose elements are row vectors, a spectral decomposition of $\mathcal{A}$ in the space $\mathcal{C}$ is achieved. $\mathcal{A}^*$ is defined by

$$\mathcal{A}^*\psi(\theta) = \begin{cases} -\dfrac{d\psi(\theta)}{d\theta}, & \theta \in [0, \tau) \\ \psi(0)A_0 + \psi(\tau)A_1, & \theta = 0 \end{cases} \tag{11}$$

It can be shown that $\sigma(\mathcal{A}^*) = \sigma(\mathcal{A})$ [6].

Any given $\lambda \in \sigma(\mathcal{A})$ determines a finite-dimensional generalized eigenspace $\mathcal{M}_\lambda(\mathcal{A})$ given by $\mathcal{M}_\lambda(\mathcal{A}) = \mathrm{Ker}(\mathcal{A} - \lambda I)^{k(\lambda)}$, where $k(\lambda)$ is an integer. $\mathcal{M}_\lambda(\mathcal{A})$, with dimension $d(\lambda)$, will be called the <u>modal subspace corresponding to λ</u>. The modal subspace $\mathcal{M}_\lambda(\mathcal{A})$ is isomorphic with $\underline{E}^{d(\lambda)}$.

Let $\Lambda = \{\lambda\colon \lambda \in \sigma(\mathcal{A}) \cap C^*\}$ and let the elements of Λ be ordered as $\lambda_1, \ldots, \lambda_N$, where N is the number of distinct elements in Λ, and let $N_\Lambda = \Sigma_{\lambda \in \Lambda} d(\lambda)$. A subspace $\mathcal{C}_+(\mathcal{a})$ of $\mathcal{C}$ is defined by

$$\mathcal{C}_+(\mathcal{a}) = \bigoplus_{\lambda \in \Lambda} \mathcal{m}_\lambda(\mathcal{a}) \tag{12}$$

The space $\mathcal{C}$ (resp. $\mathcal{C}^*$) has now been decomposed into the direct sum of the finite-dimensional subspaces $\mathcal{C}_+$ and $\mathcal{C}_-$ (resp. $\mathcal{C}^*_+$ and $\mathcal{C}^*_-$), or

$$\mathcal{C} = \mathcal{C}_+ \oplus \mathcal{C}_- \qquad (\text{resp. } \mathcal{C}^* = \mathcal{C}^*_+ \oplus \mathcal{C}^*_-) \tag{13}$$

<u>Remark</u>. All the unstable modes of the system are contained in $\mathcal{C}_+$ and the stable modes in $\mathcal{C}_-$.

For a given $\lambda \in \Lambda$, let $\varphi^1_\lambda, \varphi^2_\lambda, \ldots, \varphi^{d(\lambda)}_\lambda$ be a basis for $\mathcal{m}_\lambda(\mathcal{a})$. Define a matrix Φ_Λ (resp. ψ_Λ) by taking as its columns (resp. rows) all basis vectors for all $\lambda \in \Lambda$, and thus the columns (resp. rows) of Φ_Λ (resp. Ψ_Λ) form a basis for $\mathcal{C}_+$ (resp. $\mathcal{C}^*_+$). The matrix Ψ_Λ can be scaled in such a way that

$$\langle \Psi_\Lambda, \Phi_\Lambda \rangle = [\langle \psi_i, \varphi_j \rangle] = I \tag{14}$$

where ψ_i and φ_j are the ith row and jth column of Ψ_Λ and Φ_Λ, respectively.

Since $\mathcal{a}\mathcal{m}_\lambda(\mathcal{a}) \subset \mathcal{m}_\lambda(\mathcal{a})$, there exists an $N_\Lambda \times N_\Lambda$ matrix A_Λ such that $\mathcal{a}\Phi_\Lambda = \Phi_\Lambda A_\Lambda$. The matrix A_Λ has the following properties:

$$\text{(i)} \quad \sigma(A_\Lambda) = \Lambda \tag{15}$$

$$\text{(ii)} \quad A^*_\Lambda = A_\Lambda \tag{16}$$

$$\text{(iii)} \quad \mathcal{a}\Phi_\Lambda = \Phi_\Lambda A_\Lambda \tag{17}$$

$$\text{(iv)} \quad \mathcal{a}^*\Psi_\Lambda = A_\Lambda \Psi_\Lambda \tag{18}$$

The matrices Φ_Λ and Ψ_Λ also satisfy

$$\text{(v)} \quad \Phi_\Lambda(\xi) = \Phi_\Lambda(0)e^{A_\Lambda \xi}, \qquad \xi \in [-\tau, 0] \tag{19}$$

$$\Psi_\Lambda(\xi) = e^{A_\Lambda \theta}\Psi_\Lambda(0), \qquad \theta \in [0, \tau] \tag{20}$$

and Φ_Λ can be arranged so that A_Λ is in the Jordan form.

The solution x_t given by Eq. (4) can now be decomposed as

$$x_t = x_t^1 + x_t^2, \qquad x_t^1 \in \mathcal{C}_+, \; x_t^2 \in \mathcal{C}_- \tag{21}$$

where $x_t^1 = \Phi_\Lambda \langle \Psi_\Lambda, x_t \rangle$. The projection operator $E_\Lambda : \mathcal{C} \to \mathcal{C}_+$ is characterized by

$$E_\Lambda(\cdot) = \Phi_\Lambda \langle \Psi_\Lambda, \cdot \rangle \tag{22}$$

If $x_1(t)$ is further defined by $x_1(t) = \langle \Psi_\Lambda, x_t \rangle$, then $x_1(t)$ is the solution of

$$\dot{x}_1(t) = A_\Lambda x_1(t) + B_\Lambda u(t) \tag{23}$$

with the initial condition

$$x_1(0) = \langle \Psi_\Lambda, g \rangle \tag{24}$$

and $B_\Lambda = \Psi_\Lambda(0)B$. Because of the above decomposition, the output $y(t)$ decomposes as

$$y(t) = C\hat{P}x_t^1 + C\hat{P}x_t^2 \tag{25}$$

or

$$y(t) = C\Phi_\Lambda(0)\langle \Psi_\Lambda, x_t \rangle + C\hat{P}x_t^2 \tag{26}$$

For convenience, define $C_\Lambda = C\Phi_\Lambda(0)$ and $C_2 = C\hat{P} \mid \mathcal{C}_-$, thus

$$y(t) = C_\Lambda x_1(t) + C_2 x_t^2 \tag{27}$$

The triplet $(C_\Lambda, A_\Lambda, B_\Lambda)$ now defines a finite-dimensional system with respect to the decomposition and is henceforth called the <u>reduced system corresponding to Λ</u>.

<u>Remark.</u> The particular decomposition chosen here is by no means the only one possible. Any vertical line in the complex plane could have served the purpose and the decomposition could still have been carried out. The one chosen here is especially useful in designing <u>stable</u> control systems.

A variation of parameter formula for system (1) will also be used. This is derived in [6] to be

$$x(t) = X(t,0)x(0) + \int_{-\tau}^{0} X(t, s+\tau)A_1 g(s)\, ds + \int_{0}^{t} X(t,s)$$
$$\times\, Bu(s)\, ds \qquad (28)$$

where the fundamental matrix $X(t,s)$ satisfies

$$\frac{\partial X(t,s)}{\partial t} = A_0 X(t,s) + A_1 X(t-\tau, s)$$
$$X(t,s) \equiv 0 \qquad \text{for } s > t \text{ and } X(s,s) = I \qquad (29)$$

The "adjoint" equation corresponding to (1) is defined by

$$\dot{p}(t) = -A_0' p(t) - A_1' p(t+\tau) \qquad (30)$$

$$p(t) = \theta(t), \qquad t \geq T \qquad (31)$$

The variation of parameter formula for a nonhomogeneous adjoint equation [add a forcing term, $-f(t)$, to the right-hand side of Eq. (30)] is given by

$$p(t) = X'(T,t)p(t) + \int_{t}^{T+\tau} X'(s+\tau, t)A_1'\theta(s)\, ds$$
$$+ \int_{t}^{T} X'(s,t)f(s)\, ds \qquad (32)$$

where $X(t,s)$ is defined by (29).

II. OPTIMAL CONTROL

In this section we first pose an optimization problem and state the corresponding maximum principle. The maximum principle is then utilized to develop a theory for a linear plant-quadratic cost problem. After the optimization theory has been presented, the computational aspects are discussed. The section concludes with an extension of the observer theory for time-delay systems.

A. Maximum Principle

The maximum principle in its basic form results in optimal open-loop controls. Combining the maximum principle with the Fredholm theory for integral equations, a unified treatment for the design of optimal controllers can be achieved.

Only the most fundamental results concerning the maximum principle for time-delay systems are presented here. An excellent survey paper by Banks and Manitius [7] provides a more thorough account of the variational theory applied to hereditary systems for an interested reader.

Consider the differential-difference equation

$$\dot{x}(t) = f(x(t),x(t-\tau),u(t),t), \qquad t > t_0 \tag{33}$$

$$x(t) = g(t), \qquad t \in [t_0-\tau,\ t_0] \tag{34}$$

where x is an n-vector, u is an m-vector, g a given continuous time function, and the delay $\tau = 1$. Choose a cost function to be minimized as

$$C(u) = \int_{t_0}^{T} f^o(x(s),x(s-1),u(s))\, ds \tag{35}$$

It is assumed that the functions $f(x,z,u,t)$ and $f^o(x,z,u,t)$ together with f_x, f_z, f_x^o, f_z^o are continuous functions of their arguments. A typical optimization problem is then to choose a measurable control u from a restraint set Ω so as to steer the initial function g(t) to a target set in $\underline{E}^n$ while minimizing the cost.

Define the Hamiltonian function H to be

$$H(x,z,u,p,t) = p'(t)f(x,z,u,t) + f^o(x,z,u,t) \tag{36}$$

where p(t) is an n-vector. Then the maximum (minimum, in this case) principle [8] can be stated as

Theorem 1. If u* is an optimal control minimizing the cost (35) and x* the corresponding optimal trajectory, it is necessary that there exists a nonzero vector p*(t) such that

1. p*(t) corresponds to u*(t) and x*(t), where p*t and x*t are a solution of

$$\begin{aligned} \dot{x}^*(t) &= f(x^*(t),x^*(t-1),u^*(t),t) \\ \dot{p}^*(t) &= -H_x(x^*(t),x^*(t-1),u^*(t),p^*(t),t) \end{aligned} \tag{37}$$

$$- H_z(x^*(t+1), x^*(t), u^*(t+1), p^*(t+1), t+1),$$

$$t \in [t_o, T] \qquad (34)$$

$$p(t) = 0 \quad \text{for } t > T \qquad (38)$$

$p(T)$ is given by the appropriate transversality condition which depends on the target set,

2. the Hamiltonian H achieves its minimum, that is,

$$\min_{u \in \Omega} H(x^*(t), x^*(t-1), u(t), p^*(t), t)$$

$$= H(x^*(t), x^*(t-1), u^*(t), p^*(t), t), \quad t \in [t_o, T] \qquad (39)$$

The apparent difference between the maximum principle here and that with no delays is in the adjoint equation. Since the adjoint equation turns out to be a differential equation of advanced type, it can only be solved backwards in time.

Remark. More general system equations than (33) have been considered in [8], where the target set could also be chosen in a function space rather than in $\underline{E}^n$.

In the next section we will need a special case of the minimum principle which applies to linear systems with quadratic cost:

$$\dot{x}(t) = A_o x(t) + A_1 x(t-1) + B_o u(t) + B_1 u(t-1), \quad t > t_o \qquad (40)$$

$$x(t) = g(t), \quad t \in [t_o-1, t_o]$$

$$u(t) = v(t), \quad t \in [t_o-1, t_o)$$

with quadratic cost

$$C(u) = \int_{t_o}^{T} [x'(s)Qx(s) + u'(s)Wu(s)] \qquad (41)$$

Here, all matrices are constant and of appropriate dimensions, Q (resp. W) is a symmetric positive semidefinite (resp. definite) matrix. The initial functions are given continuous functions. The control vector u is admissible, if $u \in L_2[t_o, T]$.

Remark. The control delay is chosen to be equal to the state delay for convenience only.

Let the cost coordinate be defined by

$$\dot{x}^o(t) = x'(t)Qx(t) + u'(t)Wu(t), \qquad t > t_o$$
$$x^o(t_o) = 0 \tag{42}$$

Then the cost functional to be minimized is $C(u) = x^o(T)$.

The maximum principle for the problem is furnished in [9] for the case when u is constrained, but the same approach is also applicable here. The set of attainability K(T) is the fundamental concept utilized in [9]. This is defined to be the set of all response end points $\hat{x}_u(T) = (x_u^o(T), x_u(T)) \in \underline{E}^{n+1}$ for all admissible controls $u \in L_2[t_o, T]$. Here $\hat{x}_u$ is the response corresponding to an admissible control u. The following necessary and sufficient conditions can then be established (for further details see [10,11]).

Theorem 2. Consider the system (40) with the cost C(u). Then there exists a unique hypersurface among the family S: $x^o = c$ such that S_m is tangent to $\hat{K}(T)$, and hence m is the optimal cost. Also, there exists a unique optimal controller, namely the extremal controller $u^*(t)$ which steers the response to a single point at which S_m touches $\hat{K}(T)$. Furthermore, there is a unique solution of the equation

$$\dot{x}(t) = \begin{cases} A_o x(t) + A_1 x(t-1) - H_{oo}p(t) - H_{o1}p(t+1) + B_1 v(t-1), \\ t \in [t_o, t_o+1] \\ A_o x(t) + A_1 x(t-1) - [H_{oo} + H_{11}]p(t) - H_{o1}p(t+1) \\ - H_{1o}p(t-1), \qquad t \in [t_o+1, T-1] \\ A_o x(t) + A_1 x(t-1) - [H_{oo} + H_{11}]p(t) - H_{1o}p(t-1), \\ t \in [T-1, T] \end{cases} \tag{43}$$

$$\dot{p}(t) = -A_o'p(t) - A_1'p(t+1) - Qx(t), \qquad t \in [t_o, T] \tag{44}$$

satisfying the boundary conditions $x(t) = g(t)$ on $[t_o-1, t_o]$ and $p(t) = 0$ for $t \geq T$. Here $H_{oo} = B_o W^{-1} B_o'$, $H_{o1} = B_o W^{-1} B_1'$, $H_{1o} = B_1 W^{-1} B_o'$, and $H_{11} = B_1 W^{-1} B_1'$. That is, the optimal response $x^*(t)$ and $p^*(t)$ are such that

$$u^*(t) = -W^{-1}B_0'p(t) - W^{-1}B_1'p(t+1), \qquad t \in [t_0, T] \tag{45}$$

Equipped with the maximum principle, we will now combine it with the Fredholm theory for integral equations and solve the synthesis problem for linear systems.

B. Optimal Control in LPQ Problem

Solving the problem of a linear plant-quadratic cost (LPQ) is a well-known design technique for ordinary differential equations (ODEs). As pointed out by Wonham [12], it really is just another pole-assignment or stabilization technique rather than solving any basic structural problem of synthesis. However, a lot of experience has been accumulated in solving LPQ problems and, based on this, reasonable cost matrices Q and W can be chosen to achieve good overall dynamic performance. It is therefore important to solve the LPQ problem for time-delay systems, although some more fundamental problems, e.g., the internal model principle [12] for time-delay systems, are still open problems.

The so-called Carathéodory lemma was used by Kalman [13] to solve the LPQ problem for ODEs, resulting in a feedback control where the feedback gain satisfies the matrix Riccati equation. Krasovskii [14] was the first one to discuss the LPQ problem for time-delay systems, but more thorough accounts of the theory have been presented in [10,15-18], of which [10,15,16] use the (extended) Carathéodory lemma.

In the ODE case, the optimal feedback law for the LPQ problem can be found by several methods, e.g., by the sweep method [17]. This technique proceeds roughly as follows: Starting with the maximum principle, necessary conditions for optimality are obtained. Then a linear relationship between the adjoint variable and the state variable is postulated. The two-point boundary value problem resulting from the maximum principle can thus be decoupled. As a result, the optimal control law together with a matrix Riccati equation for the feedback gain is determined. This is in contrast with the Carathéodory lemma, which is similar to the dynamic programming approach.

In time-delay systems, difficulties may arise (particularly in numerical computations) due to the fact that the state equation involves variables with delayed arguments and the adjoint equation contains terms with advanced arguments.

A possible approach is to transform the differential-difference equations into Fredholm integral equations and then to use the theory of Fredholm's resolvents. This was done by Manitius [20] and later extended by Koivo and Lee [11] to the problem discussed at the end of Section II.A. For some further comments about the method, the reader is encouraged to consult [21,22].

Remark. The Fredholm theory has been used extensively by Kailath [23,24] in communication theory. The connection between the matrix Riccati equations and Fredholm resolvents is also discussed in [25].

The LPQ problem is to minimize the quadratic cost (41) subject to (40). Moreover, we would like to find the optimal controller in the feedback form. We will now follow the procedure of [11] suggested above.

Starting with Theorem 2, Eq. (43) is integrated forward from r to t and (44) backward from T to t, resulting in

$$x(t) = X(t,r)x(r) + \int_{r-1}^{r} X(t, s+1)[A_1x(s) + B_1u(s)]\, ds$$

$$- \int_{r}^{t-1} X(t, s+1)H_{10}p(s)\, ds - \int_{r}^{t} X(t,s)H_{00}p(s)\, ds$$

$$- \int_{t+1}^{t} X(t,s)H_{11}p(s)\, ds - \int_{r}^{t+1} X(t, s-1)H_{01}p(s)\, ds \qquad (46)$$

$$p(t) = \int_{t}^{T} X'(t,s)Qx(s)\, ds \qquad (47)$$

where the fundamental matrix $X(t,s)$ satisfies Eq. (29). Here we have used the variation of parameter formulae (28) and (32).

Now substituting $x(t)$ from Eq. (46) into Eq. (47), defining a matrix

$$M(t,s) = \int_{\max(t,s)}^{T} X'(\alpha,t)QX(\alpha,s)\, d\alpha$$

and taking into account the properties of X(t,s), integral equation (47) can be written as

$$p(t) = q(t) - \int_{r}^{T} K(t,s)p(s)\, ds \tag{48}$$

where

$$q(t) = M(t,r)x(r) + \int_{r-1}^{r} M(t,\ s+1)[A_1x(s) + B_1u(s)]\, ds$$

and the kernel K(t,s)

$$K(t,s) = \begin{cases} K_1(t,s) = M(t,\ s+1)H_{10} + M(t,s)H_{00}, & s \in [r,\ r+1] \\ K_2(t,s) = M(t,\ s+1)H_{10} + M(t,s)(H_{00} + H_{11}) & \\ \qquad + M(t,\ s-1)H_{01}, & s \in [r+1,\ T-1] \\ K_3(t,s) = M(t,s)(H_{00} + H_{11}) + M(t,\ s-1)H_{01}, & \\ \qquad s \in [T-1,\ T] & \end{cases} \tag{49}$$

The matrix M(t,s) can be shown to possess the following properties [12]:

1. M(t,s) is symmetric in its arguments; that is, $M'(t,s) = M(s,t)$.
2. M(t,s) is continuously differentiable on each of the sets $\{t > s\}$ and $\{t = s\}$, except for the lines $t = T + 1$ and $s = T + 1$.
3. M(t,s) satisfies the system of equations

$$\frac{\partial M(t,s)}{\partial t} = -A_0'M(t,s) - A_1'M(t+1,\ s), \quad t < s$$

$$\frac{\partial M(t,s)}{\partial s} = -M(t,s)A_0 - M(t,\ s+1)A_1, \quad t > s$$

$$\frac{dM(t,t)}{dt} = A_0'M(t,t) - M(t,t)A_0 - A_1'M(t+1,\ t) - M(t,\ t+1)A_1 - Q$$

where $t,s \in [r,T]$.

The following lemma can then be stated:

Lemma 3. $p(t)$ is a solution of the integral equation (48) if and only if it is a solution of the system of Eqs. (43) and (44) together with the boundary conditions.

Fredholm's first theorem and Lemma 3 lead to Lemma 4:

Lemma 4. A unique solution exists to the Fredholm integral equation (48). The solution $p(t)$ to Eq. (48) can be written as

$$p(t) = q(t) - \int_r^T R(t,\alpha,r)q(\alpha)\, d\alpha$$

where the resolvent $R(t,s,r)$ satisfies

$$R(t,s,r) = K(t,s) - \int_r^T R(t,\alpha,r)K(\alpha,s)\, d\alpha$$

$$= K(t,s) - \int_r^T K(t,\alpha)R(\alpha,s,r)\, d\alpha \tag{50}$$

It is convenient to define a modified resolvent

$$P(t,s,r) = M(t,s) - \int_r^T R(t,\alpha,r)M(\alpha,s)\, d\alpha \tag{51}$$

and thus $p(t)$ can be written as

$$p(t) = P(t,r,r)x(r) + \int_r^T P(t, s+1, r)[A_1x(s) + B_1u(s)]\, ds \tag{52}$$

From Eq. (45) the optimal control now becomes

$$u(t) = -W^{-1}\{[B_0'P(t,r,r) + B_1'P(t+1, r, r)]x(r)$$

$$+ \int_{r-1}^{r} [B_0'P(t, s+1, r) + B_1'P(t+1, s+1, r)][A_1x(s)$$

$$+ B_1u(s)]\, ds\} \tag{53}$$

Note that $P(t,s,r) \equiv 0$ for $t,s \geq T$.

We summarize the above results in Theorem 5.

<u>Theorem 5.</u> The optimal control $u^*(t)$ which minimizes the cost $C(u)$ subject to the system constraint (40) exists, is unique, and is given by (53). The feedback law $\hat{u}(x_t,t) = u(x_t,u_t,t)$ is obtained from Eq. (53) by setting $r = t$.

The basic structural feature in the feedback control is that it is not only a function of $x(t)$, but of the whole state x_t and also of the history of the control $u_t(\theta)$, $\theta \in [-1,0)$. If $A_1 = B_1 = 0$, then the familiar control law for ODEs is evident. If $B_1 = 0$, then the typical feedback controller for time-delay systems is achieved.

The LPQ problem has now been solved, but the properties of the modified resolvent need to be investigated more closely. An integral equation will first be established for the modified resolvent $P(t,s,r)$. Starting with Eq. (50) for $R(t,s,r)$, we substitute the value of the kernel $K(t,s)$ from Eq. (49) into (50), and use the definition for $P(t,s,r)$, Eq. (51). This results in

$$P(t,s,r) = M(t,s) - \int_r^T P(t,\alpha,r)K'(s,\alpha)\, d\alpha \tag{54}$$

The integral equation representation for the modified resolvent is useful; however, differential equation representation is often preferable for computational purposes.

It is easy to show that the modified resolvent is symmetric with respect to the arguments t and s, i.e., $P'(t,s,r) = P(s,t,r)$. The symmetric property together with the integral equation (54) imply that $P(t,s,r)$ also satisfies

$$P(t,s,r) = M(t,s) - \int_r^T K(t,\alpha)P(\alpha,s,r)\, d\alpha \tag{55}$$

By observing that $P(t,s,r)$ is continuous in all its arguments, we are now ready to obtain the differential characterization for $P(t,s,r)$.

<u>Theorem 6.</u> $P(t,s,r)$ is continuous with respect to its arguments in respective intervals $t,s \in [r,T]$, $r \in [t_0,T]$. Moreover, it is piecewise continuously differentiable with respect to r and

$$\frac{\partial P(t,s,r)}{\partial r} = \sum_{i,j=0}^{1} P(t,\ r+i,\ r)\, H_{ij} P(r+j,\ s,\ r) \tag{56}$$

$$\frac{dP(t,r,r)}{dr} = -P(t,r,r)A_0 - P(t,\ r+1,\ r)A_1 + \sum_{i,j=0}^{1} P(t,\ r+i,\ r) \times H_{ij} P(r+j,\ r,\ r) \tag{57}$$

$$\frac{dP(r,r,r)}{dr} = -A_0' P(r,r,r) - P(r,r,r)A_0 - A_1' P(r+1,\ r,\ r) - P(r,\ r+1,\ r)A_1 - Q - \sum_{i,j=0}^{1} P(r,\ r+1,\ r) H_{ij} \times P(r+j,\ r,\ r) \tag{58}$$

with the boundary conditions

$$P(t,s,r) = 0 \quad \text{for } t,s \geq T, \qquad P(T,T,T) = 0$$

Remark. The equations derived in Theorem 6 are the extended Riccati equations. They reduce to the usual matrix Riccati when we set $A_1 = B_1 = 0$.

Although the extended Carathéodory lemma was used advantageously for the posed problem when $B_1 = 0$ (no control delay), there were some questions concerning the applicability of the Carathéodory lemma to the case of $B_1 \neq 0$. This problem was solved formally in [22] and is discussed more thoroughly in a forthcoming paper [26]. The key observation is established by substituting the optimal control law (53) into the cost functional (41). The optimal cost then becomes

$$C(u) = x'(t_0)P(t_0,t_0,t_0)x(t_0) + x'(t_0)\int_{t_0-1}^{t_0} P(t,\ s+1,\ t)z(s)\, ds + \int_{t_0-1}^{t_0} z'(r)P(r+1,\ t,\ t)\, dr\, x(t_0) + \int_{t_0-1}^{t_0}\int_{t_0-1}^{t_0} z'(r) \times P(r+1,\ s+1,\ t)z(s)\, dr\, ds \tag{59}$$

where $z(s) = A_1 x(s) + B_1 u(s)$, $s \in [t-1,\ t]$. When $B_1 = 0$, this is exactly the form of the "V-function" which was used, for example, in [10].

<u>Remark.</u> Another interesting extension of the posed problem is to the infinite-time case, that is, $T \to \infty$. It has been discussed in [15] and more thoroughly in [18], when $B_1 = 0$. For $B_1 \neq 0$ the problem appears to be open.

If we let

$$\tilde{E}_0(t) = P(t,t,t)$$

$$\tilde{E}_1(t,\theta) = P(t, t+\theta+1, t), \qquad \theta \in [-1,0]$$

$$\tilde{E}_2(t,\xi,\theta) = P(t+\xi+1, t+\theta+1, t), \qquad \xi,\theta \in [-1,0]$$

then it was shown in [18] that as $T \to \infty$, $\tilde{E}_0(t) \to E_0$ (a constant matrix), $\tilde{E}_1(t,\theta) \to E_1(\theta)$, and $\tilde{E}_2(t,\xi,\theta) \to E_2(\xi,\theta)$; that is, the feedback gains are no longer time dependent. The optimal feedback control becomes

$$u(x_t) = -W^{-1}B_0'\left[E_0 x(t) + \int_{-1}^{0} E_1(\theta)x(t+\theta)\, d\theta\right]$$

where E_0, $E_1(\theta)$, $E_2(\xi,\theta)$ satisfy the relations

$$A_0'E_0 + E_0A_0 + A_1'E_1'(0) + E_1(0)A_1 + Q - E_0B_0W^{-1}B_0'E_0 = 0$$

$$\dot{E}_1(\theta) = A_0'E_1(\theta) + A_1'E_2(0,\theta) - E_0B_0W^{-1}B_0'E_1(\theta), \qquad \theta \in [-1,0] \tag{60}$$

$$\frac{\partial E_2(\xi,\theta)}{\partial \xi} + \frac{\partial E_2(\xi,\theta)}{\partial \theta} = -E_1'(\xi)H_{00}E_1(\theta), \qquad \xi,\theta \in [-1,0]$$

together with

$$E_1(-1) = E_0$$

$$E_2(-1,\theta) = E_1(\theta), \qquad \xi,\theta \in [-1,0] \tag{61}$$

Under certain stabilizability hypotheses, the existence of a stable closed-loop control is proved in [18].

On the above basis, we can ask for a stabilizing control for the system (40) when $B_1 = 0$. It appears feasible, but the theory is not yet conclusive. This question has been investigated in [27,28] where the form of a stabilizing controller has been found to be the same as was obtained in the infinite-time LPQ problem, that is,

$$u(x_t) = P_0 x(t) + \int_{-1}^{0} P_1(\xi)x(t+\xi)\, d\xi \qquad (62)$$

It is also worthwhile to mention that the equivalence between the modified resolvent $P(t,s,r)$ obtained from the maximum principle and the Fredholm theory and the feedback gains obtained from the Carathéodory lemma was proved in [29] for the case $B_1 = 0$.

An interesting formal approach was used by Soliman and Ray [30] to find the same solution as presented above. This will be discussed in more detail in connection with computational considerations.

A more abstract approach relating the LPQ theory of hereditary systems to that of systems which are described by equations in Hilbert space was discussed in a sequence of papers [17,18,31]. In these papers the existence of a linear operator which relates the state and the adjoint operators is established using a product space. The operator serves the same purpose as the modified resolvent $P(\cdot, \cdot, \cdot)$; that is, it decouples the two-point boundary value problem resulting from the maximum principle. Systems with time-varying delays can also be treated in the framework described; only notations become more complex. Similarly, a hereditary term $\int_{t-1}^{t} A_2(t,s)x(s)\, ds$ can also be included in the state equation; such systems are investigated in [22].

C. Computation of Optimal Solution

The computational difficulties in time-delay systems are more formidable than in systems without delay. For example, computation of the optimal feedback gain $P(\cdot,\cdot,\cdot)$ appears to be such a demanding task due to the partial differential equation (PDE) that only relatively simple problems have been discussed in the literature. Various questions dealing with approximations are still partially unsolved. For example, it is not clear at what point of computations approximations should be done. Recently, computational aspects have received increased attention by many authors, but many important questions are still unanswered.

We will first discuss standard trajectory optimization techniques, which are used mainly to solve the two-point boundary value problem resulting from the maximum principle. Then we will consider various approximation schemes to solve the LPQ problem. Finally, we will discuss the computational aspects of the spectral projection method.

1. Two-point Boundary Value

In general, most of the techniques used for the trajectory optimization for systems without delay can be generalized to include delay systems. Here we consider the computational aspects only; for theoretical investigations of hill-climbing methods in Hilbert space the reader is referred to [32-34].

Consider the problem stated in Section II.A. Assume initially that $u(t)$ is unconstrained, and that the final time T is fixed. The maximum principle then yields (cf. page 257):

(i) Canonical equations

$$\dot{x}(t) = f(x(t),x(t-1),u(t)), \qquad t > t_o \tag{63}$$

$$\begin{aligned}\dot{p}(t) = &-H_x(x(t),x(t-1),u(t),p(t),t)\\ &-H_z(x(t+1),x(t),u(t+1),p(t+1),t+1),\\ &t \in [t_o,T]\end{aligned} \tag{64}$$

(ii) Boundary conditions

$$x(t) = g(t) \qquad \text{on } [t_o-1, t_o] \tag{65}$$

$$p(t) = 0 \qquad \text{for } t \geq T \tag{66}$$

(iii) Minimization of the Hamiltonian

$$H_u(x(t),x(t-1),u(t),p(t),t) = 0 \tag{67}$$

Choose any two of the necessary conditions and iterate on the third condition to satisfy it to a desired degree of accuracy.

For example, assume that (i) and (iii) hold, and iterate on the third condition; that is, the new control u_{i+1} is given by

$$u_{i+1} = u + \epsilon_1 g_i \tag{68}$$

where ϵ_1 is a constant to be chosen by a one-dimensional minimization

$$\min_{\epsilon} C(u_i + \epsilon g_i) = C(u_i + \epsilon_1 g_i) \tag{69}$$

and

$$g_i = H_u(x_i(t), x_i(t-1), u_i(t), p_i(t), t) \tag{70}$$

Choose a small constant δ to determine a stopping condition. A standard first-order gradient algorithm for time-delay systems in the policy space can then be stated as

(a) Algorithm (gradient method).

1. Guess $u_0(t)$ for the interval $[t_0, T]$ and set $i = 0$.
2. Compute $x_i(t)$ by integrating Eq. (63) forward from t_0 to T with the given initial function (65) and control $u_i(t)$.
3. Compute $p_i(t)$ by integrating Eq. (64) backward from T to t_0 with the final function (66), $x_i(t)$, and $u_i(t)$.
4. Compute g_i from Eq. (70). If $\|g_i\| < \delta$, stop; otherwise go to Step 5.
5. Find ϵ_1 from the one-dimensional minimization (69) and set $u_{i+1} = u_i + \epsilon_1 g_i$.
6. Set $i \to i + 1$ and go to Step 2.

Similarly, we can construct a somewhat more efficient conjugate gradient algorithm for time-delay systems:

(b). Algorithm (conjugate gradient).

Perform Steps 1-4 as in the gradient method.

5. Compute $\beta_i = \langle g_i, g_i \rangle / \langle g_{i-1}, g_{i-1} \rangle$, where $\langle a, b \rangle = \int_{t_0}^{T} a'(t) \times b(t)\, dt$. If $i = 0$, set $\beta_0 = 0$.
6. Set $s_i = -g_i + \beta_{i-1} s_{i-1}$.
7. Find ϵ_1 from the one-dimensional search $\min C(u_i + \epsilon s_i) = C(u_i + \epsilon_1 s_i)$. Set $u_{i+1} = u_i + \epsilon_1 s_i$.

8. Set $i \rightarrow i+1$ and go to Step 2.

A method using second variations was constructed in [35], and a similar method for linear systems was discussed in [36].

The aforementioned methods have features which are typical in case of no delays. But in general they tend to be more time-consuming. If u is constrained, this might introduce singular arcs. These aspects were studied in [37,38] by using both gradient and conjugate gradient methods. A clipping-off gradient algorithm of [39] was tested in [40] and was found to be quite effective. This method is essentially the same as the conjugate-gradient method except for the Step 7, which is changed to

$$\epsilon_I(t) = \begin{cases} 0 & \text{if } t \text{ belongs to any saturated interval} \\ \epsilon_1 & \text{otherwise} \end{cases}$$

Other approaches that have been suggested are those based on the singular perturbation method [41,42]. A closely related approach is a parameter imbedding method [43] and a technique based on asymptotic series solution for systems with small delays [44].

2. LPQ Problem

The methods of the previous section can also be applied to the computation of the solution of the LPQ problem; the resulting control will be in the open-loop form. On the other hand, feedback gains could be computed using hill-climbing methods, which yield reasonable results [21].

After finding the structure of the optimal feedback controller, we still have to solve either the generalized Riccati equations, which are a set of coupled PDEs (56) to (58), henceforth called S, or an integral equation (55). A straightforward finite difference scheme is proposed in [45] to solve S. Another method suggested in [16] is to consider $P(t,s,r)$ as a function of only one variable r by fixing t and s; and thus treating S as a set of ODEs. Utilizing the characteristics of S, the two methods are combined and a more efficient method can be obtained [46].

The numerical solution of the integral equation (55) has not been investigated to any great degree. The problem is that we first need the fundamental solution $X(t,s)$ in order to compute $M(t,s)$; if the time interval is long, the standard discretization methods [47] yield a high-order algebraic equation. A simple Bernstein quadrature rule was used in [22] to discretize the right-hand side of equations

$$P(t,s,t) = M(t,s) - \int_t^T P(t,\alpha,t)K'(s,\alpha)\, d\alpha$$

$$P(t+1,\ s,\ t) = M(t+1,\ s) - \int_t^T P(t+1,\ \alpha,\ t)K'(s,\alpha)\, d\alpha$$

Then, giving different values for t and s, $t \in [t_0,T]$, $s \in [t,\ t+1]$, the resulting set of linear equations can be solved by any standard linear equation solver routine. The method works well when $B_1 = 0$ and the time interval is short. When $B_1 \neq 0$, the kernel $K(s,\alpha)$ is no longer continuous, and great care has to be exercised in computations.

Here we have started from one extreme, where no approximations are made before the final result is known. On the other hand, we could approximate the delay term (or terms) immediately. Perhaps the first one to suggest this was Salukvadze [48]. We will now investigate this question following the presentation of [30].

Let $t_0 = 0$ for simplicity and consider

$$\frac{\partial\eta(t,s)}{\partial t} + \frac{1}{\tau}\frac{\partial\eta(t,s)}{\partial s} = 0, \qquad t > 0,\ s \in [0,1]$$

Then a solution to the PDE is any function $\eta(t,s) = h(t-\tau s)$. If we consider the boundaries $s = 1$ and $s = 0$, we have $\eta(t,1) = h(t-\tau)$ and $\eta(t,0) = h(t)$, respectively. In particular, if we let $h(t) = x(t)$, we can write [30] the delay-differential equation (40), with $B_1 = 0$ for convenience, as

$$\dot{x}(t) = A_0x(t) + A_1y(t) + B_0u(t)$$

$$\frac{\partial\eta(t,s)}{\partial t} + \frac{1}{\tau}\frac{\partial\eta(t,s)}{\partial s} = 0, \qquad t > 0,\ s \in [0,1]$$

where

$$\eta(0,s) = h(-s\tau), \qquad s \in [0,1]$$

$$\eta(t,0) = x(t)$$

$$\eta(t,1) = x(t-\tau) = y(t)$$

If we approximate only the spatial derivatives by a finite-difference scheme, we obtain

$$\frac{\partial \eta(t,s_i)}{\partial t} = -\frac{1}{\tau}\,\frac{\eta(t,s_i) - \eta(t,s_{i-1})}{1/N}, \qquad i = 1,2, \ldots ,N$$

where $s_i = i/N$. We have now approximated the infinite-dimensional system with a finite-dimensional one. As N is made larger, the accuracy of the approximation is improved. The finite-dimensional system can be written as

$$\dot{\hat{x}} = A_N\hat{x} + B_N u$$

where $\hat{x} = [x, \eta_1, \ldots, \eta_i, \ldots, y(t)]'$

$$A_N = \begin{bmatrix} A_0 & 0 & \cdots & 0 & A_1 \\ \frac{NI_n}{\tau} & -\frac{NI_n}{\tau} & \cdots & 0 & 0 \\ \vdots & \vdots & & \vdots & \vdots \\ 0 & 0 & \cdots & \frac{NI_n}{\tau} & -\frac{NI_n}{\tau} \end{bmatrix} \qquad B_N = \begin{bmatrix} B \\ 0 \\ \vdots \\ 0 \end{bmatrix}$$

The LPQ problem for delay systems is treated by this method in [30], when $B_1 \neq 0$. A heuristic result is obtained which is the same as the one we derived in Section II.B. This construction is immediately applicable for computations. It should also be noted that the Nth approximation offers good starting values for the (N+1)st. The same approximation has also been used in [49], where a fourth-order infinite-time LPQ problem is solved. Recently, Banks and Burns [50] gave some convergence results concerning this approximation.

Delfour [51] has gone a step further by discretizing not only the spatial derivative, but also the time derivative (when $B_1 = 0$). He is able to prove the convergence of the method, which was lacking in [30]. Previous investigators [52-54] have also discretized the system equations and treated the design problem completely as a discrete-time problem. The problem is then computationally straightforward, and results of good convergence have been reported experimentally.

For linear time-invariant problems, the spectral projection method is quite useful. As will be indicated in the following section, the computation of the zeros of $\det(sI - A_0 - A_1e^{-s\lambda}) = 0$ in $\mathcal{C}_+$ is required. Using this method, the LPQ problem was treated in [55] and the convergence is also proved; that is, if the vertical line in $\underline{C}$ is pushed farther to the left, more modes are included; and finally, in the limit, the modal approximation can be made as close to the original infinite-dimensional solution as desired. In simpler cases, the eigenvalues can be computed, e.g., by Newton's method.

3. Numerical Examples

Example 1. The Zn-flotation circuit at Lake Dufault (Noranda, Quebec, Canada) is described by the following dynamic equations [40]

$$\dot{x}_1(t) = -0.43x_1(t) + 2.15x_3(t-T) - u_1(t)x_1(t) + u_2(t)x_3(t-T) - 27.2u_1(t) + 2.34u_2(t)$$

$$\dot{x}_2(t) = -0.21x_2(t) + 0.03x_4(t-T) - 0.20u_1(t)x_2(t) + 0.01u_2(t) \times x_4(t-T) - 471u_1(t) - 13.3u_2(t)$$

$$\dot{x}_3(t) = 0.2x_1(t) - 2.31x_3(t) - u_2(t)x_3(t) - 2.34u_2(t)$$

$$\dot{x}_4(t) = 0.2x_2(t) - 0.20x_4(t) - 0.01u_2(t)x_4(t) - 13.26u_2(t)$$

with the initial conditions $[x_1(t),x_2(t),x_3(t),x_4(t)]' = [8.2,-11.7,0.7,-12.0]'$ for $t \le 0$. The initial function is to be steered to zero by minimizing the cost

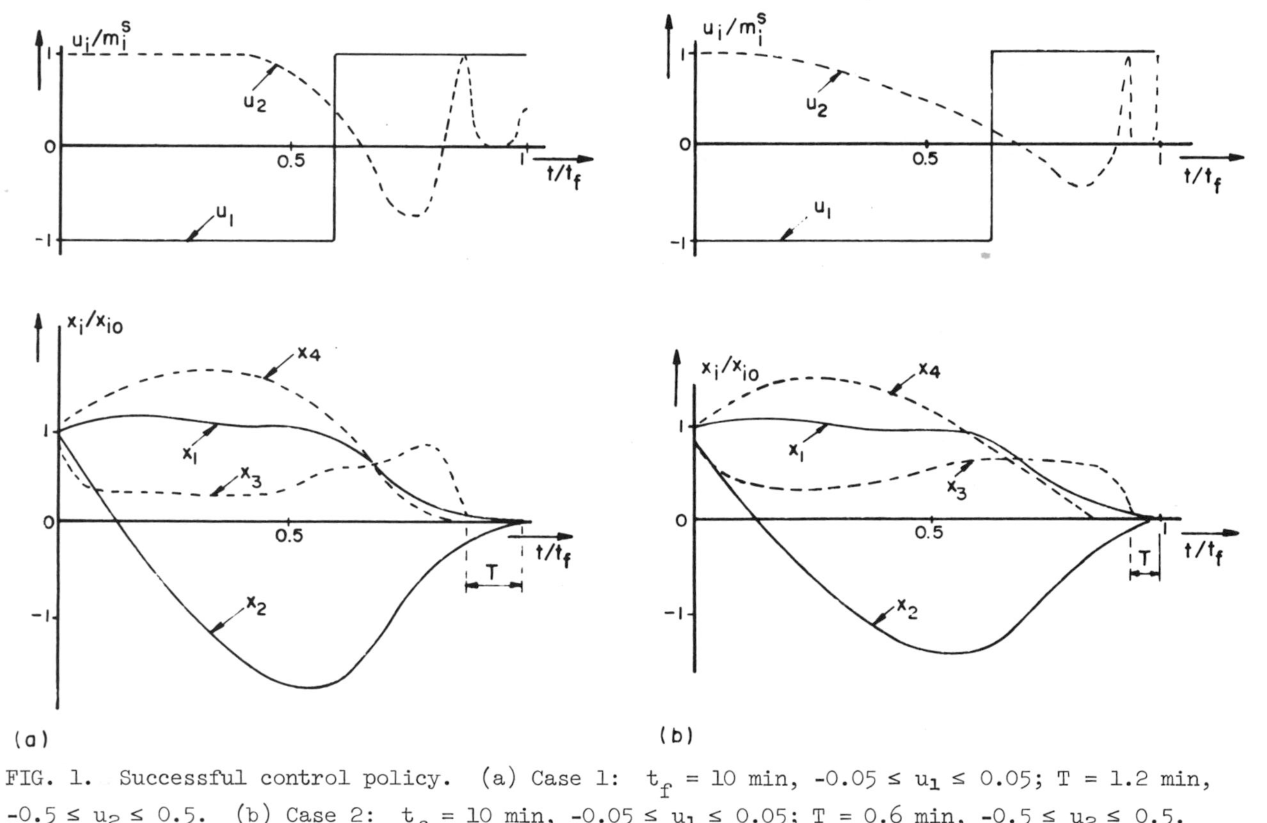

FIG. 1. Successful control policy. (a) Case 1: $t_f = 10$ min, $-0.05 \leq u_1 \leq 0.05$; $T = 1.2$ min, $-0.5 \leq u_2 \leq 0.5$. (b) Case 2: $t_f = 10$ min, $-0.05 \leq u_1 \leq 0.05$; $T = 0.6$ min, $-0.5 \leq u_2 \leq 0.5$.

$$C(u) = \int_0^{10} (x_3(s) + \{0.01x_2(s) + [27.2 + x_2(s)]0.05u_1(s)\})\ ds$$

The cost function was constructed so as to minimize losses. The control components were bounded as follows: $|u_1| \le 0.05m_1$, $|u_2| \le 0.5m_2$. The delay was estimated to be 0.6 min, but another run was performed with 1.2 min. The terminal constraint was handled by adding a penalty function to the cost.

When the clipping-off gradient method was used, the results of Figure 1 were attained. All components of x-vector are fairly close to zero at the final time. Control u_1 is a pure bang-bang control, whereas u_2 exhibits a singular arc. The slight fuzziness at the end is explained by the penalty term in the cost.

Example 2 [16]. Consider the system

$$\dot{x}(t) = x(t) + x(t-1) + u(t)$$

$$x(t) = 1, \quad t \in [1,0]$$

subject to the performance cost

$$C(u) = \int_0^2 (x^2 + u^2)\ dt$$

The optimal feedback is given by

$$u(x_t) = -P(t,t,t)x(t) - \int_{-1}^{0} P(t,\ s+\tau,\ t)x(s)\ ds$$

In [16] the generalized Riccati equations (60) are solved by changing them into ODEs with parameters. On the other hand, the integral equation (54) for P(t,s,t) is solved in [21] and the results are identical (Fig. 2).

D. Observers

In most practical problems, all components in the state vector cannot be measured directly; for example, they are not accessible for measurements. In such cases, an observer may be used to reconstruct the

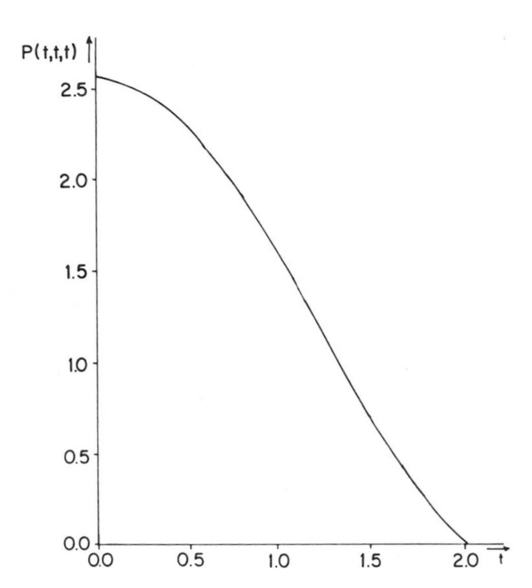

FIG. 2. The feedback gain P(t,t,t) vs time.

state vector if certain conditions are fulfilled. An observer yields asymptotically estimates of the state vector.

As pointed out in Section II.B, the stabilizing feedback controller takes the form

$$u(x_t) = P_0 x(t) + \int_{-1}^{0} P_1(\xi) A_1 x(t+\xi)\, d\xi$$

Thus, the control law depends on the state x_t. Often the state x_t is not available for measurements and must be estimated from the available data by some device, commonly called an observer. Roughly speaking, an observer is a device which estimates the state (in this case x_t) with an error which approaches zero asymptotically. Observers were first discussed by Luenberger [56]; excellent summaries of the current state of the knowledge appear in [4,57].

The theory of observers for linear constant ODEs is well developed, but attempts to extend the observer theory for time-delay systems have only produced partial results. By formally mimicking the procedure for ODEs, the observer for the time-delay system (1), (13) (with $u \equiv 0$) becomes

$$\dot{z}(t) = A_0 z(t) + A_1 z(t-1) + K[-y(t) + Cz(t)]$$

where the matrix K has to be chosen so that $x(t) - z(t) \to 0$ as $t \to \infty$. Unfortunately, it appears that this is possible only in some trivial cases. Indeed, the error $e(t) = x(t) - z(t)$ evolves according to

$$\dot{e}(t) = (A_0 + KC)e(t) + A_1 e(t-1)$$

It is difficult in general to choose the matrix K so that $x(t) - z(t) \to 0$ asymptotically, since the stability is determined by the roots of $\det(\lambda I - A_0 - KC - A_1 e^{-\lambda}) = 0$. It is now fairly obvious that using only a matrix operation on $y - Cz$ is not sufficient. As is seen in Section II on the LPQ problem, the optimal feedback control needs the entire state x_t. Hence, it seems intuitively that the operator K also needs all the information about the state over the previous delay interval. In order to construct a stable observer, another more transparent formalism is chosen, which tends to "lump" the matrices A_0 and A_1 into one. We can accomplish this by the infinitesimal generator $\mathcal{A}$ mentioned in Section I.C.

By using the decomposition of the state space given in Section I.C, Bhat and Koivo [58] were able to develop a fairly satisfactory observer theory for time-delay systems. It is encouraging to note that the Kalman filter discussed in Section III.A has a structure similar to that of an observer.

1. Observer Structure

Recall from Section I.C the equations determining the reduced system corresponding to Λ (with $u \equiv 0$):

$$\dot{x}_1(t) = A_\Lambda x_1(t) \tag{71}$$

$$x_1(0) = \langle \Psi_\Lambda, g \rangle \tag{72}$$

$$y(t) = C_\Lambda x_1(t) + C_2 x_t^2 \tag{73}$$

If the data for only the finite-dimensional part are used, the observability of the pair (C_Λ, A_Λ) implies that there exists a matrix K such that the spectrum of $A_\Lambda + KC_\Lambda$ agrees with any given symmetric set of N_Λ complex numbers [4]. What is needed now is to extend the operator K for the whole state space $\mathcal{C}$. The extension is constructed so that the unstable modes are stabilized and the stable ones are left unperturbed.

We can achieve this by defining the operator $\mathcal{K}: \underline{E}^p \to \mathcal{C}$ by $\mathcal{K}(y(t)) = \Phi_\Lambda Ky(t)$, where the matrix K stabilizes $A_\Lambda + KC_\Lambda$. The operator $\mathcal{K}$ is exactly the proper extension of K that is being sought.

Theorem 7. (Bhat and Koivo [58].) The operator $\mathcal{A} + \mathcal{K}CP$ defined on $\mathcal{D}(\mathcal{A})$ is the infinitesimal generator of a strongly continuous asymptotically stable semigroup.

Remark 1. The assumption needed in the proof of Theorem 7 is the observability of the modal matrix pair (C_Λ, A_Λ). It would be much more satisfactory to have this assumption in terms of the original data A_0, A_1, and C. Recently, such a link has been obtained by Bhat and Koivo [59]. It can be stated as the following theorem:

Theorem 8. The reduced system corresponding to Λ is observable if and only if $\operatorname{Ker}(A_0 + A_1e^{-\lambda} - \lambda I) \cap \operatorname{Ker} C = 0$, $\lambda \in \Lambda$, or equivalently

$$\operatorname{Rank} \begin{bmatrix} A_0 + A_1e^{-\lambda} - \lambda I \\ C \end{bmatrix} = n, \qquad \lambda \in \Lambda \tag{74}$$

Of course, the result is not totally satisfactory, since we are forced to compute all λ in Λ in order to check the rank condition, but it is esthetically more pleasing and a bit easier than the modal matrix assumption.

Using Theorem 7, an observer for system (9) can now be constructed. It has the following structure

$$\dot{z}_t = (\mathcal{A} + \mathcal{K}CP)z_t - \mathcal{K}y(t), \qquad z_0 = h \in \mathcal{D}(\mathcal{A})$$

or, alternatively,

$$\dot{z}_t = \mathcal{A}z_t - \Phi_\Lambda K(y - CPz_t), \qquad z_0 = h \in \mathcal{D}(\mathcal{A}) \tag{75}$$

The error $e_t = x_t - z_t$ is governed by the equation

$$\dot{e}_t = (\mathcal{A} + \mathcal{K}CP)e_t, \qquad e_0 = g - h \in \mathcal{D}(\mathcal{A})$$

and by Theorem 8, $e_t \to 0$ as $t \to \infty$, proving that z_t is an observer.

For computational purposes the form of Eq. (75) is not very useful. In order to establish a better starting point for computations a rather formal approach is followed. It is first observed that

$$(\mathcal{A} + \mathcal{K}CP)z_t(\xi) = (\mathcal{A} + \Phi_\Lambda KCP)z_t(\xi) = \begin{cases} \dfrac{dz_t(\xi)}{d\xi} + \Phi_\Lambda(\xi)KCz_t(0), & \xi \in [-1,0) \\ (A_0 + \Phi_\Lambda(0)KC)z_t(0) + A_1 z_t(-1), & \xi = 0 \end{cases} \tag{76}$$

When $\tilde{z}(t,\xi) := z_t(\xi)$, we see that a first-order hyperbolic differential equation

$$\frac{\partial\tilde{z}(t,\xi)}{\partial t} = \frac{\partial\tilde{z}(t,\xi)}{\partial\xi} + \Phi_\Lambda(\xi)KC\tilde{z}(t,0) - \Phi_\Lambda(\xi)Ky(t), \qquad \xi \in [-1,0) \tag{77}$$

with the boundary condition

$$\frac{\partial\tilde{z}(t,0)}{\partial t} = (A_0 + \Phi_\Lambda(0)KC)\tilde{z}(t,0) + A_1\tilde{z}(t,-1) - \Phi_\Lambda(0)Ky(t), \qquad \xi = 0 \tag{78}$$

and with the initial condition

$$\tilde{z}(0,\xi) = h(\xi), \qquad \xi \in [-1,0]$$

obtains. A more thorough treatment of this and some other related topics is given in [60].

The characteristics of Eq. (77) are given by $dt/d\xi = -1$ or $t = -\xi + c$, where c is a constant. Along the characteristics Eq. (77) first yields

$$\frac{d\tilde{z}}{d\xi}(c-\xi, \xi) = -\Phi_\Lambda(\xi)KC\tilde{z}(c-\xi, 0) + \Phi_\Lambda(\xi)Ky(c-\xi)$$

or, integrating this from 0 to θ, we have

$$\tilde{z}(c-\theta, \theta) = \tilde{z}(c,0) - \int_0^\theta \Phi_\Lambda(\xi)[KC\tilde{z}(c-\xi, 0) - Ky(c-\xi)]\, d\xi$$

Now setting $c = \theta + t$ and $\theta = -1$ results in

$$\tilde{z}(t,-1) = \tilde{z}(t-1, 0) - \int_0^{-1} \Phi_\Lambda(\xi)[KC\tilde{z}(t-1-\xi, 0) - Ky(t-1-\xi)]\, d\xi$$

Substituting this expression for $\tilde{z}(t,-1)$ in Eq. (78) and setting $\tilde{z}(t+\theta,0) = z(t+\theta)$ results in

$$\dot{z}(t) = A_0 z(t) + A_1 z(t-1) + \Phi_\Lambda(0)K[Cz(t) - y(t)]$$

$$+ A_1 \int_0^1 \Phi_\Lambda(\xi-1)K[Cz(t-\xi) - y(t-\xi)]\, d\xi \qquad (79)$$

The hereditary type of equation for the observer clearly confirms the earlier conjecture that the correction term has to include all the data over the interval [t-1, t], that is, an integral correction term is needed. If no delay term appears, i.e., $A_1 = 0$, then the observer reduces to the one for the corresponding ordinary linear constant differential equation system.

Remark 2. If the observation is given by $y(t) = C_1 x(t) + C_2 x(t-1)$, the observer equation changes to

$$\dot{z}(t) = A_0 z(t) + A_1 z(t-1) + \Phi_\Lambda(0)K[C_1 z(t) + C_2 z(t-1) - y(t)]$$

$$- [A_1 + \Phi_\Lambda(0)KC_2] \int_0^1 \Phi_\Lambda(\xi-1)K[C_1 z(t-\xi) + C_2 z(t-1-\xi)$$

$$- y(t-\xi)]\, d\xi$$

Remark 3. An observer theory has also been developed for general systems characterized by semigroups [61] using spectral decomposition techniques.

The above form of the observer (79) requires the computation of the unstable modes; that is, the roots of $\det(\lambda I - A_0 - A_1 e^{-\lambda}) = 0$, which lie in the right half-plane. Then the matrix Φ_Λ can be constructed with some further computations. Computational aspects related to the spectral projection method were already discussed in Section II.C.

Example. Let the system be described by

$$\dot{x}(t) = \begin{bmatrix} 0 & 0 \\ 1 & 1 \end{bmatrix} x(t) + \begin{bmatrix} -\frac{\pi}{2} & 0 \\ 0 & 0 \end{bmatrix} x(t-1)$$

$$y(t) = [0 \quad 1]\, x(t)$$

The eigenvalues in $\mathcal{C}_+$ are

$$\lambda_{1,2,3} = \left\{ i\,\frac{\pi}{2},\ -i\,\frac{\pi}{2},\ 1 \right\}$$

The corresponding eigenfunctions

$$\varphi_1(\xi) = \left[\left(\cos \frac{\pi}{2}\xi + \frac{\pi}{2} \sin \frac{\pi}{2}\xi \right),\ -\cos \frac{\pi}{2}\xi \right]',$$

$$\varphi_2(\xi) = \left[\left(\sin \frac{\pi}{2}\xi - \frac{\pi}{2} \cos \frac{\pi}{2}\xi \right),\ -\sin \frac{\pi}{2}\xi \right]',\quad \varphi_3(\xi) = [0, e^{\xi}]'$$

form a basis to $\oplus_{i=1}^{3} \operatorname{Ker}(\mathcal{A} - \lambda_i I)$. Thus, the matrix $\Phi_\Lambda(\xi) = [\varphi_1(\xi), \varphi_2(\xi), \varphi_3(\xi)]$. The matrices A_Λ and C_Λ can be computed:

$$A_\Lambda = \begin{bmatrix} 0 & \frac{\pi}{2} & 0 \\ -\frac{\pi}{2} & 0 & 0 \\ 0 & 0 & 0 \end{bmatrix}, \qquad C_\Lambda = [-1 \ 0 \ 1]$$

The observability of the pair (C_Λ, A_Λ) is obvious, and thus a matrix K can be found for stability of $(A_\Lambda + KC_\Lambda)$. For example, let $K = [2 \ 1 \ -3]'$. It is also interesting to check the rank condition with the initial data A_0, A_1, and C, and the computed eigenvalues. Now

$$\operatorname{Rank} \begin{bmatrix} A_0 + A_1 e^{-\lambda} - \lambda I \\ C \end{bmatrix} = \operatorname{Rank} \begin{bmatrix} -\frac{\pi}{2} e^{-\lambda} - \lambda & 0 \\ 1 & 1-\lambda \\ \hline 0 & 1 \end{bmatrix} = 2 \text{ for all } \lambda$$

so that, according to Theorem 8, this implies the observability of the pair (C_Λ, A_Λ).

The complete observer equation is now given by

$$\dot{z}(t) = \begin{bmatrix} 0 & 0 \\ 1 & 1 \end{bmatrix} z(t) + \begin{bmatrix} -\frac{\pi}{2} & 0 \\ 0 & 0 \end{bmatrix} z(t-1) \begin{bmatrix} 2 - \frac{\pi}{2} \\ -5 \end{bmatrix} [y(t) - z_2(t)]$$

$$+ \begin{bmatrix} -\frac{\pi}{2} \int_0^1 [(2 - \frac{\pi}{2}) \cos \frac{\pi}{2} (\xi - 1) + (1+\xi) \sin \frac{\pi}{2} (\xi-1)][z_2(t-\xi) - y(t-\xi)] \, d\xi \\ 0 \end{bmatrix}$$

2. Observer in the Closed-Loop System

Consider the system

$$\dot{x}(t) = A_0 x(t) + A_1 x(t-1) + Bu(t) \tag{80}$$

Suppose it is desired to realize a dynamic behavior corresponding to a control

$$u(x_t) = P_0 x(t) + \int_{-1}^{0} P_1(\xi) A_1 x(t+\xi) \, d\xi \tag{81}$$

Suppose that the state x_t is not directly available, but $y(t) = Cx(t)$ can be measured. Then the control has to be synthesized by means of an observer.

Since for the observer $x(t) - z(t) \to 0$ as $t \to \infty$, the control u in (81) is set to be

$$u(z_t) = P_0 z(t) + \int_{-1}^{0} P_1(\xi) A_1 z(t+\xi) \, d\xi \tag{82}$$

Using the observer (75), the control (82), and setting $e = x - z$, the combined closed-loop system becomes

$$\dot{x}(t) = (A_0 + BP_0) x(t) + A_1 x(t-1) + \int_{-1}^{0} BP_1(\xi) A_1 x(t+\xi) \, d\xi$$

$$+ BP_0 e(t) + B \int_{-1}^{0} P_1(\xi) A_0 e(t+\xi) \, d\xi$$

$$\dot{e}(t) = (A_0 + \Phi_\Lambda(0)KC)e(t) + A_1e(t-1) + A_1 \int_0^1 \Phi_\Lambda(\xi-1)Ke(t-\xi)\, d\xi$$

Thus the spectrum of the combined system matrix can be separated into

$$\sigma(A_0 + BP_0 + A_1e^{-s} + B \int_{-1}^{0} P_1(\xi)A_1e^{s\xi}\, d\xi) \,\dot{\cup}\, \sigma(A_0 + \Phi_\Lambda(0)KC$$

$$+ A_1e^{-s} + A_1 \int_{-1}^{0} \Phi_\Lambda(\theta)Ke^{-s\theta}\, d\theta)$$

where $\dot{\cup}$ denotes the union with any common elements repeated. A similar result was proven by Gressang [62] for the observer he developed.

III. STATE ESTIMATION

When the measurements are noise-corrupted, a state estimator is needed to compute the value of the state vector. Most results reported on the state estimation problem have been obtained for the case where system equations are linear and noise processes in the model can be described by Gaussian characteristics. The well-known result of Kalman and Bucy [63] specifying the optimal filter when the plant and the observation equations are linear and the additive noise processes are white and Gaussian can be extended to linear time-delay systems.

The state estimation problem in systems with time delays has been studied by several authors [64-67]. When a time-delay system is described by linear plant and observation equations containing additive white Gaussian noise processes, the optimal estimator equations have been derived by various approaches, e.g. [64-66]. The method of orthogonal projections is used in [64] to obtain the optimal estimator equations for linear time-delay systems. An alternative derivation of the optimal estimator equations by means of the matrix minimum principle is presented in [65]. The method of Fredholm's resolvent is applied in [66] to the linear problem with time-varying delays to derive the equations for the least squares estimator (or the maximum likelihood estimator). A general approach to the state estimation problem is presented in [68] to obtain the estimator that minimizes the expected value of the squared estimation error for infinite-dimensional systems.

The estimation of state variables in a nonlinear system represents a difficult problem both theoretically and computationally. A common approach to the nonlinear estimation problem is to linearize the nonlinear functions of the system model about a nominal (usually mean) value and assume that the Gaussian characteristics of the noise processes are maintained when the signals pass through the system. The resulting filter is often referred to as the extended Kalman filter, which is applicable in a neighborhood of the nominal trajectory. A general optimal filter is presented in [69], but computational difficulties impose severe restrictions.

A. Linear Systems with Time Delay

When the mathematical model of a system is described by ODEs with additive white Gaussian noise processes, the Kalman-Bucy filter is the optimal filter in the sense that it minimizes the expected value of the squared estimation error as well as the variance of the estimation error; it also represents the maximum likelihood estimator. Thus, the linear (additive white) Gaussian problem can be posed and accordingly solved by several approaches. These methods can also be applied to the linear (additive white) Gaussian problem in which the model is described by linear differential-difference equations.

We present here a formal derivation of the optimal estimator for linear time-delay systems involving additive white Gaussian noise processes. The optimal estimator is derived so as to minimize the expected value of the squared estimation error. This criterion of optimality is used in [64,65,68]; our derivation here is similar to [64,68].

1. Problem Statement

Let the system be described by the (formal) vector differential-difference equation

$$\dot{x}(t) = \sum_{i=0}^{1} A_i(t)x(t-\tau_i) + D(t)\xi(t) \tag{83}$$

$$z(t) = \sum_{i=0}^{1} C_i(t)x(t-\tau_i) + \eta(t) \tag{84}$$

where $t \geq t_0$; the delays are constants: $\tau_0 = 0$, $\tau_1 = \tau > 0$; $x(t)$ is an n-vector evaluated at time t. The matrices $A_i(t)$, $D(t)$, $C_i(t)$, $i = 0,1$ have continuous and bounded elements for all $t \geq t_0$; the dimensions are $n \times n$, $n \times r$, and $m \times n$, respectively. The plant noise $\xi(t)$ is an r-vector of Gaussian random forcing function satisfying

$$E\{\xi(t)\} = 0; \qquad E\{\xi(t_2)\xi'(t_1)\} = Q(t_2)\delta(t_2-t_1) \tag{85}$$

where $Q(t) > 0$ $\forall t$, $\delta(t)$ is the Dirac delta function, and $E\{\cdot\}$ signifies the expectation operation. The measurement vector z is of dimension m. The measurement noise $\eta(t)$ represents a white Gaussian noise process with

$$E\{\eta(t)\} = 0; \qquad E\{\eta(t_2)\eta'(t_1)\} = R(t_2)\delta(t_2-t_1) \tag{86}$$

The plant and observation noises are mutually independent and also independent upon the initial function $x(t)$, $t \in [t_0-\tau, t_0]$.

The problem is to determine the optimal estimate $\bar{x}_t(\theta) = \bar{x}(t,\theta)$ given the measurement $z(t)$, $t \geq t_0$, so as to minimize the expected value of the squared estimation error.

2. Determination of the Optimal Estimator

We will look for the optimal estimate by performing a linear operation on the measured data:

$$\bar{x}(t,\theta) = \int_{t_0}^{t} K(t,\theta,\sigma)z(\sigma)\, d\sigma \tag{87}$$

where $K(t,\theta,\sigma)$ a kernel function defined on $t \geq t_0, \sigma, \theta \geq 0$ is to be specified. It will be determined so that we minimize the expected value of the squared error conditioned on the observations:

$$\epsilon^2(t) = E\{\ \|x(t-\theta) - \bar{x}(t,\theta)\|^2\ \} \tag{88}$$

where $\|y\|^2 = y'y$; $E\{\cdot\}$ denotes the expectation operation conditioned on the measurements.

We observe first that using the trace of a matrix, Eq. (88) can be rewritten as

$$\epsilon^2(t) = E\{tr[x(t-\theta) - \bar{x}(t,\theta)][x(t-\theta) - \bar{x}(t,\theta)]'\} \tag{89}$$

To minimize $\epsilon^2(t)$, it is convenient to define formally an operator $\mathcal{L}$ and a column vector w by

$$\mathcal{L}[K(t,\theta,\sigma)] := K(t,\theta,\sigma)R(\sigma) + \int_{t_0}^{t} K(t,\theta,\rho) \sum_{i,j} C_i(\rho) \times E[x(\rho-\tau_i)x'(\sigma-\tau_j)]C_j'(\sigma)\, d\rho \tag{90}$$

$$w := \sum_i E[x(t-\theta)x'(\sigma-\tau_i)]C_i'(\sigma)$$

and denote a scalar product by $\langle\cdot, \cdot\rangle$. The substitution of (87) for $\bar{x}(t,\theta)$ and (84) for $z(\sigma)$ in (89) leads to

$$\epsilon^2(t) = tr\{E[x(t,\theta)x'(t-\theta)] + \langle\mathcal{L}[K(\cdot)],K(\cdot)\rangle - 2\langle w,K(\cdot)\rangle\} \tag{91}$$

Since $\mathcal{L}$ is positive definite, the minimization of ϵ^2 is now a standard quadratic variational problem, for which the unique solution is

$$\mathcal{L}[K^*(t,\theta,\sigma)] = w$$

or equivalently,

$$K^*(t,\theta,\sigma)R(\sigma) + \int_{t_0}^{t} d\rho K^*(t,\theta,\rho) \sum_{i,j} C_j(\rho)E[x(\rho-\tau_j)x'(\sigma-\tau_i)]C_i'(\sigma) = \sum_i E[x(t-\theta)x'(\sigma-\tau_i)]C_i'(\sigma) \tag{92}$$

where the star signifies optimum. It will be dropped in the sequel. It is postulated that $K(\cdot)$ can be written

$$K(t,\theta,\sigma) = \sum_{n=0}^{1} M(t, \theta, t+\tau_n-\sigma)C_n'(\sigma)R^{-1}(\sigma) \tag{93}$$

Substituting Eq. (93) into (92), we obtain after setting $\theta = \theta_1$, $\theta_2 = t-\sigma+\tau_i$

$$M(t,\theta_1,\theta_2) + \int_{t_0}^{t} d\rho \left[\sum_n M(t,\theta_1,t+\tau_n-\rho) \sum_j C_n'(\rho)R^{-1} \right]$$
$$\times\, C_j(\rho)E[x(\rho-\tau_j)x'(t-\theta_2)] = E[x(t-\theta_1)x'(t-\theta_2)] \tag{94}$$

In order to relate $M(t,\cdot,\cdot)$ to the covariance of the estimation error, we form

$$E\{[x(t-\theta_1) - \bar{x}(t,\theta_1)][x(t-\theta_2) - \bar{x}(t,\theta_2)]'\} = E\{x(t-\theta_1)x'(t-\theta_2)\}$$
$$- \int_{t_0}^{t} K(t,\theta_1,\rho) \sum_j C_j(\rho)E\{x(\rho-\tau_j)x'(t-\theta_2)\}\, d\rho \tag{95}$$

where $K(t,\theta,\rho)$ is specified by expression (93). A comparison of Eqs. (95) and (94) reveals that $M(t,\theta_1,\theta_2)$ is equal to the covariance in question. Moreover, the minimum value of ϵ^2 in Eq. (88) can be manipulated using Eqs. (92) to (94) to yield $\epsilon^2(t) = \mathrm{tr}[M(t,\theta,\theta)]$.

The equations specifying the evolution of the error covariance in the form of PDEs can be obtained from Eq. (94). If we set $\theta_1 = \theta_2 = 0$ in Eq. (94), we obtain (96) for $dM(t,0,0)/dt$ by making use of (83) and substituting expression (94) for $E\{x(t)x'(t)\}$, collecting the terms, and utilizing the arguments of positive definiteness of the matrices [63,64]. In a similar fashion, we can form $\partial M(t,\theta_1,0)/\partial t + \partial M(t,\theta_1,0)/\partial\theta$ and $\partial M(t,\theta_1,\theta_2)/\partial t + \partial M(t,\theta_1,\theta_2)/\partial\theta_1 + \partial M(t,\theta_1,\theta_2)/\partial\theta_2$. The resulting equations are:

$$\frac{dM(t,0,0)}{dt} = \sum_{i=0}^{1} [A_i M(t,\tau_i,0) + M(t,0,\tau_i)A_i']$$
$$- \sum_{i,k=0}^{1} M(t,0,\tau_i)C_i'R^{-1}C_k M(t,\tau_k,0) + D(t)QD'(t) \tag{96}$$

$$\frac{\partial M(t,\theta_1,0)}{\partial t} + \frac{\partial M(t,\theta_1,0)}{\partial\theta_1} = \sum_{i=0}^{1} M(t,\theta_1,\tau_i)A_i'$$
$$- \sum_{i,k=0} M(t,\theta_1,\tau_k)C_k'R^{-1}C_i M(t,\tau_i,0) \tag{97}$$

$$\frac{\partial M(t,\theta_1,\theta_2)}{\partial t} + \frac{\partial M(t,\theta_1,\theta_2)}{\partial \theta_1} + \frac{\partial M(t,\theta_1,\theta_2)}{\partial \theta_2}$$

$$= - \sum_{i,k=0}^{1} M(t,\theta_1,\tau_k)C_k'R^{-1}C_i M(t,\tau_i,\theta_2) \tag{98}$$

By the definition of $M(\cdot)$, it is clear that $M'(t,\theta,0) = M(t,0,\theta)$; therefore the equation for $M(t,\theta,\theta_2)$ is similar to that of $M(t,\theta_1,0)$.

The optimal estimate $\overline{x}(t,\theta)$ of $x(t-\theta)$ is specified by Eq. (87). For an on-line computation, it is desirable to have a (partial) differential equation for $\overline{x}(t,\theta)$. It can readily be obtained if we first form the (partial) differential equations for $K(\cdot)$ from Eq. (93):

$$\frac{\partial K(t,0,\sigma)}{\partial t} = \sum_i A_i K(t,\tau_i,\sigma) - K(t,0,t) \sum_i C_i K(t,\tau_i,\sigma) \tag{99}$$

$$\frac{\partial K(t,\theta,\sigma)}{\partial t} + \frac{\partial K(t,\theta,\sigma)}{\partial \theta} = -K(t,\theta,t) \sum_i C_i K(t,\tau_i,\sigma) \tag{100}$$

Having Eqs. (99) and (100), the optimal estimator equations are obtained directly from (87). If we set $\theta = 0$ and differentiate with respect to t, we obtain (after replacing $\overline{x}(t,0)$ by $\hat{x}(t, 0 \mid t)$ to emphasize that the measurement is available up to t):

$$\frac{d\hat{x}(t, 0 \mid t)}{dt} = \sum_{i=0}^{1} A_i \hat{x}(t, \tau_i \mid t) + K(t,0,t) \Big[z(t) - \sum_{i=0}^{1} C_i \hat{x}(t, \tau_i \mid t) \Big] \tag{101}$$

where

$$K(t,0,t) = \sum_{n=0}^{1} M(t,0,\tau_n)C_n'(t)R^{-1}(t)$$

By differentiating Eq. (87) with respect to t and θ, we obtain (again changing the notation to $(\hat{x}(t, \theta \mid t))$

$$\frac{\partial \hat{x}(t, \theta \mid t)}{\partial t} + \frac{\partial \hat{x}(t, \theta \mid t)}{\partial \theta} = K(t,\theta,t)\left[z(t) - \sum_{i=0}^{1} C_i \hat{x}(t, \tau_i \mid t)\right] \tag{102}$$

where $K(t,\theta,t) = \Sigma_{n=0}^{1} M(t,\theta,\tau_n)C_n'(t)R^{-1}(t)$. Thus, Eqs. (101) and (102) specify the optimal linear unbiased [see (87) and (101)] estimator. Equation (101) produces the optimal filtered estimate and (102) the smoothed estimate. For an on-line computation of the estimate $\hat{x}(t, \theta \mid t)$, the covariance equations, i.e., the PDEs, must also be solved. The determination of $M(t,\theta_1,\theta_2)$ indeed poses a computational burden.

Equations (101) and (102) correspond to the observer equations (77) and (78).

Let us now establish the equation corresponding to Eq. (79) for closer comparison. We first observe in Eq. (101) a term $\hat{x}(t, \tau \mid t)$, that is, an estimate of $x(t-\tau)$ at time t, which we do not know. What we do have is an estimate of $x(t-\tau)$ at time $t-\tau$ or $\hat{x}(t-\tau, 0 \mid t-\tau)$. Equation (102) now provides a way to obtain an estimate of $\hat{x}(t, \tau \mid t)$, and we manipulate it so as to solve it in terms of the known data.

The characteristics of Eq. (102) are defined by $dt/d\theta = 1$ and thus $t = \theta + d$, where d is a constant. Substituting this for t in (102) and integrating it from 0 to θ results in

$$\hat{x}(\theta+d, \theta \mid \theta+d) = \hat{x}(d, 0 \mid d) - \int_0^\theta K(\xi+d, \xi, \xi+d)\left[z(\xi+d) - \sum_{i=0}^{1} C_i \hat{x}(\xi+d, \tau_i \mid \xi+d)\right] d\xi$$

Since $d = t - \theta$, we obtain

$$\hat{x}(t, \theta \mid t) = \hat{x}(t-\theta, 0 \mid t-\theta) - \int_0^\theta K(\xi+t-\theta, \xi, \xi+t-\theta)\left[z(\xi+t-\theta)\right.$$

$$- \sum_{i=0}^{1} C_i \hat{x}(\xi+t-\theta, \tau_i \mid \xi+t-\theta) \Big] d\xi$$

Replacing $\xi-\theta$ by $-\xi$, letting $\theta=\tau= : \tau_1$ and substituting $\hat{x}(t, \tau_1 \mid t)$: into Eq. (101) yields (when $C_1 = 0$)

$$\dot{\hat{x}}(t, 0 \mid t) = A_0\hat{x}(t, 0 \mid t) + A_1\hat{x}(t-\tau, 0 \mid t-\tau)$$
$$+ K(t,0,t)[C_0\hat{x}(t, 0 \mid t) - z(t)]$$
$$+ A_1 \int_0^{\tau} K(t-\xi, \tau-\xi, t-\xi)[C_0\hat{x}(t-\xi, 0 \mid t-\xi) - z(t-\xi)]\, d\xi$$

When this equation is compared with (79), we immediately recognize the similarities: $\hat{x}(t, 0 \mid t)$ corresponds to $z(t)$, $\hat{x}(t-\tau, 0 \mid t-\tau)$ to $z(t-1)$, and $K(t-\xi, \tau-\xi, t-\xi)$ to $\Phi_\Lambda(\xi-1)K$. In fact, it seems plausible to conjecture that if a Kalman gain were chosen in the projected finite-dimensional space instead of a stabilizing matrix K, this would in the limit (when the vertical line is pushed to $-\infty$) approach the kernel $K(\cdot,\cdot,\cdot)$.

B. Nonlinear Systems with Time Delay

The estimation of state variables in nonlinear systems is a complex problem due to the fact that a signal possessing Gaussian characterization at the input of a general nonlinear element will not possess the properties of a Gaussian process (i.e., sufficient description is the first two moments) at the output of the nonlinearity. Because of this basic difficulty, solutions to a nonlinear estimation problem are usually only locally optimum, i.e., about some nominal (often the average) trajectory; general globally optimum solutions in most cases are, at least at the present, untractable. A common approach is to linearize the input-output relation of the nonlinear system about an average trajectory, and then construct a filter for the linearized system. Such a filter is often referred to as the extended Kalman filter. The basic assumptions are that (1) the linearized model is sufficiently accurate, and (2) the Gaussian noise processes are assumed to remain Gaussian after passing through the nonlinear system. These (somewhat vague) assumptions may be justified if the signals stay in a small region, which unfortunately is not readily defined.

These basic assumptions also apply to nonlinear systems involving time delays. We will first present a first-order estimator for a system which results from the linearization of a nonlinear system. Then we describe a least squares approach for the construction of an estimator for nonlinear systems.

1. First-order Estimator

The state transition in a nonlinear dynamical system with pure time delay may be described by the following differential-difference equation

$$\dot{x}(t) = f[x(t),x(t-\tau),t] \tag{103}$$

$$z(t) = h[x(t),x(t-\tau),t] \tag{104}$$

where $x(t)$ and $x(t-\tau)$ are n-dimensional vectors evaluated at time t and $t-\tau$, respectively, and $t \in [t_o,T]$; $\tau \geq 0$ represents a constant time delay; $f[x(t),x(t-\tau),t]$ and $h[x(t),x(t-\tau,t]$ are vector-valued functions and $f, h \in C^2[t_o,T]$. It is assumed that Eq. (103) possesses a unique solution for $t \geq t_o$ and for a given continuous initial function.

Suppose that a nominal trajectory $X(t)$ of the state $x(t)$ has been computed. For example, it may have been computed using the equation $X = f[X(t),X(t-\tau),t]$ with the initial function $X(t_o+\sigma)$, $\sigma \in [0,-\tau]$, if such an expression is available. Assuming now that a nominal trajectory $X(t)$ is known, we may linearize the right-hand side of Eq. (103) about the nominal trajectory:

$$f[x(t),x(t-\tau),t] = f[X(t),X(t-\tau),t] + \sum_{i=0}^{1} A_i[t][x(t-\tau_i) - X(t-\tau_i)] + \text{higher-order terms} \tag{105}$$

where $\tau_o = 0$, $\tau = \tau_1$, and $A_i[\cdot] = \partial f/\partial x(t-\tau_i)$ is the Jacobian matrix of $f[x(t),x(t-\tau),t]$ evaluated about $X(t),X(t-\tau)$, i.e., for example,

$$A_0(t) = \frac{\partial f[X(t),X(t-\tau),t]}{\partial x(t)} = \begin{bmatrix} \frac{\partial f_1}{\partial x_1(t)}, & \cdots, & \frac{\partial f_1}{\partial x_n(t)} \\ \cdot & & \\ \cdot & & \\ \cdot & & \\ \frac{\partial f_n}{\partial x_1(t)}, & \cdots, & \frac{\partial f_n}{\partial x_n(t)} \end{bmatrix} \tag{106}$$

where the derivatives are evaluated at $X(t)$, $X(t-\tau)$. It is noted that the higher-order terms in Eq. (105) represent products containing terms $[x(t) - X(t)]$ and $[x(t-\tau) - X(t-\tau]$ to at least second power and their cross-products. If only the terms up to the first order are retained in the construction of the filter, the result is a first-order (extended Kalman) filter. If the second-order terms are also included in the derivation of the estimator equations [70], a second-order filter is obtained.

In a similar fashion, the right-hand side of Eq. (104) is linearized about the nominal trajectory

$$h[x(t),x(t-\tau),t] = h[X(t),X(t-\tau)] + \sum_{i=0}^{1} C_i[t][x(t-\tau_i) - X(t-\tau_i)] + \text{higher-order terms} \tag{107}$$

where $C_i[t] = h(\cdot)/\partial x(t-\tau_i)$ is the Jacobian matrix of $h(\cdot)$ evaluated at $X(t),X(t-\tau)$. When $x(t)$, $t \in [t_0,T]$ is sufficiently close to the nominal trajectory $X(t)$ so that the contribution of the higher-order terms is negligible, the system Eqs. (103) and (104) for the problem may be written as follows:

$$\dot{x}(t) = f[X(t),X(t-\tau),t] + \sum_{i=0}^{1} A_i[t][x(t-\tau_i) - X(t-\tau_i)] + D(t)\xi(t) \tag{108}$$

$$z(t) = h[X(t),X(t-\tau)] + \sum_{i=0}^{1} C_i[t][x(t-\tau_i) - X(t-\tau_i)] + \eta(t) \tag{109}$$

where the terms $D(t)\xi(t)$ and $\eta(t)$ account for modeling and measurement errors. It is now assumed that they represent independent white Gaussian zero-mean noise processes with variances

$$E\{\xi(t)\xi'(s)\} = Q(t)\delta(t-s) \text{ and } E\{\eta(t)\eta'(s)\} = R(t)\delta(t-s)$$

One observes that if $X(t)$ satisfies the equation $\dot{X} = f(X(t),X(t-\tau),t)$, then Eq. (108) represents a differential-difference equation for $\delta x(t-\tau_i) := [x(t-\tau_i) - X(t-\tau_i)]$, which has the same form as (83). Similarly, Eq. (103) with observations $\delta z(t) := \{z(t) - h[X(t), X(t-\tau)]\}$ corresponds to (84).

The estimation problem is now posed in the same fashion as the one specified by Eqs. (83) to (86): Determine the estimate $\hat{x}(t, \theta \mid t)$ on the basis of the measurements $z(t)$, $t \in [t_0,T]$ so as to minimize the expected value of the squared error

$$\epsilon^2(t) = E\{ \|x(t-\theta) - \hat{x}(t, \theta \mid t)\|^2 \}$$

subject to the differential-difference Eqs. (108), (109) constraint.

The problem posed is the same as the estimation problem solved in Section III.A.2. The solution is provided by the following equations which correspond to Eqs. (101) and (102):

$$\begin{aligned}\frac{d\hat{x}(t, 0 \mid t)}{dt} &= f[X(t),X(t-\tau),t] + \sum_{i=0}^{1} A_i[t][\hat{x}(t,\tau_i \mid t) \\ &\quad - X(t-\tau_i)] + \sum_{i=0}^{1} M(t,0,\tau_i)C_i'(t)R^{-1}\{z(t) \\ &\quad - h[X(t),X(t-\tau)] - \sum_{j=0}^{1} C_j(t)[\hat{x}(t,\tau_j \mid t) \\ &\quad - X(t-\tau_j)]\}\end{aligned} \tag{110}$$

$$\begin{aligned}\frac{\partial\hat{x}(t, \theta \mid t)}{\partial t} + \frac{\partial\hat{x}(t, \theta \mid t)}{\partial\theta} &= -\sum_{n=0}^{1} M(t,\theta,\tau_n)C_n'(t)R^{-1}\{z(t) \\ &\quad -h[X(t),X(t-\tau)] - \sum_{i=0}^{1} C_i(t) \\ &\quad \times [\hat{x}(t,\tau_i \mid t) - X(t-\tau_i)]\}\end{aligned} \tag{111}$$

where $C_i(\cdot) = C_i[X(t),X(t-\tau)]$.

The equations for $M(t,\theta_1,\theta_2)$, $M(t,\theta_1,0)$, and $M(t,0,0)$ are the same as Eqs. (96) to (98).

Equations (110) and (111) specify the filtering and smoothing equations for the time-delay system resulting from the linearization of the nonlinear equations (103) and (104). We observe that if $\hat{x}(t-\tau_i \mid t)$ is the same as the nominal trajectory $X(t-\tau_i)$ computed on the basis of the measurement $z(t)$, then the terms involving $[\hat{x}(t-\tau_i \mid t) - X(t-\tau_i)]$ do not appear in Eqs. (110) and (111). This would be the case, for example, when $f[X(t),X(t-\tau),t] = E\{f[x(t), x(t-\tau),t] \mid z\}, \dot{X}(t) = E[\dot{x}(t) \mid z]$, and $h[X(t),X(t-\tau),t] = E\{h[x(t), x(t-\tau),t] \mid z\}$.

2. Nonlinear Estimator by Deterministic Data Fitting

An alternative approach [67,72] to nonlinear estimation problem is to attempt to match the measured data with the values generated by the mathematical model involving some unknown deterministic time functions. These disturbances account for modeling and measurement errors; their time averages are assumed to be known. The problem is then to minimize the integral of the sum of the squared errors representing the measurement and modeling errors subject to the differential-difference equation constraint.

Specifically, suppose that the nonlinear system is specified by Eqs. (103) and (104).

It is desired to determine an estimate of the state $x(t-\sigma)$, $\sigma \in [0,\tau]$ so that the estimator generates a curve which is as close as possible to the observed data $z(t)$ on $(t_0,T]$, i.e., to obtain the least integrated squared error. Specifically, the problem may be stated as follows: Minimize

$$J = \frac{1}{2}\int_{t_0}^{T} \{ \|z(t) - h[\bar{x}(t),\bar{x}(t-\tau),t]\|^2_{R^{-1}(t)} + \|\bar{\xi}(t)\|^2_{Q^{-1}(t)}\} \, dt \tag{112}$$

subject to the constraint

$$\dot{\bar{x}} = f[\bar{x}(t),\bar{x}(t-\tau),t] + \Gamma(t)\bar{v}(t) \tag{113}$$

where f[·] and h[·] are the same functions as defined in Eqs. (103) and (104), with $\bar{x}(\cdot)$ substituted for $x(\cdot)$; $\Gamma(t)$ has continuous and bounded elements for all $t \geq t_0$. The input $v(t) \in L_2(t)$ is a vector-valued error function to be determined. It is assumed that the time averages of $(f+\Gamma v)$ and $(z-h)$ are zero. Known inputs such as a nonzero time average should be included as explicit (separable) time functions on the right-hand side of Eq. (113). The weighting matrices $Q^{-1}(t)$ and $R^{-1}(t)$ are positive definite and semidefinite matrices, respectively, for all $t \geq t_0$. The sought estimate is obtained by minimizing (112) subject to the constraint (109).

The constraint (113) can be described equivalently by

$$\frac{\partial\bar{x}(t,\sigma)}{\partial t} = -\frac{\partial\bar{x}(t,\sigma)}{\partial\sigma}, \qquad \sigma \in [0,\tau],\ t \in [t_0,T] \tag{114}$$

$$\frac{d\bar{x}(t,0)}{dt} = f(t,\bar{x}(t,0),\bar{x}(t,\tau)) + \Gamma(t)\bar{v}(t), \qquad t \in [t_0,T] \tag{115}$$

where $\bar{x}(t,0) = \bar{x}(t)$ and $\bar{x}(t,\tau) = \bar{x}(t-\tau)$.

The calculus of variations can be applied to determine necessary conditions for the minimum of the performance index. With the introduction of a costate variable $p(t,\sigma)$ defined on $[t_0,T] \times [0,\tau]$, the variational calculus yields the following necessary conditions for the optimality (optimal values are indicated by the *):

$$\frac{dx^*(t,0)}{dt} = f(t,x^*(t,0),x^*(t,\tau)) - \Gamma(t)Q(t)\Gamma'(t)p^*(t,0) \tag{116}$$

$$\frac{\partial\bar{x}^*(t,\sigma)}{\partial t} + \frac{\partial\bar{x}^*(t,\sigma)}{\partial\sigma} = 0 \tag{117}$$

$$\frac{dp^*(t,0)}{dt} = -f'_{\bar{x}^*(t,0)}\, p^*(t,0) - p^*(t,0)$$

$$+ h'_{\bar{x}^*(t,0)}\, R^{-1}(t)[y(t) - h(t,\bar{x}^*(t,0),\bar{x}^*(t,\tau))] \tag{118}$$

$$\frac{\partial p^*(t,\sigma)}{\partial t} + \frac{\partial p^*(t,\sigma)}{\partial\sigma} = 0 \tag{119}$$

$$p^*(t,\tau) = f'_{\bar{x}^*(t,\tau)} p^*(t,0) - h'_{\bar{x}^*(t,\tau)} R^{-1}(t)[y(t)$$
$$- h(t,\bar{x}^*(t,0),\bar{x}^*(t,\tau))] \tag{120}$$

$$p^*(T,\sigma) = 0, \quad p^*(t_0,\sigma) = 0 \tag{121}$$

for $t \in [t_0,T]$, $\sigma \in [0,\tau]$.

Equations (120) and (121) represent the boundary conditions. The term in row i and column j of the partial derivative $f_{\bar{x}(t,0)}$ is $\partial f_i(t,\bar{x}(t,0),\bar{x}(t,\tau))/\partial \bar{x}_j(t,0)$. The other partial derivatives are similarly defined. The optimal solution to the split boundary value problem defined by Eqs. (116) to (121) provides the "least squares fit" to the observations over the interval $[t_0,T]$. An iterative technique can be used to obtain the solution. However, if a real-time estimate is needed, then it is desirable to have a sequential method of computing the solution. In the nonlinear case under consideration it can be accomplished by the following approximate relationship (the * notation is deleted hereafter):

$$\bar{x}(t,\sigma) = -\int_0^\tau M(t,\sigma,\alpha)p(t,\alpha)\, d\alpha - M(t,\sigma,0)p(t,0)$$
$$+ s(t,\sigma), \quad t \in [t_0,T],\ \sigma \in [0,\tau] \tag{122}$$

where $M(\cdot,\cdot,\cdot)$ is a differentiable $(n \times n)$-dimensional matrix and $s(\cdot,\cdot)$ is a differentiable n-dimensional vector. Equations for $M(\cdot,\cdot,\cdot)$ and $s(\cdot,\cdot)$ in (122) are so determined that the necessary conditions are satisfied. It is noted that (122) is an approximate relation between $\bar{x}(t,\cdot)$ and $p(t,\cdot)$ in the nonlinear case; it represents an exact relation in a linear case.

To determine equations for $M(\cdot,\cdot,\cdot)$ and $s(\cdot,\cdot)$, Eq. (122) is differentiated with respect to t and also with respect to σ. The results are added and equated to zero to satisfy (117). Through the use of (116) to (121), the terms in the resulting expression are expanded about $(t,s(t,0),s(t,\tau))$. Coefficients of the first powers of $p(t,\alpha)$ and $p(t,0)$ are equated to zero yielding three equations. Finally, the derivative of (122) with respect to t evaluated at

$\sigma = 0$ is set equal to (116). In a manner similar to that described above, three more equations for $M(\cdot,\cdot,\cdot)$ and $s(\cdot,\cdot)$ are obtained. The resulting equations are as follows ($\tau_0 = 0$, $\tau_1 = \tau$):

$$\frac{\partial M(t,\sigma,\alpha)}{\partial t} + \frac{\partial M(t,\sigma,\alpha)}{\partial \sigma} + \frac{\partial M(t,\sigma,\alpha)}{\partial \alpha}$$

$$= \sum_{i=0}^{1} \sum_{j=0}^{1} M(t,\sigma,\tau_i)\{-h'_{s(t,\tau_i)} R^{-1}(t) h_{s(t,\tau_j)} + h'_{s(t,\tau_i)s(t,\tau_j)} R^{-1}(t)[y(t)-h]\} M(t,\tau_j,\alpha) \tag{123}$$

$$\frac{\partial M(t,\sigma,0)}{\partial t} + \frac{\partial M(t,\sigma,0)}{\partial \sigma} = \sum_{i=0}^{1} M(t,\sigma,\tau_i) f'_{s(t,\tau_i)} + \sum_{i=0}^{1} \sum_{j=0}^{1} M(t,\sigma,\tau_i)\{-h'_{s(t,\tau_i)} R^{-1}(t) \times h_{s(t,\tau_j)} + h'_{s(t,\tau_i)s(t,\tau_j)} R^{-1}(t) \times [y(t)-h]\} M(t,\tau_j,0) \tag{124}$$

$$\frac{dM(t,0,0)}{dt} = \sum_{i=0}^{1} [f_{s(t,\tau_i)} M(t,\tau_i,0) + M(t,0,\tau_i) f'_{s(t,\tau_i)}] + \Gamma(t)Q(t)\Gamma'(t) + \sum_{i=0}^{1} \sum_{j=0}^{1} M(t,0,\tau_i)\{-h'_{s(t,\tau_i)} \times R^{-1}(t) h_{s(t,\tau_j)} + h'_{s(t,\tau_i)s(t,\tau_j)} R^{-1}(t) \times [y(t)-h]\} M(t,\tau_j,0) \tag{125}$$

$$\frac{\partial s(t,\sigma)}{\partial t} + \frac{\partial s(t,\sigma)}{\partial \sigma} = \left[\sum_{i=0}^{1} M(t,\sigma,\tau_i) h'_{s(t,\tau_i)} \right] R^{-1}(t)[y(t)-h] \tag{126}$$

$$\frac{ds(t,0)}{dt} = f + \left[\sum_{i=0}^{1} M(t,0,\tau_i) h'_{s(t,\tau_i)} \right] R^{-1}(t)[y(t)-h] \tag{127}$$

where $f = f(t,s(t,0),s(t,\tau))$; $h = h(t,s(t,0),s(t,\tau))$; the partial derivative matrices are as defined previously, and

$$h'_{s(t,\tau_i)s(t,\tau_j)} = \left[\frac{\partial}{\partial s(t,\tau_j)} (h'_{s(t,\tau_i)}), \ldots, \frac{\partial}{\partial s_n(t,\tau_j)} \times (h'_{s(t,\tau_i)}) \right] \tag{128}$$

is an $(n \times mn)$-dimensional matrix (tensor). It is noted that $M(t,0,\alpha) = M'(t,\alpha,0)$ for $t \geq t_0$.

One observes that Eqs. (123) to (127) specify an approximate solution to the original problem; it is valid in a (sufficiently) small neighborhood of the actual trajectory.

By Eq. (121) the boundary conditions of the split boundary value problem are $p(t_0,\sigma) = 0$ and $p(T_f,\sigma) = 0$, $\sigma \in [0,\tau]$. At $t = T$, it follows from (122) and (121) that $\overline{x}(T,\sigma = s(T,\sigma)$, $\sigma \in [0,\tau]$. Thus, with the use of the $\hat{x}(\cdot \mid \cdot)$ notation to emphasize the estimator dependence on the data interval, replacing T by t as the running variable, (126) and (127) can be rewritten:

$$\frac{\partial \hat{x}(t-\sigma \mid t)}{\partial t} + \frac{\partial \hat{x}(t-\sigma \mid t)}{\partial \sigma} = [M(t,\sigma,0)h'_{x(t \mid t)}(t,\hat{x}(t \mid t),\hat{x}(t-\tau \mid t)) + M(t,\sigma,\tau)h'_{\hat{x}(t-\tau \mid 0)}(t,\hat{x}(t \mid t), \hat{x}(t-\tau \mid t))]R^{-1}(t)[y(t) - h(t,\hat{x}(t \mid t), \hat{x}(t-\tau \mid t))] \tag{129}$$

$$\frac{d\hat{x}(t \mid t)}{dt} = f(t,\hat{x}(t \mid t),\hat{x}(t-\tau \mid t)) + [M(t,0,0)h'_{\hat{x}(t \mid t)}(t,\hat{x}(t \mid t),\hat{x}(t-\tau \mid t)) + M(t,0,\tau)h'_{\hat{x}(t-\tau \mid t)}(t,\hat{x}(t \mid t),\hat{x}(t-\tau \mid t))] R^{-1}(t)[y(t) - h(t,\hat{x}(t \mid t),\hat{x}(t-\tau \mid t))] \tag{130}$$

Thus, Eqs. (129), (130), and (123) to (125) specify an approximately optimal estimator for the nonlinear system.

C. Numerical Solutions to Estimator Equations

The existence of a unique solution to the estimator Eqs. (96) to (98), (101), and (102) with some specified initial conditions in a finite domain can be established (see, e.g., [45]). Specific algorithms for the numerical solution of the estimator equations are scarce, although numerical examples are presented in the literature (e.g. [67,72]).

We usually want to compute the optimal estimates on-line. If the system equations are linear and noise processes additive Gaussian, Eqs. (101) and (102) apply. In order to compute the optimal estimates, we also have to solve for the estimator gains, which requires a numerical solution to the PDEs (96) to (98) specifying the evolution of the error covariance. A possible approach is to use a finite difference approximation for the partial derivatives (Euler's method) [45,73], although the accuracy of the method is quite limited. Since the PDEs for the error covariance and for the optimal estimates are of first order, the method of characteristics often appears more appealing.

This computational technique leads to ODEs which are valid along the characteristic curves. The characteristics for the estimator equations turn out to be straight lines. Indeed, the characteristics [74] for Eqs. (101) and (102) are defined by $d\theta/dt = 1$; similarly, the characteristics for (96) to (98) are defined in terms of three independent variables t, θ_1, θ_2: $d\theta_1/dt = 1$ and $d\theta_2/dt = 1$. When the resulting equations are solved along the characteristics, only ODEs need to be considered, e.g., the left-hand side of Eq. (98) is the total derivative $dP(t,\theta_1,\theta_2)/dt$ along the characteristic curve. An example illustrates the approach.

Example. A stirred tank chemical reactor and separator unit shown schematically in Figures 3a and b. Its behavior can be described by equations which represent the material balance and energy balance [75] in the reaction. They can be written in the form of Eq. (103) by defining $(n = 2)$, $f = [f_1(\cdot), f_2(\cdot)]'$, where

$$f_1(t,x(t),x(t-\tau)) = -\frac{v_0 + v_r}{V} x_1(t) - x_1(t)e^{\alpha-(\beta/x_2(t))}$$

$$+ \frac{v_r}{V} e^{-\mu\tau} x_1(t-\tau) + \frac{v_r}{V}(1 - e^{-\mu\tau}) + \frac{v_0}{V} X_0$$

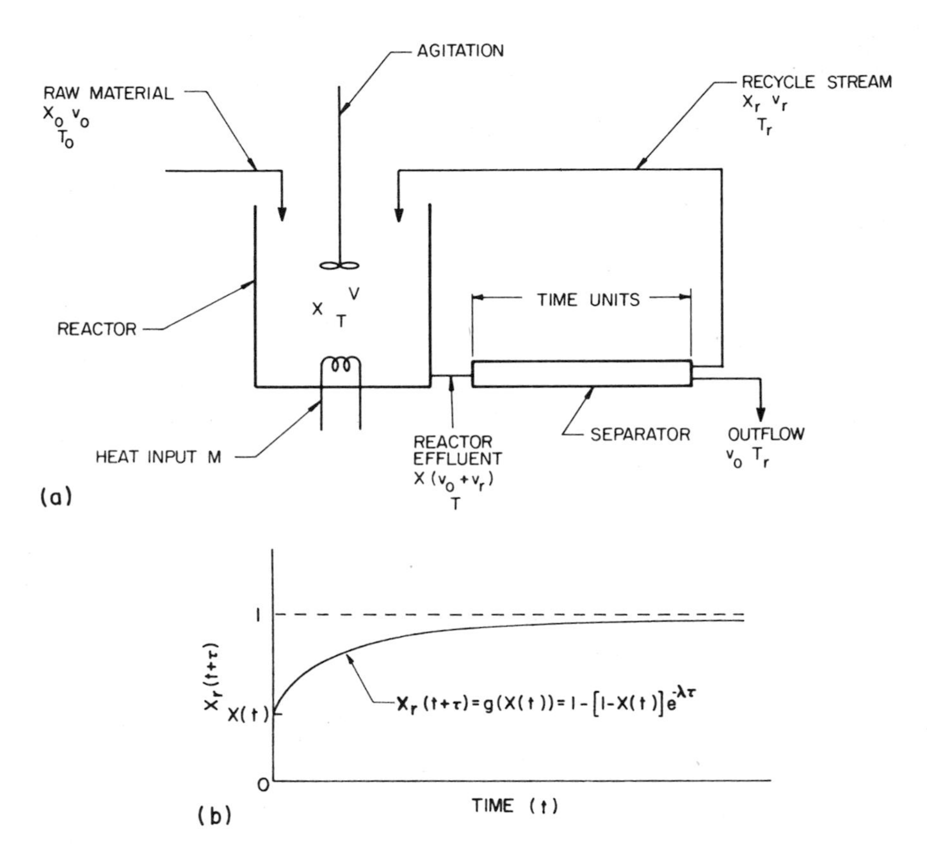

FIG. 3. (a) Schematic diagram of a stirred tank chemical reactor and separator unit. (b) Return concentration $X_r(t-\tau)$ vs transport time for separator unit with input concentration X(t) as parameter.

$$f_2(t,x(t),x(t-\tau)) = -\left(\frac{\Delta H_r X_N}{\rho C_p T_N} x_1(t)e^{\alpha-(\beta/x_2(t))}\right) - \frac{v_0 + v_r}{V} x_2(t)$$

$$+ \frac{v_0}{V} T_0 + \frac{v_r}{V} T_r + \frac{M(t)}{\rho C_p V T_N} \tag{131}$$

where

$x_1 = X, X_0, X_r$ = concentration of component A in reactor, feed stream, and recycle stream, respectively, normalized to concentration reference X_N

$x_2 = T, T_0, B_r$	= temperature of reactor, feed stream, and recycle stream, respectively, normalized to temperature reference T_N
V	= volume of reactor tank (constant)
v_0, v_r	= volumetric flow rate of input feed and outflow, and of recycle stream, respectively
ΔH_r	= heat of reaction
C_p	= heat capacity of mixture in reactor
ρ	= density of mixture in reactor
α	= reaction rate parameter
β	= activation energy parameter, normalized to X_N
M	= heat added to mixture in reactor

The functions $\overline{v}_1(t)$ and $\overline{v}_2(t)$ in Eq. (113) represent unknown plant disturbances.

Parameter values chosen for the reactor and separator unit are

$t \geq t_0 = 0$ min; $T_N = 353°$ K; $X_N = 12$ g mol/liter

$V = 500$ liters; $T_0 = 363°$ K; $X_0 = 1.0$; $V_0 = 100$ liters/min

$T_r = 353°$ K; $\mu = 4.0$; $v_r = 50$ liters/min; $C_p = 1$ kcal/kg° K

$\rho = 1$ kg/liter; $\alpha = 15.915$; $\beta = 15{,}500$; $M = 33{,}500$ kcal/min

$\tau = 0.25$ min

Two cases will be studied.

In Case 1, it is assumed that a measuring device performs the measurements according to

$$y(t) = x_2(t) - 0.1x_2^2(t) \tag{132}$$

and the heat of reaction is known to be

$$\Delta H_r = 55 \text{ kcal/g mol} \tag{133}$$

In Case 2, it is assumed that measurements are performed according to (132), but that the heat of reaction ΔH_r is an unknown constant to be estimated. The following then applies. Let

$$\Delta H_r = 100x_3(t) \tag{134}$$

and augment the plant equations by

$$\dot{x}_3(t) = 0 \tag{135}$$

Cases 1 and 2 both fit to the form of (113) and (112). The least squares filtered and smoothed estimates are thus given by the solution of (125) to (130) for the two cases [72]. A possible approximate solution to the equations can be obtained by using the characteristics and discretizing the set of differential equations obtained with discretization interval Δ. The dimensionality of the resulting difference equations is dependent upon the value of the time delay τ, and the discretization interval.

Case 1. Plant trajectories and observations are generated for

$$x(t_0 - \sigma) = \begin{bmatrix} 0.57 \\ 0.93 \end{bmatrix}, \qquad \tau = 0.25 \text{ min}$$

The above-described method is used to obtain an approximate solution to (131) to (135) for the following chosen values:

$$\hat{x}(t_0 - \sigma \mid t_0) = \begin{bmatrix} 0.56 \\ 0.89 \end{bmatrix}$$

$$Q = \begin{bmatrix} 0.056 & 0.000 \\ 0.000 & 0.005 \end{bmatrix}, \qquad R = [0.2 \times 10^{-3}]$$

for $\Delta = 0.05$ min, $t_0 = 0$, $\sigma \in [0,\tau]$, $\alpha \in [0,\tau]$.

A sample trajectory is given in Figure 4 for the actual state $x(t)$, the filtered estimate $\hat{x}(t \mid \tau)$, and the smoothed estimate $\hat{x}(t \mid t+\tau)$.

Case 2. Plant trajectories and observations are generated for the case where

$$x(t_0 - \sigma) = \begin{bmatrix} 0.58 \\ 0.93 \\ 0.50 \end{bmatrix}, \qquad \tau = 0.25 \text{ min}$$

Values chosen for this example are

$$\hat{x}(t_0 - \sigma \mid t_0) = \begin{bmatrix} 0.53 \\ 0.89 \\ 0.40 \end{bmatrix}$$

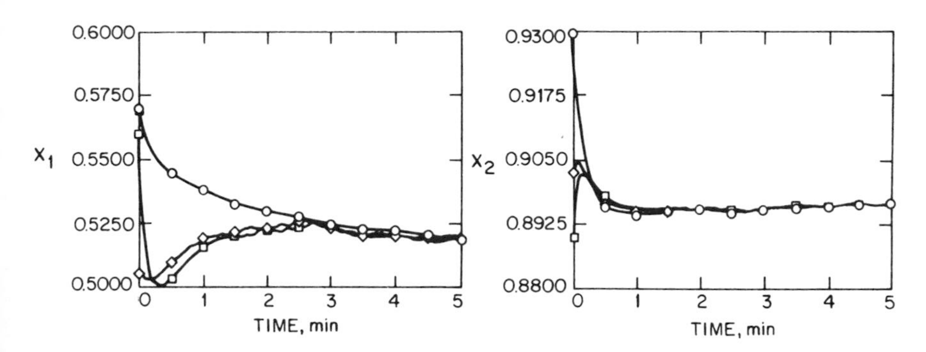

FIG. 4. Sample trajectories $x_1(t)$ and $x_2(t)$ of the nonlinear plant with delay (○), filtered estimates (□), and smoothed estimates (◇).

$$Q = \begin{bmatrix} 0.056 & 0 & 0 \\ 0 & 0.005 & 0 \\ 0 & 0 & 0.056 \end{bmatrix}, \qquad R = [0.2 \times 10^{-3}]$$

for $\Delta = 0.05$ min, $t_0 = 0$, $\sigma \in [0,\tau]$, $\alpha \in [0,\tau]$

A sample trajectory of the plant and filtered and smoothed estimates is shown in Figure 5.

Remark. The computation of the numerical solution to the first-order PDE can be very demanding, particularly in view of computer memory. Although we need not store the values of the entire solution to the estimator equations over the entire domain, the array of the covariance of the estimation error often causes storage difficulties when the time delays are large. In fact, a hybrid computer appears to be one of the most powerful tools to handle the dimensionality problem.

The approaches presented for solving nonlinear estimation problems in systems with time delay work reasonably well in a sufficiently small neighborhood of a nominal trajectory. Unfortunately, such a region is difficult to define in advance.

The numerical solutions are often computed by discretizing the system equations using either finite differences for the partial derivatives or performing the discretization along the characteris-

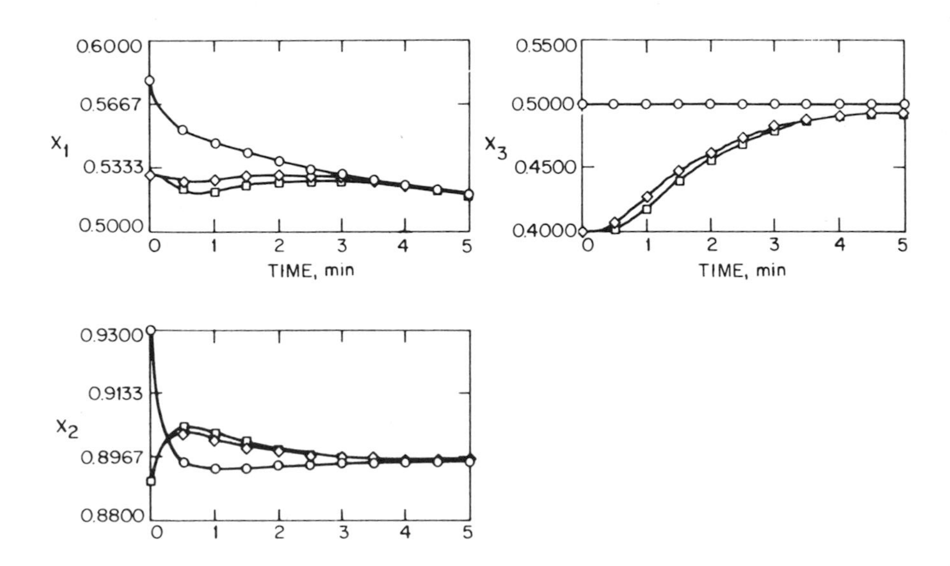

FIG. 5. Sample trajectories $x_1(t)$, $x_2(t)$, and $s_3(t)$ of the nonlinear plant (○) and an unknown parameter, filtered estimates (□), and smoothed estimates (Δ).

tics. It should be emphasized that very little is established on the discretization problem itself.

Example 2. A common procedure for solving PDEs is based on expressing the dependent variable as a product of unknown functions which are then specified by ODEs. This approach has been applied to Eqs. (96) to (98), when the plant equation (83) does not contain the noise term, i.e., DQD' in Eq. (96) is absent [76].

The equations for the characteristics of (97), (98) are: $t_1 - \theta_1$ = constant and $t - \theta_2$ = constant. A family of these characteristics give rise to an integral surface [74] on which the left-hand sides of Eqs. (97) and (98) become total derivatives.

We may then postulate the following expression:

$$M(t,\theta_1,\theta_2) = \gamma_1(t-\theta_1, t_o)\gamma_2(t)\gamma_1'(t-\theta_2, t_o) \tag{136}$$

where the equations specifying $\gamma_1(t,t_o)$ and $\gamma_2(t)$ are now to be determined. Substituting (136) into (96) and (97), (98) yields

$$\frac{d}{dt}[\gamma_1(t,t_0)] = \sum_{i=0}^{1} A_i(t)\gamma_1(t-\tau_i,t_0) \qquad \text{for } t \geq t_0$$

$$\gamma_i(t_0,t_0) = I \tag{137}$$

$$\frac{d}{dt}[\gamma_2(t)] = -\gamma_2(t)\beta(t)\gamma_2(t)$$

$$\gamma_2(t_0) = M(t_0,0,0) \tag{138}$$

where

$$\beta(t) = \sum_{i,j=0}^{1} \gamma_1'(t\ \tau_i,\ t_0)C_i'(t)R^{-1}(t)C_j(t)\gamma_1(t-\tau_j,\ t_0) \tag{139}$$

Equations (137) to (139) can be computed on-line; thus, $M(t,\theta_1,\theta_2)$ in (136) and the estimator gains are obtained directly by a forward integration. Moreover, Eqs. (100) and (102) for the estimator can also be solved by a forward integration. Thus, separating the independent variables in $M(t,\theta_1,\theta_2)$ as shown in (136) simplifies the computations considerably. A numerical example illustrates this aspect.

Suppose a process is described by the following state transition equations

$$\dot{x}_1(t) = ax_1(t) + cx_2(t-\tau)$$

$$\dot{x}_2(t) = bx_2(t) + ds(t) \tag{140}$$

where $s(t)$ is the unit step input to the system and $a = -0.015$, $b = -0.02$, $c = 0.005$, and $d = 0.40$. The measurement $z(t)$ is given by

$$z(t) = x_1(t) + x_1(t-\tau) + \eta(t) \tag{141}$$

where the noise is characterized by a zero-mean Gaussian process with

$$E[\eta(t)\eta(t_1)] = r\ \delta(t-t_1), \qquad r = 0.05$$

The estimator equations (101), (102) take on the following form:

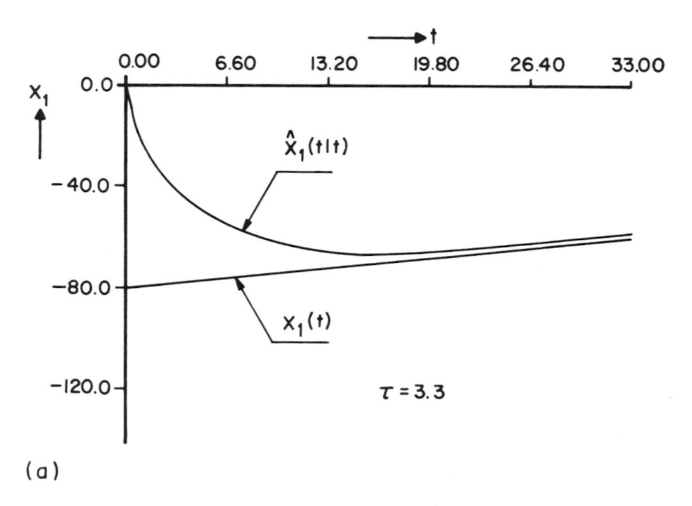

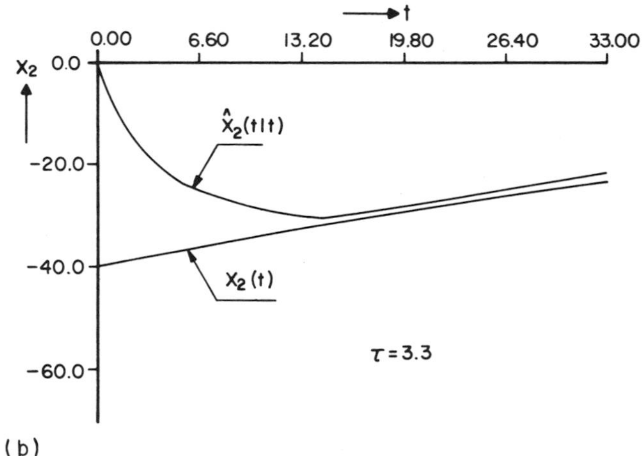

FIG. 6. (a) Sample trajectory of $x_1(t)$ and the filter response $\hat{x}_1(t \mid t)$ vs time; $\tau = 3.3$. (b) Sample trajectory $x_2(t)$ and the filter response $\hat{x}_2(t \mid t)$ vs time; $\tau = 3.3$.

$$\begin{bmatrix} \dot{\hat{x}}_1(t, 0 \mid t) \\ \dot{\hat{x}}_2(t, 0 \mid t) \end{bmatrix} = \begin{bmatrix} a & 0 \\ 0 & b \end{bmatrix} \begin{bmatrix} \hat{x}_1(t, 0 \mid t) \\ \hat{x}_2(t, 0 \mid t) \end{bmatrix} + \begin{bmatrix} 0 & c \\ 0 & 0 \end{bmatrix} \begin{bmatrix} \hat{x}_1(t, \tau \mid t) \\ \hat{x}_2(t, \tau \mid t) \end{bmatrix}$$
$$+ \begin{bmatrix} 0 \\ ds(t) \end{bmatrix} + r^{-1} \begin{bmatrix} M_{11}(t,0,0) + M_{11}(t,0,\tau) \\ M_{12}(t,0,0) + M_{12}(t,0,\tau) \end{bmatrix}$$
$$\times [z(t) - \hat{x}_1(t, 0 \mid t) - \hat{x}_1(t, \tau \mid t)] \quad (142)$$

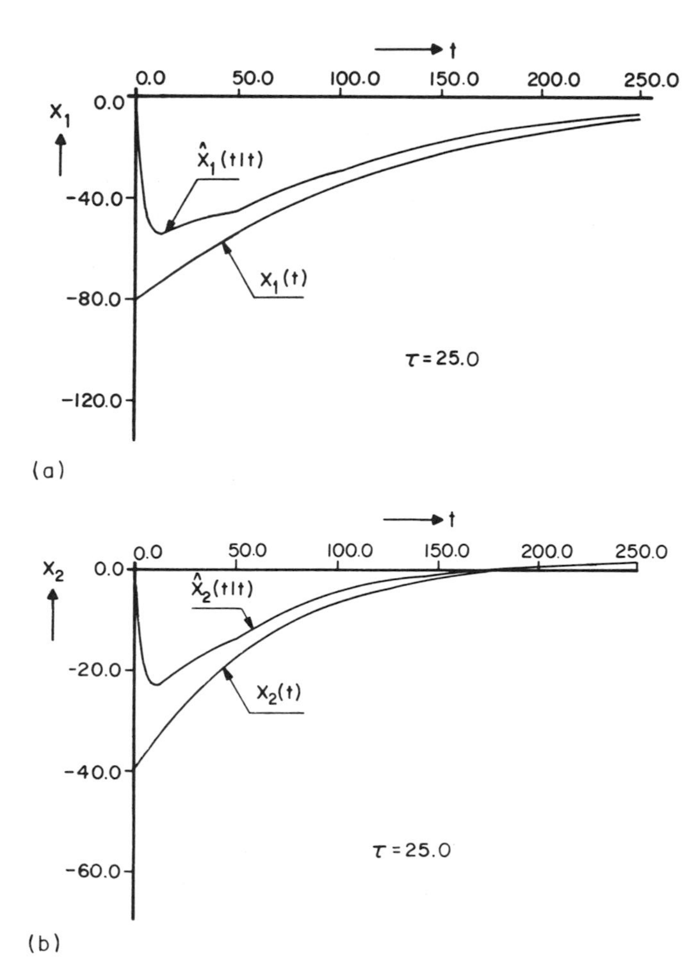

FIG. 7. (a) Sample trajectory $x_1(t)$ and the filter response $\hat{x}_1(t \mid t)$ vs time; $\tau = 25.0$. (b) Sample trajectory $x_2(t)$ and the filter response $\hat{x}_2(t \mid t)$ vs time; $\tau = 25.0$.

Equation (102) can be written similarly. The numerical values $M(t,\theta_1,\theta_2)$ are computed using Eqs. (136), (137). The simulation results on CDC 6500 are shown in Figures 6 and 7 for $\tau = 3.3$ and $\tau = 25$. The estimator tracks the "true" values quite well after a certain time.

IV. OPTIMAL CONTROL AND ESTIMATION IN STOCHASTIC DELAYED SYSTEMS

The optimization of stochastic linear dynamical systems described by difference equations or ODEs is well established when the cost functional to be minimized is quadratic in the control and state variables and the noise processes are Gaussian. The separation principle for this class of problems both in discrete time and in continuous time is well known. Basically it states that the generation of the optimal state estimates which constitute the input to the feedback controller and the determination of the optimal feedback gain matrices which operate on the optimal estimates of the states can be treated as independent problems. These solutions combined establish the optimal solution.

The optimal control is determined here for linear systems with time delay when the cost functional is quadratic in the control and state variables and the noises represent Gaussian processes. The configuration of the feedback loop within unspecified gain matrices is assumed, although the optimal feedback controller could also be derived.

The feedback loop consists of the state estimator and the controller. The state estimator operates on the noise corrupted measurements. It generates the input to the controller. The mathematical description of the estimator as well as that of the controller is assumed within unspecified gain matrices. The optimal gain matrices of the estimator are so determined that the mean-squared value of the estimation error is minimized. The optimal gain matrices of the controller are then so determined that the cost functional is minimized.

A. Statement of the Problem

The plant in a given system is described by the following equation:

$$\dot{x}(t) = \sum_{i=0}^{1} A_i x(t-\tau_i) + B(t)u(t) + D(t)\xi(t) \tag{143}$$

where the n-vector $x(t+\sigma)$, $t \geq t_0$, $-\tau \leq \sigma \leq 0$, describes the state of the system. The delay times are specified by $\tau_0 = 0$, $\tau_1 = \tau$ (a positive constant). The n x n matrices $A_0(t)$ and $A_1(t)$ have continuous and bounded elements on $t \geq t_0$. The n x m matrix $B(t)$ has continuous and bounded elements for $t \geq t_0$. The control $u(t)$ is a measurable m-vector in a compact convex set $\Omega \in R^m$ for all t. $D(t)$, the n x r matrix, has continuous and bounded elements for all t. The noise $\xi(t)$ represents a white Gaussian process with zero mean and covariance

$$E[\xi(t)\xi'(t_1)] = Q_1\delta(t-t_1), \qquad Q_1 > 0 \tag{144}$$

The output of the system is described by

$$z(t) = \sum_{i=0}^{1} H_i x(t-\tau_i) + \eta(t) \tag{145}$$

where $H_0 = H_0(t)$ and $H_1 = H_1(t)$ are $p \times \eta$ matrices with continuous and bounded elements in $t \geq t_0$. The noise is represented by a white Gaussian process $\eta(t)$ with zero mean and covariance

$$E[\eta(t)\eta'(t_1)] = Q_2\delta(t-t_1), \qquad Q_2 > 0 \tag{146}$$

For the system described, it is desired to design a feedback controller so as to minimize the performance functional

$$C_c(u) = E\left\{\int_{t_0}^{t_f} [x'(t)Wx(t) + u'(t)Ru(t)]\, dt\right\} \tag{147}$$

where t_f is a specified final time of the process, $E\{\cdot\}$ signifies the expectation operation conditioned on the measurements, and the matrices W and R are positive semidefinite and positive definite, respectively. Thus, the optimal feedback controller $u^*(t,\cdot)$ is to be determined when the measurements are available up to time t.

B. Configuration of the Feedback Loop

The feedback loop consists of a state estimator operating on the measurements $z(t)$ and a controller which operates on the output of the estimator generating the control variable.

It is assumed that the state of the estimator, denoted by $\hat{x}(t+\sigma \mid t)$, $-\tau \le \sigma \le 0$, furnishes the input to the controller. The input-output relation of the controller is specified by

$$u(t,\hat{x}_t) = K(t,0)\hat{x}(t \mid t) + \int_{-\tau}^{0} K(t,\sigma)\hat{x}(t+\sigma \mid t)\, d\sigma \tag{148}$$

where the subscripted argument emphasizes the functional dependence. Admissible expressions for the gain matrices $K(t,0)$ and $K(t,\sigma)$ are continuous and bounded; moreover, they must be stabilizing, i.e., they are to be such that the matrices

$$\{A_0 + BK(t,0),\ A_1,\ BK(t,\cdot)\} \tag{149}$$

represent a stable system, where $K(t,\cdot)\hat{x} = \int_{-\tau}^{0} K(t,\sigma)\hat{x}(t+\sigma \mid t)\, d\sigma$. The gain matrices $K(t,0)$ and $K(t,\sigma)$, $-\tau \le \sigma < 0$, $t_0 \le t \le t_f$, will be determined so that the cost functional (147) is minimized.

The input to the estimator is comprised of the measurement $z(t)$. The evolution of the state in the estimator is described by

$$\frac{d\hat{x}(t \mid t)}{dt} = \sum_{i=0}^{1} A_i\hat{x}(t-\tau_i \mid t) + G(t,0)[z(t) - \sum_{i=0}^{1} H_i\hat{x}(t-\tau_i \mid t)] + B(t)u \tag{150}$$

$$\frac{\partial\hat{x}(t+\theta \mid t)}{\partial t} = \frac{\partial\hat{x}(t+\theta \mid t)}{\partial\theta} + G(t,\theta)[z(t) - \sum_{i=0}^{1} H_i\hat{x}(t-\tau_i \mid t)] \tag{151}$$

where $\hat{x}(t+\theta \mid t)$, $-\tau \le \theta \le 0$, signifies the state of the estimator based on the measurement $z(t)$. The initial function for the estimator is $\hat{x}(t+\theta \mid t) \equiv 0$ (null vector). Admissible expressions for the gain matrices $G(t,0)$ and $G(t,\theta)$ are continuous and bounded and such that Eqs. (150) and (151) represent a stable system.

The estimate for the state of the system (143) obtained from Eqs. (150) and (151) is unbiased.

The gain matrices $G(t,0)$ and $G(t,\theta)$ will now be determined so as to minimize the mean-squared error (given the measurements) or, equivalently, the trace of the covariance $M(t,\sigma,\alpha)$ of the estimation error

$$C_e(G) = \mathrm{tr}[M(t_f,0,0) + \int_{-\tau}^{0} [M(t_f,0,\alpha) + M(t_f,\alpha,0)]\, d\alpha$$

$$+ \int_{-\tau}^{0} d\sigma \int_{-\tau}^{0} d\alpha [M(t_f,\sigma,\alpha)] \quad (152)$$

where

$$M(t,\sigma,\alpha) = E\{[x(t+\sigma) - \hat{x}(t+\sigma \mid t)][x(t+\alpha) - \hat{x}(t+\alpha \mid t)]' \mid z(t)\}$$

It is noted that the existence of the admissible gain matrices $K(t,0)$, $K(t,\cdot)$, and $G(t,0)$, $G(t,\cdot)$ has been assumed.

Using the feedback control (148) in (143), the plant equation becomes

$$\dot{x}(t) = \sum_{i=0}^{1} A_i x(t-\tau_i) + B(t)[K(t,0)\hat{x}(t \mid t)$$

$$+ \int_{-\tau}^{0} K(t,\sigma)\hat{x}(t+\sigma \mid t)\, d\sigma] + D(t)\xi(t) \quad (153)$$

Now the problem is to determine the gain matrices of the estimator and of the controller so as to achieve

$$\min_K C_c(K) \quad \text{and} \quad \min_G C_e(G)$$

while fulfilling the constraint equations (150) to (153). It may be noticed that a problem of the same type is obtained by replacing C_e by C_c and minimizing only $C_c(K,G)$ relative to both the gain matrices K and G.

C. Optimal Solution

The solution to the posed problem can be obtained by different methods, e.g., by the matrix minimum principle [70] or by using innovations [77]. If the former approach is applied, the problem is first reformulated by expressing the cost functional (147) and the constraint equations (153), (150), (151) in terms of the second-order moments, i.e., the covariance matrices. Then the optimization problem can be solved as an application of the matrix minimum principle. The details of the derivation can be found in [77]. The in optimal solution is only summarized here.

The optimal gain matrices are specified by

$$K^*(t,\sigma) = R^{-1}B'(t)P(t,\sigma,0), \qquad -\tau \le \sigma \le 0 \tag{154}$$

$$G(t,\theta_1) = \sum_{i=0}^{1} M((t,\theta_1,\tau_i)H_i'Q_2^{-1}, \qquad -\tau \le \theta_1 \le 0 \tag{155}$$

where $P(t,\sigma,\alpha)$ and $M(t,\theta_1,\theta_2)$ are determined by Eqs. (56) to (58) and (96) to (98), respectively. It is emphasized here that the optimal gain of the feedback controller is the same as in the deterministic case. Moreover, the introduction of the feedback controller does not change the equations which specify the evolution of the covariance of the estimation error.

The optimal system evolves according to Eqs. (153) and (154), and the optimal estimator is specified by (150), (151), (148), (154), and (155). An example illustrates the approach.

Example. The plant and measurement equations of a system are described by

$$\begin{aligned} \dot{x}(t) &= -0.2x(t) - 0.4x(t-0.3) + u(t) + \xi(t) \\ z(t) &= x(t) + x(t-0.3) + \eta(t) \end{aligned} \tag{156}$$

where $\xi(t)$ and $\eta(t)$ represent white Gaussian noises with zero mean and variances $Q_1\delta(t-t_1)$ and $Q_2\delta(t-t_1)$, respectively; the values of Q_1 and Q_2 are positive constants.

The problem is to determine the optimal feedback controller $u^*(t,\hat{x}_t)$ so as to minimize

$$C_c(u) = E\left\{\int_0^{10} [x^2(t) + u^2(t)]\, dt\right\}$$

when the measurements $z(t)$, $t_0 \le t$, are given.

The configuration of the feedback loop is specified by (148), (150), and (151). The optimal gain matrices of the controller and estimator are determined by Eqs. (154) and (155):

$$K^*(t,0) = P(t,0,0), \qquad K^*(t,\sigma) = P(t,\sigma,0)$$

$$G^*(t,0) = [M(t,0,0) + M(t,0,0.3)]Q_2^{-1}$$

$$G^*(t,\theta_1) = [M(t,\theta_1,0) + M(t,\theta_1,0.3)]Q_2^{-1}$$

where $t_0 \le t$, $-\tau \le \sigma < 0$, and $-\tau \le \theta_1 < 0$. The matrices $P(t,\sigma,\alpha)$ and $M(t,\theta_1,\theta_2)$, for $t_0 \le t$, $-\tau \le \sigma$, $\alpha \le 0$, and $-\tau \le \theta_1$, $\theta_2 \le 0$, are computed from Eqs. (56) to (58) and (96) to (98).

The equations for the optimal estimator are written from Eqs. (151) and (152):

$$\frac{d\hat{x}(t \mid t)}{dt} = -0.2\hat{x}(t \mid t) - 0.4\hat{x}(t-0.3 \mid t) + G^*(t,0)[z(t) - \hat{x}(t \mid t) - \hat{x}(t-0.3 \mid t)] + u^*(t,\hat{x}_t)$$

$$\frac{\partial\hat{x}(t+\theta \mid t)}{\partial t} = \frac{\partial\hat{x}(t+\theta \mid t)}{\partial\theta} + G^*(t,\theta)[z,t - \hat{x}(t \mid t) - \hat{x}(t-0.3 \mid t)] \tag{157}$$

where $G^*(t,\theta)$, $t \ge t_0$, and $-\tau \le \theta \le 0$ are specified by Eq. (155).

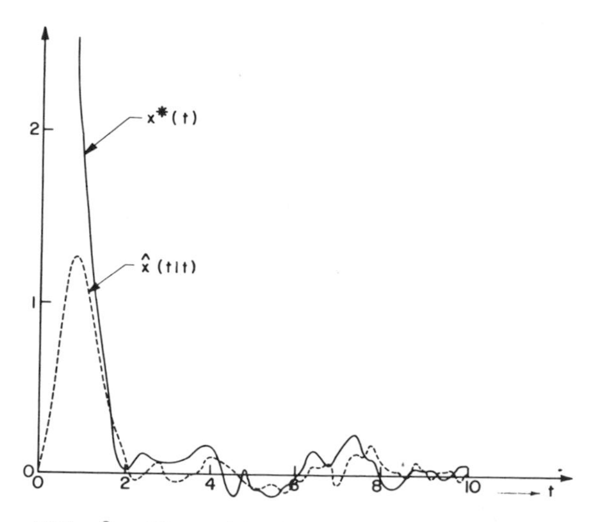

FIG. 8. The optimal performance of the system: $x^*(t)$, sample trajectory of the state; $\hat{x}(t \mid t)$, the optimal estimate.

Thus, the optimal feedback loop of the system is determined by cascading the optimal estimator specified by (157) and the controller given by

$$u^*(t,\hat{x}_t) = K^*(t,0)\hat{x}(t \mid t) + \int_{-\tau}^{0} K^*(t,\sigma)\hat{x}(t+\sigma \mid t)\, d\sigma$$

To simulate the performance of the optimal system, the PDEs are solved by means of characteristics. The performance of the optimal system for the values $Q_1 = 0.2$ and $Q_2 = 0.2$ is displayed in Figure 8.

V. CONCLUSIONS

In this chapter we have summarized those results of the optimal control theory for time-delay systems which to us seem most relevant in design. The presentation has been rather formal in certain sections, and the reader is urged to read more detailed proofs in the references provided.

We first discuss the pertinent mathematical theory of time-delay systems emphasizing the spectral decomposition, which is essential in the theory of observers. A maximum principle is stated for the posed optimization problem and then applied to the linear case. In fact, a large portion of the chapter is devoted to the linear systems, because of its well-understood structure and ease of application. The LPQ problem is treated in detail utilizing the Fredholm theory and summarizing the various other approaches that have been used to solve it. Computational considerations are very important in connection with the practical implementation, and a brief up-to-date survey of recent techniques is provided. The design of feedback controllers leads to the theory of observers and state estimation. An observer theory is presented by utilizing the spectral decomposition technique and an example illustrates the highlights of the theory.

A formal derivation of the optimal estimator equations for linear systems involving arguments with constant delays has been presented. The noise processes in the plant and observation equations are assumed to be white and Gaussian with known covariances.

Two approaches are given to nonlinear estimation problems in which the plant and observation equations contain variables with delayed arguments. A first-order estimator is obtained for the system which results from the linearization of the system equations about a known nominal trajectory. Then, an alternative estimation is presented by considering the modeling and measurement errors as unknown time functions; the estimator is so constructed that the best fit to the measured data is achieved in the sense of the least value of the integrated squared (estimation) errors.

Numerical examples illustrate computational aspects of the estimation. Then, the optimal control of linear stochastic time-delay systems is presented; it is demonstrated that the optimal gain matrices for the controller and the optimal gain matrices for the estimator can be computed independently. Examples are used throughout the chapter to illustrate the applicability of the underlying theory.

REFERENCES

1. P. Harriott, Process Control, McGraw-Hill, New York, 1964.
2. C. L. Smith, Digital Computer Process Control, International Textbook Company, Scranton, Pa., 1972.
3. O. J. M. Smith, A controller to overcome dead time, I.S.A. J., 6: 28-33 (1959).
4. H. H. Rosenbrock, Computer-aided Control System Design, Academic, London, 1974.
5. J. K. Hale, Functional Differential Equations, Springer-Verlag, New York, 1971.
6. A. Halanay, Differential Equations, Oscillations, Stability, Time Lags, Academic, New York, 1966.
7. H. T. Banks and A. Manitius, Application of abstract variational theory to hereditary systems — a survey, IEEE Trans. Auto. Contr., AC-19: 524-533 (1974).
8. G. L. Kharatishvili, Maximum principle in the theory of optimum time-delay processes, Dokl. Akad. Nauk, 136: 39-42 (1961).
9. E. B. Lee, Variational problems for systems having delay in the control action, IEEE Trans. Auto. Contr., AC-13: 697-699 (1968).
10. Y. Alekal, P. Brunovsky, D. H. Chyung, and E. B. Lee, The quadratic problem for systems with time delays, IEEE Trans. Auto. Contr., AC-16: 673-687 (1971).

11. H. N. Koivo and E. B. Lee, Controller synthesis for linear systems with time delays, Automatica, 8: 203-208 (1972).

12. W. M. Wonham, Linear Multivariable Control: A Geometric Approach, Springer-Verlag, New York, 1974.

13. R. E. Kalman, The theory of optimal control and calculus of variations, in Mathematical Optimization Techniques (R. Bellman, ed.), Univ. of California Press, Berkeley, California, 1963.

14. N. N. Krasovskii, On analytical design of optimum regulators in time-delay systems, J. Appl. Math. Mech., 26: 50-67 (1962).

15. D. W. Ross and Flügge-Lotz, An optimal control problem for systems with differential-difference equation dynamics, SIAM J. Contr., 7: 609-623 (1969).

16. D. H. Eller, J. K. Aggarwal, and H. T. Banks, Optimal control of linear time-delay systems, IEEE Trans. Auto. Contr., AC-14, 678-687 (1969).

17. M. C. Delfour and S. K. Mitter, Controllability, observability, and optimal feedback control of affine hereditary differential systems, SIAM J. Contr., 10: 298-328 (1972).

18. M. C. Delfour, C. McCalla, and S. K. Mitter, Stability and the infinite-time quadratic cost problem for linear hereditary differential systems, SIAM J. Contr., 13: 48-88 (1975).

19. A. E. Bryson and Y. C. Ho., Applied Optimal Control, Blaisdell, Waltham, Mass., 1969.

20. A. Manitius, Optimal control of time-lag systems with quadratic performance indices, Proceedings of the Fourth IFAC Congress, Warsaw, Poland, 1969.

21. A. Manitius, "Optimal control of hereditary systems," Research Report CRM-472, Centre de Recherches Mathematique, Univ. Montreal, 1974.

22. H. N. Koivo, "Fredholm resolvent in the optimization of linear systems with state and control retardations," Ph.D. thesis, Univ. of Minnesota, Minneapolis, Minn., 1971.

23. T. Kailath, Fredholm resolvents, Wiener-Hopf equations, and Riccati differential equations, IEEE Trans. Inform. Theory, IT-15: 665-672 (1969).

24. T. Kailath, Application of a resolvent identity to a linear smoothing problem, SIAM J. Contr., 7: 68-74 (1969).

25. A. Schumitzky, On the equivalence between matrix Riccati equations and Fredholm resolvents, J. Compu. Syst. Sci., 2: 76-87 (1968).

26. H. N. Koivo, On the existence of feedback controllers for hereditary differential systems, to be published.

27. N. N. Krasovskii and Y. S. Osipov, Stabilization of a controlled system with time delay, Eng. Cyb., no. 6, 1-11 (1963).

28. H. F. Vandevenne, Controllability and stabilizability properties of delay systems, Proceedings of the 1972 IEEE Conference on Decision and Control, 1972, pp. 370-377.

29. H. N. Koivo, On the equivalence of maximum principle openloop controllers and Caratheodory feedback controllers for time-delay systems, J. Optim. Theory Appl., 14: 163-178 (1974).

30. M. A. Soliman and W. H. Ray, Optimal feedback control for linear-quadratic systems having time delays, Int. J. Contr., 15: 609-627 (1972).

31. M. C. Delfour and S. K. Mitter, Hereditary differential systems with constant delays. I. General case, J. Diff. Eqs., 12: 213-235 (1972).

32. J. W. Daniel, Approximate Minimization of Functionals, Prentice-Hall, Englewood Cliffs, N. J., 1971.

33. A. P. Wierzbicki and A. Hatko, Computational methods in Hilbert space for optimal control problems with delays, in The Fifth Conference on Optimization Techniques, Part I (R. Conti and A. Ruberti, eds.), Springer-Verlag, New York, 1973, pp. 304-318.

34. E. Polak, Computational Methods in Optimization, Academic, New York, 1971.

35. D. Mackinnon, Optimal control of systems with pure time delays using a variational programming approach, IEEE Trans. Auto. Contr., AC-12: 255-262 (1967).

36. T. E. Mueller, Optimal control of linear systems with time lag. Proceedings of the Third Annual Allerton Conference on Circuit System Theory, Univ. of Illinois, Urbana, Ill., 1965, pp. 339-345.

37. M. A. Soliman and W. H. Ray, On the optimal control of systems having pure time delays and singular arcs. I. Some necessary conditions for optimality, Int. J. Contr., 16: 963-976 (1973).

38. M. A. Soliman and W. H. Ray, On the optimal control of systems having pure time delays and singular arcs. II. Computational considerations, Int. J. Contr., 18: 773-783 (1973).

39. V. H. Quintana and E. J. Davison, Clipping-off gradient algorithm to compute optimal controls with constrained magnitude, Int. J. Contr., 20: 243-257 (1974).

40. H. N. Koivo and R. Cojocariu, An optimal control for flotation circuits, Preprints of the Sixth IFAC, Boston, Mass., 1975.

41. P. Sannuti, Near optimum design of time-lag systems by singular perturbation method, Proc. 1970 JACC, 489-496 (1970).

42. P. B. Reddy and P. Sannuti, Optimal control of singularly perturbed time delay systems with an application to a coupled core nuclear reactor, Proceedings of the 1974 Conference on Decision and Control, 1974, pp. 793-803.

43. H. C. Chan and W. R. Perkins, Optimization of time delay systems using parameter imbedding, Automatica, 9: 257-261 (1973).

44. P. Sannuti and P. B. Reddy, Asymptotic series solution of optimal systems with small time delay, IEEE Trans. Auto. Contr., AC-18: 250-259 (1973).

45. Y. Alekal, "Synthesis of feedback controllers for systems with time delay," Ph.D. Dissertation, Univ. of Minnesota, Minneapolis, Minn., 1969.

46. J. K. Aggarwal, Computation of optimal control for time-delay systems, IEEE Trans. Auto. Contr., AC-15: 683-685 (1970).

47. F. B. Hildebrand, Methods of Applied Mathematics, Prentice-Hall, Englewood Cliffs, N. J., 1965.

48. M. E. Salukvadze, Concerning the synthesis of an optimal controller in linear delay systems subject to constantly acting perturbations, Auto. Remote Contr., 23: 1495-1501 (1962).

49. D. W. Ross, Controller design for time-lag systems via a quadratic criterion, IEEE Trans. Auto. Contr., AC-16: 664-673 (1971).

50. H. T. Banks and J. A. Burns, An abstract framework for approximate solutions to optimal control problems governed by hereditary systems, Proceedings of the International Conference on Differential Equations, Univ. of Southern California, Sept. 3-7, 1974.

51. M. C. Delfour, "Numerical solution of the optimal control problem for linear hereditary differential systems with a linear quadratic cost function and approximation of the Riccati differential equation," Research Report CRM-408, Centre de Recherches Mathematiques, Univ. of Montreal, 1974.

52. R. W. Koepcke, On the control of linear systems with pure time-delay, Trans. ASME, J. Basic Eng., Series D, 87: 74-80 (1965).

53. F. Kurzweil, The control of multivariable processes in the presence of pure transport delays, IEEE Trans. Auto. Contr., AC-10: 27-34 (1965).

54. H. W. Smith, Dynamic control of a two-stand cold mill, Automatica, 5: 183-190 (1969).

55. H. T. Banks and A. Manitius, Projection series for functional differential equations with applications to optimal control problems, J. Diff. Eqs., 18: 296-332 (1975).

56. D. G. Luenberger, Observing the state of a linear system, IEEE Trans. Mil. Electron., MIL-8: 74-80 (1964).

57. D. G. Luenberger, An introduction to observers, IEEE Trans. Auto. Contr., AC-16: 596-602 (1971).

58. K. P. M. Bhat and H. N. Koivo, An observer theory for time-delay systems, IEEE Trans. Auto. Contr., to be published.

59. K. P. M. Bhat and H. N. Koivo, "Modal characterizations of controllability and observability for time-delay systems," Control Systems Report No. 7508, Department of Electrical Engineering, Univ. of Toronto, May 1975.

60. M. C. Delfour, "State theory of linear hereditary differential systems," Research Report CRM-395, Centre de Recherches Mathematiques, Univ. of Montreal, 1974.

61. R. V. Gressang and G. B. Lamont, Observers for systems characterized by semigroups, IEEE Trans. Auto Contr., AC-20: 523-528 (1975).

62. R. V. Gressang, Observers and pseudo observers for linear time-delay systems, Proceedings of the 12th Allerton Conference on Circuit and System Theory, Allerton House, Monticello, Ill., Oct. 2-4, 1974.

63. R. E. Kalman and R. S. Bucy, New results in linear filtering and prediction theory, J. Basic Eng., Trans. ASME, 83: 55-108 (1961).

64. H. Kwakernaak, Optimal filtering in systems with time delays, IEEE Trans. Auto. Contr., AC-12: 169-173 (1967).

65. A. J. Koivo, Optimal estimation for linear stochastic systems described by functional differential equations, Inform. Contr., 19: 232-245 (1971).

66. H. N. Koivo, Least-squares estimator for hereditary systems with time varying delay, IEEE Trans. Syst., Man. Cyb., SMC-4: 275-283 (1974).

67. T. K. Yu, J. H. Seinfeld, and W. H. Ray, Filtering in nonlinear time delay systems, IEEE Trans. Auto. Contr., AC-19: 324-333 (1974).

68. A. V. Balakrishnan and J. L. Lions, State estimation for infinite-dimensional systems, J. Comp. Syst. Sci., 1: 391-403 (1967).

69. H. J. Kushner, Approximations to optimal nonlinear filters, IEEE Trans. Auto. Contr., AC-12: 546-556 (1967).

70. R. P. Wishner, J. A. Tabaczynski, and M. Athans, A comparison of three nonlinear filters, Automatica, 6: 487-496 (1969).

71. A. Lindquist, Optimal control of linear stochastic systems with applications to time lag systems, Inform. Sci., 5: 81-126 (1973).

72. A. J. Koivo and R. L. Stoller, Least-squares estimator for nonlinear systems with transport delay, J. Dynam. Syst., Meas. Contr., Trans. ASME, 96: 301-306 (1974).

73. D. G. Lainiotis, Estimation: A brief survey, Inform. Sci., 7: 191-202 (1974).

74. R. Courant and D. Hilbert, Methods of Mathematical Physics, Wiley (Interscience), New York, 1962.

75. L. A. Gould and W. Kipiniak, Dynamic optimization and control of stirred-tank chemical reactor, Trans. AIEE, Part 1, Commun. Electron., 79: (1960).

76. D. W. Repperger and A. J. Koivo, On stable forward filtering and fixed-lag smoothing in a class of systems with time delays, IEEE Trans. Auto. Contr., AC-19: 266-267 (1974).

77. A. J. Koivo, Optimal control of linear stochastic systems described by functional differential equations, J. Optim. Theory Appl., 9: 161-175 (1972).

Chapter 6

SECOND-ORDER COMPUTATIONAL METHODS FOR DISTRIBUTED PARAMETER OPTIMAL CONTROL PROBLEMS

Kun Soo Chang

Department of Chemical Engineering
University of Waterloo
Waterloo, Ontario, Canada

I. INTRODUCTION

A considerable number of computational techniques for distributed parameter optimal control and optimization problems have appeared in recent years. Unlike lumped parameter problems, distributed parameter problems can be highly involved depending on the form of partial differential equations that describe the state of the system. The state equations may be broadly classified as linear, nonlinear, low- or high-order partial differential equations for the general case, but the partial differential operators, even if linear, may appear in the form of hyperbolic, parabolic, elliptic, or a mixture of these types. Lumped parameter dynamics may also play a role in the part of the system characterization. The controls are usually distributed over the domain of all independent variables, but are often partially distributed, making some of them independent of time or of some space variables. A few may appear only on the boundary. Sometimes the state of the system may be given in the form of integral equations or may be converted into this form if it is more tractable. All these complications point to the fact that there can hardly be a truly unified computational approach to distributed parameter optimal control and optimization problems. Thus, one technique may work well for one type of problem, but the same technique may not be suitable for another type. Furthermore, if a numerical solution is sought, the success of a particular computational technique depends not only on the particular control algorithm used, but also on the numerical scheme employed for the solution of partial differential equations. The fact that there are many numerical solution schemes available, some superior to others for a given problem, compounds the complexities of the task of solving the optimal control or optimization problems.

Fortunately, in spite of the difficulties, many computational techniques for distributed parameter control systems have evolved in the past. Publications dealing with the techniques are collected in the reference section and briefly surveyed under the headings of various methods. Although it is generally difficult to draw a clear line of distinction between computational algorithms and pertinent optimal control theory, very little effort has been made to include comprehensive publications on control theory of distributed parameter systems. Thus, the cited references [1-204] include only those publications directly dealing with control computations of distributed parameter problems. In each group of methods classified, the references are arranged and cited roughly in chronological order. This arrangement, it is hoped, should provide some degree of historical perspective in the development of computational techniques for distributed parameter control problems.

A. Gradient and Steepest Descent Methods

Gradient and steepest descent or ascent methods are very simple to program and easy to implement. For distributed parameter (DP) control problems, Volin and Ostrovskii [173,174] developed a gradient procedure for hill climbing in function space which required the solution of the system and adjoint equations at each iteration. In the work of Denn, Gray, and Ferron [47] and Denn [45], a steepest descent technique was developed for second-order nonlinear parabolic partial differential equations (PDEs). A gradient method for a system described by vector PDE was treated by Sage and Chaudhuri [146]. In the work of Seinfeld and Lapidus [155,156], a method of steepest ascent was used for hyperbolic and parabolic systems and for a singular control problem in connection with a method of direct search. A gradient algorithm for a jacketed tubular reactor system described by nonlinear first-order hyperbolic equations was formulated in the work of Chang and Bankoff [36,37]. In Kim's work [84], an algorithm based on the first variation for a multidimensional linear diffusion process was proposed. A gradient method for feedback control was used in Baldwin, Ruedrich, and Durbin [10]. An extension of the Volin and Ostrovskii's gradient method to a multibed reactor system problem is available in Paynter [134]. In Lee and Shen [101], a steepest descent method for a linear boundary control problem is available. Algorithms based on gradient methods were formulated and used extensively in Ogunye [121] and Ogunye and Ray [122-126] for nonlinear systems described by a set of first-order hyperbolic PDEs with orthogonal characteristics. An application of gradient method for nonlinear diffusion systems with interacting boundary conditions was presented in Bansal [14] and Bansal and Chang [15]. A gradient technique was also used in Hahn, Fan, and Hwang [74] for a jacketed tubular reactor problem. A comparison of a gradient method with penalty methods and Galerkin's method can be found in Lions [106, 107]. A gradient method for a nonlinear control problem was also presented in Chaudhuri [39]. The formulation of a first-order algorithm for nonlinear problems also appears in Holliday and Storey [78]. A gradient algorithm for dispersion-type packed bed reactors experiencing catalyst decay can be found in Ajinkya and Ray [1].

B. Conjugate Gradient Methods

Methods of conjugate gradient for lumped systems have been extended to distributed parameter problems. Development of a conjugate gradient algorithm for one-dimensional linear DP systems was given in Lee and Shen [101]. In Kenneth, Sibony, and Yvon [81], a conjugate gradient method for the solution of penalized convex programming in Hilbert space is available. Applications of conjugate gradient method to linear distributed parameter systems (DPS) with quadratic functional can be found in Michel and Cornick [42,43,115]. The formulation and application of modified conjugate gradient method for optimal heating problems can be found in Yang [191] and Yang and Chang [192]. In Ball and Hewit [12,13], a conjugate gradient algorithm for both constrained and unconstrained linear systems is available.

C. Methods of Weighted Residuals

The methods commonly referred to as the method of moments, the method of least squares, the method of Galerkin, and the method of projections are in a general sense all methods of weighted residuals. An applied pioneering work of the method of harmonics to a DP control problem was carried out by McCausland [112], who used a Fourier series representation in the control of temperature distribution in a solid. Butkovskii [27] developed and presented the method of moments in the general theoretical framework of DP optimal control problems. Subsequent papers by Butkovskii and Poltavskii [29-31] demonstrated the application of the method to wave equations. An application of the method of moments to a bounded energy one-dimensional heat conduction problem can be found in Yavin and Rasis [194]. A system translation of a linear parabolic equation to a problem of moments was treated in Gal'chuk [61], and an approximate optimal feedback control by Galerkin's method was carried out in Zahradnik and Lynn [200]. A time-optimal control of linear diffusion processes by Galerkin's method was treated in Prabhu and McCausland [139], while Galerkin's method as a numerical algorithm was treated in Lions [106,107]. In Parkin and Zahradnik [131], a method of weighted residuals to trajectory approximation was applied. A suboptimal control algorithm based on weighted residuals was presented in Neuman and Sen [118,119]. A design procedure for linear processes using Galerkin's method of projection is found in Orner and Foster [129]. A numerical approximation approach for a parabolic system using a Ritz-Galerkin procedure is available in McKnight and Bosarge [114]. A method of boundary control of temperature distribution can be found in Fattorini [57] in which the problem was reduced to moment problems.

D. Linear and Nonlinear Programming Methods

Distributed parameter control problems can be converted into programming problems. Sakawa [147,148] used linear, nonlinear, and quadratic programming techniques to solve a heat conduction problem and integral equation problems. A method of linear programming to sampled-data DP systems was used in Lorchirachoonkul and Pierre [108]. A method of penalization for convex programming in a Hilbert space was proposed in Kenneth, Sibony, and Yvon [81]. In Koivo and Kruh [89], a simplex method of linear programming in the design of an approximate optimal feedback controller of a heat conduction system was given. A dynamic programming approach can also be used as in the work of Graham and D'Souza [72]. A modified Frank-Wolfe algorithm of quadratic programming was used for a temperature control problem by Barnes [16,17]. In Lions [106,107], the optimal control of DPS in the framework of linear quadratic and duality problems was treated and compared with the results of penalty methods. A method of linear programming was also applied in Rosen [144] for an iterative graphical solution. An extension of penalty methods to linear DPS with inequality constraints was presented in Sasai [150]. Tarng and Nardizzi [166] applied a method of penalty function to PDEs. For discrete-time models, the use of a mathematical programming method can be found in Sheirah and Hamza [157].

E. Direct Search and Quasilinearization Methods

A quasilinearization technique, first developed by Bellman and Kalaba as a tool for solving nonlinear two-point boundary value problems in both ordinary differential equations (ODEs) and PDEs is well known. Sage and Chaudhuri [146] and Chaudhuri [39] applied the technique to DP control problems. In the work of Seinfeld and Lapidus [155,156], a method of direct search on the system performance index was proposed as a successful computational tool for hyperbolic and parabolic DPS. They also incorporated a method of quasilinearization in the solution of two-point boundary value problems. Zone [202] also formulated a quasilinearization algorithm for periodic DP processes.

F. Functional Analysis Methods

Techniques of functional analysis have found wide applications in DP control problems, especially in those cases where integral equation representation of the state was possible. There are many publications dealing with the theoretical aspects of functional analysis approach to DP systems which deserve considerable attention on their own merits. However, we restrict the scope here to those publications dealing directly with computational aspects as applied to distributed parameter control systems. Axelband [5] applied a method of functional analysis to the control problem of a class of DPS using a steepest descent technique in function space. A Banach space approach for linear DPS was proposed by Chaudhuri [38]. Uzgiris and D'Souza [171] dealt with the optimum time control of a parabolic system with nonlinear boundary conditions. A linear problem with bounded inputs was treated in Axelband [6]. An integral equation formulation and a successive approximation solution method were treated in Sakawa [148,149]. In another work by Axelband [7], a solution algorithm for linear optimal pursuit problem can be found. Numerical algorithms for solving linear DP control problems were proposed and well described in Brogan [23,24]. Solution techniques for linear systems with constrained inputs were presented in Weigand and D'Souza [185], and a technique of obtaining explicit solutions for an abstract minimum norm problem was dealt with in Fahmy [55]. A bounded energy control problem of a nonlinear integral equation by means of approximation by a sequence of degenerate kernels can be found in Yavin and Sivan [196]. An excellent treatment of function space methods for linear DPS is available in Axelband [8]. In Lee and Shen [101], a one-dimensional conduction problem was treated in a functional analysis setting. A treatment of an improved iteration technique using a modified steepest descent in Hilbert space can be found in Chang [35]. A computational algorithm based on functional analysis and Green's identity is available in Tzafestas [167], while a suggested computational scheme based on integrodifferential equation for input derivative constraint problems can be found in Yavin [193]. In Yang [191] and Yang and Chang [192], a method of conjugate gradient in the numerical computation of optimal control in function space is available. A minimum time control of a system with multiple-norm constraints was dealt with in Singh and Rajamani [160]. A method of conjugate gradient in function space was also applied in

Ball and Hewit [12]. A treatment of optimal boundary control for a linear stochastic system in Hilbert space can be found in Omatu, Shibata, and Hata [128]. For systems with discrete constrained inputs, a treatment is available in Hamza and Rasmy [76].

G. Second-order Methods

Second-order methods in a general sense are all methods of gradient, but they possess excellent convergence properties which are superior to those of ordinary gradient methods. For this reason, second-order methods have been widely investigated for lumped parameter systems (LPS). However, only fairly recently has any such technique been developed for DP control systems. Tzafestas and Nightingale [170] developed an algorithm for DP control systems based on the second-order expansion of the DP Hamilton-Jacobi equation in the differential dynamic programming formulation. The technique was an extension of Jacobson and Mayne's differential dynamic programming approach for LPS. In the subsequent paper [167], Tzafestas derived a set of necessary conditions for final value control of nonlinear composite DPS and LPS in functional analysis setting. He proposed an iterative computational algorithm based on successive improvement by sensitivity functions of the initial conditions of the adjoint equations. Tzafestas [168] also put forth another algorithm based on variational calculus which would exhibit a second-order convergence property. Zone and Chang [203] formulated an algorithm based on the second-order expansion of the Hamiltonian function for a general class of nonlinear DPS. Subsequently, a successive approximation algorithm based on the direct second-order expansion of the performance index was developed in Zone and Chang [204] and Zone [202]. In this algorithm, the integration of adjoint equations is not necessary. This algorithm and Tzafestas and Nightingale's algorithm [170] seem to possess many desirable features of efficient computation. Recently, Holliday and Storey [78] have applied a second-order gradient method in the numerical solution of parabolic and first-order hyperbolic DP problems.

H. Discretization and Numerical Schemes

There are many numerical solution techniques available for PDEs. Some have appeared in book form, others in journals. To name a few roughly in chronological order, we have Lapidus [98], Saul'yev [152], Forsythe and Wasow [59], Smith [162], Mitchell [116], Ames [3], Bush [25], Fairweather and Mitchell [56], Lee [99], Dorri [48], Dupont [49], Rachford [140], Dupont, Fairweather, and Johnson [50], Paul and Strainge [132], Walsh [177], and many others. It was Butkovskii [26,27] who initially suggested a spatial discretization of the DP model in order to apply well-developed techniques for LPS. Wang and Tung [183] and Wang [178] discussed in some detail the difficulties associated with approximating DPS by spatial and time discretization. It was believed preferable to preserve the distributed nature of the problem and its solution as far as possible before the discretization for computation was performed. Various discretization and solution schemes for DP optimal control problems can be found in Gould and Murray-Lasso [68], Yeh and Tou [197,198], Vostrova [175], Kim and Erzberger [85], Yavin and Sivan [195], Sage [145], Wismer [188,189], Kusic [96], Alvarado and Mukundan [2], Cea and Malanowski [34], Graham [69], Sasai and Shimemura [151], Casti and Rao [33], Bégis and Glowinski [18], Davis and Perkins [44], Kogan, Krtolica, and Lapina [88], Chaudhuri [39], Johnson [79], Pavlov [133], Friedman and Yavin [60], and Butkovskii, Darinskii, and Pustyl'nikov [28]. Many other papers mentioned elsewhere in the chapter have also dealt in varying degrees with numerical solution techniques of PDEs.

I. Analytical Controller Design: Feedback and Feedforward Controllers

For some linear DPS, analytical controller design is possible. The techniques were used in Koppel [92], Koppel and Shih [93], Lim [103, 104], and L'vova [110]. Feedback and feedforward controller design schemes were also put forward for adaptive and stochastic as well as deterministic control problems. They may be found in Khatri and Goodson [83], Wang [179], Wiberg [187], Denn [46], Tzafestas and

Nightingale [169], Shih [158], Paraskos and McAvoy [130], Greenberg [73], Hahn, Fan, and Hwang [75], Lim and Fang [105], Baker and Brosilow [9], Davis and Perkins [44], Chu and Shih [41], Pedersen and Nardizzi [136], and Mutharasan and Coughanowr [117].

J. Trajectory Approximations, Modal Control, and Other Techniques

A variety of techniques such as rational approximation, trajectory approximation, eigenfunction expansion, transform technique, modal control, approximate expansion, perturbation technique, suboptimal control, discrete-time control, and Wiener-Hopf spectral analysis have been applied in the past for DP control problems. These techniques may be found in various research publications; they are Ball [11], Kadymov and Listengarten [80], Pierre [137], Khatri and Goodson [82], Egorov [52], Raspopov [142], Kolb and Pierre [90], Golub' [64], Erzberger and Kim [53], Goldwyn, Sriram, and Graham [62], Wang [180], Seinfeld and Kumar [154], Wang [181], Butkovskii and Poltavskii [32], Andreev and Orkin [4], Weigand and D'Souza [186], Golub' [67], Paynter, Dranoff, and Bankoff [135], Hassan and Solberg [77], Stafford and Nightingale [164], Singh [159], Graham [70], Goldwyn, Sriram, and Graham [63], Lynn and Zahradnik [111,201], Graham [71], Klein and Hughes [86], Porter and Bradshaw [20,138], Leibowitz and Surendran [102], Yu and Seinfeld [199], Kotchenko and Solomin [95], Vermeychuk [172], Bradshaw [19], and McGlothin [113].

K. Applications

Applications of optimal control computation for DP processes can be found in many areas. The processes include thermal regenerators and heat exchangers, nuclear rockets and reactors, massive-body heating, crystal growing, drying, tubular reactors, packed beds with catalyst decay, gas-solid reactions, countercurrent processes, two-phase flow processes, natural gas pipelines, river aeration, continuous strip heating, beam vibration, plasma, and magnetohydrodynamics. They can be found in the work of Wismer and Lefkowitz [190], Briggs and Shen [21,22], Golub' [65,66], Sirazetdinov [161], Komkow [91], Raspopov

[143], Tarassov, Perlis, and Davidson [165], Fjeld and Kristiansen [58], Kwakernaak, Strijbos, and Tijssen [97], Vyrk [176], Ogunye and Ray [122,125,126], Weigand [184], Seinfeld, Gavalas, and Hwang [153], Rapoport [141], Sood, Funk, and Delmastro [163], Nishida, Ichikawa, and Tazaki [120], Lee, Koppel, and Lim [100], Klestov and Sirazetdinov [87], Chaudhuri [40], Evans et al. [54], Ajinkya and Ray [1], Wang [182], Olivei [127], and Earp and Kershenbaum [51]. In these publications, examples of solution techniques can also be found.

With this overall view of computational techniques for DPS, we now present the main topic of this chapter, the second-order computational techniques for DP control systems. Two different approaches are available and they are treated in detail in Sections II and III. In the first approach, a successive approximation algorithm based on the direct second-order expansion of the performance index is used for the computation of a general class of nonlinear DP control systems with nonlinear functional boundary conditions. This is treated in Section II. In the second approach, an iterative algorithm based on the expansion of the Hamilton-Jacobi equation is used in the framework of the differential dynamic programming approach for nonlinear DP control problems. This is treated in Section III.

II. SECOND-ORDER METHOD OF SUCCESSIVE APPROXIMATIONS

We derive here a successive approximation algorithm based on the direct expansion of the performance index up to the second order for a general class of nonlinear DPS having nonlinear functional boundary conditions. The nonlinear system contains both domain and boundary controls and also, as is usually the case with many physical processes, the boundary conditions contain a lumped parameter vector which is driven by its own dynamics. Although a method based on the second-order expansion of the system Hamiltonian function is also available [203], the present method is different from the standard method in that the need for the solution of the system adjoint equations is eliminated. The development follows the work of Zone and Chang [204]. We begin first by explaining the model of PDEs used in the treatment.

A. Description of the Mathematical Model

A general state equation model which describes a wide class of DP control processes may be given by the following system of PDEs:

$$\frac{\partial^{n_i} w_i}{\partial t^{n_i}} = h_i \left[t, x_1, \ldots, x_s, w_1, \ldots, w_m, \ldots, \left(\frac{\partial^K w_j}{\partial t^{k_0}\, \partial x_1^{k_1} \ldots \partial x_s^{k_s}} \right)_i, \right.$$

$$\left. u_1, \ldots, u_r \right] \tag{1}$$

$$i,j = 1,2, \ldots, m, \qquad \sum_{\ell=0}^{s} k_\ell = K, \qquad k_0 < n_i \text{ (for each } i \text{ and } j)$$

where $w_i(t,x)$ $(i = 1,2, \ldots, m)$ are the distributed state variables; $u_i(t,x)$ $(i = 1,2, \ldots, r)$ are the domain control variables; h_i $(i = 1,2, \ldots, m)$ are the specified functions of their arguments; t is time; and $x = (x_1, x_2, \ldots, x_s)^T$ (T for transpose) is the spatial coordinate vector of domain Ω. The term

$$\left(\frac{\partial^K w_j}{\partial t^{k_0}\, \partial x_1^{k_1} \ldots \partial x_s^{k_s}} \right)_i$$

denotes all the possible partial derivatives of w_j up to order K $(K = 1,2, \ldots, N_i)$, where N_i is the highest-order partial derivative on the right-hand side of the ith equation. The inequality $k_0 < n_i$ for all j implies that $\partial^{n_i} w_i / \partial t^{n_i}$ on the left-hand side is the highest time-derivative term in the ith equation. We assume that the problem given by Eq. (1) is well-posed. By introducing new variables $v_1, v_2, \ldots$, and v_n as

$$v_1 = w_1, \qquad v_2 = \frac{\partial w_1}{\partial t}, \ \ldots, \qquad v_{n_1} = \frac{\partial^{n_1 - 1} w_1}{\partial t^{n_1 - 1}},$$

$$v_{n_1+1} = w_2, \qquad v_{n_1+2} = \frac{\partial w_2}{\partial t}, \ \ldots, \qquad v_{n_1+n_2} = \frac{\partial^{n_2 - 1} w_2}{\partial t^{n_2 - 1}},$$

$$\ldots \qquad \ldots \qquad \ldots \qquad \ldots$$

$$v_{p+1} = w_m, \qquad v_{p+2} = \frac{\partial w_m}{\partial t}, \quad \ldots , \qquad v_{p+n_m} = \frac{\partial^{n_m-1} w_m}{\partial t^{n_m-1}} \tag{2}$$

where $p = \Sigma_{i=1}^{m-1} n_i$ and $p + n_m = n$, and thus enlarging the number of state variables, we can reduce Eqs. (1) to a system of equations with only first-order partial derivatives with respect to time

$$\frac{\partial v_i}{\partial t} = f_i \left[t, x_1, \ldots, x_s, v_1, \ldots, v_n, \left(\frac{\partial^m v_j}{\partial x_1^{k_1} \ldots \partial x_s^{k_s}} \right)_i, \right.$$

$$\left. u_1, \ldots, u_r \right] \tag{3}$$

$$i,j = 1,2, \ldots, n, \qquad \sum_{\ell=1}^{s} k_\ell = M \text{ (for each i and j)}$$

where $(\;)_i$ in Eq. (3) denotes all the possible spatial partial derivatives of v_j up to order M ($M = 1,2, \ldots, \bar{N}_i$) in the ith equation, where $\bar{N}_i$ is the highest-order spatial partial derivative on the right-hand side of the ith equation. Equation (3) can be written more compactly as

$$\frac{\partial v(t,x)}{\partial t} = f \left[t, x, v(t,x), \frac{\partial v}{\partial x_1}, \ldots, \right.$$

$$\left. \times \frac{\partial^m v}{\partial x_1^{k_1} \ldots \partial x_s^{k_s}}, \ldots, u(t,x) \right] \tag{4}$$

where $v(t,x)$ is an n-dimensional state vector, f is an n-dimensional vector-valued function, and $u(t,x)$ is an r-dimensional control vector. Many high-order nonlinear PDEs can be reduced by the introduction of new variables in this fashion to a system of first-order quasilinear PDEs of the form

$$A(v,x,t) \frac{\partial v}{\partial t} + \sum_{i=1}^{s} B_i(v,x,t) \frac{\partial v}{\partial x_i} + C(v,u,x,t) = 0 \tag{5}$$

which is a special form or a further reduced form of Eq. (4).

However, the reduction depends on the way the variables are redefined, and therefore is not unique. The symbols are explained in the next section where the optimal control problem is formulated around this form of state equation.

B. Description of the Optimal Control Problem

With such a reduction in mind, we now state a general nonlinear control problem for the following reduced system of PDEs:

$$A(v,x,t)v_t + \sum_{i=1}^{s} B_i(v,x,t)v_{x_i} + C(v,u,m,x,t) = 0 \tag{6}$$

where $x = (x_1,x_2, \ldots ,x_s)^T$ is the s-dimensional Euclidean space coordinate vector; $t \in [t_0,t_f]$ is the time variable; $v(x,t)$ is the n-dimensional distributed state of the system; $u(x,t)$ is the r-dimensional distributed control; and $m(t)$ is the q-dimensional spatially independent distributed control. Here and in the sequel, the short notations $v_t = \partial v/\partial t$, $v_{x_i} = \partial v/\partial x_i$, etc. are often used. The vector functions v, u, and m are defined for all $t \in [t_0,t_f]$ and for all $x \in \Omega$ where $\Omega = \{x \mid x_i^0 < x_i < x_i^f,\ i = 1,2, \ldots ,s\}$. The variables t_0 and t_f are the initial and final times, respectively, and, similarly, x_i^0 and x_i^f are the initial and final coordinate points of the ith space coordinate. Since Eq. (6) may be a reduced form of hyperbolic, parabolic, or elliptic PDEs or a mixture of these equations, the n x n matrices A and B_i $(i = 1,2, \ldots ,s)$ are not necessarily of the maximal rank and thus are not necessarily invertible; C is an n-dimensional vector-valued function of its arguments as specified. The initial and functional boundary conditions are specified as

$$Gv(x,t_0) = a(x) \tag{7}$$

where G is a coefficient matrix and a(x) is a known vector of x, and

$$\left[g_i[v(x,t),b_i(x,t),z(t)]\right]_{x_i=x_i^0} = 0 \tag{8}$$

$$\left[\hat{g}_i[v(x,t),d_i(x,t),z(t)]\right]_{x_i=x_i^f} = 0 \tag{9}$$

for $t \in [t_o,t_f]$ and $i = 1,2, \dots ,s$. The boundary controls b_i and d_i $(i = 1,2, \dots ,s)$ have appropriate dimensions and are not a function of x_i since they are defined at x_i^o and x_i^f, respectively. The vector $z(t)$ is an α-dimensional spatial coordinate-free parameter which may be considered as a spatially independent boundary control, but is driven by another control $k(t)$ through the dynamic equation:

$$\dot{z} = \varphi[z(t),k(t),t], \qquad z(t_o) = z_o \tag{10}$$

where φ is an α-dimensional vector-valued function of its arguments and $k(t)$ is a β-dimensional time-dependent control vector (here $\dot{z} = dz/dt$). The general form of performance index which is to be minimized is given by

$$\begin{aligned} P = & \int_{t_o}^{t_f}\int_{\Omega} Q(v,v_x,u,m,x,t)\, d\Omega\, dt \\ & + \sum_{i=1}^{s} \int_{t_o}^{t_f}\int_{\partial\Omega_i} \Big\{ L_i[v(x,t),b_i(x,t)] \Big|_{x_i=x_i^o} \\ & + \hat{L}_i[v(x,t),d_i(x,t)] \Big|_{x_i=x_i^f} \Big\}\, d\partial\Omega_i\, dt \\ & + \int_{\Omega} M[v(x,t_f)]\, d\Omega + \int_{t_o}^{t_f} N[z(t),k(t)]\, dt \end{aligned} \tag{11}$$

where

$$\Omega = \{x \mid x_i^o < x_i < x_i^f,\ i = 1,2, \dots ,s\} \tag{12}$$

$$\partial\Omega_i = \{\xi_i' \mid x_i = x_i^o \quad \text{or} \quad x_i = x_i^f\} \tag{13}$$

and

$$\xi_i' = (x_1, \dots ,x_{i-1},x_{i+1}, \dots ,x_s)^T \tag{14}$$

The integrands Q, L_i, $\hat{L}_i$, M, and N in the performance index are scalar-valued functions of their arguments and defined on the respective spaces, and $d\Omega$ and $d\partial\Omega_i$ denote the product of differentials $dx_1dx_2 \dots dx_s$ and $dx_1 \dots dx_{i-1},dx_{i+1} \dots dx_s$, respectively.

The control problem is then to find a set of control functions $u(x,t), m(t), b_i, d_i$ $(i=1,2, \ldots ,s)$, and $k(t)$ so as to minimize P in a specified time period, $t_f - t_o$. In this general setting, the proof of well-posedness of the problem is a difficult task. However, to proceed to our main aim of developing the algorithm, we assume here that any given problem in this general framework is well-posed in the usual sense. Some conditions for well-posedness will appear in the form of convergence conditions in the later sections.

C. Development of Equations

Suppose that a nominal set of controls u, m, b_i, d_i $(i=1,2, \ldots ,s)$, k is available. Then it is possible to integrate the state equations (6) and (10) in the forward time direction and obtain the nominal solution surface (or trajectory). We now expand the performance index up to second-order terms around this nominal solution surface. The increment in the performance index around this solution surface can be expressed by

$$\Delta P = \delta P + \frac{1}{2} \delta^2 P + o(\delta^2 \cdot) \tag{15}$$

where $o(\delta^2 \cdot)$ represents the third and higher-order terms in the Taylor series expansion. The evaluation from Eq. (11) gives

$$\begin{aligned} \delta P = & \int_{t_o}^{t_f} \int_{\Omega} \left\{ Q_v^T \, \delta v + \sum_{i=1}^{s} Q_{v_{x_i}}^T \, \delta v_{x_i} + Q_u^T \, \delta u + Q_m^T \, \delta m \right\} d\Omega \, dt \\ & + \sum_{i=1}^{s} \int_{t_o}^{t_f} \int_{\partial\Omega_i} \left\{ \left(L_v^T \, \delta v + L_b^T \, \delta b \right)_{x_i = x_i^o} \right. \\ & \left. + \left(\hat{L}_v^T \, \delta v + \hat{L}_d^T \, \delta d \right)_{x_i = x_i^f} \right\} d\partial\Omega_i \, dt + \int_{\Omega} M_v^T \, \delta v \Big|_{t=t_f} d\Omega \\ & + \int_{t_o}^{t_f} \left(N_z^T \, \delta z + N_k^T \, \delta k \right) dt \end{aligned} \tag{16}$$

$$\delta^2 P = \int_{t_0}^{t_f}\int_{\Omega} \begin{bmatrix} \delta v \\ \delta u \\ \delta m \\ \delta v_x \end{bmatrix}^T \begin{bmatrix} Q_{vv} & Q_{vu} & Q_{vm} & Q_{vv_x} \\ Q_{vu}^T & Q_{uu} & Q_{um} & Q_{uv_x} \\ Q_{vm}^T & Q_{um}^T & Q_{mm} & Q_{mv_x} \\ Q_{vv_x}^T & Q_{uv_x}^T & Q_{mv_x}^T & Q_{v_x v_x} \end{bmatrix} \begin{bmatrix} \delta v \\ \delta u \\ \delta m \\ \delta v_x \end{bmatrix} d\Omega\, dt$$

$$+ \sum_{i=1}^{s} \int_{t_0}^{t_f} \int_{\partial\Omega_i} \left\{ \begin{bmatrix} \delta v \\ \delta b \end{bmatrix}^T \begin{bmatrix} L_{vv} & L_{vb} \\ L_{vb}^T & L_{bb} \end{bmatrix} \begin{bmatrix} \delta v \\ \delta b \end{bmatrix}_{x_i = x_i^0} \right.$$

$$\left. + \begin{bmatrix} \delta v \\ \delta b \end{bmatrix}^T \begin{bmatrix} \hat{L}_{vv} & \hat{L}_{vd} \\ \hat{L}_{vd}^T & \hat{L}_{dd} \end{bmatrix} \begin{bmatrix} \delta v \\ \delta d \end{bmatrix}_{x_i = x_i^f} \right\} d\partial\Omega_i\, dt$$

$$+ \int_{\Omega} \delta v^T M_{vv}\, dv \Big|_{t=t_f} d\Omega + \int_{t_0}^{t_f} \begin{bmatrix} \delta z \\ \delta k \end{bmatrix}^T \begin{bmatrix} N_{zz} & N_{zk} \\ N_{zk}^T & N_{kk} \end{bmatrix} \begin{bmatrix} \delta z \\ \delta k \end{bmatrix} dt \quad (17)$$

Here we used the notations $Q_v = (Q_{v_1}, Q_{v_2}, \ldots, Q_{v_s})^T$, $(Q_{vu})_{i,j} = \partial^2 Q/(\partial v_i\, \partial u_j)$ for the ith row and the jth column, and $v_x = (v_{x_1}^T, \ldots, v_{x_s}^T)^T$, etc. In the above equations and in the remainder of the development, the subscript index i of L_i, $\hat{L}_i$, b_i, d_i, and also of g_i, $\hat{g}_i$ is often suppressed for notational convenience. In the subsequent development, we assume that

(i) $Q_{uv_x} = 0$

(ii) $Q_{v_x v_x} = 0$

(iii) $Q_{vv_{x_i}}$ is symmetric $(i = 1,2, \ldots, s)$

(iv) $Q_{um} = 0$ and $Q_{mv_x} = 0$

The first three conditions allow the first integral of Eq. (17) to be expressed in terms of δv, δu, and δm, the first-order deviations in the state and control vectors from their nominal values. The term δv_{x_i} $(i=1,2, \ldots ,s)$ can then be eliminated through integration by parts. The fourth condition is not essential but simplifies the development and may be relaxed if necessary. We now linearize the state equations (6), (10) and the initial and boundary conditions (7) to (9) around the nominal solution surface to obtain the accessory state equations

$$A\,\delta v_t + \sum_{i=1}^{s} B_i\,\delta v_{x_i} + H_v^T\,\delta v + C_u^T\,\delta u + C_m^T\,\delta m = 0 \tag{18}$$

$$\delta \dot{z} = \varphi_z^T\,\delta z + \varphi_k^T\,\delta k \tag{19}$$

and the linearized initial and boundary conditions

at $t = t_o$, $$G\,\delta v(x,t_o) = 0 \tag{20}$$

$$\delta z(t_o) = 0 \tag{21}$$

at $x_i = x_i^o$, $t > t_o$, $$g_v^T\,\delta v + g_b^T\,\delta b + g_z^T\,\delta z = 0 \tag{22}$$

at $x_i = x_i^f$, $t > t_o$, $$\hat{g}_v^T\,\delta v + \hat{g}_d^T\,\delta d + \hat{g}_z^T\,\delta z = 0 \tag{23}$$

$(i = 1,2, \ldots ,s)$

where in Eq. (18)

$$H = Av_t + \sum_{i=1}^{s} B_i v_{x_i} + C \tag{24}$$

The variables δv and δz are now the state variables in the accessory problem. The scheme is then to choose a set of accessory control vectors δu, δm, δb_i, δd_i $(i = 1,2, \ldots ,s)$, and δk so as to minimize ΔP subject to the accessory state equations (18), (19) and the initial and boundary conditions (20) to (23). To achieve this, we introduce Lagrange multipliers $\delta\lambda$, $\delta\pi$, $\delta\psi_i$, $\delta\hat{\psi}_i$ $(i = 1,2, \ldots ,s)$ and adjoin the constraining equations (18), (19), (22), (23) to ΔP_1, where

$$\Delta P_1 = \epsilon\,\delta P + \tfrac{1}{2}\delta^2 P \tag{25}$$

The adjustable parameter ϵ, $0 < \epsilon \le 1$, is introduced to ensure that a fraction of the step may be taken when a full step invalidates the second-order approximation to the perturbation in the performance index and leads to a set of controls which produces an increase in the performance index, a likely occurrence when the nominal solution surface is far from the optimum. The adjusted increment in the performance index, ΔP_1, reduces to the second-order Taylor series expansion of the perturbation in the performance index, ΔP, when ϵ is set equal to unity. Thus the augmented performance index becomes

$$\Delta P_1^* = \Delta P_1 + \int_{t_o}^{t_f} \int_{\Omega} \delta\lambda^T \Big\{ A\ \delta v_t + \sum_{i=1}^{s} B_i\ \delta v_{x_i} + H_v^T\ \delta v$$

$$+ C_u^T\ \delta u + C_m^T\ \delta m \Big\}\ d\Omega\ dt + \sum_{i=1}^{s} \int_{t_o}^{t_f} \int_{\partial\Omega_i} \Big\{ \Big[\delta\psi_i^T \Big(g_v^T\ \delta v$$

$$+ g_b^T\ \delta b + g_z^T\ \delta z \Big) \Big]_{x_i = x_i^o} + \Big[\delta\hat{\psi}_i^T \Big(\hat{g}_v^T\ \delta v + \hat{g}_d^T\ \delta d$$

$$+ \hat{g}_z^T\ \delta z \Big) \Big]_{x_i = x_i^f} \Big\}\ d\delta\Omega_i\ dt + \int_{t_o}^{t_f} \delta\pi^T \Big(\varphi_z^T\ \delta z + \varphi_k^T\ \delta k - \delta\dot{z} \Big)\ dt \qquad (26)$$

The time and space derivatives of the accessory state variables in ΔP_1^* are integrated by parts, and the first variation of it is set equal to zero for arbitrary variations in the accessory state and control vectors. After a considerable amount of algebraic manipulations and rearrangement, we obtain the following set of stationary conditions and the accessory controls:

$$(A^T\ \delta\lambda)_t + \sum_{i=1}^{s} (B_i^T\ \delta\lambda)_{x_i} - H_v\ \delta\lambda - \Big[Q_{vv} - \sum_{i=1}^{s} (Q_{vv_{x_i}})_{x_i} \Big] \delta v$$

$$- Q_{vu}\ \delta u - Q_{vm}\ \delta m - \epsilon \Big[Q_v - \sum_{i=1}^{s} (Q_{v_{x_i}})_{x_i} \Big] = 0 \qquad (27)$$

$$\delta\dot{\pi} + \varphi_z\ \delta\pi + N_{zz}\ \delta z + N_{zk}\ \delta k + \epsilon N_z + \sum_{i=1}^{s} \int_{\partial\Omega_i} \Big(g_z\ \delta\psi_i \Big|_{x_i = x_i^o}$$

$$+ \hat{g}_z\ \delta\hat{\psi}_i \Big|_{x_i = x_i^f} \Big)\ d\partial\Omega_i = 0 \qquad (28)$$

subject to the transversality conditions

at $t = t_f$, $$\epsilon M_v + M_{vv}\,\delta v + A^T\,\delta\lambda = 0 \tag{29}$$

$$\delta\pi(t_f) = 0 \tag{30}$$

at $x_i = x_i^o$, $$t < t_f,\ (L_{vv} - Q_{vv_{x_i}})\,\delta v - B_i^T\,\delta\lambda + g_v\,\delta\psi_i + L_{vb}\,\delta b + \epsilon(L_v - Q_{v_{x_i}}) = 0 \tag{31}$$

at $x_i = x_i^f$, $$t < t_f,\ (\hat{L}_{vv} + Q_{vv_{x_i}})\,\delta v + B_i^T\,\delta\lambda + \hat{g}_v\,\delta\hat{\psi}_i + \hat{L}_{vd}\,\delta d + \epsilon(\hat{L}_v + Q_{v_{x_i}}) = 0 \tag{32}$$

and the accessory control vectors

$$\delta u = -Q_{uu}^{-1}(\epsilon Q_u + Q_{vu}^T\,\delta v + C_u\,\delta\lambda \tag{33}$$

$$\delta m = -\left(\int_\Omega Q_{mm}\,d\Omega\right)^{-1}\int_\Omega(\epsilon Q_m + Q_{vm}^T\,\delta v + C_m\,\delta\lambda)\,d\Omega \tag{34}$$

$$\delta b_i = -L_{bb}^{-1}(\epsilon L_b + L_{vb}^T\,\delta v + g_b\,\delta\psi_i)_{x_i = x_i^o} \qquad (i = 1,2,\ \ldots\ ,s) \tag{35}$$

$$\delta d_i = -\hat{L}_{dd}^{-1}(\epsilon\hat{L}_d + \hat{L}_{vd}^T\,\delta v + \hat{g}_d\,\delta\hat{\psi}_i)_{x_i = x_i^f} \qquad (i = 1,2,\ \ldots\ ,s) \tag{36}$$

$$\delta k = -N_{kk}^{-1}(\epsilon N_k + N_{zk}^T\,\delta z + \varphi_k\,\delta\pi) \tag{37}$$

When we substitute the accessory control vectors of Eqs. (33) to (37) into Eqs. (18), (27), (19), (28), further simplified accessory state and adjoint equations result:

(i) $$A\,\delta v_t + \sum_{i=1}^{s} B_i\,\delta v_{x_i} + \alpha\,\delta v + \beta\,\delta\lambda + \epsilon\gamma - C_m^T\left[\int_\Omega Q_{mm}\,d\Omega\right]^{-1} \int_\Omega(\epsilon Q_m + Q_{vm}^T\,\delta v + C_m\,\delta\lambda)\,d\Omega = 0 \tag{38}$$

where

$$\alpha = H_v^T - C_u^T Q_{uu}^{-1} Q_{vu}^T$$

$$\beta = -C_u^T Q_{uu}^{-1} C_u$$

$$\gamma = -C_u^T Q_{uu}^{-1} Q_u$$

(ii) $$(A^T \delta\lambda)_t + \sum_{i=1}^{s} (B_i^T \delta\lambda)_{x_i} + \sigma\ \delta v - \alpha^T \delta\lambda + \epsilon\nu$$

$$+ Q_{vm} \left[\int_\Omega Q_{mm}\, d\Omega \right]^{-1} \int_\Omega (\epsilon Q_m + Q_{vm}^T \delta v + C_m \delta\lambda)\, d\Omega = 0 \quad (39)$$

where

$$\sigma = -Q_{vv} + \sum_{i=1}^{s} \left[Q_{vv_{x_i}} \right]_{x_i} + Q_{vu} Q_{uu}^{-1} Q_{vu}^T$$

$$\nu = -Q_v + \sum_{i=1}^{s} \left[Q_{v_{x_i}} \right]_{x_i} + Q_{vu} Q_{uu}^{-1} Q_u$$

(iii) $$\delta\dot{z} = \eta_1\ \delta z + \eta_2\ \delta\pi + \epsilon\eta_3 \quad (40)$$

where

$$\eta_1 = \varphi_z^T - \varphi_K^T N_{kk}^{-1} N_{zk}^T$$

$$\eta_2 = -\varphi_k^T N_{kk}^{-1} \varphi_k$$

$$\eta_3 = -\varphi_k^T N_{kk}^{-1} N_k$$

(iv) $$\delta\dot{\pi} = \mu_1\ \delta z + \mu_2\ \delta\pi + \epsilon\mu_3 - \sum_{i=1}^{s} \int_{\partial\Omega_i} \Big(g_z\ \delta\psi_i \Big|_{x_i = x_i^o}$$

$$+ \hat{g}_z\ \delta\hat{\psi}_i \Big|_{x_i = x_i^f} \Big)\, d\partial\Omega_i \quad (41)$$

where

$$\mu_1 = N_{zk}N_{kk}^{-1}N_{zk}^T - N_{zz}$$

$$\mu_2 = N_{zk}N_{kk}^{-1}\varphi_k - \varphi_z$$

$$\mu_3 = N_{zk}N_{kk}^{-1}N_k - N_z$$

In the above equations, the parameter ϵ appears explicitly. The presence of ϵ can be eliminated if we introduce the following new variables [202]:

$$\begin{aligned} &\delta v^* = \delta v/\epsilon && \delta b_i^* = \delta b_i/\epsilon && \delta k^* = \delta k/\epsilon \\ &\delta\lambda^* = \delta\lambda/\epsilon && \delta d_i^* = \delta d_i/\epsilon && \delta\psi^* = \delta\psi/\epsilon \\ &\delta z^* = \delta z/\epsilon && \delta u^* = \delta u/\epsilon && \delta\hat{\psi}^* = \delta\hat{\psi}/\epsilon \\ &\delta\pi^* = \delta\pi/\epsilon && \delta m^* = \delta m/\epsilon && \end{aligned} \tag{42}$$

Then the accessory controls in Eqs. (33) to (37) become

$$\delta u^* = -Q_{uu}^{-1}(Q_u + Q_{vu}^T\ \delta v^* + C_u\ \delta\lambda^*) \tag{43}$$

$$\delta m^* = -\left[\int_\Omega Q_{mm}\ d\Omega\right]^{-1}\int_\Omega (Q_m + Q_{vm}^T\ \delta v^* + C_m\ \delta\lambda^*)\ d\Omega \tag{44}$$

$$\delta b_i^* = -L_{bb}^{-1}(L_b + L_{vb}^T\ \delta v^* + g_b\ \delta\psi_i^*)_{x_i=x_i^o} \tag{45}$$

$$\delta d_i^* = -\hat{L}_{dd}^{-1}(\hat{L}_d + \hat{L}_{vd}\ \delta v^* + \hat{g}_d\ \delta\hat{\psi}_i^*)_{x_i=x_i^f} \tag{46}$$

$$\delta k^* = -N_{kk}^{-1}(N_k + N_{zk}^T\ \delta z^* + \varphi_k\ \delta\pi^*) \tag{47}$$

The accessory systems in Eqs. (38) to (41) also become

$$A\ \delta v_t^* + \sum_{i=1}^{s} B_i\ \delta v_{x_i}^* + \alpha\ \delta v^* + \beta\ \delta\lambda^* + \gamma$$
$$-C_m^T\left[\int_\Omega Q_{mm}\ d\Omega\right]^{-1}\int_\Omega (Q_m + Q_{vm}^T\ \delta v^* + C_m\ \delta\lambda^*)\ d\Omega = 0 \tag{48}$$

$$(A^T \delta\lambda^*)_t + \sum_{i=1}^{s} (B_i^T \delta\lambda^*)_{x_i} + \sigma\, \delta v^* - \alpha^T \delta\lambda^* + \nu$$

$$+ Q_{vm} \left[\int_\Omega Q_{mm}\, d\Omega \right]^{-1} \int_\Omega (Q_m + Q_{vm}^T \delta v^* + C_m \delta\lambda^*)\, d\Omega = 0 \quad (49)$$

$$\delta\dot{z}^* = \eta_1 \delta z^* + \eta_2 \delta\pi^* + \eta_3 \quad (50)$$

$$\delta\dot{\pi}^* = \mu_1 \delta z^* + \mu_2 \delta\pi^* + \mu_3 - \sum_{i=1}^{s} \int_{\delta\Omega_i} \left(g_z \delta\psi_i^* \Big|_{x_i = x_i^o} \right.$$

$$\left. + \hat{g}_z \delta\hat{\psi}_i^* \Big|_{x_i = x_i^f} \right) d\delta\Omega_i \quad (51)$$

subject to the ϵ-free transversality conditions:

at $t = t_o$, $\quad \delta z^*(t_o) = 0 \quad (52)$

$$G\, \delta v^*(x, t_o) = 0 \quad (53)$$

at $t = t_f$, $\quad M_v + M_{vv} \delta v^* + A^T \delta\lambda^* = 0 \quad (54)$

$$\delta\pi^*(t_f) = 0 \quad (55)$$

at $x_i = x_i^o$, $\quad g_v^T \delta v^* + g_b^T \delta b^* + g_z^T \delta z^* = 0 \quad (56)$

$$(L_{vv} - Q_{vv_{x_i}})\, \delta v^* - B_i^T \delta\lambda^* + g_v \delta\psi_i^* + L_{vb} \delta b^*$$

$$+ (L_v - Q_{v_{x_i}}) = 0 \quad (57)$$

at $x_i = x_i^f$, $\quad \hat{g}_v^T \delta v^* + \hat{g}_d^T \delta d^* + \hat{g}_z^T \delta z^* = 0 \quad (58)$

$$(\hat{L}_{vv} + Q_{vv_{x_i}})\, \delta v^* + B_i^T \delta\lambda^* + \hat{g}_v \delta\hat{\psi}_i^* + \hat{L}_{vd} \delta d^*$$

$$+ (\hat{L}_v + Q_{v_{x_i}}) = 0 \qquad (i = 1, 2, \ldots, s) \quad (59)$$

The advantage of using ϵ-free equations (43) to (59) is that it is not necessary to reintegrate these accessory equations when a readjustment of ϵ is made in the implementation of algorithm. The effect of ϵ-readjustment is only on the choice of the control set $\{\delta u, \delta m, \delta b_i, \delta d_i, \delta k\}$ through the transformation in Eq. (42). Consequently, if Eqs. (43) to (59) admit a finite solution, then we can reduce the set of control improvement vectors $\{\delta u, \delta m, \delta b_i, \delta d_i, \delta k\}$ linearly to zero by simply reducing ϵ to zero in Eq. (42).

D. Algorithm

We now outline the iterative scheme of algorithm based on the equations developed.

1. Select a set of nominal control functions.
2. Solve state equations (10) and (6) in forward time. If this is the initial iteration, evaluate the performance index P and go to Step 4.
3. Evaluate the performance index P. If an increase over the previous value occurs, reduce ϵ and go to Step 6; if the improvement in P is less than some preassigned small positive quantity, stop the iteration.
4. Solve the distributed two-point boundary value problem of Eqs. (48) to (59) by a distributed or lumped Riccati transformation (see Section II.F).
5. Set $\epsilon = 1$.
6. Compute the new controls from $u^{new} = u^{old} + \epsilon\delta u^*$, $m^{new} = m^{old} + \epsilon\delta m^*$, $b_i^{new} = b_i^{old} + \epsilon\delta b_i^*$, $d_i^{new} = d_i^{old} + \epsilon\delta d_i^*$, $(i = 1, 2, \ldots, s)$, $k^{new} = k^{old} + \epsilon\delta k^*$ where δu^*, δm^*, etc. are calculated from Eqs. (43) to (47). Go to Step 2.

E. Sufficiency Conditions for Convergence

In order to ensure a reduction in the performance index at each iteration until eventual convergence is attained, a set of conditions is sought which, if satisfied, guarantees that $\Delta P < 0$ for each iteration. For an appropriately small value of ϵ $(0 < \epsilon \le 1)$, hence for an appropriately small perturbation in the system controls, if we analyze δP and $\delta^2 P$ in Eq. (15), then we arrive at the conditions that make $\Delta P < 0$ (for detail see Zone [202]). The sufficiency conditions thus obtained are given below. If they are satisfied along the nominal solution surface, a reduction in the performance index occurs.

1. Q_{uu} is positive definite, $\forall t \in [t_o,t_f]$, $\forall x \in \Omega$.
2. $\int_\Omega Q_{mm}\, d\Omega$ is positive definite, $\forall t \in [t_o,t_f]$.
3. $(L_i)_{bb}$ and $(\hat{L}_i)_{dd}$ are positive definite for $i = 1,2, \ldots ,s$,

$$\forall t \in [t_o,t_f], \quad \forall \xi_i' \in \partial\Omega_i.$$

4. N_{kk} is positive definite, $\forall t \in [t_o,t_f]$.
5. $\bar{\bar{\lambda}} > \|(\int_\Omega Q_{mm}\, d\Omega)^{-1}\| \int_\Omega \|Q_{vm}^T\|^2\, d\Omega$, $\forall t \in [t_o,t_f]$

   where $\bar{\bar{\lambda}} = \min_{x\in\Omega} \bar{\lambda}\,(x,t)$ at each $t \in [t_o,t_f]$

   and $\bar{\lambda}\,(x,t)$ is the smallest eigenvalue of

$$Q_{vv} - \sum_{i=1}^{s} (Q_{vv_{x_i}})_{x_i} - Q_{vu}Q_{uu}^{-1}Q_{vu}^T$$

   Here $\|\ \|$ is a conveniently chosen norm.
6. $M_{vv}\,|_{t=t_f}$ is positive semidefinite, $\forall x \in \Omega$.
7. $(L_{vv} - Q_{vv_{x_i}} - L_{vb}L_{bb}^{-1}L_{vb}^T)_{x_i=x_i^o}$ and

   $(\hat{L}_{vv} + Q_{vv_{x_i}} - \hat{L}_{vd}\hat{L}_{dd}^{-1}\hat{L}_{vd}^T)_{x_i=x_i^f}$ are

   positive semidefinite for $i = 1,2, \ldots ,s$,

$$\forall t \in [t_o,t_f], \quad \forall \xi_i' \in \partial\Omega_i.$$

8. $(N_{zz} - N_{zk}N_{kk}^{-1}N_{zk}^T)$ is positive semidefinite, $\forall t \in [t_o,t_f]$.

It is noted that the first four conditions are analogous forms of the strengthened Legendre-Clebsch conditions for lumped systems and must be satisfied in order for a local minimum to exist for the accessory problem.

F. Solution of Two-point Boundary Value Problem

The set of PDEs (48) to (51) together with (52) to (59) form a linear two-point boundary value problem which may be converted to a final-time problem through the introduction of a set of Riccati transformations. To achieve this, we first eliminate the Lagrange multipliers $\delta\psi_i^*$ and $\delta\hat{\psi}_i^*$ $(i = 1,2, \ldots ,s)$ from Eq. (51) by making use of the transversality conditions of (57) and (59). Then the transformation (asterisk dropped for convenience)

$$\delta\lambda(x) = \int_\Omega W(x,\xi)A(\xi)\ \delta v(\xi)\ d\xi + \Gamma(x)\ \delta z + Z(x) \tag{60}$$

is introduced in the accessory PDEs (48) and (49). Here $\xi = (\xi_1,\xi_2, \ldots ,\xi_s)^T$ denotes a spatial coordinate vector and $d\xi$ represents the product $d\xi_1 \cdots d\xi_s$. The time dependence of all variables is implicit. Similarly, the transformation (asterisk dropped)

$$\delta\pi = \int_\Omega R(\xi)A(\xi)\ \delta v(\xi)\ d\xi + D\ \delta z + Y \tag{61}$$

is introduced into Eq. (50) and into the equation obtained from (51) by eliminating $\delta\psi_i^*$ and $\delta\hat{\psi}_i^*$ as described above. The time derivatives of the accessory state variables are then eliminated. Since the equations must hold for arbitrary δv and δz, the resulting Riccati PDEs and ODEs are arrived at by equating like terms. Appropriate final-time and boundary conditions for $W(x,\xi)$, $\Gamma(x)$, $Z(x)$, and $R(\xi)$, D, Y are then imposed to solve the resulting equations. The final expressions are quite complicated; the detailed development can be found in Zone [202]. Because of the complicated nature of the distributed Riccati equations, particularly the increase in the number of spatial variables, it may sometimes be preferable to discretize the distributed accessory equations in space or time, whichever is more amenable, and apply a lumped parameter Riccati transformation to the resulting system of ODEs.

G. Computational Example

We present here a computational example taken from Zone and Chang [204] and Zone [202]. This example is intended to show how the algorithmic equations are formed and how a discretization scheme is employed to numerically solve optimal control problems.

1. Description of Problem

We consider a problem of a first-order, reversible chemical reaction $A \rightleftarrows B$ occurring in a continuous tubular reactor with axial diffusion. Both forward and reverse rate constants are of the temperature-dependent Arrhenius form. The system equation representing the concentration, $v(x,t)$, of A and the initial and boundary conditions (IC and BC) are given in dimensionless form as:

$$v_t = v_{xx} - 2Kv_x - K\{[F_1 \exp(-E_1/Ru) + F_2 \exp(-E_2/Ru)]v - F_2 \exp(-E_2/Ru)\} \quad (62)$$

IC $\quad v(x,0) = v_{oo}(x), \quad 0 \le x \le 1,\ t = 0 \quad (63)$

BC $\quad v_x = 2K(v-v^o) \quad$ at $x = 0,\ t > 0 \quad (64)$

$\quad v_x = 0 \quad$ at $x = 1,\ t > 0 \quad (65)$

where $v(x,t)$ is the mass fraction of A; K and F_i $(i = 1,2)$ are dimensionless parameters ($K = Bo/2$, Bo = Bodenstein number); E_1 and E_2 are activation energies; R is the gas constant; u is temperature; $v_{oo}(x)$ and v^o are the initial and inlet mass fractions of A; and, finally, $x \in [0,1]$ and t are the dimensionless distance and time variables, respectively. The objective is to manipulate the reactor temperature $u(x,t)$ in both time and space so as to minimize the performance index:

$$P = \frac{1}{2}\int_0^{t_f}\int_0^1 \left\{v^2 + \left(\frac{u - u_d}{u^o}\right)^2\right\} dx\, dt \quad (66)$$

where u_d and u^o are constants.

The expressions of Eqs. (62) to (65) can now be recast into the general first-order form of (6) to (9) by letting

$$v_1 = v \qquad v_2 = v_x \quad (67)$$

Then Eqs. (62) to (65) become (we let K = 1 for convenience)

$$\begin{pmatrix} 0 & 0 \\ 1 & 0 \end{pmatrix} \begin{pmatrix} v_1 \\ v_2 \end{pmatrix}_t + \begin{pmatrix} 1 & 0 \\ 0 & -1 \end{pmatrix} \begin{pmatrix} v_1 \\ v_2 \end{pmatrix}_x + \begin{pmatrix} -v_2 \\ [F_1 \exp(-E_1/Ru) \\ + F_2 \exp(-E_2/Ru)]v_1 + 2v_2 - F_2 \exp(-E_2/Ru) \end{pmatrix} = \begin{pmatrix} 0 \\ 0 \end{pmatrix} \tag{68}$$

IC $$(1 \quad 0) \begin{pmatrix} v_1 \\ v_2 \end{pmatrix} = v_{00}(x), \qquad 0 \le x \le 1,\ t = 0 \tag{69}$$

BC $$\begin{aligned} & v_2 - 2(v_1 - v^0) = 0 && \text{at } x = 0,\ t > 0 \\ & v_2 = 0 && \text{at } x = 1,\ t > 0 \end{aligned} \tag{70}$$

2. Algorithmic Equations

We now derive the pertinent equations in the method of second-order successive approximations. Comparing Eqs. (11) with (66), we have

$$Q = \frac{1}{2}\left\{v_1^2 + \left(\frac{u - u_d}{u^0}\right)^2\right\} \tag{71}$$

The full set of sufficiency conditions for convergence of the algorithm given in Section II.E is satisfied since

$$Q_{uu} = \left(\frac{1}{u^0}\right)^2 > 0 \tag{72}$$

and

$$Q_{vv} = \begin{pmatrix} 1 & 0 \\ 0 & 0 \end{pmatrix} \tag{73}$$

is positive semidefinite. Thus, by a cursory examination of only the system performance index one can be assured of the convergence of the second-order algorithm to a local optimum from any nominal solution surface. The accessory state and adjoint equations (48) and (49) for this example are

$$\begin{pmatrix} 0 & 0 \\ 1 & 0 \end{pmatrix} \begin{pmatrix} \delta v_1^* \\ \delta v_2^* \end{pmatrix}_t + \begin{pmatrix} 1 & 0 \\ 0 & -1 \end{pmatrix} \begin{pmatrix} \delta v_1^* \\ \delta v_2^* \end{pmatrix}_x - \begin{pmatrix} 0 & 1 \\ \theta_1 & -2 \end{pmatrix} \begin{pmatrix} \delta v_1^* \\ \delta v_2^* \end{pmatrix}$$

$$-\begin{pmatrix} 0 & 0 \\ 0 & \theta_2 \end{pmatrix}\begin{pmatrix} \delta\lambda_1^* \\ \delta\lambda_2^* \end{pmatrix} - \begin{pmatrix} 0 \\ \theta_3 \end{pmatrix} = \begin{pmatrix} 0 \\ 0 \end{pmatrix} \tag{74}$$

and

$$\begin{pmatrix} 0 & 1 \\ 0 & 0 \end{pmatrix}\begin{pmatrix} \delta\lambda_1^* \\ \delta\lambda_2^* \end{pmatrix}_t + \begin{pmatrix} 1 & 0 \\ 0 & -1 \end{pmatrix}\begin{pmatrix} \delta\lambda_1^* \\ \delta\lambda_2^* \end{pmatrix}_x - \begin{pmatrix} 1 & 0 \\ 0 & 0 \end{pmatrix}\begin{pmatrix} \delta v_1^* \\ \delta v_2^* \end{pmatrix}$$

$$+ \begin{pmatrix} 0 & \theta_1 \\ 1 & -2 \end{pmatrix}\begin{pmatrix} \delta\lambda_1^* \\ \delta\lambda_2^* \end{pmatrix} - \begin{pmatrix} v_1 \\ 0 \end{pmatrix} = \begin{pmatrix} 0 \\ 0 \end{pmatrix} \tag{75}$$

where

$$\theta_1 = -[F_1 \exp(-E_1/Ru) + F_2 \exp(-E_2/Ru)] \tag{76}$$

$$\theta_2 = (cu^o)^2 \tag{77}$$

$$\theta_3 = (u - u_d)c \tag{78}$$

and

$$c = \frac{\{[E_1F_1 \exp(-E_1/Ru) + E_2F_2 \exp(-E_2/Ru)]v_1 - E_2F_2 \exp(-E_2/Ru)\}}{Ru^2} \tag{79}$$

The transversality conditions are

$$\delta v_1^* = 0 \qquad \text{at } t = 0 \tag{80}$$

$$\delta\lambda_2^* = 0 \qquad \text{at } t = 1 \tag{81}$$

$$\delta v_{1_x}^* = 2\delta v_1^* \qquad \text{at } x = 0,\ t > 0 \tag{82}$$

$$\delta\lambda_{2_x}^* = 0 \qquad \text{at } x = 0,\ t < t_f \tag{83}$$

$$\delta v_{1_x}^* = 0 \qquad \text{at } x = 1,\ t > 0 \tag{84}$$

$$\delta\lambda_{2_x}^* = -2\delta\lambda_2^* \qquad \text{at } x = 1,\ t < t_f \tag{85}$$

The control correction is from Eq. (43)

$$\delta u^{*(j)} = -[(u - u_d) + (u^o)^2 c\ \delta\lambda_2^*] \tag{86}$$

and

$$u^{(j+1)}(x,t) = u^{(j)}(x,t) + \epsilon\ \delta u^{*(j)}(x,t) \tag{87}$$

where the superscript j is the iteration number and $\delta u^{*(j)}$ is calculated along the solution surface of the jth iteration.

3. Solution Scheme of Equations

We now have to solve the state equation (68) and the accessory state and adjoint equations (74) and (75), together with the control function iterative scheme of (86) and (87) using the second-order algorithm outlined in Section II.D. Since Eq. (68) is equivalent to (62), which is of the parabolic form

$$v_t = v_{xx} + av_x + bv + q \tag{88}$$

with appropriate a, b, and q and the corresponding initial and spatial boundary conditions:

$$v = v_{oo}(x) \qquad \text{at } t = 0 \tag{89}$$

$$v_x = -a(v-v^o) \qquad \text{at } x = 0 \tag{90}$$

$$v_x = 0 \qquad \text{at } x = 1 \tag{91}$$

we shall consider the numerical solution scheme of this equation. There are many numerical solution schemes available for this type of parabolic PDEs (for example, see Section I.H). Based on the criteria of consistency, numerical stability, and numerical convergence, it was found for the type of the present problem that the implicit Crank-Nicholson technique worked well and permitted rapid integration. A grid of 25 space nodes and 101 time nodes provided a sufficiently fine discretization.

Following the Crank-Nicholson discretization scheme, we replace Eq. (88) with the following difference formula:

$$\frac{V_{i,k+1} - V_{i,k}}{\Delta t}$$

$$= \frac{1}{(\Delta x)^2}\left[\frac{(V_{i+1,k+1} - 2V_{i,k+1} + V_{i-1,k+1}) + (V_{i+1,k} - 2V_{i,k} - V_{i-1,k})}{2}\right]$$

$$+ \frac{a}{2\Delta x}\left[\frac{(V_{i+1,k+1} - V_{i-1,k+1}) + (V_{i+1,k} - V_{i-1,k})}{2}\right]$$

$$+ \frac{(b_{i,k+1} + b_{i,k})}{2}\frac{(V_{i,k+1} + V_{i,k})}{2} + \frac{(q_{i,k+1} + q_{i,k})}{2} \tag{92}$$

$$(i = 1,2, \ldots ,L;\ k = 1,2, \ldots ,100)$$

where $\Delta t = t_f/100$ is the time increment; $\Delta x = 1/(L-1)$ is the space increment; L is the total number of space nodes and taken to be 25; and $V_{i,k}$ is the numerical value of v at the ith space node and kth time node. We let

$$b'_{i,k} = \frac{1}{2}(b_{i,k} + b_{i,k+1}) \tag{93}$$

$$q'_{i,k} = \frac{1}{2}(q_{i,k} + q_{i,k+1}) \tag{94}$$

and regroup the variables such that the left-hand side of the equation contains only variables at the (k+1)th time node. This gives

$$\alpha V_{i+1,k+1} + \beta_{i,k}V_{i,k+1} + \gamma V_{i-1,k+1}$$

$$= -\alpha V_{i+1,k} + \rho_{i,k}V_{i,k} - \gamma V_{i-1,k} + \psi_{i,k} \tag{95}$$

$$(i = 1,2, \ldots ,L;\ k = 1,2, \ldots ,100)$$

where the coefficients are

$$\alpha = \frac{\Delta t}{2(\Delta x)^2} + \frac{a\Delta t}{4\Delta x}, \qquad \beta_{i,k} = \frac{b'_{i,k}\Delta t}{2} - \frac{\Delta t}{(\Delta x)^2} - 1$$

$$\gamma = \frac{\Delta t}{2(\Delta x)^2} - \frac{a\Delta t}{4\Delta x}, \qquad \rho_{i,k} = -2 - \beta_{i,k}$$

$$\psi_{i,k} = -q'_{i,k}\Delta t$$

At i = 1 and i = L, Eq. (95) contains the fictitious variables $V_{0,k}$ and $V_{L+1,k}$. These variables can be eliminated if we first discretize the boundary conditions of Eqs. (90), (91) by the central-difference formula

$$\frac{V_{2,k} - V_{0,k}}{2\Delta x} = -a(V_{1,k} - v_k^o) \tag{96}$$

$$\frac{V_{L+1,k} - V_{L-1,k}}{2\Delta x} = 0 \tag{97}$$

and then substitute the expressions for $V_{0,k}$ and $V_{L+1,k}$ of Eqs. (96), (97) into (95) when i = 1 and i = L. The initial condition in (89) is discretized to yield

$$V_{i,1} = v_{00i} \qquad (i = 1,2, \ldots ,L) \tag{98}$$

where v_{00i} is the value of $v_{00}(x)$ at the ith space node. Equation (98) initializes the Crank-Nicholson scheme by providing the state variable values along the first time grid. Equation (95) provides L equations in L unknowns which are solved for each successive time grid until the final time is reached.

The accessory state and adjoint equations (74) and (75) can be written as

$$\delta v^*_{1_t} = \delta v^*_{1_{xx}} - 2\delta v^*_{1_x} + \theta_1\, \delta v^*_1 + \theta_2\, \delta\lambda^*_2 + \theta_3 \tag{99}$$

$$\delta\lambda^*_{2_t} = -\delta\lambda^*_{2_{xx}} - 2\delta\lambda^*_{2_x} + \delta v^*_1 - \theta_1\, \delta\lambda^*_2 + v_1 \tag{100}$$

These two equations are discretized spatially into six segments by the application of the central-difference approximations

$$\delta V_x = \frac{\delta V_{i+1} - \delta V_{i-1}}{2\Delta x} \tag{101}$$

$$\delta V_{xx} = \frac{\delta V_{i+1} - 2\delta V_i + \delta V_{i-1}}{(\Delta x)^2} \qquad (i = 1,2, \ldots ,7) \tag{102}$$

where $\delta V_i(t)$ is the time-dependent value of the state variable at the ith spatial node. The discretization of Eqs. (99) and (100) leads to

$$\delta \dot{V}_i = \eta_1\, \delta V_{i-1} + \eta_{2i}\, \delta V_i + \eta_3\, \delta V_{i+1} + \theta_{2i}\, \delta\Lambda_i + \theta_{3i} \tag{103}$$

$$\delta \dot{\Lambda}_i = -\eta_3\, \delta\Lambda_{i-1} - \eta_{2i}\, \delta\Lambda_i - \eta_1\, \delta\Lambda_{i+1} + \delta V_i + V_i \tag{104}$$

where

$$\eta_1 = \frac{1}{(\Delta x)^2} + \frac{1}{\Delta x}, \qquad \eta_{2i} = \theta_{1i} - \frac{2}{(\Delta x)^2}$$

$$\eta_3 = \frac{1}{(\Delta x)^2} - \frac{1}{\Delta x}, \qquad \Delta x = \frac{1}{(N-1)}, \; N = 7$$

In the process of discretization, we dropped the asterisk superscript as well as the subscripts 1 and 2 from δv_1^* and $\delta\lambda_2^*$ for notational convenience. The new subsrcipt i indicates that the variable to which it is appended is evaluated at the ith spatial node, and $\delta\Lambda_i$ represents the numerical value of $\delta\lambda_2^*(t)$. Here again at i = 1 and i = N, Eqs. (103) and (104) contain the fictitious variables δV_0, $\delta\Lambda_0$, δV_{N+1}, and $\delta\Lambda_{N+1}$. In order to eliminate these variables, we discretize the boundary conditions in Eqs. (82) to (85), applying the formula (101) to both δv_x^* and $\delta\lambda_x^*$ to obtain

$$\delta V_0 = \delta V_2 - (4\Delta x)\, \delta V_1 \tag{105}$$

$$\delta\Lambda_0 = \delta\Lambda_2 \tag{106}$$

$$\delta V_{N+1} = \delta V_{N-1} \tag{107}$$

$$\delta\Lambda_{N+1} = \delta\Lambda_{N-1} - (4\Delta x)\, \delta\Lambda_N \tag{108}$$

and then substitute them into Eqs. (103) and (104) when i = 1 and i = N. Now we also obtain from the initial and final time conditions in Eqs. (80) and (81)

$$\delta V_i(0) = 0 \tag{109}$$

$$\delta\Lambda_i(1) = 0 \qquad (i = 1,2, \ldots ,N) \tag{110}$$

The resulting set of ODEs (N = 7) constitutes a two-point boundary value problem in time and can be solved either by applying a lumped parameter Riccati transformation or by the method of superposition. Generally, the former technique is preferable because of the inherent instability of the discretized equations when the integration is performed simultaneously in either forward or reverse direction in time. The lumped two-point boundary value problem posed by the discretized accessory equations (103) and (104) subject to (109) and (110) is now solved by the Riccati transformation

$$\delta\Lambda(t) = F(t)\ \delta V(t) + s(t) \tag{111}$$

The (7 x 1) vectors δV and $\delta\Lambda$ represent the accessory state and adjoint variables as a function of time at the seven (N = 7) discrete space nodes. Here the (7 x 7) square matrix F and the vector s are Riccati variables. In the subsequent computations, the Riccati equations were integrated by an explicit fourth-order Runge-Kutta technique employing 100 time increments. The DP accessory equations could not be too finely discretized in space for a given time increment or instability would result in the solution of the Riccati equations, indicating that a stability requirement, $\Delta t/(\Delta x)^2 \leq$ constant, similar to that obtained for certain explicit finite-difference techniques for the solution of parabolic PDEs, was prevailing.

4. Computational Results

To carry out the computations, the following parameter values were assigned:

$$K = 1.0, \qquad F_1 = 5.02 \times 10^3, \qquad F_2 = 3.99 \times 10^5$$

$$t_f = 1.0, \qquad u^0 = 300^\circ R, \qquad u_d = 800^\circ R$$

$$E_1 = 10^4 \text{ Btu/lb mol}, \qquad E_2 = 2 \times 10^4 \text{ Btu/lb mol}$$

$$v^0(t) = \begin{cases} 0.6 - 0.4 \cos(2\pi t) & 0 \leq t \leq 0.5 \\ 1.0 & 0.5 \leq t \leq 1.0 \end{cases}$$

The initial condition $v_{00}(x)$ was taken as the steady-state profile for the inlet mass fraction of A of 0.2 such that the temperature was spatially optimum and minimized the steady-state exit mass fraction of A. This temperature policy was also used as the initial control guess in the control iteration. With the numerical data, the state equation and the accessory state and adjoint equations were numerically solved by the discretization schemes of Section II.G.3. The result of the improvement of control policy with iterations at the reactor entrance is presented in Figure 1. Similar

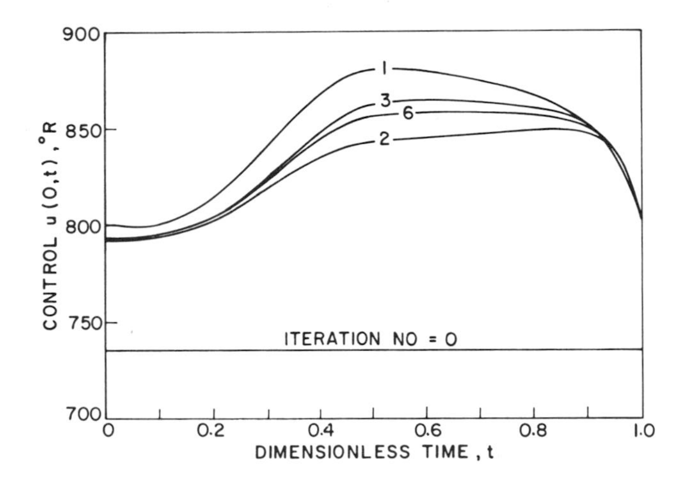

FIG. 1. Improvement of control policy with iterations at $x = 0$.

behavior existed at other spatial positions. Only three iterations were required to reach within 0.1% of the optimal performance index (30.377×10^{-3}). However, three additional iterations were needed to have the convergence within one degree of the optimal temperature policy, although there was very little improvement in the performance index during the additional iterations. An entirely different initial guess for the control policy was also tried, but it was found that the iterated control converged to the same optimal policy after six iterations. The efficiency of the technique in terms of reduction in the performance index with iterations is presented in Figure 2. The result of the first-order gradient method is also presented for comparison. It is to be noted that the present second-order method provided a reasonably good suboptimal policy in only one iteration. Figure 3 shows the optimal concentration profiles along the reactor at various times. The time required for each iteration of the present second-order algorithm for this problem was 10.2 sec central processor time on an IBM 360 Model 75. This time includes the complete numerical solution of the state equation and the accessory state and adjoint equations.

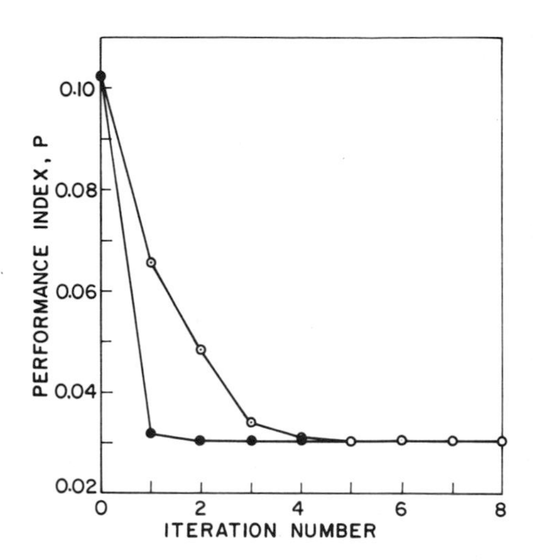

FIG. 2. Reduction in performance index with iterations. ○, first variation; ●, successive spproximation.

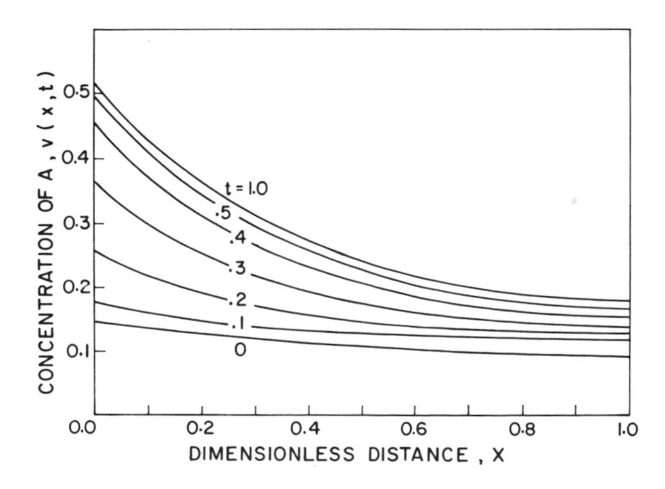

FIG. 3. Final optimal concentration profiles in the reactor.

III. SECOND-ORDER METHOD OF DIFFERENTIAL DYNAMIC PROGRAMMING APPROACH

We now present another second-order algorithm which was developed by Tzafestas and Nightingale [170]. The method is based on a differential dynamic programming approach and requires the solution of equations resulting from a DP Hamilton-Jacobi equation. It is suitable for DP systems with distributed and boundary controls and nonlinear performance index, but is more conveniently used for problems where the nonlinearity of the system equations appears only in the algebraic part, for it is then possible to apply the Green's identity.

A. Description of the Optimal Control Problem

The processes we consider here are assumed to be represented by the following well-posed system of PDEs:

$$\frac{\partial v}{\partial t} = M(v,u,x,t) \qquad x \in \Omega \tag{112}$$

where x is the s-dimensional Euclidean space coordinate vector; $t \in [t_o,t_f]$ is the time variable (t_o and t_f are the fixed initial and final times); $v(x,t)$ is the n-dimensional distributed state vector; $u(x,t)$ is the r-dimensional distributed control vector; and M is a well-posed spatial operator defined over the s-dimensional Euclidean space Ω. The initial and functional boundary conditions are given by

$$v(x,t_o) = v_o(x) \tag{113}$$

$$N(v,b,x,t) = 0 \qquad x \in \partial\Omega \tag{114}$$

where N is a well-posed spatial operator over the boundary $\partial\Omega$ of Ω ($\bar{\Omega} = \Omega \cup \partial\Omega$); v_o is the initial state, and $b(x,t)$ is the boundary control of appropriate dimension defined on $\partial\Omega$. The performance index is given by

$$J = \int_{t_o}^{t_f} L\, dt + J_f \tag{115}$$

where L is a nonlinear spatial functional of $v(x,t)$, $u(x,t)$, and

$b(x,t)$, and L_f is a functional of $v(x,t_f)$. The control problem can now be stated as follows: Find a pair of control functions $u(x,t)$ and $b(x,t)$ within an admissible control function space so as to minimize the performance index J.

B. Differential Dynamic Programming Formulation

First we define a set $S_t = \{v(x,t) \mid x \in \Omega\}$. In this notation, the end states are given by $S_{t_o} = \{v(x,t_o) \mid x \in \Omega\}$ and $S_{t_f} = \{v(x,t_f) \mid x \in \Omega\}$. Suppose we used the pair of optimal controls $u^o(x,t)$ and $b^o(x,t)$ for the time interval $[t_o,t_f]$ and drove the system from the initial state S_{t_o} to the terminal state S_{t_f}. The corresponding trajectory (or the solution surface) is then obtained from Eqs. (112) to (114). Let $J^o(S_t,t)$ denote the optimum performance index thus obtained by using u^o and b^o. Obviously, J^o depends on the state set S_t, and this is indicated in the notation. For an appropriately posed problem of this type, it was shown that the following DP Hamilton-Jacobi equation would be satisfied for the optimal performance index $J^o(S_t,t)$ [170]:

$$-\frac{\partial J^o(S_t,t)}{\partial t} = \min_{u,b} \left\{ L(S_t;u,b,t) + \int_\Omega \langle J^o_v(S_t,t), M(v,u,x,t) \rangle \, d\Omega \right\} \quad (116)$$

where $J^o_v(S_t,t) = \delta J^o(S_t,t)/\delta v$ denotes the first-order functional derivative of $J^o(S_t,t)$ with respect to $v(x,t)$ and $\langle \, , \, \rangle$ denotes the inner product. However, the optimal controls u^o and b^o are not known. But now we suppose that a nominal control pair $\bar{u}(x,t)$ and $\bar{b}(x,t)$ in the time interval $[t_o,t_f]$ is available. Then by integrating Eqs. (112) to (114) we can find the corresponding nominal trajectory and nominal performance index

$$\frac{\partial \bar{v}(x,t)}{\partial t} = M(\bar{v},\bar{u},x,t) \quad (117)$$

$$\bar{v}(x,t_o) = v_o(x) \quad (118)$$

$$N(\bar{v},\bar{b},x,t) = 0 \quad (119)$$

$$\bar{J} = \int_{t_o}^{t_f} L(\bar{S}_t;\bar{u},\bar{b},t)\, dt + J_f(\bar{S}_{t_f}) \tag{120}$$

Equations (112) to (115) may now be written in terms of the nominal trajectory by setting

$$v(x,t) = \bar{v}(x,t) + \delta v(x,t) \tag{121}$$

$$u(x,t) = \bar{u}(x,t) + \delta u(x,t) \tag{122}$$

$$b(x,t) = \bar{b}(x,t) + \delta b(x,t) \tag{123}$$

$$S_t = \bar{S}_t + \delta S_t \tag{124}$$

where δv, δu, δb, and δS_t are measured with respect to the nominal quantities $\bar{v}$, $\bar{u}$, $\bar{b}$, and $\bar{S}_t$. They are

$$\frac{\partial\{\bar{v}(x,t) + \delta v(x,t)\}}{\partial t} = M(\bar{v} + \delta v,\ \bar{u} + \delta u,\ x,\ t) \tag{125}$$

$$\bar{v}(x,t_o) + \delta v(x,t_o) = v_o(x) \tag{126}$$

$$N(\bar{v} + \delta v,\ \bar{b} + \delta b,\ x,\ t) = 0 \tag{127}$$

$$J = \int_{t_o}^{t_f} L(\bar{S}_t + \delta S_t;\ \bar{u} + \delta u,\ \bar{b} + \delta b,\ t)\, dt + J_f(\bar{S}_{t_f} + \delta S_{t_f}) \tag{128}$$

The corresponding Hamilton-Jacobi equation is

$$-\frac{\partial J^o(\bar{S}_t + \delta S,\ t)}{\partial t} = \min_{\substack{\bar{u}+\delta u \\ \bar{b}+\delta b}} \Big\{L(\bar{S}_t + \delta S_t;\ \bar{u} + \delta u,\ \bar{b} + \delta b,\ t) + \int_\Omega \langle J_v^o(\bar{S}_t + \delta S_t,\ t),\ M(\bar{v} + \delta v,\ \bar{u} + \delta u,\ x,\ t)\rangle\, d\Omega\Big\} \tag{129}$$

We shall now work with Eq. (129) and derive a set of equations for a second-order algorithm.

C. Derivation of Iteration Equations

We expand $J^o(\bar{S}_t + \delta S_t, t)$ in a Volterra-Taylor series about $\bar{S}_t$:

$$J^o(\bar{S}_t + \delta S_t, t) = J^o(\bar{S}_t, t) + \int_\Omega \langle J_v^o(\bar{S}_t, t), \delta v\rangle \, d\Omega$$

$$+ \frac{1}{2}\int_\Omega\int_\Omega \langle \delta v, J^o_{vv_1}(\bar{S}_t, t)\, \delta v_1\rangle \, d\Omega \, d\Omega_1 + o(\delta^2 v) \qquad (130)$$

where $o(\delta^2 v)$ stands for all terms of order higher than two; $J^o_{vv_1}$ denotes $\delta^2 J^o/(\delta v\, \delta v_1)$ with $v_1 = v(x_1, t)$ ($x_1 \in \Omega$ is still a vector); and $d\Omega_1$ refers to x_1. We also expand $J_v^o(\bar{S}_t + \delta S_t, t)$ about $\bar{S}_t$ to get

$$J_v^o(\bar{S}_t + \delta S_t, t) = J_v^o(\bar{S}_t, t) + \int_\Omega \langle J^o_{vv_1}(\bar{S}_t, t), \delta v_1\rangle \, d\Omega_1$$

$$+ o(\delta v) \qquad (131)$$

If we define $I(\bar{S}_t, t)$ to represent the difference between the optimal performance index at $(\bar{S}_t, t)$ obtained by using the optimal controls $u^o = \bar{u} + \delta u^o$, $b^o = \bar{b} + \delta b^o$ and the nominal performance index $\bar{J}$ obtained by using $\bar{u}$, $\bar{b}$, then we have

$$J^o(\bar{S}_t, t) = \bar{J}(\bar{S}_t, t) + I(\bar{S}_t, t) \qquad (132)$$

Substituting Eqs. (132) into (130) and then (130), (131) into (129), and making use of the fact that the minimization with respect to $\bar{u} + \delta u$ and $\bar{b} + \delta b$ is equivalent to the minimization with respect to δu and δb, we have

$$-\frac{\partial \bar{J}}{\partial t} - \frac{\partial I}{\partial t} - \int_\Omega \langle \frac{\partial}{\partial t} J_v^o, \delta v\rangle \, d\Omega - \frac{1}{2}\int_\Omega\int_\Omega \langle \delta v, \frac{\partial}{\partial t} J^o_{vv_1}\, \partial v_i\rangle \, d\Omega \, d\Omega_1$$

$$+ o(\delta^2 v) = \min_{\delta u, \delta b} \Big\{ L(\bar{S}_t + \delta S_t; \bar{u} + \delta u, \bar{b} + \delta b, t) + \int_\Omega \langle [J_v^o(\bar{S}_t, t)$$

$$+ \int_\Omega \langle J^o_{vv_1}(\bar{S}_t, t), \delta v_1\rangle \, d\Omega_1 + o(\delta v)],$$

$$[M(\bar{v} + \delta v, \bar{u} + \delta u, x, t)]\rangle \, d\Omega \Big\} \qquad (133)$$

In Eq. (133), if δv produced by δu, δb acting through (125) to (127) is small, then the terms $o(\delta^2 v)$ and $o(\delta v)$ can be neglected. But then J^o is optimal if only δv is sufficiently small. Hence, the superscript of J^o will subsequently be dropped. Equation (133) is rewritten as

$$-\frac{\partial \bar{J}}{\partial t} - \frac{\partial I}{\partial t} - \int_\Omega \langle \frac{\partial}{\partial t} J_v, \delta v \rangle \, d\Omega - \frac{1}{2} \int_\Omega \int_\Omega \langle \delta v, \frac{\partial}{\partial t} J_{vv_1} \delta v_1 \rangle \, d\Omega \, d\Omega_1$$

$$= \min_{\delta u, \delta b} \Big\{ L(\bar{S}_t + \delta S_t; \bar{u} + \delta u, \bar{b} + \delta b, t) + \int_\Omega \langle [J_v(\bar{S}_t, t)$$

$$+ \int_\Omega \langle J_{vv_1}(\bar{S}_t, t), \delta v_1 \rangle \, d\Omega_1], [M(\bar{v} + \delta v, \bar{u} + \delta u, x, t)] \rangle \, d\Omega \Big\} \quad (134)$$

We now expand the following quantities up to the second order about the nominal set $\bar{u}$, $\bar{b}$, and $\bar{v}$, and $\bar{S}_t$.

$$J_f(\bar{S}_{t_f} + \delta S_{t_f}, t_f) = J_f(\bar{S}_{t_f}, t_f) + \int_\Omega \langle (J_f)_v, \delta v \rangle \, d\Omega$$

$$+ \frac{1}{2} \int_\Omega \int_\Omega \langle \delta v, (J_f)_{vv_1} \delta v_1 \rangle \, d\Omega \, d\Omega_1 \quad (135)$$

$$L(\bar{S}_t + \delta S_t; \bar{u} + \delta u, \bar{b} + \delta b, t)$$

$$= L(\bar{S}_t; \bar{u}, \bar{b}, t) + \int_\Omega \langle L_v, \delta v \rangle \, d\Omega + \int_\Omega \langle L_u, \delta u \rangle \, \delta\Omega$$

$$+ \int_{\partial\Omega} \langle L_b, \delta b \rangle \, d\partial\Omega + \frac{1}{2} \int_\Omega \int_\Omega \langle \delta v, L_{vv_1} \delta v_1 \rangle \, d\Omega \, d\Omega_1$$

$$+ \int_\Omega \int_\Omega \langle \delta u, (L_{uv_1} + L^T_{v_1 u}) \delta v_1 \rangle \, d\Omega \, d\Omega_1 + \int_{\partial\Omega} \int_\Omega \langle \delta b, (L_{bv_1}$$

$$+ L^T_{v_1 b}) \delta v_1 \rangle \, d\partial\Omega \, d\Omega_1 + \frac{1}{2} \int_\Omega \int_\Omega \langle \delta u, L_{uu_1} \delta u_1 \rangle \, d\Omega \, d\Omega_1$$

$$+ \frac{1}{2} \int_{\partial\Omega} \int_{\partial\Omega} \langle \delta b, L_{bb_1} \delta b_1 \rangle \, d\partial\Omega \, d\partial\Omega_1 \quad (136)$$

where $L_{vv_1} = \delta^2 L/[\delta v(x,t) \, \delta v(x_1,t)]$ is a second-order functional derivative and $v_1 = v(x_1,t)$, $u_1 = u(x_1,t)$ $(x_1 \in \Omega)$, $b_1 = b(x_1,t)$ $(x_1 \in \partial\Omega)$, etc.

The operators M and N are also expanded up to order two:

$$M(\bar{v} + \delta v, \bar{u} + \delta u, x, t) = M + M_v \, \delta v + M_u \, \delta u + \frac{1}{2} \delta^2 M \quad (137)$$

$$N(\bar{v} + \delta v, \bar{b} + \delta b, x, t) = N + N_v \, \delta v + N_b \, \delta b + \frac{1}{2} \delta^2 N \quad (138)$$

where M_v denotes the first-order operator derivative in the Gateaux sense of M with respect to v (and similar notations apply for others), and $\delta^2 M$ and $\delta^2 N$ are second-order differentials. The quantities on the right-hand side are all evaluated along $\bar{u}$, $\bar{b}$, and $\bar{v}$.

We assume that each of the operators M and N consists of a linear combination of a partial differential part and a nonlinear algebraic part:

$$M(v,u,x,t) = \alpha v + f(v,u) \tag{139}$$

$$N(v,b,x,t) = \beta v + g(v,b) \tag{140}$$

where α and β are linear operators and f and g are the algebraic parts. Then the second differentials $\delta^2 M$ and $\delta^2 N$ are purely algebraic operators (if they are not algebraic, only the algebraic parts are retained [170]). This subsequently allows us to use the Green's identity. To proceed, we assume a specific form of operators such that M_v and N_v are of the standard form

$$M_v[\cdot] = \sum_{i=1}^{s} \sum_{j=1}^{s} \delta A_{i,j} \frac{\partial^2[\cdot]}{\partial x_i \, \partial x_j} + \sum_{i=1}^{s} \delta B_i \frac{\partial[\cdot]}{\partial x_i} + \delta C[\cdot] \tag{141}$$

$$N_v[\cdot] = \delta G[\cdot] + \delta W \frac{\partial[\cdot]}{\partial \sigma} \tag{142}$$

and that M_u and N_b are pure algebraic operators. Here $\partial[\cdot]/\partial\sigma$ is the derivative with respect to the conormal, σ, of the surface, $\partial\Omega$, relative to the operator M. The compact Green's identity for Eqs. (141), (142) with functions X and Y is

$$\int_\Omega \left[\langle Y, M_v X\rangle - \langle M_v^* Y, X\rangle \right] d\Omega = - \int_{\partial\Omega} \left[\langle Y, A_\sigma \frac{\partial X}{\partial\sigma}\rangle - \langle \frac{\partial Y}{\partial\sigma}, A_\sigma X\rangle \right.$$

$$\left. + \sum_{i=1}^{s} \langle Y, Q_i \cos(\nu, x_i) X\rangle \right] d\partial\Omega \tag{143}$$

where M_v^* is the adjoint of M_v; (ν, x_i) is the angle between the normal, ν, of $\partial\Omega$ and the x_i-coordinate axis; and

$$A_\sigma(x,t) = \left\{ \sum_{j=1}^{s} \left[\sum_{i=1}^{s} A_{i,j} \cos(\nu, x_i) \right]^2 \right\}^{1/2} \tag{144}$$

$$Q_i(x,t) = B_i - \sum_{j=1}^{s} \frac{\partial A_{i,j}}{\partial x_j} \tag{145}$$

When these expansions are substituted into the right-hand side of Eq. (134) and the Green's identity is applied, we obtain

$$- \frac{\partial \bar{J}}{\partial t} - \frac{\partial I}{\partial t} - \int_\Omega \langle \frac{\partial}{\partial t} J_v, \delta v\rangle \, d\Omega - \frac{1}{2} \int_\Omega \int_\Omega \langle \delta v, \frac{\partial}{\partial t} J_{vv_1} \, \delta v_1\rangle \, d\Omega \, d\Omega_1$$

$$= \min_{\delta u, \delta b} \Big\{ L + \int_\Omega \langle J_v, M\rangle \, d\Omega + \int_\Omega \langle L_v + M_v^* J_v, \delta v\rangle \, d\Omega$$

$$+ \int_\Omega \langle L_u + M_u^T J_v, \delta u\rangle \, d\Omega + \int_{\partial\Omega} \langle L_b + N_b^T (\delta W^{-1})^T \, \delta A_\sigma^T J_v, \delta b\rangle \, d\partial\Omega$$

$$+ \frac{1}{2} \int_\Omega \int_\Omega \langle \delta v, [L_{vv_1} + (M_{vv_1})^T J_v + M_v^* J_{vv_1}$$

$$+ J_{vv_1} (M_{v_1}^*)^T] \, \delta v_1\rangle \, d\Omega \, d\Omega_1 + \int_\Omega \int_\Omega \langle \delta u, [L_{uv_1} + M_u^T J_{vv_1}$$

$$+ M_{uv_1}^T J_{v_1}] \, \delta v_1\rangle \, d\Omega \, d\Omega_1 + \int_{\partial\Omega} \int_\Omega \langle \delta b, \{L_{bv_1} + \delta W^{-1} \, \delta A_\sigma [N_b^T J_{vv_1}$$

$$+ N_{bv_1}^T J_{v_1}]\} \, dv_1\rangle \, d\partial\Omega \, d\Omega_1 + \frac{1}{2} \int_\Omega \int_\Omega \langle \delta u, L_{uu_1} \, \delta u_1\rangle \, d\Omega \, d\Omega_1$$

$$+ \frac{1}{2} \int_{\partial\Omega} \int_{\partial\Omega} \langle \delta b, L_{bb_1} \, \partial b_i\rangle \, d\partial\Omega \, d\Omega_1 \Big\} \tag{146}$$

subject to the boundary condition for J_v

$$N_v^* J_v \equiv \delta A_\sigma^T \frac{\partial J_v}{\partial \sigma} + \Big[\delta G^T \, \delta W^{-1} \, \delta A_\sigma^T - \sum_{i=1}^{s} \delta Q_i^T \cos(\nu, x_i) \Big] J_v = 0 \tag{147}$$

where N_v^* is the adjoint of N_v. In Eq. (146) the terms up to second-order in δv, δu, and δb have been retained.

Defining a function $\hat{H}$ by

$$\hat{H} = L + \int_\Omega \langle J_v, M\rangle \, d\Omega + \int_{\partial\Omega} \langle J_v, \delta A_\sigma \, \delta W^{-1} N\rangle \, d\partial\Omega \tag{148}$$

we may express Eq. (146) as

$$- \frac{\partial \bar{J}}{\partial t} - \frac{\partial I}{\partial t} - \int_{\Omega} \langle \frac{\partial}{\partial t} J_v, \delta v \rangle \, d\Omega - \frac{1}{2} \int_{\Omega}\int_{\Omega} \langle \delta v, \frac{\partial}{\partial t} J_{vv_1} \delta v_1 \rangle \, d\Omega \, d\Omega_1$$

$$= \min_{\delta u, \delta b} \Big\{ L + \int_{\Omega} \langle J_v, M \rangle \, d\Omega + \int_{\Omega} \langle \hat{H}_v, \delta v \rangle \, d\Omega + \int_{\Omega} \langle \hat{H}_u, \delta u \rangle \, d\Omega$$

$$+ \int_{\partial\Omega} \langle \hat{H}_b, \delta b \rangle \, d\partial\Omega + \frac{1}{2} \int_{\Omega}\int_{\Omega} \langle \delta v, [\hat{H}_{vv_1} + M_v^* J_{vv_1}$$

$$+ J_{vv_1} (M_{v_1}^*)^T] \, \delta v_1 \rangle \, d\Omega \, d\Omega_1 + \int_{\Omega}\int_{\Omega} \langle \delta u, [\hat{H}_{uv_1} + M_u^T J_{vv_1}] \, \delta v_1 \rangle \, d\Omega \, d\Omega_1$$

$$+ \int_{\partial\Omega}\int_{\Omega} \langle \delta u, \{\hat{H}_{bv_1} + \delta W^{-1} \, \delta A_\sigma N_b^T J_{vv_1}] \, \delta v_1 \rangle \, d\partial\Omega \, d\Omega_1$$

$$+ \frac{1}{2} \int_{\Omega}\int_{\Omega} \langle \delta u, \hat{H}_{uu_1} \, \delta u_1 \rangle \, d\Omega \, d\Omega_1$$

$$+ \frac{1}{2} \int_{\partial\Omega}\int_{\partial\Omega} \langle \delta b, \hat{H}_{bb_1} \, \delta b_1 \rangle \, d\partial\Omega \, d\Omega_1 \Big\} \tag{149}$$

Taking the functional derivatives with respect to δu and δb of the right-hand side of Eq. (149) and setting them equal to zero, we obtain the optimal incremental controls δu^o and δb^o:

$$\delta u^o(x,t) = \sigma_u(x,t) + \int_{\Omega} \Lambda_u(x,x_1,t) \, \delta v(x_1,t) \, d\Omega_1 \qquad (x \in \Omega) \tag{150}$$

$$\delta b^o(x,t) = \sigma_b(x,t) + \int_{\Omega} \Lambda_b(x,x_1,t) \, \delta v(x_1,t) \, d\Omega_1 \qquad (x \in \partial\Omega) \tag{151}$$

where

$$\sigma_u(x,t) = - \int_{\Omega} \hat{H}^+_{uu_1} \hat{H}_{u_1} \, d\Omega_1 \tag{152}$$

$$\sigma_b(x,t) = - \int_{\partial\Omega} \hat{H}^+_{bb_1} \hat{H}_{b_1} \, d\partial\Omega_1 \tag{153}$$

$$\Lambda_u(x,x_1,t) = - \int_{\Omega} \hat{H}^+_{uu_2} [\hat{H}_{u_2 v_1} + M^T_{u_2} J_{v_2 v_1}] \, d\Omega_2 \tag{154}$$

$$\Lambda_b(x,x_1,t) = - \int_{\partial\Omega} \hat{H}_{bb_2} [\hat{H}_{b_2 v_1} + \delta W^{-1} \, \delta A_\sigma N^T_{b_2} J_{v_2 v_1}] \, d\partial\Omega_2 \tag{155}$$

Here, $\hat{H}_{uu_2} = \partial^2\hat{H}/(\partial u\ \partial u_2)$ and $u_2 = u(x_2,t)$, etc. ($x_2 \in \Omega$ is a spatial vector different from x or x_1, and $d\Omega_2$ goes with x_2); $\hat{H}^+_{uu_1}$ and $\hat{H}^+_{bb_1}$ are the functional inverses of $\hat{H}_{uu_1}$ and $\hat{H}_{bb_1}$, respectively. To ensure convergence, we introduce small adjustable numbers ϵ_1 and ϵ_2 ($0 < \epsilon_1$, $\epsilon_2 \le 1$) into the control vector iteration equations (150) and (151) to have

$$\delta u^o(x,t) = \epsilon_1\sigma_u(x,t) + \int_\Omega \Lambda_u(x,x_1,t)\ \delta v(x_1,t)\ d\Omega_1 \qquad (x \in \Omega) \tag{156}$$

$$\delta b^o(x,t) = \epsilon_2\sigma_b(x,t) + \int_\Omega \Lambda_b(x,x_1,t)\ \delta v(x_1,t)\ d\Omega_1 \qquad (x \in \partial\Omega) \tag{157}$$

Finally, introducing the optimal incremental controls, given by Eqs. (156) and (157), into (149) and equating the kernels of like terms, we obtain the equations for J_{vv_1}, J_v, and I:

$$\begin{aligned}
-\frac{\partial}{\partial t} J_{vv_1} = {} & \hat{H}_{vv_1} + M^*_v J_{vv_1} + J_{vv_1}(M^*_{v_1})^T - \int_\Omega\int_\Omega [\hat{H}_{uv2} + M^T_u J_{vv_2}]^T \\
& \times \hat{H}^+_{u_2u_3}[\hat{H}_{u_3v_1} + M^T_{u_3} J_{v_3v_1}]\ d\Omega_2\ d\Omega_3 \\
& - \int_{\partial\Omega}\int_{\partial\Omega} [\hat{H}_{bv_2} + \delta W^{-1}\ \delta A_\sigma N^T_b J_{vv_2}]^T \hat{H}^+_{b_2b_3}[\hat{H}_{b_3v_1} \\
& + \delta W^{-1}\ \delta A_\sigma N^T_{b_3} J_{v_3v_1}]\ d\partial\Omega_2\ d\partial\Omega_3
\end{aligned} \tag{158}$$

$$-\frac{\partial J_{v_1}}{\partial t} = \hat{H}_{v_1} + \int_\Omega \Lambda^T_u(x,x_1,t)\hat{H}_u\ d\Omega + \int_{\partial\Omega} \Lambda^T_b(x,x_1,t)\hat{H}_b\ d\partial\Omega \tag{159}$$

$$\begin{aligned}
-\frac{\partial I}{\partial t} = {} & -\epsilon_1\left(1 - \frac{\epsilon_1}{2}\right) \int_\Omega\int_\Omega \langle \hat{H}_b, \hat{H}^+_{bb_1}\hat{H}_{u_1}\rangle\ d\partial\Omega\ d\partial\Omega_1 \\
& -\epsilon_2\left(1 - \frac{\epsilon_2}{2}\right) \int_{\partial\Omega}\int_{\partial\Omega} \langle \hat{H}_b, \hat{H}^+_{bb_1}\hat{H}_{u_1}\rangle\ d\partial\Omega\ d\partial\Omega_1
\end{aligned} \tag{160}$$

where all quantities are evaluated along the nominal trajectory, $\bar{v}$, obtained by using $\bar{u}$ and $\bar{b}$. The spatial boundary conditions for J_{vv_1} and J_v are determined from Eqs. (130) to (132) (with superscript dropped) and (147)

$$N_v^* J_{vv_1} = 0 \qquad (x \in \partial\Omega,\ x_1 \in \bar{\Omega}) \tag{161}$$

$$N_v^* J_v = 0 \qquad (x \in \partial\Omega) \tag{162}$$

The terminal conditions are

$$[J_{vv_1}]_{t=t_f} = (J_f)_{vv_1} \qquad (x, x_1 \in \Omega) \tag{163}$$

$$[J_v]_{t=t_f} = (J_f)_v \qquad (x \in \Omega) \tag{164}$$

$$I(S_{t_f}, t_f) = 0 \tag{165}$$

D. Algorithm

The second-order algorithm based on the differential dynamic programming technique can now be outlined.

1. Select a pair of nominal control functions $\bar{u}(x,t)$ $(x \in \Omega)$ and $\bar{b}(x,t)$ $(x \in \partial\Omega)$.
2. Solve the state equations (117) to (119) in forward time to obtain the nominal trajectory $\bar{v}(x,t)$ for $x \in \Omega$, $t \in [t_0, t_f]$ and calculate the nominal performance index $\bar{J}$ from Eq. (120).
3. Solve Eqs. (158) to (160) in backward time; calculate σ_u, σ_b, Λ_u, and Λ_b from (152) to (155) and store them.
4. Improve the controls by $u^{new} = u^{old} + \delta u^o$ and $b^{new} = b^{old} + \delta b^o$, where δu^o and δb^o are obtained from Eqs. (156) and (157) with appropriate ϵ_1 and ϵ_2, and $\delta v = v^{new} - v^{old}$.
5. If a reasonable decrease in the performance index J does not occur, modify ϵ_1 and ϵ_2 and go to Step 2.
6. If the decrease in the performance index J is less than some preassigned value, then stop the iteration. Otherwise go to Step 2.

It is to be noted that at the beginning of the iterations, the first δv in Step 4 is not available. One may repeat Steps 1 and 2 for a set of neighboring $\bar{u}$ and $\bar{b}$ or set $\delta v = 0$ in Step 4 for the first iteration.

In order for each iteration to be successful so that a decrease in the performance index occurs, the backward equations must have bounded solutions and the matrix functions $\hat{H}^+_{uu}$ and $\hat{H}^+_{bb}$ must be positive definite. This latter condition follows from Eq. (160), where a backward integration in time provides negative I if H^+_{uu} and H^+_{bb} are positive definite.

E. Example

So far, no result of extensive numerical application has been reported. Tzafestas and Nightingale [170] gave an example for illustration which is taken up here again.

1. Description of Problem

We consider a nonlinear boundary control system

$$\frac{\partial v(x,t)}{\partial t} = (v+1)^2 \frac{\partial^2 v}{\partial x^2} \qquad x \in (0.1) \tag{166}$$

with boundary conditions

$$v(0,t) = 0 \tag{167}$$

$$v(1,t) = kb(t) \qquad t \in [t_0, t_f] \tag{168}$$

and an appropriate initial condition. The minimizing performance index is

$$J = \frac{1}{2}\int_{t_0}^{t_f} \left\{ \frac{1}{2}\int_0^1\int_0^1 v(x,t)Q(x,x_1,t)v(x_1,t)\,dx\,dx_1 + R(t)b^2(t)\right\} dt \tag{169}$$

The boundary control of the system is b(t).

2. Algorithmic Equations

We identify for this example

$$M(v,x,t) = (v+1)^2 \frac{\partial^2 v}{\partial x^2} \tag{170}$$

$$N(v,b,x,t) = 0 : \{v(0,t) = 0,\ v(1,t) - kb(t) = 0\} \tag{171}$$

This gives

$$M_v\,\delta v = A\frac{\partial^2 \delta v}{\partial x^2} + C\,\delta v \tag{172}$$

$$N_v\,\delta v = 0 : \{\delta v(0,t) = 0.\ \delta v(1,t) - k\,\delta b(t) = 0\} \tag{173}$$

and

$$M_v^* Y = \frac{\partial^2(AY)}{\partial x^2} + CY \tag{174}$$

$$N_v^* Y = 0 : \{Y(0,t) = 0,\ Y(1,t) = 0\} \tag{175}$$

for a function Y. Here

$$A = (\bar{v} + 1)^2, \qquad C = 2(\bar{v} + 1)\,\frac{\partial^2 \bar{v}}{\partial x^2} \tag{176}$$

Green's theorem gives for function Y

$$\int_0^1 \left[YA\,\frac{\partial^2 \delta v}{\partial x^2} - \delta v\,\frac{\partial^2(AY)}{\partial x^2} \right] dx = \left[YA\,\frac{\partial \delta v}{\partial x} - \delta v\,\frac{\partial (AY)}{\partial x} \right]_0^1$$

$$= -k \left(A\,\frac{\partial Y}{\partial x} \right)_{x=1} \delta b(t) \tag{177}$$

The optimal boundary control is calculated by

$$\delta b^o(t) = \epsilon_2 \sigma_b(t) + \int_0^1 \Lambda_b(x_1,t)\,\delta v(x_1,t)\,dx_1 \tag{178}$$

where σ_b and Λ_b are given by

$$\sigma_b(t) = kR^{-1}(t) \left[(\bar{v} + 1)^2\,\frac{\partial J_v(x,t)}{\partial x} \right]_{x=1} \tag{179}$$

$$\Lambda_b(x_1,t) = kR^{-1}(t) \left[(\bar{v} + 1)^2\,\frac{\partial J_{vv_1}}{\partial x} \right]_{x=1} \tag{180}$$

The backward equations for $J_{vv_1}(x,x_1,t)$ and $J_{v_1}(x,t)$ are

$$-\frac{\partial (J_{vv_1})}{\partial t} = Q(x,x_1,t) + [\bar{v}(x,t) + 1]\,\frac{\partial^2 (J_{vv_1})}{\partial x^2}$$

$$+ [\bar{v}(x_1,t) + 1]\,\frac{\partial^2 (J_{vv_1})}{\partial x_1^2}$$

$$+ 2\,\frac{\partial (\bar{v} + 1)^2}{\partial x} \cdot \frac{\partial (J_{vv_1})}{\partial x} + 2\,\frac{\partial (\bar{v} + 1)^2}{\partial x_1} \cdot \frac{\partial (J_{vv_1})}{\partial x_1}$$

$$+ 2\left[(\bar{v} + 1)\frac{\partial^2 \bar{v}}{\partial x^2} + (\bar{v} + 1)\frac{\partial^2 \bar{v}}{\partial x_1^2}\right] J_{vv_1}$$

$$- k(\bar{v} + 1)^4_{x=1}\left[\frac{\partial J_{vv_1}(x,x',t)}{\partial x'}\right]_{x'=1}$$

$$\times \left[\frac{\partial J_{vv_1}(x'',x_1,t)}{\partial x''}\right]_{x''=1} \tag{181}$$

$$-\frac{\partial(J_{v_1})}{\partial t} = \int_0^1 Q(x,x_1,t)\bar{v}(x_1,t)\,dx_1 + (\bar{v} + 1)^2 \frac{\partial^2(J_{v_1})}{\partial x^2}$$

$$+ 2(\bar{v} + 1)\frac{\partial^2(\bar{v}(x,t))}{\partial x^2} J_{v_1}$$

$$+ \left[kR^{-1}\left\{(\bar{v} + 1)^2 \frac{\partial(J_{vv_1}(x_1,x,t))}{\partial x_1}\right\}_{x_1=1}\right]$$

$$\left[R\bar{b}(t) - k\left\{(\bar{v} + 1)^2 \frac{\partial(J_{v_1}(x_1,t))}{\partial x_1}\right\}_{x_1=1}\right] \tag{182}$$

subject to

$$J_{vv_1}(0,x_1,t) = 0 \qquad x_1 \in (0,1) \tag{183}$$

$$J_{vv_1}(1,x_1,t) = 0 \qquad x_1 \in (0,1) \tag{184}$$

$$J_{v_1}(0,t) = 0 \tag{185}$$

$$J_{v_1}(1,t) = 0 \tag{186}$$

The decrease in the performance index for each iteration is obtained by integrating Eq. (160).

$$I(\bar{S}_t,t) = -\epsilon_2\left(1 - \frac{\epsilon_2}{2}\right)\int_{t_f}^{t} R^{-1}\left[R\bar{b} - k\left\{(\bar{v} + 1)^2 \frac{\partial(J_{v_1})}{\partial x}\right\}_{x=1}\right]^2 d\tau \tag{187}$$

The authors have reported [170] that the convergence criterion $|\ I(S_{t_o},t_o)\ | < 0.01$ was satisfied after three iterations.

IV. CONCLUDING REMARKS

We have briefly surveyed the computational techniques available for distributed parameter optimal control problems and then presented two second-order computational methods available in the literature. The first method employs a successive approximation algorithm based on the direct expansion of the performance index and the second method employs an iterative algorithm based on the expansion of the Hamilton-Jacobi equation of distributed parameter systems.

In the first method, a nonlinear system of partial differential equations is first reduced to a system of first-order quasilinear partial differential equations by a set of proper substitutions. The algorithmic equations are then derived for the system of equations by directly expanding the performance index about the nominal solution surface. Aside from the generality of the formulated control problem accrued from the reduction process, other advantages result. The need for the solution of the system adjoint equations in the two-point boundary value problem can be eliminated, the set of sufficiency conditions for convergence can easily be tested, and an excellent near-optimal control can usually be obtained after only one or two iterations. The last point makes this second-order technique quite attractive over the ordinary gradient methods because of the general tendency of slow convergence and premature termination associated with these methods. In addition, the accessory state and adjoint equations need not be integrated as nearly accurately as the state equations, and the same solutions can be reused for different values of adjustable parameter ϵ since the accessory equations are free of ϵ.

In the second method, a differential dynamic programming approach is used and the algorithmic equations are derived for the system of equations by expanding the Hamilton-Jacobi equation. The method is good for nonlinear distributed parameter systems, but is most conveniently used for the systems where the nonlinearity occurs only in the algebraic part of the partial differential operators. It is then possible to make use of the Green's identity relationship and cast the iterative equations into a more tractable form retaining the full power of second-order convergence properties. For the cases where the nonlinearity occurs in the main operator part, no extensive computational experience has been reported in the literature. It is expected, however, to possess an excellent convergence property and other advantages associated with a second-order method.

REFERENCES

1. M. B. Ajinkya and W. H. Ray, The optimization of axially dispersed packed bed reactors experiencing catalyst decay, Chem. Eng. Sci., 28: 1719-1729 (1973).

2. F. L. Alvarado and R. Mukundan, An optimization problem in distributed parameter systems, Int. J. Contr., 9: 665-677 (1969).

3. W. F. Ames, Numerical Network for Partial Differential Equations, Barnes & Noble, New York, 1969.

4. Yu. N. Andreev and V. M. Orkin, Approximation of optimal control for a distributed system, Auto. Remote Contr., 30: 681-690 (1969).

5. E. I. Axelband, Function space methods for the optimal control of a class of distributed parameter control systems, Proceedings of the 1965 JACC, Troy, New York, pp. 374-380.

6. E. I. Axelband, An approximation technique for the optimal control of linear distributed parameter systems with bounded inputs, IEEE Trans. Auto. Contr., AC-11: 42-45 (1966).

7. E. I. Axelband, A solution to the optimal pursuit problem for distributed parameter systems, J. Compu. Syst. Sci., 1: 261-286 (1967).

8. E. I. Axelband, Optimal control of linear distributed parameter systems, in Advances in Control Systems, vol. 7 (C. T. Leondes, ed.), Academic, New York, 1969, pp. 257-310.

9. T. E. Baker and C. B. Brosilow, Controller design for distributed systems via Bass' technique, AIChE J., 18: 734-738 (1972).

10. J. T. Baldwin, R. A. Ruedrich, and L. D. Durbin, Optimal boundary control signal synthesis for the backflow cell-network model of a nonlinear heat exchange reactor with axial mixing, Proceedings of the 1969 JACC, Boulder, Colorado, pp. 803-814.

11. S. J. Ball, Approximate models for distributed-parameter heat-transfer systems, Proceedings of the 1963 JACC, Minneapolis, Minn., pp. 131-139.

12. D. J. Ball and J. R. Hewit, An optimal control problem for a class of distributed parameter systems, Automatica, 9: 263-267 (1973).

13. D. J. Ball and J. R. Hewit, An alternative approach to a distributed parameter optimal control problem, Int. J. Syst. Sci., 5: 309-316 (1974).

14. J. G. Bansal, "Optimal control of a class of nonlinear diffusional systems--A distributed parameter approach," M.A.Sc. thesis, Univ. of Waterloo, 1970.

15. J. G. Bansal and K. S. Chang, Optimal control of dispersion type distributed parameter systems with time-delay in interacting two-point boundary conditions, Int. J. Contr., 16: 481-500 (1972).

16. E. R. Barnes, An extension of Gilbert's algorithm for computing optimal controls, JOTA, 7: 420-443 (1971).

17. E. R. Barnes, Computing optimal controls in systems with distributed parameters, IFAC Symposium on the Control of Distributed Parameter Systems, Banff, Alberta, Canada, 1971.

18. D. Bégis and R. Glowinski, Dual numerical techniques for some variational problems involving biharmonic operator application to an optimal control problem in thin plates theory, in Technique of Optimization (A. V. Balakrishnan, ed.), Academic, New York, 1972, pp. 159-172.

19. A. Bradshaw, Modal control of distributed-parameter vibratory systems, Int. J. Contr., 19: 957-968 (1974).

20. A. Bradshaw and B. Porter, Modal control of a class of distributed-parameter systems: Multi-eigenvalue assignment, Int. J. Contr., 16: 277-285 (1972).

21. D. L. Briggs and C. N. Shen, Distributed parameter optimum control of a nuclear rocket with thermal stress constraints, Trans. ASME, J. Basic Eng., Series D89, 300-306 (1967).

22. D. L. Briggs and C. N. Shen, Switching analysis for constrained bilinear distributed parameter system with applications, Trans. ASME, J. Basic Eng., Series D91, 277-283 (1969).

23. W. L. Brogan, Theory and application of optimal control for distributed parameter systems — II, Automatica, 4: 121-137 (1967).

24. W. L. Brogan, Optimal control theory applied to systems described by partial differential equations, in Advances in Control Systems, vol. 6 (C. T. Leondes, ed.), Academic, New York, 1968, pp. 221-316.

25. S. F. Bush, An iterative method for the solution of sets of first-order hyperbolic differential equations, SIAM J. Appl. Math., 15: 193-206 (1967).

26. A. G. Butkovskii, Some approximate methods for solving problems of optimal control of distributed parameter systems, Auto. Remote Contr., 22: 1429-1438 (1961).

27. A. G. Butkovskii, Distributed Control Systems, American Elsevier, New York, 1969.

28. A. G. Butkovskii, Yu. V. Darinskii, and L. M. Pustyl'nikov, Control of distributed systems by displacement of the source, Auto. Remote Contr., 35: 701-719 (1974).

29. A. G. Butkovskii and L. N. Poltavskii, Optimal control of a distributed oscillatory system, Auto. Remote Contr., 26: 1835-1848 (1965).

30. A. G. Butkovskii and L. N. Poltavskii, Optimal control of a two-dimensional distributed oscillatory system, Auto. Remote Contr., 27: 553-563 (1966).

31. A. G. Butkovskii and L. N. Poltavskii, Optimal control of wave processes, Auto. Remote Contr., 27: 1542-1547 (1966).

32. A. G. Butkovskii and L. N. Poltavskii, Finite control of systems with distributed parameters, Auto. Remote Contr., 30: 491-501 (1969).

33. J. Casti and H. S. Rao, An initial value method for a class of distributed control processes, IEEE Trans. Auto. Contr., AC-16, 513-515 (1971).

34. J. Cea and K. Malanowski, An example of a max-min problem in partial differential equations, SIAM J. Contr., 8: 305-316 (1970).

35. K. S. Chang, On the numerical computation of a class of distributed parameter systems, IEEE Trans. Auto. Contr., AC-15: 514-516 (1970).

36. K. S. Chang and S. G. Bankoff, Oscillatory operation of jacketed tubular reactors, IEC Fundamen., 7: 633-639 (1968).

37. K. S. Chang and S. G. Bankoff, Optimal control of tubular reactors, AIChE J., 15: 410-418 (1969).

38. A. K. Chaudhuri, Concerning optimum control of linear distributed parameter systems, Int. J. Contr., 2: 365-384 (1965).

39. S. P. Chaudhuri, Optimal control computational techniques for a class of nonlinear distributed parameter systems, Int. J. Contr., 15: 419-432 (1972).

40. S. P. Chaudhuri, Distributed optimal control in a nuclear reactor, Int. J. Contr., 16: 927-937 (1972).

41. C. K. Chu and Y. P. Shih, Low sensitivity optimal control of a class of linear distributed systems, Int. J. Contr., 16: 325-336 (1972).

42. D. E. Cornick and A. N. Michel, Numerical optimization of linear distributed-parameter systems, JOTA, 14: 73-98 (1974).

43. D. E. Cornick and A. N. Michel, Numerical optimization of distributed parameter systems by the conjugate gradient method, IEEE Trans. Auto. Contr., AC-17: 358-362 (1972).

44. J. M. Davis and W. R. Perkins, Optimal control of distributed parameter systems with separable controllers, Automatica, 8: 187-193 (1972).

45. M. M. Denn, Optimal boundary control for a nonlinear distributed system, Int. J. Contr., 4: 167-178 (1966).

46. M. M. Denn, Optimal linear control of distributed systems, IEC Fundamen., 7: 410-413 (1968).

47. M. M. Denn, R. D. Gray, and J. R. Ferron, Optimization in a class of distributed-parameter systems, IEC Fundamen., 5: 59-66 (1966).

48. M. K. Dorri, Approximate solution on an analog computer of certain partial differential equations of heat exchange, Auto. Remote Contr., 27: 1445-1451 (1966).

49. T. Dupont, L-estimates for Galerkin methods for second order hyperbolic equations, SIAM J. Numer. Anal., 10: 880-889 (1973).

50. T. Dupont, G. Fairweather, and J. P. Johnson, Three-level Galerkin methods for parabolic equations, SIAM J. Numer. Anal., 11: 392-410 (1974).

51. R. G. Earp and L. S. Kershenbaum, Optimal temperature profiles in tubular reactors for several catalyst decay laws, Chem. Eng. Sci., 30: 35-45 (1975).

52. A. I. Egorov, Optimal processes in systems containing distributed parameter plants. II, Auto. Remote Contr., 26: 1178-1187 (1965).

53. H. Erzberger and M. Kim, Optimum boundary control of distributed parameter systems, Inform. Contr., 9: 265-278 (1966).

54. J. W. Evans, J. Szekely, W. H. Ray, and Y. K. Chuang, On the optimum temperature progression for irreversible non-catalytic gas-solid reactions, Chem. Eng. Sci., 28: 683-690 (1973).

55. M. M. Fahmy, Optimal control of distributed-parameter systems, Proc. IEE, 115: 572-576 (1968).

56. G. Fairweather and A. R. Mitchell, A new computational procedure for A.D.I. methods, SIAM J. Numer. Anal., 4: 163-171 (1967).

57. H. O. Fattorini, Boundary control of temperature distributions in a parallelepipedon, SIAM J. Contr., 13: 1-13 (1975).

58. M. Fjeld and T. Kristiansen, Optimization of a nonlinear distributed parameter system using periodic boundary control, Int. J. Contr., 10: 601-624 (1969).

59. G. E. Forsythe and W. R. Wasow, Finite-Difference Methods for Partial Differential Equations, Wiley, New York, 1967.

60. M. Friedman and Y. Yavin, Computation of optimal controls for two classes of nonlinear distributed parameter systems, Int. J. Contr., 18: 705-712 (1973).

61. L. I. Gal'chuk, Optimal control of systems described by parabolic equations, SIAM J. Contr., 7: 546-558 (1969).

62. R. M. Goldwyn, K. P. Sriram, and M. H. Graham, Time optimal control of a linear diffusion process, SIAM J. Contr., 5: 295-308 (1967).

63. R. M. Goldwyn, K. P. Sriram, and M. H. Graham, Time optimal control of a linear hyperbolic system, Int. J. Contr., 12: 645-656 (1970).

64. N. N. Golub', Control of the heating of "linearly" viscoelastic plates in the case of thermal stress limitations, Auto. Remote Contr., 27: 195-205 (1966).

65. N. N. Golub', Optimal control of the heating of massive bodies with various phase constraints, Auto. Remote Contr., 28: 562-578 (1967).

66. N. N. Golub', Optimal control of heating in massive bodies with internal heat sources, Auto Remote Contr., 28: 1881-1889 (1967).

67. N. N. Golub', Optimum control of nonlinear distributed-parameter systems, Auto. Remote Contr., 30: 1572-1581 (1969).

68. L. A. Gould and M. A. Murray-Lasso, On the modal control of distributed systems with distributed feedback, IEEE Trans. Auto. Contr., AC-11: 729-737 (1966).

69. J. W. Graham, Time-optimal control of a class of linear-distributed parameter systems with nonlinear boundary conditions, Int. J. Contr., 12: 297-303 (1970).

70. J. W. Graham, A Hamilton-Jacobi approach to the optimal control of distributed parameter systems, Int. J. Contr., 12: 479-487 (1970).

71. J. W. Graham, Optimal boundary control of a class of distributed parameter systems, Int. J. Contr., 14: 937-949 (1971).

72. J. W. Graham and A. F. D'Souza, Optimal control of distributed parameter systems subject to quadratic loss, Proceedings of the 1969 JACC, Boulder, Colorado, pp. 703-709.

73. S. G. Greenberg, Pointwise regulation of distributed parameter systems, Proceedings of the 1970 JACC, Atlanta, Georgia, pp. 584-589.

74. D. R. Hahn, L. T. Fan, and C. L. Hwang, Optimal startup control of a jacketed tubular reactor, Proceedings of the 1970 JACC, Atlanta, Georgia, pp. 451-461.

75. D. R. Hahn, L. T. Fan, and C. L. Hwang, Feedforward-feedback control of distributed parameter systems, Int. J. Contr., 13: 363-382 (1971).

76. M. H. Hamza and M. E. Rasmy, Optimal control of distributed parameter systems with discrete constrained inputs, Int. J. Contr., 20: 159-168 (1974).

77. M. A. Hassan and K. O. Solberg, Discrete time control of linear distributed parameter systems, Automatica, 6: 409-417 (1970).

78. J. H. Holliday and C. Storey, Numerical solution of certain nonlinear distributed parameter optimal control problems, Int. J. Contr., 18: 817-825 (1973).

79. T. L. Johnson, Minimum-energy terminal state control of first order linear hyperbolic systems in one spatial variable using the method of characteristics, SIAM J. Contr., 11: 119-129 (1973).

80. Y. B. Kadymov and B. A. Listengarten, An approximation method for calculating transients in automatic control systems which include components with distributed parameters, Auto. Remote Contr., 25: 450-457 (1964).

81. P. Kenneth, M. Sibony, and J. P. Yvon, Penalization techniques for optimal control problems governed by partial differential equations, in Computing Methods in Optimization Problems, vol. 2 (L. A. Zadeh et al., eds.), Academic, New York, 1969, pp. 177-186.

82. H. C. Khatri and R. E. Goodson, Optimal control of systems with distributed parameters, Proceedings of the 1965 JACC, Troy, New York, pp. 390-397.

83. H. C. Khatri and R. E. Goodson, Optimal feedback solutions for a class of distributed systems, Trans. ASME, J. Basic Eng., Series D88: 337-342 (1966).

84. M. Kim, Successive approximation method in optimum distributed-parameter systems, JOTA, 4: 40-43 (1969).

85. M. Kim and H. Erzberger, On the design of optimum distributed parameter system with boundary control function, IEEE Trans. Auto. Contr., AC-12: 22-28 (1967).

86. R. E. Klein and R. O. Hughes, The distributed parameter control of torsional bending in seagoing ships, Proceedings of the 1971 JACC, St. Louis, Missouri, pp. 867-875.

87. E. A. Klestov and T. K. Sirazetdinov, Response-optimum control of the angular and torsional oscillations of an elastic flying wing, Auto. Remote Contr., 33: 1611-1619 (1972).

88. B. Y. Kogan, R. Krtolica, and I. A. Lapina, Computing techniques in automatic control solution of nonstationary boundary-value problems with variable boundary conditions on a hybrid computer, Auto. Remote Contr., 33: 287-295 (1972).

89. A. J. Koivo and P. Kruh, On the design of approximately optimal feedback controllers for a distributed parameter system, Int. J. Contr., 10: 53-63 (1969).

90. R. C. Kolb and D. A. Pierre, Averaged integral-square error and an approximation to the optimal control of distributed parameter systems, Proceedings of the 1966 JACC, Seattle, Washington, pp. 811-822.

91. V. Komkov, The optimal control of a transverse vibration of a beam, SIAM J. Contr., 6: 401-421 (1968).

92. L. B. Koppel, Optimum control of a class of distributed-parameter processes, IEC Fundamen., 6: 299-303 (1967).

93. L. B. Koppel and Y. P. Shih, Optimal control of a class of distributed-parameter systems with distributed controls, IEC Fundamen., 7: 414-422 (1968).

94. L. B. Koppel, Y. P. Shih, and D. R. Coughanowr, Optimal feedback control of a class of distributed-parameter systems with space-independent controls, IEC Fundamen., 7: 286-295 (1968).

95. F. F. Kotchenko and A. V. Solomin, Calculation of distributed parameter systems by means of logarithmic frequency responses, Auto. Remote Contr., 34: 2015-2019 (1973).

96. G. L. Kusic, Finite differences to implement the solution for optimal control of distributed parameter systems, IEEE Trans. Auto. Contr., AC-14: 397-400 (1969).

97. H. Kwakernaak, R. C. W. Strijbos, and P. Tijssen, Optimal operation of thermal regenerators, IEEE Trans. Auto. Contr., AC-14: 728-731 (1969).

98. L. Lapidus, Digital Computation for Chemical Engineers, McGraw-Hill, New York, 1962.

99. E. S. Lee, A generalized Newton-Raphson method for nonlinear partial differential equations — Packed-bed reactors with axial mixing, Chem. Eng. Sci., 21: 143-157 (1966).

100. H. H. Lee, L. B. Koppel, and H. C. Lim, Integrated approach to design and control of a class of countercurrent processes, IEC Process Des. Develop., 11: 376-382 (1972).

101. H. E. Lee and D. W. C. Shen, Optimal control of a class of distributed-parameter systems using gradient methods, Proc. IEE, 116: 1237-1244 (1969).

102. M. A. Leibowitz and K. Surendran, Optimal control of a class of distributed systems, IEEE Trans. Auto. Contr., AC-18: 69-71 (1973).

103. H. C. Lim, Time-optimal control of a class of linear distributed-parameter processes, IEC Fundamen., 8: 757-765 (1969).

104. H. C. Lim, Time-optimal output control computations for a class of linear tubular processes, Can. J. Chem. Eng., 48: 301-307 (1970).

105. H. C. Lim and R. J. Fang, Optimal feedback control of a class of linear tubular processes, AIChE J., 18: 282-286 (1972).

106. J. L. Lions, Optimal control of deterministic distributed parameter systems, IFAC Symposium on the Control of Distributed Parameter Systems, Banff, Alberta, Canada, 1971.

107. J. L. Lions, On the optimal control of distributed parameter systems, in Techniques of Optimization (A. N. Balakrishnan, ed.), Academic, New York, 1972, pp. 137-158.

108. V. Lorchirachoonkul and D. A. Pierre, Optimal control of multivariable distributed-parameter systems through linear programming, Proceedings of the 1967 JACC, pp. 702-710.

109. J. Løvland, Optimal control of two-stream plus-flow processes, IEC Fundamen., 11: 566-569 (1972).

110. N. N. L'vova, Optimal control of a certain distributed nonhomogeneous oscillatory system, Auto. Remote Contr., 34: 1550-1559 (1973).

111. L. L. Lynn and R. L. Zahradnik, The use of orthogonal polynomials in the near-optimal control of distributed systems by trajectory approximation, Int. J. Contr., 12: 1079-1087 (1970).

112. I. McCausland, On optimal control of temperature distribution in a solid, Proc. IEE (Sci. Gen.), 112: 543-548 (1965).

113. G. E. McGlothin, Optimal control of one-dimensional distributed parameter systems with mixed boundary conditions, Int. J. Contr., 20: 945-954 (1974).

114. R. S. McKnight and W. E. Bosarge, The Ritz-Galerkin procedure for parabolic control problems, SIAM J. Contr., 11: 510-524 (1973).

115. A. N. Michel and D. E. Cornick, Numerical optimization of distributed parameter systems, Proceedings of the 1971 JACC, St. Louis, Missouri, pp. 183-191.

116. A. R. Mitchell, Computational Methods in Partial Differential Equations, Wiley, New York, 1969.

117. R. Mutharasan and D. R. Coughanowr, Feedback direct digital control algorithms for a class of distributed-parameter systems, IEC Process Des. Develop., 13: 168-176 (1974).

118. C. P. Neuman and A. Sen, A rapid sub-optimal control algorithm for distributed parameter regulator problems, Int. J. Contr., 16: 539-548 (1972).

119. C. P. Neuman and A. Sen, Weighted residual methods and the suboptimal control of distributed parameter systems, Int. J. Contr., 18: 1291-1301 (1973).

120. N. Nishida, A. Ichikawa, and E. Tazaki, Optimal design and control in a class of distributed parameter systems under uncertainty — Application to tubular reactor with catalyst deactivation, AIChE J., 18: 561-568 (1972).

121. A. F. Ogunye, "General optimal control synthesis for single and multiple bed reactors experiencing catalyst decay," Ph.D. Dissertation, Univ. of Waterloo, 1969.

122. A. F. Ogunye and W. H. Ray, Optimization of a vinyl chloride monomer reactor, IEC Process Des. Develop., 9: 619-624 (1970).

123. A. F. Ogunye and W. H. Ray, Optimal control policies for tubular reactors experiencing catalyst decay. Part I, Single bed reactors, AIChE J., 17: 43-51 (1971).

124. A. F. Ogunye and W. H. Ray, Optimal control policies for tubular reactors experiencing catalyst decay. Part II, Multiple-bed, AIChE J., 17: 365-370 (1971).

125. A. F. Ogunye and W. H. Ray, Optimization of cyclic tubular reactors with catalyst decay, IEC Process Des. Develop., 10: 410-416 (1971).

126. A. F. Ogunye and W. H. Ray, Optimization of recycle reactors having catalyst decay, IEC Process Des. Develop., 10: 416-420 (1971).

127. A. Olivei, Boundary model control of the temperature of the melt for crystal-growing in crucibles, Int. J. Contr., 20: 129-157 (1974).

128. S. Omatu, H. Shibata, and S. Hata, Optimal boundary control for a linear stochastic distributed parameter system using functional analysis, Inform. Contr., 24: 264-278 (1974).

129. P. A. Orner and A. M. Foster, A design procedure for a class of distributed parameter control systems, Trans. ASME, J. Dynam. Syst. Meas. Contr., 93: 86-93 (1971).

130. J. A. Paraskos and T. J. McAvoy, Feedforward computer control of a class of distributed parameter processes, AIChE J., 16: 754-761 (1970).

131. E. S. Parkin and R. L. Zahradnik, Computation of near-optimal control policies by trajectory approximation: Hyperbolic distributed parameter systems with space-independent controls, AIChE J., 17: 409-414 (1971).

132. R. J. A. Paul and K. R. Strainge, Simulation of second-order hyperbolic equations with split boundary conditions, Int. J. Contr., 15: 433-450 (1972).

133. V. G. Pavlov, Group properties and invariant solutions in the problem of analytical design of controllers in a process with distributed parameters, Auto. Remote Contr., 34: 1201-1207 (1973).

134. J. D. Paynter, The optimal control and design of a multi-bed deactivating catalyst system, Chem. Eng. Sci., 24: 1277-1283 (1969).

135. J. D. Paynter, J. S. Dranoff, and S. G. Bankoff, Application of a suboptimal design method to a distributed-parameter reactor problem, IEC Process Des. Develop., 9: 303-309 (1970).

136. K. C. Pedersen and L. R. Nardizzi, Optimally sensitive control for distributed parameter systems, Int. J. Contr., 16: 723-735 (1972).

137. D. A. Pierre, Minimum mean-square-error design of distributed-parameter control systems, Proceedings of the 1965 JACC, Troy, New York, 381-389.

138. B. Porter and A. Bradshaw, Modal control of a class of distributed parameter systems, Int. J. Contr., 15: 673-681 (1972).

139. S. S. Prabhu and I. McCausland, Time-optimal control of linear diffusion processes using Galerkin's method, Proc. IEE, 117: 1398-1404 (1970).

140. H. H. Rachford, Two-level discrete-time Galerkin approximations for second order nonlinear parabolic partial differential equations, SIAM J. Numer. Anal., 10: 1010-1026 (1973).

141. E. Y. Rapoport, Problem of optimal control of the heating of massive bodies, Auto. Remote Contr., 32: 629-635 (1971).

142. B. M. Raspopov, Problem of optimal fast response of uncoupled heat and mass transfer processes, Auto. Remote Contr., 26: 1791-1796 (1965).

143. B. M. Raspopov, Problems in the optimum control of heat- and mass-transfer processes during drying, Auto. Remote Contr., 29: 336-343 (1968).

144. J. B. Rosen, Iterative graphical solution of boundary value problems using linear programming, in Techniques of Optimization (A. V. Balakrishnan, ed.), Academic, New York, 1972, pp. 183-194.

145. A. P. Sage, Optimum Systems Control, Prentice-Hall, Englewood Cliffs, N.J., 1968.

146. A. P. Sage and S. P. Chaudhuri, Gradient and quasi-linearization computational techniques for distributed parameter systems, Int. J. Contr., 6: 81-98 (1967).

147. Y. Sakawa, Solution of an optimal control problem in a distributed-parameter system, IEEE Trans. Auto. Contr., AC-9: 420-426 (1964).

148. Y. Sakawa, Optimal control of a certain type of linear distributed-parameter systems, IEEE Trans. Auto. Contr., AC-11: 35-41 (1966).

149. Y. Sakawa, On a solution of an optimization problem in linear systems with quadratic performance index, SIAM J. Contr., 4: 382-395 (1966).

150. H. Sasai, A note on the penalty method for distributed parameter optimal control problems, SIAM J. Contr., 10: 730-736 (1972).

151. H. Sasai and E. Shimemura, On the convergence of approximating solutions for linear distributed parameter optimal control problems, SIAM J. Contr., 9: 263-273 (1971).

152. V. K. Saul'yev, Integration of Equations of Parabolic Type by the Method of Nets (G. J. Tee, trans.), Pergamon Press, Oxford, 1964.

153. J. H. Seinfeld, G. R. Gavalas, and M. Hwang, Control of plug-flow tubular reactors by variation of flow rate, IEC Fundamen., 9: 651-655 (1970).

154. J. H. Seinfeld and K. S. P. Kumar, Synthesis of sub-optimal feedback controls for a class of distributed parameter systems, Int. J. Contr., 7: 417-424 (1968).

155. J. H. Seinfeld and L. Lapidus, Computational aspects of the optimal control of distributed-parameter systems, Chem. Eng. Sci., 23: 1461-1483 (1968).

156. J. H. Seinfeld and L. Lapidus, Singular solutions in the optimal control of lumped- and distributed-parameter systems, Chem. Eng. Sci., 23: 1485-1499 (1968).

157. M. A. Sheirah and M. H. Hamza, Optimal control of distributed parameter systems, Int. J. Contr., 19: 891-902 (1974).

158. Y. P. Shih, Optimal control of distributed-parameter systems with integral equation constraints, Chem. Eng. Sci., 24: 671-680 (1969).

159. R. N. P. Singh, A unified approach to a state-space model for linear distributed systems, Int. J. Contr., 11: 471-478 (1970).

160. R. N. P. Singh and V. S. Rajamani, Minimal-time control of distributed-parameter systems with multiple-norm constraints on the control function, Int. J. Contr., 15: 241-254 (1972).

161. T. K. Sirazetdinov, Analytic design of regulators for magnetohydrodynamic processes, II, Auto. Remote Contr., 28: 1813-1822 (1967).

162. G. D. Smith, Numerical Solution of Partial Differential Equations, Oxford University Press, London, 1969.

163. A. K. Sood, G. L. Funk, and A. C. Delmastro, Dynamic optimization of a natural gas pipeline using a gradient search technique, Int. J. Contr., 14: 1149-1157 (1971).

164. E. M. Stafford and J. M. Nightingale, Generalised fourier methods for 1st-order distributed systems, Proc. IEE, 117: 1864-1868 (1970).

165. V. J. Tarassov, H. J. Perlis, and B. Davidson, Optimization of a class of river aeration problems by the use of multivariable distributed parameter control theory, Water Resour. Res., 5: 563-573 (1969).

166. M. Y. Tarng and L. R. Nardizzi, A computational technique for the control of systems described by partial difference equations, Int. J. Contr., 17: 1189-1199 (1973).

167. S. G. Tzafestas, Final-value control of nonlinear composite distributed- and lumped-parameter systems, J. Franklin Inst., 290: 439-451 (1970).

168. S. G. Tzafestas, Optimal distributed-parameter control using classical variational theory, Int. J. Contr., 12: 593-603 (1970).

169. S. G. Tzafestas and J. M. Nightingale, Optimal control of a class of linear stochastic distributed-parameter systems, Proc. IEE, 115: 1213-1220 (1968).

170. S. G. Tzafestas and J. M. Nightingale, Differential dynamic-programming approach to optimal nonlinear distributed-parameter control systems, Proc. IEE, 116: 1079-1084 (1969).

171. S. C. Uzgiris and A. F. D'Souza, Optimal control of distributed parameter systems with nonlinear boundary conditions, Proceedings of the 1966 JACC, Seattle, Washington, pp. 675-683.

172. J. G. Vermeychuk, New methods for suboptimal feedback control of parabolic systems, AIChE J., 20: 159-166 (1974).

173. Yu. M. Volin and G. M. Ostrovskii, A method of successive approximations for calculating optimal modes of some distributed-parameter systems, Auto. Remote Contr., 26: 1188-1194 (1965).

174. Yu. M. Volin and G. M. Ostrovskii, Optimization of a system with distributed parameters, J. Appl. Math. Mech., 29: 708-715 (1966).

175. Z. I. Vostrova, Optimal processes in sampled-data systems containing objects with distributed parameters, Auto. Remote Contr., 27: 767-779 (1966).

176. A. K. Vyrk, Optimal heating of massive bodies in continuous furnaces, Auto. Remote Contr., 31: 1132-1141 (1970).

177. R. T. Walsh, Optimization and comparison of partial difference methods, SIAM J. Numer. Anal., 10: 785-797 (1973).

178. P. K. C. Wang, Advances in Control Systems (C. T. Leondes, ed.), vol. 1, Academic, New York, 1964, pp. 75-172.

179. P. K. C. Wang, On the feedback control of distributed parameter systems, Int. J. Contr., 3: 255-273 (1966).

180. P. K. C. Wang, Control of a distributed parameter system with a free boundary, Int. J. Contr., 5: 317-329 (1967).

181. P. K. C. Wang, Confinement of thermonuclear plasmas — A distributed control problem, Proceedings of the 1968 JACC, Ann Arbor, Michigan, pp. 1119-1120.

182. P. K. C. Wang, Optimal confinement of collisionless plasmas by localized time-varying electric fields, Int. J. Contr., 19: 449-472 (1974).

183. P. K. C. Wang and F. Tung, Optimum control of distributed parameter systems, Trans. ASME, J. Basic Eng., 86: 67-79 (1964).

184. W. A. Weigand, Optimal control of plug-flow heat exchanger with control produced by wall flux or wall temperature, IEC Fundamen, 9: 641-651 (1970).

185. W. A. Weigand and A. F. D'Souza, Optimal control of linear distributed parameter systems with constrained inputs, Proceedings of the 1968 JACC, Ann Arbor, Michigan, pp. 239-252.

186. W. A. Weigand and A. F. D'Souza, Optimal control of linear distributed parameter systems with constrained inputs, Trans. ASME, J. Basic Eng., 91: 161-167 (1969).

187. D. M. Wiberg, Feedback control of linear distributed systems, Trans. ASME, J. Basic Eng., 89: 379-384 (1967).

188. D. A. Wismer, A decomposition principle for the optimization of a class of distributed parameter systems, Proceedings of the 1968 JACC, Ann Arbor, Michigan, pp. 1092-1101.

189. D. A. Wismer, An efficient computational procedure for the optimization of a class of distributed parameter systems, Trans. ASME, J. Basic Eng., 91: 190-194 (1969).

190. D. A. Wismer and I. Lefkowitz, Analysis and computer simulation for dynamic control of continuous strip processes, Proceedings of the 1963 JACC, Minneapolis, Minnesota, pp. 255-261.

191. J. T. Yang, "The solution of distributed parameter systems, and the synthesis of their optimal control using functional analysis," Ph.D. Dissertation, Univ. of Waterloo, 1972.

192. J. T. Yang and K. S. Chang, Numerical computation of optimal control for a class of distributed parameter systems, Proceedings of the 1974 JACC, 424-429, Austin, Texas, pp. 424-429.

193. Y. Yavin, Optimal control with a bounded energy input derivative for a class of distributed systems, Int. J. Contr., 14: 141-148 (1971).

194. Y. Yavin and Y. Rasis, The bounded energy optimal control for a class of heat conduction systems, Int. J. Contr., 11: 153-164 (1970).

195. Y. Yavin and R. Sivan, The optimal control of a distributed parameter system, IEEE Trans. Auto. Contr., AC-12: 758-761 (1967).

196. Y. Yavin and R. Sivan, The bounded energy optimal control for a class of distributed parameter systems, Int. J. Contr., 8: 525-536 (1968).

197. H. H. Yeh and J. T. Tou, Design of optimum control for a class of distributed-parameter systems, Proceedings of the 1966 JACC, Seattle, Washington, pp. 684-693.

198. H. H. Yeh and J. T. Tou, Optimum control of a class of distributed-parameter systems, IEEE Trans. Auto. Contr., AC-12: 29-37 (1967).

199. T. K. Yu and J. H. Seinfeld, Suboptimal control of stochastic distributed parameter systems, AIChE J., 19: 389-392 (1973).

200. R. L. Zahradnik and L. L. Lynn, Near-optimal feedback control of distributed parameter systems by trajectory approximation, Proceedings of the 1970 JACC, Atlanta, Georgia, pp. 590-595.

201. R. L. Zahradnik and L. L. Lynn, Distributed systems with distributed controls by trajectory approximation, IEC Fundamen., 10: 176-179 (1971).

202. G. Zone, "Second-order optimal control algorithms for lumped- and distributed-parameter processes," Ph.D. Dissertation, Univ. of Waterloo, 1973.

203. G. Zone and K. S. Chang, "A second-variation algorithm for a class of nonlinear distributed parameter systems," National Conference on Automatic Control, NRC, Univ. of Waterloo, 1970.

204. G. Zone and K. S. Chang, A successive approximation method for nonlinear distributed-parameter control systems, Int. J. Contr., 15: 255-272 (1972).

PART 2

APPLICATIONS

Chapter 7

THE CONTROL OF VIBRATING ELASTIC SYSTEMS

M. Köhne

Institut für Systemdynamik und Regelungstechnik
Universität Stuttgart
Stuttgart, West Germany

I. INTRODUCTION

In recent years, due to the building of larger and more flexible engineering structures, the analysis and design of flexible mechanical systems have steadily increased in importance. Further developments can be expected in the future. A few examples are:

- High buildings, long bridges, and pipe lines
- Magnetically levitated and air cushion high-speed ground vehicles
- Large ships and submersible vessels
- Ocean mining systems with long elastic hauling pipes
- Aircraft, rockets, missiles, and spacelabs
- Satellites with flexible appendages of extendable booms
- Astronomical telescopes and large antennas

As a rough approximation, these distributed parameter systems or, at least, their most essential parts, can be considered as vibrating strings, beams, membranes, plates, shells, and combinations of these well-known subjects for the theory of elasticity [1], mechanical engineering [2], and mathematical physics [3]. Systems of this nature are usually described by partial differential equations with two or more independent variables (time t and space coordinate vector z). The differential equations of the elastic systems enumerated above are characterized by the even order of the highest occurring time derivative. This is demonstrated in Table 1, where the partial differential equations of some well-known distributed parameter systems are summarized. In this chapter, mainly <u>transverse</u> or <u>biharmonic</u> vibrations of elastic systems (beams and plates, for example) are investigated.

Mechanical engineers are primarily interested in structural analysis [1] (determination of stress and displacement distributions under prescribed loads and constraints), free and forced vibrations [2] (computation of modes and natural frequencies), problems of parametric excitation and elastic stability [4-6] (buckling of columns and plates, flutter problems), and optimal structural design of elastic systems [7].

TABLE 1

Classification of Dynamic Linear Distributed Parameter Systems

Highest order of the time-/space-derivative		Example	Type of equation	Physical phenomena
1	1	$\beta\dot{w} + \nu\nabla_z w = u$	Hyperbolic (1st order)	Transport process
1	2	$\beta\dot{w} - \gamma\Delta_z w = u$	Parabolic	Diffusion, heat conduction
2	2	$\mu\ddot{w} - \gamma\Delta_z w = u$	Hyperbolic (2nd order)	Propagation of waves
2	4	$\mu\ddot{w} + \alpha\Delta_z^2 w = u$	Not classified	Transverse vibrations of beams and thin plates
2	8	$\mu\Delta_z^2\ddot{w} + \epsilon\Delta_z^4 w + \ldots$	Not classified	Vibrations of shells
4	4	$\kappa\ddddot{w} - \theta\ddot{w}'' + \alpha w^{IV} + \ldots$	Not classified	Coupled vibrations of beams

$\dot{w} = \partial w/\partial t$, $w' = \partial w/\partial z$, ∇_z = nabla operator, $\Delta_z = \nabla_z^2$ = Laplace operator (abbreviations)

Some possible objectives of control engineers are the determination of distributed or boundary-value control variables (loads, forces, bending moments) to obtain a desired dynamic behavior of the considered elastic system [8-14], and the application of Lyapunov's stability theory for solving kinetic stability problems [4,6]. In particular, the control or damping of elastic vibrations has been recognized as an important engineering problem for a long time. Various experimental and theoretical results with passive damping devices have been obtained. The use of active control systems was initiated by the development of aircraft and spacecraft systems [15]. We will briefly mention the main results of optimal control of elastic systems.

The optimal control theory of distributed parameter systems has been developed over the last ten years by the pioneering work of Butkowskii [16], Wang [17], and Lions [18]. Some of their results have been applied to vibrating elastic systems of the nature indicated above. Komkov [13] has developed Pontryagin's maximum principle for the optimal control of transversely vibrating beams and plates, and has recently extended this principle to control problems of structural mechanics [14]. With these results we are able to solve minimal time and minimal energy control problems.

Two examples of linear optimal regulator problems have been investigated by Yonkin [19]. The first example is a flexible taut string with position control at one end and the uncontrolled end fixed. The second example concerns the optimal control of a long slender column with a weight at the top due to a camera system. The flexible column is mounted on a truck and the control is provided by a torque at the base. Sirazetdinov [20] has investigated a similar problem, the optimal control of elastic aircraft with a point control force applied to the rear edge. All these, together with examples in [21-23], are boundary value control problems and can be considered as special types of distributed control problems if generalized functions (distributions) are applied [24]. Of practical importance are lumped control forces which take effect at spatially discrete points. In [10] and later in [25], the solution of this problem of pointwise control has been undertaken by the author, together with the solution of state reconstruction problems.

The purpose of this chapter is to give a summary of several investigations [10-12,25,26] on the linear optimal feedback control problem of elastic systems; in particular, transversely vibrating beams and plates. The beam hinged at both ends and the rectangular plate will serve as test cases (similar to the heat conductor in other publications on distributed parameter systems) since their solutions can easily be attained. In Section V, however, we also consider a practically oriented example, namely the optimal control of lateral vibrations of an elastic hauling pipe, an essential component of ocean mining systems.

In the following section, the basic system equations are written in state-space form and a few examples of elastic systems are given. The optimal tracking and regulator problem is solved in Section III, and in addition both the implicit and the explicit model-following problems are formulated and solved for distributed parameter systems.

The basic concept of observer theory is developed in Section IV, and distributed parameter observers of full and reduced order are designed. Conditions for the existence of such observers have not yet been proven, but several examples are given in order to demonstrate their existence. Simulation results for the optimal feedback control problem of the hauling pipe mentioned above are obtained in Section V.

II. SYSTEM DESCRIPTION

A. Basic System Equations

Consider an elastic system defined on a fixed spatial domain Ω, which is a simply connected, open subset of the r-dimensional euclidean space E_r with boundary Γ (Figure 1). The spatial coordinate vector and the time variable will be denoted by $z = (z_1, \ldots, z_r) \in \Omega \subset E_r$ and $t \in T = (t_o, t_f) \subset R_1$, respectively.

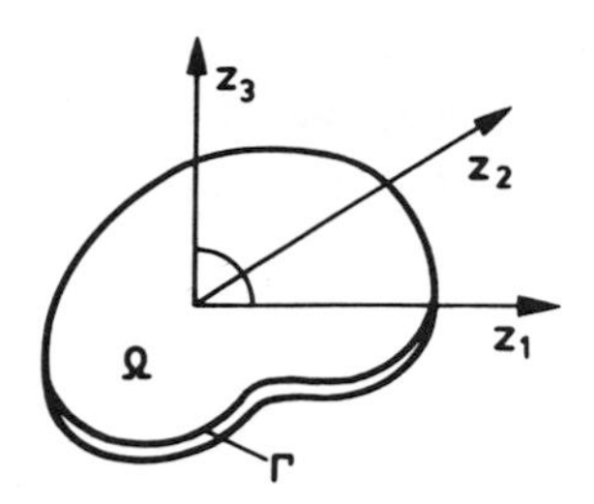

FIG. 1. A thin plate as an example of an elastic system.

Assume

$$\Pi = \Omega \times T, \qquad \bar{\Omega} = \Omega \cup \Gamma$$

$$\Sigma = \Gamma \times T, \qquad \bar{T} = [t_o, t_f]$$

for abbreviations.

It is necessary to define an equation model for the elastic systems considered. Many different models have been proposed for distributed parameter systems (DPS) [16,18,24,27,28], a state-space form similar to lumped parameter system (LPS) description seems to be the most appropriate for our problems. We describe elastic systems by the linear partial differential equation (PDE)

$$\frac{\partial}{\partial t} x(z,t) = Ax(z,t) + Bu(z,t), \qquad z,t \in \Pi \tag{1}$$

the observation equation

$$y(z,t) = Cx(z,t), \qquad z,t \in \Pi \tag{2}$$

and the boundary and initial conditions

$$A_\Gamma x(z,t) = B_\Gamma v(z,t), \qquad z,t \in \Sigma \tag{3}$$

and

$$x(z,t_o) = x_o(z), \qquad z \in \Omega \tag{4}$$

where $u \in U$ is a k-vector distributed control input, $v \in V$ is an ℓ-vector boundary control input (or disturbance input), $y \in Y$ is an m-vector observation output, and $x \in X$ is an n-vector state. A and A_Γ are linear matrix differential operators with respect to the spatial coordinates (sometimes indicated by the subscript z, e.g., A_z and $A_{\Gamma z}$). B, B_Γ, and C are matrix integral operators or matrix functions. (See Sections II.B and C for examples.)

U, V, Y, and X are given Hilbert spaces; the distributed input space $U = L_k^2(\Pi)$, the boundary input space $V = L_\ell^2(\Sigma)$, the output space $Y = L_m^2(\Pi)$, and the state space $X = L_n^2(\Pi)$. Here $H_n(\Omega)$ denotes the Hilbert space of n-vector functions square integrable over Ω, with the inner product defined by

$$(x_1,x_2)_{H_n(\Omega)} = \int_\Omega x_1^T(z,t)x_2(z,t)\,dz, \qquad x_1,x_2 \in H_n(\Omega) \tag{5}$$

The superscript T denotes transposed vectors. The corresponding inner product of $L_n^2(\Pi)$ is then defined by

$$(x_1,x_2)_{L_n^2(\Pi)} = \int_T (x_1,x_2)_{H_n(\Omega)}\,dt \tag{6}$$

$H_n(\Gamma)$ and $L_n^2(\Sigma)$ are defined in a similar manner [29].

Assuming that a unique solution of the mathematical model (1) to (4) exists and is dependent upon the given distributed, boundary, and initial data, it can be written as

$$x(z,t) = G_\Pi u(z,t) + G_\Sigma v(z,t) + G_\Omega x_o(z) \tag{7}$$

where G_Π, G_Σ, and G_Ω are matrix integral operators whose kernels are the Green's matrix $G(z,\zeta,t,\tau)$ and its derivatives with respect to space and time [30,31]. The theory of elasticity and vibrations deals with the solution of the system equations (1) to (4) to obtain (7).

Here, however, we want to solve control problems, especially the output or state regulator problems, where the control vector u (or possibly v) has to be determined as a function of the output y or the state x. In applications, the initial state x_o may not be known precisely. The parameters of the deterministic mathematical model are assumed to be known exactly, i.e., modeling and measurement errors are not taken into consideration. To illustrate the category of systems (1) to (4), we will first present a few examples of vibrating elastic systems in state-space form.

B. State Equations of Elastic Systems

1. Transverse Vibrating Slender Beams

In general, the one-dimensional elastic system vibrating transversely derives its restoring force from two sources: axial tension and bending stiffness. We ignore the axial elastic displacement but consider the effect of an axial tensile force $f(z,t)$ upon the bending vibration of the beam (Figure 2).

The PDE of motion is developed in [2]:

$$\mu(z)\ddot{w}(z,t) + \delta_E\dot{w}(z,t) + \delta_I\dot{w}^{IV}(z,t) \qquad z \in (0,L),\ t \in T\ ,$$

$$- (f(z,t)w'(z,t))' + (\alpha(z)w''(z,t))'' = p(z,t)^{\dagger} \qquad (8)$$

where

w	= lateral deflection (assumed to be small)
p	= external transverse force per unit length (distributed load)
μ	= mass per unit length
δ_E, δ_I	= external and internal damping coefficients
f	= longitudinal tensile force
α	= flexural rigidity

$^{\dagger}\dot{w}(z,t) = \partial w(z,t)/\partial t$ and $w'(z,t) = \partial w(z,t)/\partial z$ are used for simplicity.

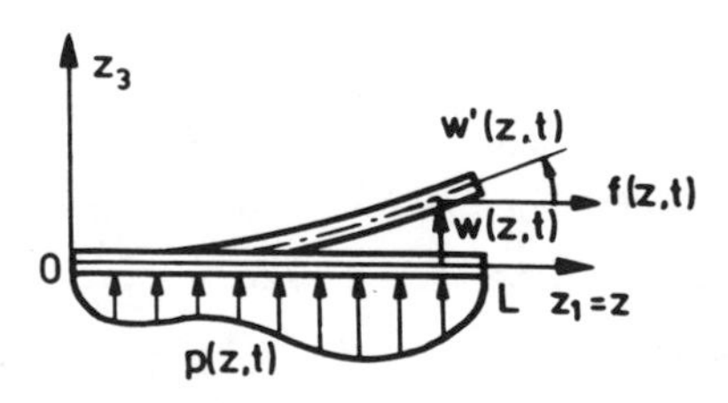

FIG. 2. Transverse vibrating slender beam.

Sometimes the tension f is due to gravity. Consider a hanging beam (e.g., the hauling pipe in Section V) with lumped mass m_L at the end $z = L$ to obtain the time-independent force

$$f \equiv f(z) = [m_L + \mu(L - z)]g, \qquad z \in [0,L] \tag{9}$$

The boundary conditions result from geometric compatibility, bending moment balance, or shearing force balance. We distinguish between four cases (Figure 3), with the end

$$\text{Clamped} \quad w(z,t) = 0, \quad w'(z,t) = 0 \tag{10a}$$
$$\text{Hinged} \quad w(z,t) = 0, \quad \alpha w''(z,t) = 0, \qquad z \in \{0,L\} \tag{10b}$$
$$\text{Sliding} \quad w'(z,t) = 0, \quad -(\alpha w''(z,t))' = 0, \tag{10c}$$
$$\text{Free} \quad \alpha w''(z,t) = 0, \quad -(\alpha w''(z,t))' + fw'(z,t) = 0 \tag{10d}$$

Several other combinations are possible. In general, the following boundary conditions must be satisfied:

$$\sum_{k=0}^{3} \alpha_{i,k}(z,t) \frac{\partial^k w(z,t)}{\partial z^k} = v_i(t), \qquad i = 1, \ldots ,4, \qquad z,t \in \Sigma \tag{11}$$

We choose $u = p$ and the state vector

$$x(z,t) = [w(z,t),\dot{w}(z,t)]^T \tag{12}$$

to obtain the state equation of transverse vibrating elastic systems

$$\frac{\partial}{\partial t} x(z,t) = \begin{bmatrix} 0 & 1 \\ A_{21} & A_{22} \end{bmatrix} x(z,t) + \begin{bmatrix} 1 \\ \frac{1}{\mu(z)} \end{bmatrix} u(z,t) \tag{13}$$

The elements A_{21} and A_{22} are the differential operators

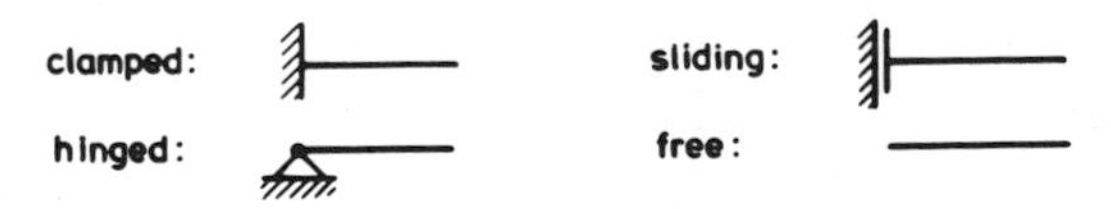

FIG. 3. Simple supports of the ends of a beam.

$$A_{21} = \frac{1}{\mu(z)}\left[\frac{\partial}{\partial z}\left(f(z,t)\frac{\partial}{\partial z}\right) - \frac{\partial^2}{\partial z^2}\left(\alpha(z)\frac{\partial^2}{\partial z^2}\right)\right] \tag{13a}$$

and

$$A_{22} = \frac{-1}{\mu(z)}\left(\delta_E + \delta_I\frac{\partial^4}{\partial z^4}\right) \tag{13b}$$

If the coefficients μ, f, and α are constant, we introduce the parameters $a = \sqrt{\alpha/\mu}$, $c = \sqrt{f/\mu}$, $2\beta_E = \delta_E/\mu$, $2\beta_I = \delta_I/\mu$.

Choosing the state vector

$$x(z,t) = [aw''(z,t) + cw'(z,t), \dot{w}(z,t)]^T \tag{14}$$

instead of (12), we obtain the system equation (1) in the form

$$\frac{\partial}{\partial t}x(z,t) = \begin{bmatrix} 0 & a\frac{\partial^2}{\partial z^2} + c\frac{\partial}{\partial z} \\ -a\frac{\partial^2}{\partial z^2} + c\frac{\partial}{\partial z} & 2\beta_E + 2\beta_I\frac{\partial^4}{\partial z^4} \end{bmatrix} x(z,t) + \begin{bmatrix} 0 \\ \frac{1}{\mu} \end{bmatrix} u(z,t) \tag{15}$$

If the lateral velocity $\dot{w}(z,t)$ is observed at the point $z = z_o$, Eq. (2) reads

$$y(t) = \left[0, \int_0^L \delta(z-z_o)(\cdot)\, dz\right] x(z,t) \tag{16}$$

where $\delta(z-z_o)$ represents the Dirac delta distribution. As an example of boundary conditions (3), consider a beam with hinged ends, where external bending moments $v_o(t)$ and $v_L(t)$ may act at the points z=0 and z=L:

$$\begin{bmatrix} a & 0 \\ 0 & 1 \end{bmatrix} x(z,t) = \begin{bmatrix} \frac{1}{\mu} \\ 0 \end{bmatrix} v_z(t), \qquad z \in \Gamma = \{0,L\},\ t \in T \tag{17}$$

In this case the boundary operators A_Γ and B_Γ reduce to a simple matrix and a column vector, respectively.

Assuming the initial conditions of the beam $w(z,0) = w_o(z)$ and $\dot{w}(z,0) = w_1(z)$, the corresponding condition (4) can easily be derived from the state vector (12) or (14), for example,

$$x(z,0) = [aw_o''(z) + cw_o'(z), w_1(z)]^T \tag{18}$$

Equations (15) to (18) describe the transverse motion of slender uniform beams. They are specific examples of the general system description (1) to (4).

Remark 1. Whether or not the elastic system is well-posed depends not only on the choice of the state variables (12) or (14), but also on the choice of the norm of the Hilbert space and whether damping effects are considered. An example is given in [12].

Remark 2. The consideration of small internal damping corresponds to parabolic regularization of hyperbolic equations [18].

Remark 3. The tensile force f(z,t) may be an external control force or an additional input variable. We are then confronted with a bilinear DPS.

Remark 4. If the restoring forces are entirely due to the axial tension and the bending stiffness is negligible ($a=0$, $\beta_I=0$), Eq. (15) describes the lateral vibration of flexible strings, hanging ropes, heavy chains, and cables. The coefficient c is the wave propagation velocity. Longitudinal vibrations of rods and torsional vibrations of circular shafts are also governed by Eq. (15) if the corresponding variables and parameters are used [2].

2. Timoshenko Beam

In the preceding section we did not consider the effects of shearing and rotatory inertia. In general, the total lateral deflection w(z,t) of a beam consists of two components, one caused by bending and the other by shearing

$$w(z,t) = w_B(z,t) + w_S(z,t)$$

Considering these effects we obtain two coupled PDEs [2]

$$\begin{aligned} &\mu\ddot{w} - (\theta\ddot{w}_B')' + (\alpha_B w_B'')'' = p \\ &\theta\ddot{w}_B' - \alpha_S(w'-w_B') - (\alpha_B w_B'')' = 0 \end{aligned} \tag{19}$$

where θ is the mass moment of inertia. If the state vector

$$x(z,t) = (w'-w_B',\ \dot{w},\ w_B'',\ \dot{w}_B')^T \tag{20}$$

is chosen, the state equation (2) reads in this case

$$\frac{\partial}{\partial t} x(z,t) = \left[\begin{array}{cc|cc} 0 & \frac{\partial}{\partial z} & 0 & -1 \\ \frac{1}{\mu}\frac{\partial}{\partial z}(\alpha_S(\cdot)) & 0 & 0 & 0 \\ \hline 0 & 0 & 0 & \frac{\partial}{\partial z} \\ \frac{\alpha_S}{\theta} & 0 & \frac{1}{\theta}\frac{\partial}{\partial z}(\alpha_B(\cdot)) & 0 \end{array}\right] x(z,t)$$

$$+ \left[\begin{array}{c} 0 \\ \frac{1}{\mu} \\ \hline 0 \\ 0 \end{array}\right] u(z,t) \tag{21}$$

Boundary and initial conditions can be given in a similar manner. Equation (21) is one possible example of an elastic system of the fourth order. A second example is the coupled transverse and torsional vibration of beams, which is of particular importance in the dynamics of aircraft structures.

3. Coupled Transverse and Torsional Vibrations of Beams

We denote by d(z) the distance from the neutral axis of the beam to the center of torsion. The motion is then governed by the pair of differential equations [13]:

$$\left.\begin{array}{ll} \mu\ddot{w} - \mu d\ddot{\varphi} + (\alpha_B w'')'' = p(z,t), & w = w(z,t) \\ \rho I_o\ddot{\varphi} - \mu d\ddot{w} - (\alpha_T\varphi')' = m(z,t), & \varphi = \varphi(z,t) \end{array}\right\} \quad z,t \in \Pi \tag{22}$$

where

- φ = angle of rotation
- p = force per unit length
- m = moment per unit length
- ρ = mass density
- I_o = polar moment of inertia with respect to the shear center ($I_o = I_p + Ad^2$)

I_p = polar moment of the cross section ($\rho I_p = \rho I_o - \mu d^2$)
A = cross-sectional area

Choose the state

$$x(z,t) = [w''(z,t),\ \dot{w}(z,t),\ \varphi'(z,t),\ \dot{\varphi}(z,t)]^T \tag{23a}$$

and the control vector

$$u(z,t) = [p(z,t),\ m(z,t)]^T \tag{23b}$$

to obtain the following system equation:

$$\frac{\partial}{\partial t}x(z,t) = \left[\begin{array}{cc|cc} 0 & \frac{\partial^2}{\partial z^2} & 0 & 0 \\ \frac{-I_o}{\mu I_p}\frac{\partial}{\partial z^2}(\alpha_B(\cdot)) & 0 & \frac{d}{\rho I_p}\frac{\partial}{\partial z}(\alpha_T(\cdot)) & 0 \\ \hline 0 & 0 & 0 & \frac{\partial}{\partial z} \\ \frac{-d}{\rho I_p}\frac{\partial^2}{\partial z^2}(\alpha_B(\cdot)) & 0 & \frac{1}{\rho I_p}\frac{\partial}{\partial z}(\alpha_T(\cdot)) & 0 \end{array}\right] x(z,t)$$

$$+ \left[\begin{array}{cc} 0 & 0 \\ \frac{I_o}{\mu I_p} & \frac{d}{\rho I_p} \\ \hline 0 & 0 \\ \frac{d}{\rho I_p} & \frac{1}{\rho I_p} \end{array}\right] u(z,t) \tag{24}$$

The coupling between both the transverse vibrations and torsional vibrations is caused by the distance d(z). For boundary and initial conditions see [13].

4. Membranes and Thin Plates

Now we consider two-dimensional systems (r=2) and, specifically, membranes and thin plates. The membranes have no bending resistance and the restoring forces are due exclusively to tension as opposed to plates, where the bending stiffness is responsible for the restoring forces.

Assume that the two-dimensional system, in the equilibrium position, lies in a plane (reference plane). The differential equation for the lateral displacement $w(z,t) = w(z_1,z_2,t)$ subject to uniform tension f per unit length is similar to the equation of strings [2]:

$$\mu\ddot{w}(z,t) + \delta_E\dot{w}(z,t) - f\,\Delta w(z,t) = p(z,t), \qquad z,t \in \Pi \tag{25}$$

where

p = external pressure

μ = mass per unit area

δ_E = damping coefficient per unit length

$\Delta = \nabla^2$ = Laplace operator

∇ = nabla operator

A possible boundary condition is

$$a_o w(z,t) + a_1 \frac{\partial w(z,t)}{\partial n} = v(z,t), \qquad z,t \in \Sigma \tag{26}$$

where the derivative is taken in the direction of the outer normal n of the boundary Γ. One of the coefficients a_o and a_1 can be zero. Instead of (25) we can use one of the following state equations [with $c = \sqrt{f/\mu}$ and $2\beta_E = \delta_E/\mu$ for abbreviation]:

$$\frac{\partial}{\partial t}\begin{bmatrix} w(z,t) \\ \dot{w}(z,t) \end{bmatrix} = \begin{bmatrix} 0 & 1 \\ c^2\Delta & -2\beta_E \end{bmatrix}\begin{bmatrix} w(z,t) \\ \dot{w}(z,t) \end{bmatrix} + \begin{bmatrix} 0 \\ \frac{1}{\mu} \end{bmatrix} p(z,t) \tag{27}$$

$$\frac{\partial}{\partial t}\begin{bmatrix} c\,\nabla w(z,t) \\ \dot{w}(z,t) \end{bmatrix} = \begin{bmatrix} 0 & c\,\nabla \\ c\,\nabla & -2\beta_E \end{bmatrix}\begin{bmatrix} c\,\nabla w(z,t) \\ \dot{w}(z,t) \end{bmatrix} + \begin{bmatrix} 0 \\ \frac{1}{\mu} \end{bmatrix} p(z,t) \tag{28}$$

Similar equations hold true if vibrating plates are considered. The equivalent equation of (25) is

$$\mu\ddot{w}(z,t) + \delta_E\dot{w}(z,t) + \alpha_P\,\Delta^2 w(z,t) = p(z,t), \qquad z,t \in \Pi \tag{29}$$

where

α_p = $Eh^3/12(1 - \nu^2)$ = plate flexural rigidity

h = thickness of the plate

E = Young's modulus

ν = Poisson's ratio, $\nu \in (0,0.5)$

Denote by n and s the coordinates in the directions normal and tangential to the boundary Γ (Figure 4) to derive the following boundary conditions [2], with the edge

$$\text{Clamped} \qquad w = 0, \qquad \frac{\partial w}{\partial n} = 0, \tag{30a}$$

$$\text{Supported} \qquad w = 0, \qquad M_n = 0, \qquad z,t \subset \Sigma \tag{30b}$$

$$\text{Free} \qquad M_n = 0, \qquad Q_n - \frac{\partial M_{ns}}{\partial s} = 0, \tag{30c}$$

The relations between moment, shearing forces, and deflections, in terms of normal and tangential coordinates, are

$$\begin{aligned} M_n &= \alpha_p \Delta w - (1 - \nu)\alpha_p \left(\frac{1}{R}\frac{\partial w}{\partial n} + \frac{\partial^2 w}{\partial s^2}\right) \\ M_{ns} &= (1 - \nu)\alpha_p \left(\frac{\partial^2 w}{\partial n\, \partial s} - \frac{1}{R}\frac{\partial w}{\partial s}\right), \qquad Q_n = -\alpha_p \frac{\partial}{\partial n}\Delta w \end{aligned} \tag{31}$$

where the Laplace operator has the form

$$\Delta = \frac{\partial^2}{\partial n^2} + \frac{1}{R}\frac{\partial}{\partial n} + \frac{\partial^2}{\partial s^2} \tag{32}$$

and R denotes the radius of curvature of the boundary curve Γ. The equations similar to (27) and (28) are thus

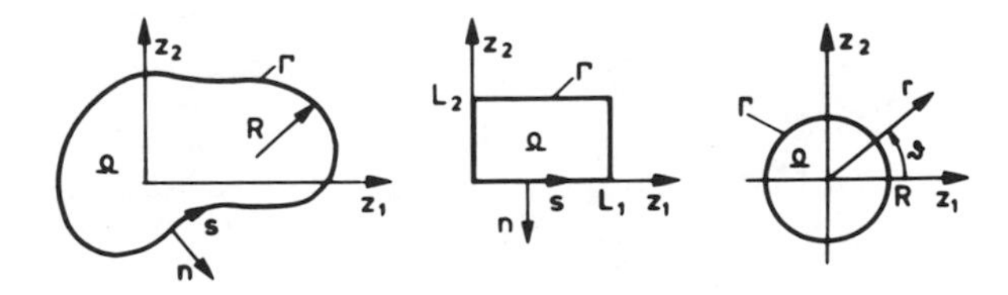

FIG. 4. Two-dimensional elastic systems (thin plates).

$$\frac{\partial}{\partial t}\begin{bmatrix} w(z,t) \\ \dot{w}(z,t) \end{bmatrix} = \begin{bmatrix} 0 & 1 \\ -a^2\Delta^2 & -2\beta_E \end{bmatrix}\begin{bmatrix} w(z,t) \\ \dot{w}(z,t) \end{bmatrix} + \begin{bmatrix} 0 \\ \frac{1}{\mu} \end{bmatrix} p(z,t) \quad (33)$$

$$\frac{\partial}{\partial t}\begin{bmatrix} a\,\Delta w(z,t) \\ \dot{w}(z,t) \end{bmatrix} = \begin{bmatrix} 0 & a\,\Delta \\ -a\,\Delta & -2\beta_E \end{bmatrix}\begin{bmatrix} a\,\Delta w(z,t) \\ \dot{w}(z,t) \end{bmatrix} + \begin{bmatrix} 0 \\ \frac{1}{\mu} \end{bmatrix} p(z,t) \quad (34)$$

We omit the description of the boundary and initial conditions in the form of (3) and (4), since this is straightforward.

Remark 5. If the thickness h of the plate is not constant and if normal forces act in the plane of the plate, the governing equations are given in [13].

Remark 6. Sometimes the Laplace operator Δ is called harmonic operator and hence the *hyperbolic* equations (25), (27), and (28) are called harmonic and Eqs. (29), (33), and (34) biharmonic.

Remark 7. Assume undamped and nondisturbed elastic systems ($\beta_E = 0$, $u = 0$). Note that the matrix differential operator A of the harmonic system (28) is *symmetric*, but the operator A of the biharmonic system (34) is *skew symmetric*. This is a characteristic property of elastic systems. In both cases the adjoint operator is $A^* = -A$. With the above assumption, $\beta_E = 0$, $u = 0$, Eqs. (27), (28), (33), and (34) must also be selfadjoint. This is the case, since $-\dot{x}=A^*x$ is equal to $\dot{x}=Ax$.

C. Control and Observation

1. Control Variables

In the preceding sections we have assumed spatially distributed and continuous control variables u(z,t), e.g., distributed loads p(z,t) or bending moments m(z,t).

An important simplification occurs if the (scalar) control variables are separable in space and time:

$$u(z,t) = \sum_{i=1}^{I} p_i(z)u_i(t) = p^T(z)u(t), \qquad z \in \bar{\Omega},\ t \in T \qquad (35)$$

where the profiles $p_i(z)$ are given a priori and only the amplitudes $u_i(t)$ are allowed to vary with time. Such control functions are always assumed in [13]. Specific examples are single forces $u_j(t)$ at discrete points a_j and moments $m_k(t)$ at discrete points b_k

$$u(z,t) = \sum_{j=1}^{J} \delta(z-a_j)u_j(t) + \sum_{k=1}^{K} \delta'(z-b_k)m_k(t) \tag{35a}$$

where $\delta(z)$ is the Dirac delta distribution and $\delta'(z)$ its first derivative. The manipulation points a_j and b_k may be time-dependent, $a_j = \nu_j t$, for example, if the control force $u_j(t)$ is moving with constant velocity ν_j.

Remark 8. The boundary control variables $v(z,t)$ can be considered as a special case of distributed controls. Thus, we restrict our investigations mainly to control problems with distributed controls since, if generalized functions or distributions are used, boundary control problems are formally included [12,24].

2. Output Variables

In general, distributed measurements are of the form

$$y(z,t) = \int_{\Omega_M} C(z,\zeta,t)x(\zeta,t)\, d\zeta \tag{36}$$

where the weighting matrix $C(z,\zeta,t)$ is assumed to be time-dependent. The integration over the spatial domain $\Omega_M \leq \Omega$ is assumed, since the real sensors may average over some portion of the spatial domain Ω.

Examples are proportional measurement

$$y(z,t) = \int_{\Omega_M} \delta(z-\zeta)Cx(\zeta,t)\, d\zeta = Cx(z,t), \qquad z \in \Omega_M,\ t \in T \tag{36a}$$

and modal observation

$$y_m(t) = \int_{\Omega} \varphi_m(\zeta)Cx(\zeta,t)\, d\zeta = Cx_m(t), \qquad m = 1,2, \ldots ,M \tag{36b}$$

where C is a constant matrix, $\varphi_m(z)$ are eigenfunctions of the investigated system, and $x_m(t)$ are generalized Fourier coefficients, since (36b) can be regarded as a Fourier integral transformation.

Of practical importance are observations at discrete points $c_n \in \Omega_M$ (idealized local observations):

$$y_n(t) = \int_{\Omega_M} \delta(\zeta - c_n) Cx(\zeta,t)\, d\zeta = Cx(c_n,t), \quad n=1,2, \ldots ,N \quad (36c)$$

In the case of moving sensors, the points c_n are time-dependent.

Remark 9. Both the controllability and observability of elastic systems are closely related to the choice of control and observation variables, since these properties depend on the type, number, and location of manipulators and sensors. Such problems appear in parameter identification, state estimation, and control of distributed systems. The reader is referred to [17,30,32,33], and Chapter 2. The controllability of elastic systems has been investigated by Fattorini [34]. The observability defined by Wang [17] can be applied to elastic systems of the form (1) to (4). This definition requires initial state recovery from measurements over some finite time interval and spatial domain.

We assume in the following sections controllable and observable systems where necessary. Common assumptions are mode controllability and mode observability, i.e., a finite number of essential modes of an elastic system are controllable and observable. Consequently, lumped manipulators and sensors should not be placed in a node of the considered eigenfunctions [30,32].

III. FEEDBACK CONTROL OF ELASTIC SYSTEMS

A. Output Regulator Problem

In output regulator problems we seek control variables $u(z,t)$ such that the corresponding output $y(z,t)$ reaches the desired vector $y_d(z,t)$. The control theory of DPS [30,35] can be applied to obtain the following dynamic control law:

$$u(z,t;y,y_d) = -\int_0^t \int_\Omega F(z,\zeta,t,\tau)[y(\zeta,\tau) - y_d(\zeta,\tau)]\, d\zeta\, d\tau$$

$$\triangleq -F[y(z,t) - y_d(z,t)] \quad (37)$$

where the kernel $F(z,\zeta,t,\tau)$ of the integral operator F is the Green's matrix of the regulator. If the system to be controlled is self-adjoint, one usual type of control is modal control. For simplicity, we regard the case with scalar input and output; then $F(z,\zeta,t,\tau)$ is the Green's function of the regulator

$$F(z,\zeta,t,\tau) = \sum_{i=1}^{\infty} \sum_{j=1}^{\infty} \varphi_i(z)\hat{f}_{i,j}(t-\tau)\varphi_j(\zeta) = \varphi^T(z)\hat{F}(t-\tau)\varphi(\zeta) \tag{38}$$

where

$$\varphi^T(z) = [\varphi_1(z),\varphi_2(z), \ldots]$$

is the infinite-dimensional vector of the complete set of orthogonal eigenfunctions of the system. For implementation of the control law (37), only a finite number of modes can be realized. The distinct modes of the closed-loop system are decoupled when a diagonal matrix $F(t-\tau)$ is chosen. For modal control problems of elastic systems the reader is referred to [8,9,11,30,36].

Of practical importance is the pointwise control of one-dimensional elastic systems with control forces and sensors at discrete points as discussed in the preceding section. A useful method of solving this problem is to divide the entire system into a finite number of coupled distributed subsystems [30]. Each subsystem is described in the frequency domain by a difference equation with transcendental coefficients. The controller is designed in such a manner that a finite number of single feedback control loops are obtained.

Frequency methods are also preferred in the case of regulator problems with pointwise or boundary observation and boundary control [37]. As an alternative to the modal analysis and modal control of bending vibrations, such vibrations can be studied in terms of the DP concepts of wave propagation, reflection, and characteristic termination. In particular, the dynamics of lateral vibration of a thin, uniform beam are factored by Vaughan [23] into a form separating the process of propagation from boundary effects. This allows the influence of various terminal impedance matrices to be described in terms of a reflection matrix, which is a generalization of the concept of reflection coefficient for the wave equation. Several specific cases of terminal impedance matrices are considered in [23]. A different form of the wave reflection matrix has been developed by Van de Vegte [22]. This form presents a clearer correlation between the values of its elements and the system response characteristics.

The aim of the following section is the application of the optimal control theory [18,27,28,38] to solve linear optimal tracking problems with state feedback controls. Closely related are problems of model-following and state reconstruction.

B. Optimal Tracking Problem

1. Problem Formulation

Let the vibrating systems be governed by Eqs. (1) to (4). Assume distributed and boundary controls $u(z,t)$ and $v(z,t)$ belonging to the sets of admissible controls $U_{ad} \subseteq U$ and $V_{ad} \subseteq V$. Here T is the time interval $(0,t_f)$ with fixed final time t_f. Denote the difference between the output y and the desired output y_d by

$$e(z,t;u,v) = y(z,t;u,v) - y_d(z,t) \tag{39}$$

We seek an optimal distributed control u^o (or a boundary control v^o) which minimizes the quadratic performance functional

$$J(u,v) = \frac{1}{2}\int_T [(e,Qe)_{H_m(\Omega)} + (u,Ru)_{H_k(\Omega)} + (e,Q_\Gamma e)_{H_m(\Gamma)} + (v,R_\Gamma v)_{H_\ell(\Gamma)}]\, dt + \frac{1}{2}(e_f,Q_f e_f)_{H_m(\Omega)} \tag{40}$$

where $(\cdot,\cdot)$ are suitable scalar products of the respective Hilbert spaces and the subscript f denotes final time. Q, R, Q_Γ, R_Γ, and Q_f are symmetric weighting matrices or operators. In our first investigation we consider only distributed control problems and choose $Q_\Gamma = R_\Gamma = 0$.

We want to apply the optimal control theory of DPS, especially the variational inequalities of Lions [18], to obtain the unique optimal solution u^o of this problem. We therefore introduce the quadratic form $a(u,u)$, the linear form $b(u)$, and the constant c:

$$a(u,u) = \int_T \{(y(u), Qy(u)) + (u,Ru)\} \, dt + (y_f(u),Q_f y_f(u))$$

$$b(u) = \int_T (y(u),Qy_d) \, dt + (y_f(u),Q_f y_{df})$$

$$c = \int_T (y_d,Qy_d) \, dt + (y_{df},Q_f y_{df})$$

and write the performance functional (40) in the form

$$J(u) = \frac{1}{2} [a(u,u) - 2b(u) + c] \tag{40a}$$

From [18] we conclude that a minimizing element $u^o \in U_{ad}$ exists, if the assumption

$$a(u,u) \geqq k \|u\|^2 \qquad \forall u \in U, \; k > 0 \tag{41}$$

holds, and that this minimizing element is characterized by the variational inequality

$$a(u^o, u-u^o) \geqq b(u-u^o), \qquad \forall u \in U_{ad} \tag{42}$$

The proofs of existence and uniqueness of u^o are given in [18] for parabolic and hyperbolic systems. They can be extended to elastic systems of the form (1) to (4). In our case we obtain the inequality

$$\int_T \{(x^o, C^* QC(x-x^o)) + (u^o, R(u-u^o))\} \, dt + (x_f^o, C^* Q_f C(x_f-x_f^o))$$

$$\geqq \int_T ((x-x^o), C^* Qy_d) \, dt + ((x_f-x_f^o), C^* Q_f y_{df}) \tag{42a}$$

where C^* is the adjoint operator and $x^o = x(z,t;u^o)$, $x_f^o = x(z,t_f;u^o)$.

In order to derive linear control laws we restrict ourselves to unconstrained square integrable controls $u \in U_{ad} = U$. Consequently, only the equals sign holds in (42a).

2. Optimal Control Law

Assume integral operators in (40) and choose $Q_\Gamma = R_\Gamma = 0$:

$$J(u) = \frac{1}{2}\int_T\int_\Omega\int_\Omega \{e^T(z,t)Q(z,\zeta,t)e(\zeta,t) + u^T(z,t)R(z,\zeta,t)u(\zeta,t)\}$$

$$d\zeta\ dz\ dt + \frac{1}{2}\int_\Omega\int_\Omega e^T(z,t_f)Q_f(z,\zeta)e(\zeta,t_f)\ d\zeta\ dz \tag{40b}$$

If $(u,Ru) \geqq k\ \|u\|^2$, $\forall u \in U$, $k > 0$, and $(e,Qe) \geqq 0$, we obtain the optimal linear control law from (42a):

$$u^o(z,t;x,q) = -\int_\Omega S(z,z',t)B^*\left\{\int_\Omega K(z',\zeta,t)x(\zeta,t)\ d\zeta - q(z',t)\right\} dz' \tag{43}$$

where $S(z,z',t)$ is the kernel of the operator $R^{-1} = S$, which is defined by

$$\int_\Omega S(z',z,t)R(z,\zeta,t)\ dz = \delta(z',\zeta)I$$

with identity matrix I. B^* is the adjoint operator of B. For simplicity, we assume space-dependent matrices $B(z)$ and $C(z)$. Extensions to operators are always possible. The kernel $K(z,\zeta,t)$ is the solution of the nonlinear PDE of Riccati type:

$$\frac{-\partial}{\partial t}K(z,\zeta,t) = A_z^*K(z,\zeta,t) + [A_\zeta^*K^T(z,\zeta,t)]^T + C^T(z)Q(z,\zeta,t)C(\zeta)$$
$$- \int_\Omega\int_\Omega K(z,\zeta',t)B(\zeta')S(\zeta',z',t)B^T(z')K(z',\zeta,t)\ dz'\ d\zeta'$$
$$z,\zeta \in \Omega,\ t \in T \tag{44}$$

with boundary conditions

$$\bar{A}_{\Gamma z}K(z,\zeta,t) = 0, \qquad z \in \Gamma$$
$$\bar{A}_{\Gamma\zeta}K(z,\zeta,t) = 0, \qquad \zeta \in \Gamma,\ t \in T \tag{44a}$$

and final condition

$$K(z,\zeta,t_f) = C^T(z)Q_f(z,\zeta)C(\zeta, \qquad z,\zeta \in \Omega \tag{44b}$$

The feedforward control vector $q(z,t)$ is a solution of the PDE

$$\frac{-\partial}{\partial t} q(z,t) = A^{*}q(z,t) - \int_{\Omega}\int_{\Omega} K(z,\zeta',t)B(\zeta')S(\zeta',z',t)B^{T}(z')q(z',t)$$

$$dz'\, d\zeta' + \int_{\Omega} C^{T}(z)Q(z,\zeta,t)y_{d}(\zeta,t)\, d\zeta, \qquad z,t \in \Pi \qquad (45)$$

with boundary and final conditions

$$\bar{A}_{\Gamma}q(z,t) = 0, \qquad z,t \in \Sigma \qquad (45a)$$

and

$$q(z,t_{f}) = \int_{\Omega} C^{T}(z)Q_{f}(z,\zeta)y_{d}(\zeta,t_{f})\, d\zeta, \qquad z \in \Omega \qquad (45b)$$

The subscripts z and ζ of the adjoint operator A^{*} denote the derivatives with respect to z or ζ. The boundary operator $\bar{A}_{\Gamma}$ depends on the boundary conditions of the considered system and can only be computed for particular examples. It is not feasible to solve the rather general nonlinear equation (44), therefore we seek solutions for some of the concrete examples of elastic systems discussed in Section II.B.

3. Optimal Regulators for Transversal Vibrating Systems

Consider Eq. (34) with homogeneous boundary conditions and nonhomogeneous initial conditions for the description of thin plates or slender beams ($\beta = \beta_{E}$).

$$\frac{\partial}{\partial t} x(z,t) = \begin{bmatrix} 0 & a\,\Delta_{z} \\ -a\,\Delta_{z} & -2\beta \end{bmatrix} x(z,t) + \begin{bmatrix} 0 \\ \frac{1}{\mu} \end{bmatrix} u(z,t), \qquad z,t \in \Pi \qquad (46)$$

$$x(z,t) = 0, \qquad z,t \in \Sigma$$

$$x(z,0) = x_{o}(z), \qquad z \in \Omega$$

Assume C = I, given $y_{d}(z,t) = x_{d}(z,t)$ and final time t_{f}. We want to minimize the performance functional (40b) with given weighting matrices

$$Q(z,\zeta,t) = \begin{bmatrix} q_{11}(z,\zeta,t) & 0 \\ 0 & q_{22}(z,\zeta,t) \end{bmatrix} \quad Q_f(z,\zeta) = \begin{bmatrix} q_{f11}(z,\zeta) & 0 \\ 0 & q_{f22}(z,\zeta) \end{bmatrix}$$

$$R(z,\zeta,t) = \delta(z-\zeta)r(z,t)$$

If the Riccati matrix

$$K(z,\zeta,t) = \begin{bmatrix} k_{11}(z,\zeta,t) & k_{12}(z,\zeta,t) \\ k_{21}(z,\zeta,t) & k_{22}(z,\zeta,t) \end{bmatrix}$$

$$\triangleq \begin{bmatrix} k_{i,j}(z,\zeta,t) \end{bmatrix}, \quad \begin{matrix} i=1,2 \\ j=1,2 \end{matrix} \tag{47}$$

is introduced, we obtain the control law

$$u^{o}(z,t) = \frac{-1}{\mu r(z,t)} \left\{ \int_{\Omega} [k_{21}(z,\zeta,t), k_{22}(z,\zeta,t)] x(\zeta,t) d\zeta - q_2(z,t) \right\} \tag{48}$$

where $k_{ij}(z,\zeta,t)$ and $q_i(z,t)$ are solutions of the following equations (arguments neglected):

$$\begin{aligned}
-\dot{k}_{11} &= -a\,\Delta_z\, k_{21} - a\,\Delta_\zeta\, k_{12} - \frac{1}{\mu^2}\int k_{12} r^{-1} k_{21}\, d\zeta' + q_{11} \\
-\dot{k}_{12} &= -a\,\Delta_z\, k_{22} + a\,\Delta_\zeta\, k_{11} - 2\beta k_{12} - \frac{1}{\mu^2}\int k_{12} r^{-1} k_{22}\, d\zeta' \\
-\dot{k}_{21} &= -a\,\Delta_\zeta\, k_{22} + a\,\Delta_z\, k_{11} - 2\beta k_{21} - \frac{1}{\mu^2}\int k_{22} r^{-1} k_{21}\, d\zeta' \\
-\dot{k}_{22} &= a\,\Delta_z\, k_{12} + a\,\Delta_\zeta\, k_{21} - 4\beta k_{22} - \frac{1}{\mu^2}\int k_{22} r^{-1} k_{22}\, d\zeta' + q_{22}
\end{aligned} \tag{49a}$$

Note the symmetry of the Riccati matrix $K(z,\zeta,t)$:

$$k_{12}(z,\zeta,t) = k_{21}(\zeta,z,t), \quad \text{therefore} \quad K(z,\zeta,t) = K^T(\zeta,z,t)$$

$$\begin{aligned}
-\dot{q}_1 &= -a\,\Delta_z\, q_2 - \frac{1}{\mu^2}\int k_{12} r^{-1} q_2\, d\zeta + \int q_{22} x_{d2}\, d\zeta \\
-\dot{q}_2 &= a\,\Delta_z\, q_1 - 2\beta q_2 - \frac{1}{\mu^2}\int k_{22} r^{-1} q_2\, d\zeta + \int q_{22} x_{d2}\, d\zeta
\end{aligned} \tag{49b}$$

with boundary and final conditions

$$K(z,\zeta,t)=0 \qquad q(z,t)=0, \qquad z,\zeta \in \Gamma,\ t \in T \quad (49c)$$

$$K(z,\zeta,t_f)=Q_f(z,\zeta), \qquad q(z,t_f)=Q_f x_d(z,t_f) \qquad z,\zeta \in \Omega,\ t = t_f \quad (49d)$$

4. Approximate Solution of the Riccati Equation

It is obvious that Eqs. (49a) to (49d) cannot be solved exactly. Only approximate solutions are possible

$$\hat{K}(z,\zeta,t) = E(z)\hat{K}(t)E^T(\zeta) \quad \text{and} \quad \hat{q}(z,t) = E(z)\hat{q}(t) \qquad (50)$$

with

$$K(t) = \begin{bmatrix} K_{11}(t) & K_{21}^T(t) \\ K_{21}(t) & K_{22}(t) \end{bmatrix}, \qquad \hat{q}(t) = \begin{bmatrix} \hat{q}_1(t) \\ \hat{q}_2(t) \end{bmatrix},$$

$$E(z) = \begin{bmatrix} e(z) & 0 \\ 0 & e(z) \end{bmatrix}$$

for example, where the elements $e_m(z)$ of the vector $e(z)$ are suitably chosen basis functions. We use solutions of the eigenvalue problem

$$\Delta_z\, e(z) = -\Lambda e(z), \qquad z \in \Omega, \qquad e(z) = 0, \qquad z \in \Gamma \qquad (51)$$

which are

$$e_m(z) = \sqrt{\frac{2}{L}}\ \sin m\pi\, \frac{z}{L}, \qquad \lambda_m^2 = \left(\frac{m\pi}{L}\right)^2, \qquad m = 1,2,\ \dots\ ,M \quad (51a)$$

for a beam of length L and

$$e_k(z) = e_{mn}(z_1,z_2) = \frac{2}{\sqrt{L_1 L_2}}\ \sin m\pi\, \frac{z_1}{L_1}\ \sin n\pi\, \frac{z_2}{L_2},$$

$$\lambda_k^2 = \lambda_{mn}^2 = \left(\frac{m\pi}{L_1}\right)^2 + \left(\frac{n\pi}{L_2}\right)^2,\ k = 1,2,\ \dots\ ,MN \qquad (51b)$$

if a rectangular plate with lengths L_1 and L_2 is considered. The modal transformation of the system of Eqs. (49a) to (49d) leads to the ordinary equations

$$-\dot{\hat{K}}(t) = \hat{\Lambda}^T\hat{K}(t) + \hat{K}(t)\hat{\Lambda} - \hat{K}(t)\hat{B}\hat{S}(t)\hat{B}^T\hat{K}(t) + \hat{Q}(t), \qquad (52a)$$

$$-\dot{\hat{q}}(t) = \hat{\Lambda}\hat{q}(t) - \hat{K}(t)\hat{B}\hat{S}(t)\hat{B}^T\hat{q}(t) + \hat{Q}(t)\hat{x}_d(t), \qquad t \in T \qquad (52b)$$

$$\hat{K}(t_f) = \hat{Q}_f, \qquad \hat{q}(t_f) = \hat{Q}_f \hat{x}_d(t_f) \tag{52c}$$

where the following abbreviations are used:

$$\hat{\Lambda} = \begin{bmatrix} 0 & -a\Lambda \\ a\Lambda & -2\beta I \end{bmatrix}, \qquad \Lambda = \text{diag}(\lambda_1^2, \lambda_2^2, \ldots, \lambda_M^2),$$

$$I = \int_\Omega e(z)e^T(z)\, dz$$

$$\hat{B} = \frac{1}{\mu}\begin{bmatrix} 0 \\ I \end{bmatrix}, \qquad \hat{Q} = \begin{bmatrix} \hat{Q}_{11} & 0 \\ 0 & \hat{Q}_{22} \end{bmatrix}, \qquad \hat{x}_d(t) = \int_\Omega e(z)x_d(z,t)\, dz$$

$$\hat{S}(t) = \int_\Omega e(z)r^{-1}(z,t)e^T(z)\, dz,$$

$$\hat{Q}_{ii}(t) = \int_\Omega\int_\Omega e(z)q_{ii}(z,\zeta,t)e^T(\zeta)\, d\zeta\, dz$$

We restrict ourselves to the case of an infinite time interval ($t_f \to \infty$). Assuming $x_d = 0$ and $Q_f = 0$, consider time-independent weighting matrices Q and R. We obtain the steady-state positive definite solution $\hat{K}$ with time-independent submatrices $\hat{K}_{ij}$. If Q_{ii} and $\hat{S}$ are diagonal matrices, the $\hat{K}_{ij}$ are also diagonal with elements $\hat{k}_{ij,m}$

$$\hat{k}_{21,m} = \frac{a\mu^2\lambda_m^2}{\hat{s}_m}\left[1 - \sqrt{1 + \frac{\hat{s}_m\hat{g}_{11,m}}{a^2\mu^2\lambda_m^4}}\right] \triangleq -\mu^2\bar{\gamma}_m < 0 \tag{53a}$$

$$\hat{k}_{22,m} = \frac{-2\beta\mu^2}{\hat{s}_m}\left(1 - \left\{1 + \frac{\hat{s}_m\hat{q}_{22,m}}{4\beta^2\mu^2} + 2\left(\frac{a\lambda_m^2}{2\beta}\right)^2 \left[\sqrt{1 + \frac{\hat{s}_m\hat{q}_{11,m}}{a^2\mu^2\lambda_m^4}} - 1\right]\right\}^{1/2}\right) \tag{53b}$$

$$\triangleq \frac{-2\mu^2}{\hat{s}_m}(\beta - \bar{\beta}_m) > 0. \qquad m = 1,2,\ldots,M$$

$$k_{11,m} = k_{22,m}\sqrt{1 + \frac{\hat{s}_m\hat{q}_{11,m}}{a^2\mu^2\lambda_m^4}} - 2\beta k_{21,m} > 0 \tag{53c}$$

The elements $q_{ii}(z,\zeta)$ and $r(z)$ must be suitably chosen,

$$q_{ii}(z,\zeta) = \frac{q_{ii}}{L^2} \begin{cases} \zeta(L-z) & \text{if } \zeta \in [0,z], \\ z(L-\zeta) & \text{if } \zeta \in [z,L], \end{cases} \qquad q_{ii} > 0, \ i = 1,2$$

and $r(z) = 1$, for example. In this case the elements of $\hat{Q}$ and $\hat{S}$ are

$$\hat{q}_{ii,m} = q_{ii} \left(\frac{L}{m\pi}\right)^2, \qquad \hat{s}_m = 1, \qquad m = 1,2, \ldots ,M$$

Note that one of the elements q_{ii} may be zero to obtain a positive definite solution $\hat{K}$. The approximate solution $\hat{K}(z,\zeta)$ approaches the exact solution as M tends to infinity $(M \to \infty)$.

Substitution of (48) into (46) leads to the equation of the closed loop

$$\frac{\partial}{\partial t} x(z,t) = \begin{bmatrix} 0 & a\Delta_z \\ -a\,\Delta_z - \frac{1}{\mu} \int_{\Omega} e^T(z)\hat{K}_{21}e(\zeta)(\cdot)\, d\zeta & -2\beta - \frac{1}{\mu} \int_{\Omega} e^T(z)\hat{K}_{22}e(\zeta)(\cdot)\, d\zeta \end{bmatrix} x(z,t)$$

The damping ratio and the rigidity of the elastic system can be influenced in a suboptimal manner, since only M modes of the elastic system are fed back by the suboptimal regulator if the approximate solution (50) is used. The choice of M is an engineering design problem and depends on the considered system. We will discuss this problem further in Section V, where the optimal control of hauling pipes in ocean mining systems is investigated.

C. Optimal Model-following Problems

In our last example we assumed $y_d = 0$, but our general problem formulation allows the consideration of space- and time-dependent command inputs $y_d(z,t)$ or $x_d(z,t)$. Thus, we are able to solve <u>optimal servo problems</u> or <u>optimal tracking problems</u>, where the command inputs are given functions. When the output or state of the plant is to follow a desirable response to command inputs of another system (the model) rather than follow the command inputs directly, the problem will be called a model-following problem [39]. We distinguish between explicit model-following, where a real physical model has to be built, and implicit model-following, where the dynamics of the model are incorporated in the performance functional [40]. The linear regulator problem with cross-terms (u,Pe) in the performance functional is closely related.

1. Tracking Problem with Additional Output Feedback Loop

Instead of the performance functional (40), we want to minimize

$$J(u) = \frac{1}{2}\int_T [(e,Qe) + 2(u,Pe) + (u,Ru)]\, dt \tag{54}$$

which can be written explicitly as

$$J(u) = \frac{1}{2}\int_T\int_\Omega\int_\Omega [e^T(z,t), u^T(z,t)] \begin{bmatrix} Q(z,\zeta,t) & P^T(z,\zeta,t) \\ P(z,\zeta,t) & R(z,\zeta,t) \end{bmatrix} \times \begin{bmatrix} e(\zeta,t) \\ u(\zeta,t) \end{bmatrix} d\zeta\, dz\, dt \tag{54a}$$

For the existence of the inverse operator $R^{-1} = S$ with kernel $S(z,\zeta,t)$, the integral operator R must be positive definite. P and Q are chosen so that the modified operator $\bar{Q} = Q - P^*SP$ is not negative definite. We obtain formally the slightly modified optimal control law

$$u^o(z,t) = -\int_\Omega S(z,z',t) \left\{B^T(z') \left[\int_\Omega \bar{K}(z',\zeta,t)x(\zeta,t)\, d\zeta - \bar{q}(z',t)\right] + P(z,z',t)(y(z',t) - y_d(z',t))\right\} dz' \tag{55}$$

where $\bar{K}$ and $\bar{q}$ are solutions of the modified equations (44) and (45) in which $\bar{A}_z = A_z - BSPC$ and $\bar{Q} = Q - P^*SP$ replace A_z and Q. Figure 5 shows the basic structure of the closed-loop system.

Remark 10. In contrast to the optimal regulator of the preceding section, we now obtain an additional feedback loop where the

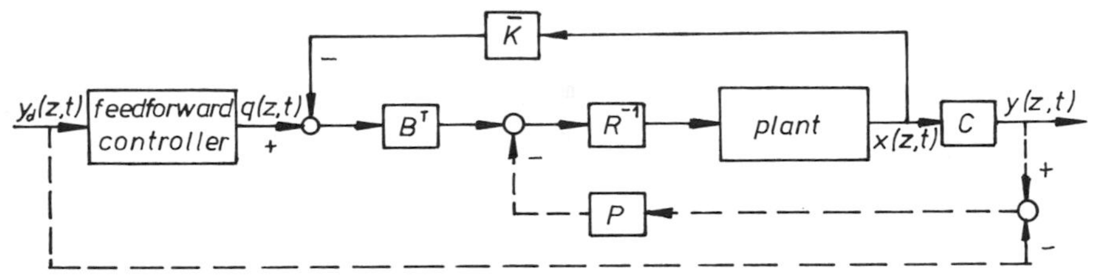

FIG. 5. Basic structure of the optimal closed-loop system.

error $e = y - y_d$ is fed back (Figure 5). The result can therefore be interpreted as the solution of an optimal regulator problem, where the plant itself includes an output feedback control loop of the type discussed in Section III.A.

2. Implicit Model-following

Consider an elastic system of the form (1) to (4) and assume that an ideal system or model is described by

$$\frac{\partial}{\partial t} x_M(z,t) = M_z x_M(z,t), \qquad z,t \in \Pi \tag{56a}$$

$$y_M(z,t) = C_M x_M(z,t) \tag{56b}$$

with boundary and initial conditions

$$A_\Gamma x_M(z,t) = B_\Gamma v(z,t), \qquad z,t \in \Sigma \tag{56c}$$

$$x_M(z,0) = x_{M0}(z), \qquad z \in \Omega \tag{56d}$$

The state x_M of the model satisfies the boundary conditions (3). For simplicity, the homogeneous conditions (3) and (56c) are assumed, $v \equiv 0$. We consider the performance criterion (40b) with $Q_f = 0$ to minimize the dynamic differences between the ideal model (56a) to (56d) and the actual system (1) to (4). Introducing the error

$$e(z,t) = \dot{y}(z,t) - C_M(z) M_z x(z,t) \tag{57}$$

the criterion (40b) is similar to (54a)

$$J(u) = \frac{1}{2} \int_T \int_\Omega \int_\Omega [x^T(z,t), u^T(z,t)] \begin{bmatrix} \tilde{Q}(z,\zeta,t) & \tilde{P}^*(z,\zeta,t) \\ \tilde{P}(z,\zeta,t) & \tilde{R}(z,\zeta,t) \end{bmatrix}$$

$$\times \begin{bmatrix} x(\zeta,t) \\ u(\zeta,t) \end{bmatrix} d\zeta \; dz \; dt \qquad (58)$$

where the abbreviations used are:

$$\tilde{Q}(z,\zeta,t) = [A_z^* C^T(z) - M_z^* C^T(z)]Q(z,\zeta,t)[C(\zeta)A_\zeta - C_M(\zeta)M_\zeta]$$

$$\tilde{P}(z,\zeta,t) = B^T(z)C^T(z)Q(z,\zeta,t)[C(\zeta)A_\zeta - C_M(\zeta)M_\zeta]$$

$$\tilde{R}(z,\zeta,t) = R(z,\zeta,t) + B^T(z)C^T(z)Q(z,\zeta,t)C(\zeta)B(\zeta)$$

From (55) we deduce the optimal control law

$$u^o(z,t) = -\int_\Omega S(z,z',t) \Big\{ B^T(z') \Big[\int_\Omega \bar{K}(z',\zeta,t)x(\zeta,t) \, d\zeta - \bar{q}(z',t) \Big] + B^T(z)C^T(z)Q(z,z',t)[C(z')A_{z'} - C_M(z')M_{z'}]x(z',t) \Big\} \, dz' \qquad (59)$$

where $\bar{K}$ and $\bar{q}$ are solutions of the modified Eqs. (44) and (45) which contain

$$\bar{A}_z = A_z - B\tilde{S}\tilde{P}C, \qquad \bar{Q} = \tilde{Q} + \tilde{P}^*\tilde{S}\tilde{P}, \qquad \text{with } \tilde{S} = \tilde{R}^{-1}$$

instead of A_z and Q. For the existence of an optimal solution of the form (59) the operator $\tilde{R}$ must be positive definite. Note that the model need not be realized, since the square of the error between the ideal model and the plant is represented by one term in the performance functional and is therefore minimized. Thus, the coefficients in the equation of motion of the elastic system are altered by the feedback loop (59) so that they approach the corresponding coefficients in Eqs. (56a) and (56b) of the model.

Consider, for example, the transverse vibrating beam or plate with system equations (46). Assume a model operator

$$M_z = \begin{bmatrix} 0 & a\,\Delta_z \\ -m_1 a\,\Delta_z & -m_2 2\beta \end{bmatrix}, \qquad m_1, m_2 > 0 \qquad (60)$$

and $C_M = C = I$, $Q(z,\zeta,t) = \delta(z-\zeta)Q$, $R(z,\zeta,t) = \delta(z-\zeta)$, where Q is a constant diagonal matrix. From (59) we obtain ($\overline{q} = 0$):

$$u^o(z,t) = \frac{1}{\mu[1 + (q_{22}/\mu^2)]}\left\{\int_\Omega [\overline{k}_{21}(z,\zeta,t),\overline{k}_{22}(z,\zeta,t)]x(\zeta,t)\, d\zeta - q_{22}[(1-m_1)a\,\Delta_z,\ (1-m_2)\,2\beta]x(z,t)\right\} \qquad (59a)$$

Of specific interest is the limit of $u^o(z,t)$ as $q_{22} \to \infty$. Since the elements k_{ij} depend on $\sqrt{q_{22}}$, we obtain from (59a)

$$\lim_{q_{22}\to\infty} u^o(z,t) = -\mu[(1-m_1)a\,\Delta_z,\ (1-m_2)2\beta]x(z,t)$$

and the equation of the closed-loop system reduces to

$$\dot{x}(z,t) = M_z x(z,t), \qquad z \in \Omega,\ t \in T$$

To approach this ideal limit case we have to choose $q_{22} \gg 1$; this is evident from our problem formulation. Sometimes building a real physical model is easier than solving the corresponding Riccati equation. We consider explicit model-following as an alternative concept.

3. Explicit Model-following

Combining the equation of the plant with the real model into a single system by introducing

$$\xi(z,t) = [x(z,t),x_M(z,t)]^T$$

gives

$$\dot{\xi}(z,t) = \begin{bmatrix} A_z & 0 \\ 0 & M_z \end{bmatrix}\xi(z,t) + \begin{bmatrix} B \\ 0 \end{bmatrix}u(z,t) \triangleq \overline{A}_z\xi(z,t) + \overline{B}u(z,t) \qquad (61a)$$

$$\begin{bmatrix} A_\Gamma & 0 \\ 0 & A_\Gamma \end{bmatrix}\xi(z,t) = \begin{bmatrix} 0 \\ 0 \end{bmatrix}, \qquad z,t \in \Sigma \qquad (61b)$$

$$\xi(z,0) = \xi_o(z), \qquad z \in \Omega \tag{61c}$$

The performance functional (40b) with $Q_f = 0$ is appropriate for achieving model-following if the error

$$e(z,t) = y(z,t) - y_M(z,t)$$

is introduced. The results of the preceding sections are applicable to this problem [12]. The optimal control law is

$$u^o(z,t) = -\int_\Omega S(z,z',t)B^T(z') \int_\Omega [\overline{K}_{11}(z',\zeta,t)x(\zeta,t)$$

$$+ \overline{K}_{12}(z',\zeta,t)x_M(\zeta,t)]\, d\zeta\, dz'$$

$$\triangleq -[Fx(z,t) + F_M x_M(z,t)] \tag{62}$$

where $K_{i,j}$ are submatrices of $\overline{K}$ (i=1,2; j=1,2), which is the solution of the modified equation of Riccati type:

$$-\dot{\overline{K}} = \overline{A_z^*}\,\overline{K} + [\overline{A_\zeta^*}\,\overline{K}^T]^T - \int_\Omega\int_\Omega \overline{KBSB^T}\,\overline{K}\, dz'\, d\zeta' + \overline{Q} \tag{63a}$$

$$\overline{A}_\Gamma \overline{K}(z,\zeta,t) = 0, \qquad z,\zeta \in \Gamma,\ t \in T \tag{63b}$$

$$K(z,\zeta,t_f) = 0, \qquad z,\zeta \in \Omega \tag{63c}$$

Writing Eq. (63a) explicitly we observe essential properties

$$\begin{aligned}
-\dot{\overline{K}}_{11} &= A_z^*\overline{K}_{11} + (A_\zeta^*\overline{K}_{11}^T)^T - \iint \overline{K}_{11}BSB^T\overline{K}_{11}\, dz'\, d\zeta' + C^TQC \\
-\dot{\overline{K}}_{12} &= A_z^*\overline{K}_{12} + (M_\zeta^*\overline{K}_{12}^T)^T - \iint \overline{K}_{11}BSB^T\overline{K}_{12}\, dz'\, d\zeta' - C^TQC_M \\
-\dot{\overline{K}}_{21} &= M_z^*\overline{K}_{21} + (A_\zeta^*\overline{K}_{21}^T)^T - \iint \overline{K}_{21}BSB^T\overline{K}_{11}\, dz'\, d\zeta' - C_M^TQC_M \\
-\dot{\overline{K}}_{22} &= M_z^*\overline{K}_{22} + (M_\zeta^*\overline{K}_{22}^T)^T - \iint \overline{K}_{21}BSB^T\overline{K}_{12}\, dz'\, d\zeta' + C_M^TQC_M
\end{aligned} \tag{64}$$

Remark 11. The submatrix $\overline{K}_{11}$ is independent of the model and not coupled with other submatrices. $\overline{K}_{11}$ is a solution of Eq. (44) with corresponding boundary and final conditions. Therefore, $\overline{K}_{11} \equiv K$, and the optimal control law (61) can really be divided into two parts, Fx and $F_M x_M$, indicating that feedforward of the model state variables is required. The model must be built as a part of the control system.

Remark 12. The combined system (61a) is not completely controllable; the model must therefore be asymptotically stable if the steady-state solution of (63a) is required.

The results of the last three sections are not restricted to elastic systems. They may be applied to other distributed systems. Applications are useful in those cases where the desired state $y_d(z,t)$ can be interpreted as a state or output of an ideal or real model. If the desired state is a prescribed function of space and time, the model-following problem reduces to the tracking problem.

4. Relation to the Tracking Problem

The desired output may be thought of as a state of a model ($C_M = I$). Replace $x_M(z,t)$ by $y_d(z,t)$ in the control law (62)

$$u^o(z,t) = -[Fx(z,t) + F_M y_d(z,t)]$$

Comparison with (43) under consideration of $B^* = B^T(z')$ and $\bar{K}_{11} \equiv K$ gives

$$q(z,t) = -\int_\Omega \bar{K}_{12}(z,\zeta,t) y_d(\zeta,t)\, d\zeta \tag{65}$$

The partial time derivative

$$-\dot{q}(z,t) = \int_\Omega [\dot{\bar{K}}_{12}(z,\zeta,t) + \bar{K}_{12}(z,\zeta,t) M_\zeta] y_d(\zeta,t)\, d\zeta$$

together with Eq. (64) is necessary for the proof of (65). We obtain

$$-\dot{q}(z,t) = \int_\Omega [-A_z^* \bar{K}_{12}(z,\zeta,t) + C^T(z) Q(z,\zeta,t) + \int_\Omega\int_\Omega K(z,\zeta,t)$$

$$\times B(\zeta')S(\zeta',z',t)B^T(z')\bar{K}_{12}(z'\zeta,t)\, dz'\, d\zeta']$$

$$\times y_d(\zeta,t)\, d\zeta$$

and conclude that the right-hand side of Eq. (65) fulfills (45). This completes the proof.

Remark 13. In applications, a time-constant deflection profile $y_d(z)$ is usually desired. These profile control problems [30] are included in our results. Examples are given in [12,18].

IV. STATE OBSERVERS

All versions of the optimal regulator, tracking, and model-following problems solved in Section III have in common the basic assumption that the complete state vector $x(z,t)$ is available. For LPS, Kalman filters or Luenberger observers [41,42] have been developed to obtain a suitable estimation or reconstruction of the state. In this decade, many papers have been published on the extension and application of optimal filter theory to estimate the state of DPS, based on noisy measurements of the output variables (the reader is referred to the Introduction and Chapter 3). However, only a few papers [25,29,43, 44] which are concerned with the development and application of observer theory to noise-free DPS are known to the author.

A straightforward application of Luenberger's theory is possible if the distributed plant is approximated as a finite number of LPS [44]. However, it is the author's opinion that the distributed nature of the plant and the observer should be retained as long as possible, e.g., until numerical results are required [45,46]. Distributed parameter observers for diffusion systems are developed in [43]. Another treatment is possible if the observer is considered as a tracking system similar to that of the preceding section [12, 25]. We will follow both lines of thought.

A. Full-order Observers

Let the considered elastic system be described by Eqs. (1) to (4) with an unknown initial state $x(z,t_o)$. Referring to the observer for LPS [41], we define the DP observer as a tracking system [25]:

$$\begin{aligned} \frac{\partial}{\partial t}\hat{x}(z,t) &= A_z\hat{x}(z,t) + Bu(z,t) - G[\hat{y}(z,t) - y(z,t)] \\ \hat{y}(z,t) &= C\hat{x}(z,t), \qquad z,t \in \Pi \\ A_\Gamma\hat{x}(z,t) &= B_\Gamma v(z,t), \qquad z,t \in \Sigma \end{aligned} \tag{66}$$

$$\hat{x}(z,t_o) = \hat{x}_o(z), \qquad z \in \Omega$$

where $\hat{x}(z,t)$ is the state of the observer and G is a suitably chosen integral operator with the kernel $G(z,\zeta,t)$. G can be considered as the distributed weight of the difference between the measured plant output $y(z,t)$ and the observer output $\hat{y}(z,t)$. The design problem is to choose the gain matrix $G(z,\zeta,t)$ in such a way that the observer output tracks the plant output as closely as possible (Figure 6a).

The observer can also be represented as

$$\frac{\partial}{\partial t}\hat{x}(z,t) = (A_z - GC)\hat{x}(z,t) + Bu(z,t) + Gy(z,t) \tag{66a}$$

This shows that the stability of the observer is determined by the behavior of A_z-GC. Introducing the reconstruction or observation error

$$e(z,t) = \hat{x}(z,t) - x(z,t) \tag{67}$$

we obtain the differential equation with associated boundary and initial conditions

$$\begin{aligned} \frac{\partial}{\partial t} e(z,t) &= (A_z-GC)e(z,t), && z,t \in \Pi \\ A_\Gamma e(z,t) &= 0, && z,t \in \Sigma \\ e(z,t) &= \hat{x}_o(z) - x(z,t_o), && z \in \Omega \end{aligned} \tag{68}$$

which has the solution (if one exists)

$$e(z,t) = \int_\Omega E(z, \zeta, t-t_o)[\hat{x}_o(\zeta) - x(\zeta,t_o)]\, d\zeta \tag{69}$$

where $E(z, \zeta, t-t_o)$ is the Green's function of the system (68).

For the system (66) to be an observer, the reconstruction error must have the property

$$\lim_{t\to\infty} e(z,t) = 0 \tag{70}$$

for all $e(z,t_o)$. From (66a) and (68) we conclude that the observer needs to be asymptotically stable. If the system (1) to (4) is completely observable (reconstructible), the design of the observer (66)

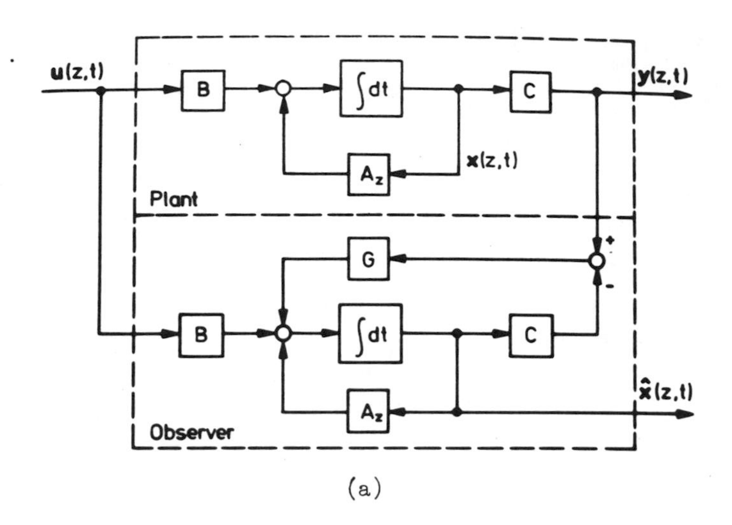

(a)

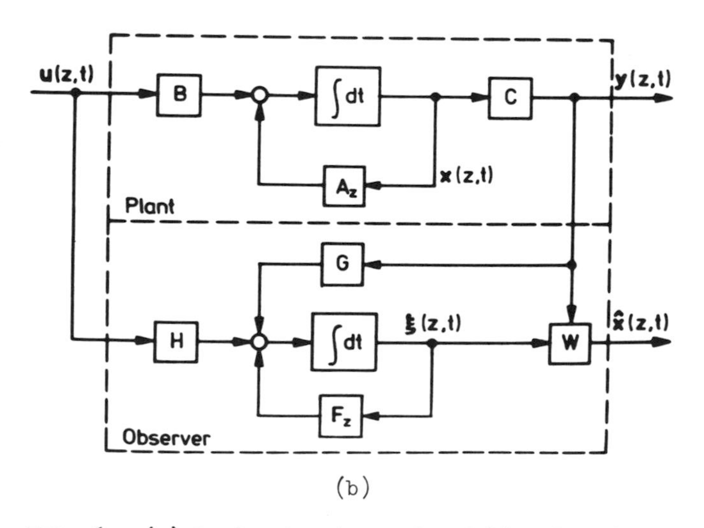

(b)

FIG. 6. (a) Basic structure of a full-order observer (identity observer); (b) basic structure of a reduced-order observer.

depends on a suitable choice of the integral operator G. In general, we consider bounded operators G and C of the form

$$GCe(z,t) = \int_{\Omega} G(z,\eta,t) \int_{\Omega} C(\eta,\zeta,t)e(\zeta,t)\, d\zeta\, d\eta \tag{71}$$

Example 1. The observer design will be demonstrated for the transverse vibrating beam described by Eq. (15) with c = 0:

$$\frac{\partial}{\partial t} x(z,t) = \begin{bmatrix} 0 & a\,\Delta_z \\ -a\,\Delta_z & -2(\beta_E + \beta_I\,\Delta_z^2) \end{bmatrix} x(z,t) + \begin{bmatrix} 0 \\ 1 \end{bmatrix} u(z,t) \quad (72a)$$

where $x = [a\,\Delta_z w,\ \dot{w}]^T$, $\Delta_z = \partial^2/\partial z^2$, $u = p/\mu$, $\Omega = (0,L)$, and $\Gamma = \{0.L\}$. Assume the velocity $x_2 = \dot{w}$ can be measured

$$y(z,t) = [0,\quad 1]\ x(z,t), \qquad z \in \Omega,\ t \geq 0 \quad (72b)$$

It is obvious that this system is observable. Homogeneous boundary conditions are considered

$$x(z,t) = 0, \qquad z \in \Gamma,\ t \geq 0 \quad (72c)$$

The initial state x(z,0) is unknown. Here the matrix operator G of (71) is a vector $g = [g_1, g_2]^T$, since y(z,t) is a scalar function. From (66) the following observer is obtained:

$$\frac{\partial}{\partial t} \hat{x}(z,t) = \begin{bmatrix} 0 & a\,\Delta_z - \int_\Omega g_1(z,\zeta,t)(\cdot)\,d\zeta \\ -a\,\Delta_z & -2(\beta_E + \beta_I\,\Delta_z^2) - \int_\Omega g_2(z,\zeta,t)(\cdot)\,d\zeta \end{bmatrix} \hat{x}(z,t)$$

$$+ \begin{bmatrix} 0 \\ 1 \end{bmatrix} u(z,t) + \begin{bmatrix} \int_\Omega g_1(z,\zeta,t)(\cdot)\,d\zeta \\ \int_\Omega g_2(z,\zeta,t)(\cdot)\,d\zeta \end{bmatrix} y(z,t) \quad (73)$$

$$\hat{y}(z,t) = [0,\quad 1]\hat{x}(z,t), \qquad z,t \in \Pi$$

$$\hat{x}(z,t) = 0, \qquad z,t \in \Sigma$$

$$\hat{x}(z,t_o) = \hat{x}_o(z), \qquad z \in \Omega$$

For simplicity we choose $g_1 \equiv 0$ and $g_2 = \delta(z-\zeta)2\gamma$, $\gamma > 0$.

Modal expansion of the state $\hat{x}(z,t)$ is used to obtain the solution (69) of the error equation (68)

$$e(z,t) = \int_{\Omega} \sum_{m=1}^{\infty} \varphi_m(z)\Phi_m(t-t_o)\varphi_m(\zeta)[\hat{x}_o(\zeta) - x(\zeta,t_o)]\, d\zeta$$

where $\Phi_m(t)$ is the fundamental matrix

$$\Phi_m(t) = \frac{e^{-\beta_m t}}{\bar{\omega}_m} \begin{bmatrix} \bar{\omega}_m \cos \bar{\omega}_m t + \beta_m \sin \bar{\omega}_m t & \omega_m \sin \bar{\omega}_m t \\ -\omega_m \sin \bar{\omega}_m t & \bar{\omega}_m \cos \bar{\omega}_m t - \beta_m \sin \bar{\omega}_m t \end{bmatrix} \tag{74}$$

The eigenfunctions $\varphi_m(z)$ are equal to the basis functions (51a). The following abbreviations are used:

$$\omega_m = a\left(\frac{m\pi}{L}\right)^2, \qquad \bar{\omega}_m = \sqrt{\omega_m^2 - \beta_m^2}, \qquad \beta_m = \beta_E + \beta_I\left(\frac{m\pi}{L}\right)^4 + \gamma$$

Since $\Phi_m(t) \to 0$ as $t \to \infty$, condition (70) holds. The observer is asymptotically stable, and the eigenvalues of the observer depend on the choice of γ. Eigenvalues somewhat more negative than those of the observed system can be chosen so that the convergence is faster than other system effects. These results can be extended to other elastic systems, e.g., the systems of Section II.B.

B. The Basic Concept of Observer Theory

The following section will show that the observer is not required to have the same order as the observed elastic system. The notion of an asymptotic DP state observer introduced earlier is now given a precise meaning similar to that of LPS [42] in the following definition.

Definition. A linear dynamic DP system of the order

$$\begin{aligned} \frac{\partial}{\partial t}\xi(z,t) &= F_z\xi(z,t) + Gy(z,t) + Hu(z,t), && z,t \in \Pi \\ F_\Gamma\xi(z,t) &= D_\Gamma v(z,t), && z,t \in \Sigma \\ \xi(z,t_o) &= \xi_o(z), && z \in \Omega \end{aligned} \tag{75}$$

is called an asymptotic state observer (reconstructor) for the system (1) to (4) if and only if there exists a matrix integral operator W satisfying

$$\lim_{t\to\infty} \left\{ x(z,t) - W \begin{bmatrix} y(z,t) \\ \xi(z,t) \end{bmatrix} \right\} = 0 \tag{76a}$$

and the reconstructed state is

$$\hat{x}(z,t) = W \begin{bmatrix} y(z,t) \\ \xi(z,t) \end{bmatrix} \tag{76b}$$

Further research must establish that an operator W with kernel $W(z,\zeta,t)$ exists for any observable elastic system studied here.

Instead of (67), we now introduce the error

$$\epsilon(z,t) \stackrel{\Delta}{=} \xi(z,t) - Tx(z,t) = \xi(z,t) - \int_\Omega T(z,\zeta,t)x(\zeta,t)\, d\zeta \tag{77}$$

where ξ need not have the same order as x, e.g., $p \leq n$. Substituting (2) and (77) into (76a), we obtain before taking the limit

$$x(z,t) - W \begin{bmatrix} y(z,t) \\ \xi(z,t) \end{bmatrix} = x(z,t) - W \begin{bmatrix} C \\ T \end{bmatrix} x(z,t) + W \begin{bmatrix} 0 \\ \epsilon(z,t) \end{bmatrix} \tag{78}$$

If the operator T is selected so that

$$\delta(z-\eta)I - \int_\Omega W(z,\zeta,t) \begin{bmatrix} C(\zeta,\eta,t) \\ T(\zeta,\eta,t) \end{bmatrix} d\zeta = 0, \qquad \lim_{t\to\infty} \epsilon(z,t) = 0 \tag{79}$$

then, as a consequence, (76a) is satisfied using (78). In the remainder of this section we assume homogeneous boundary condition (3), $v \equiv 0$. Using Green's theorem

$$\int_\Omega (T(z,\zeta,t)A_\zeta - \{A_\zeta^* T^T(z,\zeta,t)\}^T)x(\zeta,t)\, d\zeta = \int_\Gamma (T(z,\zeta,t)A_{\Gamma\zeta}$$
$$+ \{\bar{A}_{\Gamma\zeta}T^T(z,\zeta,t)\}^T)x(\zeta,t)\, d\zeta \tag{80}$$

it is not difficult to verify by direct substitution of (77) into (75) and utilization of the system equations (1) to (4) that

$$\frac{\partial}{\partial t}\epsilon(z,t) = F_z\epsilon(z,t), \qquad z,t \in \Pi$$

$$F_\Gamma\epsilon(z,t) = 0, \qquad z,t \in \Sigma \tag{81}$$

$$\epsilon(z,t) = \hat{x}_o(z) - T_o x(z,t_o) = \hat{x}_o(z) - \int_\Omega T(z,\zeta,t_o)x(\zeta,t_o)\, d\zeta$$

when

$$\int_\Omega G(z,\eta,t)C(\eta,\zeta,t)\, d\eta = \frac{\partial}{\partial t}T(z,\zeta,t) + \{A_\zeta^* T^T(z,\zeta,t)\}^T - F_z T(z,\zeta,t) \tag{82a}$$

with associated boundary and initial conditions

$$F_{\Gamma z}T(z,\zeta,t) = 0 \qquad \bar{A}_{\Gamma\zeta}(z,\zeta,t) = 0, \qquad z,\zeta \in \Gamma,\ t \in T \tag{82b}$$

$$T(z,\zeta,t_o) = T_o(z,\zeta), \qquad z,\zeta \in \Omega,\ t = t_o \tag{82c}$$

and

$$H = TB = \int_\Omega T(z,\zeta,t)B(\zeta,t)(\cdot)\, d\zeta . \tag{82d}$$

Further research must establish that the operators W and T exist for observable elastic systems studied here. Since H is easily computed from (82d), it must be shown that for any choice of T satisfying (79), observer operators F, F_Γ, and G exist satisfying (82a) to (82c).

Remark 14. The full-order observer (p=n) of the preceding section is called an identity-observer, since we obtain this observer from Eqs. (75) and (82a) to (82d) if the operator T is time-independent with the specific kernel $T(z,\zeta) = \delta(z,\zeta)I$.

Remark 15. If m linear independent state variables of the system (1) to (4) can be measured, an observer of reduced-order p=n-m can be designed, which is called a minimal-order observer.

C. Reduced-order Observer

The basic concept of observer theory can be applied to design minimal-order observers. For simplicity, we assume that the first m state variables can be measured and that the remaining ones have to be reconstructed. In this case, the operator C reduces to the matrix $C = (I,0)$ with the $m \times m$ identity matrix I and $m \times (n-m)$ zero matrix. It is then convenient to partition the state vector as $x(z,t) = (y^T(z,t), w^T(z,t))^T$ and accordingly write the system (1) to (4) in the form:

$$\frac{\partial}{\partial t}\begin{bmatrix} y(z,t) \\ --- \\ w(z,t) \end{bmatrix} = \begin{bmatrix} A_{11} & \vdots & A_{12} \\ -- & + & -- \\ A_{21} & \vdots & A_{22} \end{bmatrix} \begin{bmatrix} y(z,t) \\ --- \\ w(z,t) \end{bmatrix} + \begin{bmatrix} B_1 \\ -- \\ B_2 \end{bmatrix} u(z,t) \tag{83}$$

with associated boundary and initial conditions. Only $w(z,t)$ has to be reconstructed by the observer:

$$\frac{\partial}{\partial t}\hat{w}(z,t) = A_{22}\hat{w}(z,t) + A_{21}y(z,t) + B_2u(z,t) + L\Big[\frac{\partial}{\partial t}y(z,t) - A_{11}y(z,t) - A_{12}\hat{w}(z,t) - B_1u(z,t)\Big] \tag{84}$$

$$A_{2\Gamma}\hat{w}(z,t) = 0, \qquad z,t \in \Sigma$$

$$\hat{w}(z,t_o) = \hat{w}_o(z), \qquad z \in \Omega$$

where the error of the first equation of (83) is weighted by a suitably chosen operator L.

The required time-differentiation of the output y can be avoided by introducing the substitution

$$\hat{w}(z,t) \triangleq \xi(z,t) + Ly(z,t) = \xi(z,t) + \int_\Omega L(z,\zeta,t)y(\zeta,t)\, d\zeta \tag{85}$$

From (84) we obtain the observer of reduced order:

$$\dot{\xi}(z,t) = (A_{22} - LA_{12})\xi(z,t) + (A_{21} - LA_{11} - \dot{L})y(z,t) + (A_{22} - LA_{12})Ly(z,t) + B_2 - LB_1)u(z,t) \tag{86}$$

$$A_{2\Gamma}[\xi(z,t) + Ly(z,t)] = 0, \qquad z,t \in \Sigma$$

$$\xi(z,t_o) = \xi_o(z), \qquad z \in \Omega$$

The reconstructed state vector is defined by

$$\hat{x}(z,t) \triangleq \begin{bmatrix} y(z,t) \\ \hline Ly(z,t) + \xi(z,t) \end{bmatrix} = \left[\begin{array}{c|c} I & 0 \\ \hline L & I \end{array}\right] \begin{bmatrix} y(z,t) \\ \hline \xi(z,t) \end{bmatrix} \tag{87}$$

The integral operator L must be chosen so that the observer (86) is asymptotically stable. This should be possible if the system (1) to (4) is completely observable. Thus, the reconstruction error $\epsilon(z,t)$ approaches zero for increasing time, and we conclude from (77) and (78)

$$\hat{x}(z,t) = \begin{bmatrix} C \\ \hline T \end{bmatrix}^{-1} \begin{bmatrix} y(z,t) \\ \hline \xi(z,t) \end{bmatrix} \quad \text{and} \quad \hat{x}(z,t) = W \begin{bmatrix} y(z,t) \\ \hline \xi(z,t) \end{bmatrix}$$

In comparison with (87), an example of the introduced operators T and W can now be given:

$$T = [-L \quad I], \quad W = \left[\begin{array}{c|c} I & 0 \\ \hline L & I \end{array}\right] = \int_{\Omega} \left[\begin{array}{c|c} \delta(z-\zeta)I & 0 \\ \hline L(z,\zeta,t) & \delta(z-\zeta)I \end{array}\right] (\cdot)\, d\zeta$$

Example 2. The Timoshenko beam may serve as an example for the design of reduced-order observers. The submatrices $A_{i,j}$ of the matrix differential operator A_z can be obtained from the comparison of (83) with (21) if the first two state variables are assumed to be measurable. The observer (86) is of the second order

$$\dot{\xi}(z,t) = \begin{bmatrix} 0 & \frac{\partial}{\partial z} \\ \frac{\alpha_B}{\theta} & \ell_{21}(z) \end{bmatrix} \xi(z,t) - \begin{bmatrix} \frac{\ell_{12}(z)}{\mu} \\ 0 \end{bmatrix} u(z,t)$$

$$+ \begin{bmatrix} [\ell_{21}(z) - \ell_{12}(z)\frac{\alpha_S}{\mu}]\frac{\partial}{\partial z} & 0 \\ \frac{\alpha_S}{\theta} + \ell_{21}^2(z) & [\ell_{12}(z)\frac{\alpha_B}{\theta}] - \ell_{21}(z) \end{bmatrix} y(z,t) \tag{88}$$

$$\begin{bmatrix} 0 & 0 \\ 0 & 1 \end{bmatrix} \xi(z,t) + \begin{bmatrix} 0 & \ell_{12}(z) \\ \ell_{21}(z) & 0 \end{bmatrix} y(z,t) = \begin{bmatrix} 0 \\ 0 \end{bmatrix}, \qquad z,t \in \Sigma$$

$\xi(z,t_o) = \xi_o(z), \qquad z \in \Omega$

where a time-independent operator L with the kernel $\delta(z-\zeta)L(z)$ and a beam with clamped ends have been chosen. L(z) is a chosen matrix with space-dependent or constant elements ℓ_{12} and ℓ_{21}. This observer is asymptotically stable under the condition: $\ell_{21}(z) < 0$, $\forall z \in \Omega$, since the homogeneous part of Eq. (88) can be regarded as a wave equation, describing vibrating strings, with space-dependent damping coefficient $\beta_E = -\ell_{21}(z)$ (see (28) for comparison). A similar result can be obtained if the elastic system is governed by Eq. (24).

This example demonstrates the variety of possibilities in designing reduced-order observers for DPS. It is evident that considerable freedom exists in the choice of the operators W, T, and L. A special observer can be developed if an estimate of a linear functional is desired.

D. Functional Observer

The linear control law of an elastic system with single input is determined by a linear functional of the system state (see Eq. (48), for example). We therefore want to develop a less complex observer to yield an estimate of a given linear functional

$$f(z,t) \triangleq s^T K x(z,t) = \int_\Omega \int_\Omega s^T(z,\zeta,t)K(\zeta,\eta,t)x(\eta,t)\, d\eta\, d\zeta \tag{89}$$

As in LPS [41], the general form of the associated observer is exactly analogous to a reduced-order observer (75) for the entire state vector. The reconstructed functional $\hat{f}$ is defined by

$$\hat{f}(z,t) \triangleq m^T y(z,t) + n^T \xi(z,t) = \int_\Omega [m^T(z,\zeta,t)y(\zeta,t) + n^T(z,\zeta,t)\xi(\zeta,t)]\, d\zeta \tag{90}$$

where $\xi(z,t)$ is as defined in Section IV.C, and m and n are vector integral operators with kernels satisfying

$$m^T(z,\zeta,t)C(\zeta,\eta,t) + n^T(z,\zeta,t)T(\zeta,\eta,t) = s^T(z,\zeta,t)K(\zeta,\eta,t) \tag{91}$$

It is not difficult to apply this general result to Example 2.

E. Closed-loop Properties

It is important to study the effect induced by using the estimate $\hat{x}(z,t)$ in the control law (43), (55), or (62) instead of the true state $x(z,t)$. Suppose we have the system (1) to (4) and the control law

$$u(z,t) = -\int_\Omega\int_\Omega S(z,\eta,t)B^T(\eta)K(\eta,\zeta,t)\hat{x}(\zeta,t)\, d\zeta\, d\eta \tag{92a}$$

where an observer of the form (75) is considered. From Eq. (76b) we conclude that (92a) can also be written as

$$u(z,t) = -[Dy(z,t) + E\xi(z,t)] \tag{92b}$$

The matrix integral operators D and E are determined by the operator equation $DC + ET = SB^TK$, where the kernel of T satisfies (82a) to (82c). Introduction of the error (77) leads to the composite system

$$\frac{\partial}{\partial t}\begin{bmatrix} x(z,t) \\ \epsilon(z,t) \end{bmatrix} = \begin{bmatrix} A_z - BSB^TK & -BE \\ 0 & F_z \end{bmatrix}\begin{bmatrix} x(z,t) \\ \epsilon(z,t) \end{bmatrix}, \qquad z,t \in \Pi$$

$$\begin{bmatrix} A_\Gamma & 0 \\ 0 & F_\Gamma \end{bmatrix}\begin{bmatrix} x(z,t) \\ \epsilon(z,t) \end{bmatrix} = \begin{bmatrix} B_\Gamma \\ D_\Gamma \end{bmatrix} v(z,t), \qquad z,t \in \Sigma \tag{93}$$

Thus, the eigenvalues of the composite system are those of $A_z - BSB^TK$ and F_z. These would be obtained if the control law could be directly implemented, in addition to those of the observer itself. This result corresponds to that in LPS [41].

Another important problem is the investigation of the effect induced on the minimum value J^o of the quadratic cost functional $J(u)$ by using the estimate $\hat{x}(z,t)$ instead of $x(z,t)$. The minimum value is well known for state regulator problems ($y_d = 0$) [17].

$$J^o = \int_\Omega\int_\Omega x^T(z,t_o)K(z,\zeta,t_o)x(\zeta,t_o)\, d\zeta\, dz$$

and tracking problems ($y_d \neq 0$) [12]. In applications, however, neither the entire state vector nor its initial value is available, and an asymptotic state observer has to be employed. It is assumed that a "cost" increment will be obtained as a result of the error in the estimate of the plant state vector. Further research should be directed toward the solution of this problem.

F. Implementation of Observers

It is obvious that DP observers can be implemented only approximately, and modal series expansion or finite difference methods should be used for implementation. From Example 1 we obtain the effect of modal approximation on the error $\varepsilon(z,t)$ if a finite number M of eigen functions $\varphi_m(z)$ is assumed. The error decreases in this case as well, since a stable elastic system with external and internal damping forces ($\beta_E, \beta_I > 0$) has been considered. In general, the accuracy of the approximation depends on the number of significant modes of the observer. A similar situation occurs during the implementation of modal control [8,11,30,36] and modal simulation [26,47] of DPS.

Another problem arises since the assumed distributed observation is not available. Only local observations at discrete points $z = c_n \in \Omega$, as mentioned in Section II.C, are possible. The solution of this problem is included in our general formulation of observer theory. To demonstrate this, the observation equation (2) is written in the form

$$y(z,t) = \int_\Omega \sum_{n=1} \delta(z-c_n)\, \delta(\zeta-c_n) C(t) x(\zeta,t)\, d\zeta$$

$$= \sum_{n=1}^{N} \delta(z-c_n) y_n(t) \tag{94}$$

where the vector $y_n(t)$ is defined in Eq. (36c). Equation (71) can now be written as

$$GCe(z,t) = \int_{\Omega} G(z,\eta,t) \int_{\Omega} \sum_{n=1}^{N} \delta(\eta-c_n)\, \delta(\zeta-c_n) C(t)e(\zeta,t)\, d\zeta\, d\eta$$

$$= \sum_{n=1}^{N} G(z,c_n,t)C(t)e(c_n,t)$$

thus the observers of the preceding sections can be implemented, providing the points c_n $(n = 1, \ldots ,N)$ are suitably chosen so that the considered elastic system is observable.

Thus far, observers that take discrete sensor locations into account have been designed only for diffusion systems [43] and heat conduction processes [25] governed by a scalar state equation of parabolic type. In [25], an observer has been implemented experimentally to estimate the space- and time-dependent continuous temperature profile $\vartheta(z,t)$ using thermocouples for measurement at two or three discrete points. The observer converges rather rapidly, and the estimated scalar state $\hat{\vartheta}(z,t)$ has been used successfully in in a closed feedback loop. (See [45] for comparison with the DP filter.)

The success of the experimental study [25] should encourage the implementation of state observers suggested in this chapter for elastic systems with more than one state variable. Future application areas could be large astronomical telescopes [36] and antennas, where the deflection of the elastic primary mirror must be reconstructed from spatially discrete measurements. Applications of observers are also proposed for high-speed ground transportation systems with air cushion vehicles [48] and for magnetically levitated vehicles [49,50] on elastic rails.

V. OPTIMAL LINEAR FEEDBACK CONTROL OF LATERAL VIBRATIONS OF ELASTIC HAULING PIPES FOR DEEP SEA MINING

We would like to close this chapter with a section about the application of optimal control theory to an elastic system of future industrial importance in which simulation results are available, namely the optimal feedback control of the lateral vibrations of elastic hauling pipes, an essential feature of ocean mining systems.

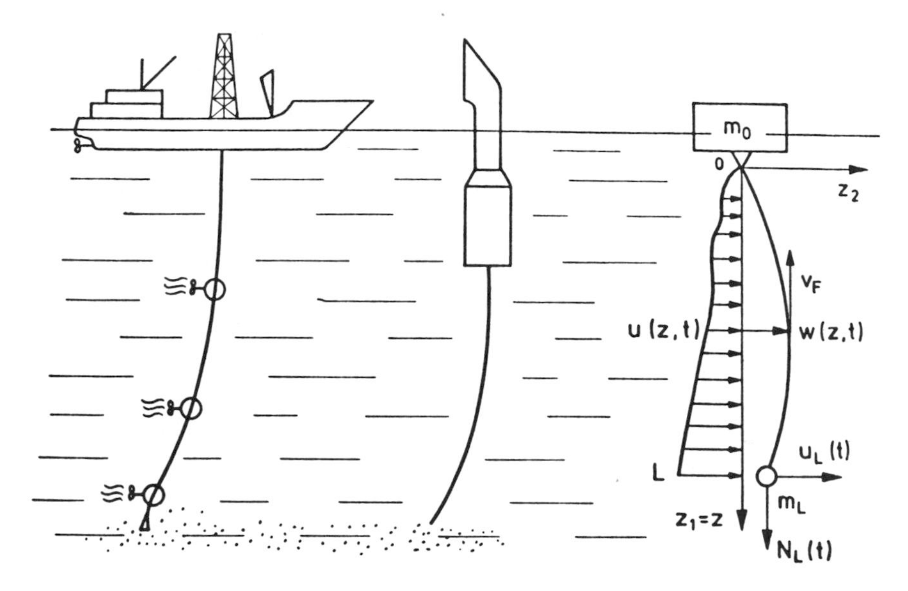

FIG. 7. Real and idealized ocean mining systems.

A. Introductory Remarks

The recovery of hot brines or manganese nodules from the sea floor represents a future problem in ocean mining. The use of long pipes assures continuous hauling of the minerals from a depth of 2,000 to 6,000 m [51-53]. A mathematical model describing the dynamic behavior of the pipe which hangs vertically on a mining vessel or a semi-submersible (Figure 7) has been developed in [11,12] on the basis of [26]. This mathematical model considers both the transverse and the longitudinal vibrations of the pipe and, moreover, the longitudinal vibrations of the fluid within the pipe. Since the transverse vibrations of the pipe string are of great industrial importance, they have explicitly been taken into consideration. The method of weighted residuals [54] has been applied to this DP system for approximation, numerical computation, and analog simulation [11,12,26].

When mining minerals from the sea floor, it is important to keep the pipe string as nearly vertical as possible, since both the drop of pressure and the abrasion of the pipe will increase sharply in horizontal hauling. This objective can be met when the lateral pipe deflection is controlled by means of ship screws or water propulsion. A modal feedback control system has been proposed and simulated [11] which facilitates essential improvement in the dynamic behavior of the pipe string if spatially distributed disturbances (sea currents, ship movement) or boundary-valued disturbances (waves at the upper end and single forces at the lower end of the pipe) are considered. The theoretical results of Section III can be applied to compute optimal feedback control systems which bring the pipe from a deflected initial position to a desired state such that a performance functional takes its minimum value [12]. The desired state may be the vertical equilibrium position or any other reachable state of the pipe string. Some simulation results of this optimal regulator problem will be presented in the following sections.

B. Problem Formulation

Consider the idealized hauling pipe of Figure 7 and assume:

The pipe string in its equilibrium state hangs vertically. Forces and motions occur only in the z_1-z_2 plane. (For simplicity, the index of the z_1-axis is omitted: $z_1 = z$.)

$w(z,t)$ and $w'(z,t)$ are small, so that the linear beam theory can be applied.

The mass m_o of the vessel is much greater than the mass $m_P = \mu_P L$ of the pipe, thus the reaction of the pipe motion on that of the ship need not be considered.

The effects of longitudinal inertia can be omitted (decoupling between longitudinal and transverse vibrations).

The parameters α, $\mu = \mu_F + \mu_P$, v_F, δ_E, and δ_F are constant.

These simplifying assumptions are made to obtain a mathematical model which contains the most essential influences on the lateral vibrations of hauling pipes, but avoids an unnecessary computational burden during our initial investigations.

If these assumptions hold, the transversal vibrations of the hauling pipe are governed by Eq. (8), where the Coriolis force

$\delta_F \dot{w}'(z,t)$ with $\delta_F = 2\mu_F v_F$ is considered instead of the neglected internal damping force ($\delta_I = 0$). The longitudinal force

$$f(z,t) = N_L(t) + \bar{m}_L g + \mu_P g(L-z) - \frac{1}{2}\mu_F v_F^2 \tag{95a}$$

depends on the vertical force $N_L(t)$ and the lumped mass m_L at the free lower end z=L, the weight of the pipe, and the velocity v_F ($\bar{m}_L$ and $\bar{\mu}_P$ are reduced values of m_L and μ_P due to buoyancy). Choosing the state vector (12) and introducing the abbreviations $a^2 = \alpha/\mu$, $2\beta_E = \delta_E/\mu$, $2\beta_F = \delta_F/\mu$, and $\gamma(z,t) = f(z,t)/\mu$, we obtain the state equation (13) with the following differential operators:

$$A_{21} = \frac{\partial}{\partial z}\left(\gamma(z,t)\frac{\partial}{\partial z}\right) - a^2\frac{\partial^4}{\partial z^4}, \qquad A_{22} = -2\left(\beta_E + \beta_F\frac{\partial}{\partial z}\right) \tag{95b}$$

Suppose that the support of the upper end of the pipe allows rotation (without external moments) and lateral (small) deflection v(t). The transversal single force $u_L(t)$ may act at the lower end of the pipe. Then the boundary conditions are of the form [12]:

$$A_{1\Gamma}x(z,t) \triangleq \begin{bmatrix} 1 & 0 \\ a^2\frac{\partial^2}{\partial z^2} & 0 \end{bmatrix} x(z,t) = \begin{bmatrix} 1 \\ 0 \end{bmatrix} v(t), \qquad z = 0,\ t \in T$$

$$A_{2\Gamma}x(z,t) \triangleq \begin{bmatrix} a^2\frac{\partial^2}{\partial z^2} & 0 \\ \gamma(z,t)\frac{\partial}{\partial z} - a^2\frac{\partial^3}{\partial z^3} & \frac{m_L}{\mu}\frac{\partial}{\partial t} \end{bmatrix} x(z,t) = \begin{bmatrix} 0 \\ \frac{1}{\mu} \end{bmatrix} u_L(t), \qquad z = L \tag{95c}$$

Note that $A_{2\Gamma}$ contains a partial derivative with respect to time if $m_L \neq 0$. For simplicity, we assume $m_L = 0$ in the remainder of this section. The given initial conditions are

$$x_o(z) = [w(z,t_o),\dot{w}(z,t_o)]^T \tag{95d}$$

In this chapter we suppose that the distributed state vector can be measured ($y = x$). In reality, the position (or velocity or acceleration) of the hauling pipe will be measurable only at discrete points (or small sections) using sonar techniques. An observer or filter would be necessary to estimate the state from the few spatially discrete measurements.

First of all, we assume distributed control forces $u(z,t)$. However, only lumped control forces, caused by ship screws or water thrusts, for example, can be realized practically. We will therefore consider lumped control forces at discrete points later on. We introduce the known disturbance vector $d(z,t)$ and the end force $u_L(t)$ in the form $u_L(z,t) = \delta(z-L)u_L(t)$ to obtain the state equation

$$\dot{x}(z,t) = \begin{bmatrix} 0 & 1 \\ A_{21} & A_{22} \end{bmatrix} x(z,t) + \begin{bmatrix} 0 \\ \frac{1}{\mu} \end{bmatrix} \times [u(z,t) + \delta(z-L)u_L(t) + d(z,t)] \quad (95e)$$

The second condition of (95c) becomes homogeneous.

We want to minimize the performance index (40b) with time-constant weighting matrices

$$Q(z,\zeta) = \begin{bmatrix} q_{11}(z) & 0 \\ 0 & q_{22}(z) \end{bmatrix} \delta(z,\zeta), \qquad R(z,\zeta) = r(z)\,\delta(z-\zeta), \quad Q_f = 0$$

This optimal control problem can be solved if the results of Section III are applied.

C. Solution of the Optimal Tracking Problem

If R is positive definite and Q is nonnegative definite, an optimal control law of the form (48) can be obtained, where $k_{ij}(z,\zeta,t)$ and $q_i(z,t)$ are solutions of the equations (arguments omitted[†])

[†] $[\cdots]_{z,\zeta}$ represents derivatives with respect to z or ζ in the brackets on the following page.

$$-\dot{k}_{11} = [(\gamma k_{21}')' - a^2 k_{21}^{IV}]_z + [(\gamma k_{12}')' - a^2 k_{12}^{IV}]_\zeta$$
$$- \frac{1}{\mu^2} \int k_{12} r^{-1} k_{21}\, dz' + q_{11}\, \delta$$
$$-\dot{k}_{12} = [(\gamma k_{22}')' - a^2 k_{22}^{IV}]_z + [k_{11} - 2\beta_E k_{12} - 2\beta_F k_{12}']_\zeta$$
$$- \frac{1}{\mu^2} \int k_{12} r^{-1} k_{22}\, dz' \tag{96a}$$
$$-\dot{k}_{21} = [(\gamma k_{22}')' - a^2 k_{22}^{IV}]_\zeta + [k_{11} - 2\beta_E k_{21} - 2\beta_F k_{21}']_z$$
$$- \frac{1}{\mu^2} \int k_{22} r^{-1} k_{21}\, dz'$$
$$-\dot{k}_{22} = [k_{12} - 2\beta_E k_{22} - 2\beta_F k_{22}']_z + [k_{21} - 2\beta_E k_{22} - 2\beta_F k_{22}']_\zeta$$
$$- \frac{1}{\mu^2} \int k_{22} r^{-1} k_{22}\, dz' + q_{22}\, \delta$$

$$-\dot{q}_1 = (\gamma q_2')' - a^2 q_2^{IV} - \frac{1}{\mu^2} \int k_{12} r^{-1} q_2\, dz' + \int q_{11} x_{d1}\, d\zeta$$
$$+ \frac{1}{\mu} \int k_{12}(d + \delta u_L)\, d\zeta \tag{96b}$$
$$-\dot{q}_2 = q_1 - 2\beta_E q_2 - 2\beta_F q_2' - \frac{1}{\mu^2} \int k_{22} r^{-1} q_2\, dz' + \int q_{22} x_{d2}\, d\zeta$$
$$+ \frac{1}{\mu} \int k_{22}(d + \delta u_L)\, d\zeta$$

with associated boundary and final conditions

$$\bar{A}_{1\Gamma} K \triangleq \begin{bmatrix} 0 & a^2 \dfrac{\partial^2}{\partial z^2} \\ 0 & 1 \end{bmatrix} K(z,\zeta,t), \qquad \bar{A}_{1\Gamma} q(z,t) = 0, \quad z = 0$$

$$\bar{A}_{2\Gamma} K \triangleq \begin{bmatrix} 0 & -\gamma(z,t) \dfrac{\partial}{\partial z} + a^2 \dfrac{\partial^3}{\partial z^3} \\ 0 & a^2 \dfrac{\partial^2}{\partial z^2} \end{bmatrix} K(z,\zeta,t), \qquad \bar{A}_{2\Gamma} q(z,t) = 0, \quad z = L \tag{96c}$$

$$K(z,\zeta,t_f) = 0, \qquad q(z,t_f) = 0, \qquad z,\zeta \in (0,L), \qquad t = t_f \tag{96d}$$

$$e_m'''(L) = 0$$

Note the significant result that (96c) defines only boundary conditions of k_{21}, k_{22}, and q_2 (also of k_{12}, since K is symmetric). This result is in fact correct since we need no boundary conditions for k_{11} and q_1, since no spatial derivatives of these functions occur in (96a) and (96b). For solving Eq. (96b), the desired state $x_d(z,t)$, the distributed disturbance $d(z,t)$, and the lumped disturbance $u_L(t)$ must be available. This will be impossible in most real ocean mining systems. The state $x_d(z,t)$ may be regarded as the state of a desired model, for example,

$$\dot{x}_d(z,t) = A_z x_d(z,t) + bu_d(z,t) \tag{97}$$

with associated boundary and initial conditions A_z and b as in Eq. (95e) and the given input $u_d(z,t)$. The main causes of the disturbance $d(z,t)$ are ship movements and sea currents. Assume constant ship velocity v_S; the corresponding disturbance can then be approximated by $d_S = cv_S^2$, where c is constant. The other disturbances are unknown and thus cannot be taken into consideration. The same difficulties occur if the boundary disturbances $v(t)$ and $u_L(t)$ have to be defined. The deflection $v(t)$ depends on the ship dynamics and sea waves, for example, and $u_L(t)$ could be a reaction force caused by the connection of the pipe to a scraper unit on the sea floor or by the lower end of the pipe touching an obstacle.

Both functions, $v(t)$ and $u_L(t)$, can also be defined as control inputs. This leads to boundary control problems and the development of the optimal control laws is a straightforward application of our general results in Section III if $u_L(z,t) = \delta(z-L)u_L$ is considered instead of $u(z,t)$. For example [12],

$$u_L^o(t) = \frac{-1}{\mu r(L)} \left\{ \int_0^L [k_{21}(L,\zeta,t), k_{22}(L,\zeta,t)]x(\zeta,t)\, d\zeta - q_2(L,t) \right\} \tag{98}$$

In Eqs. (44) and (45) or (96a) and (96b) the terms $KBR^{-1}B^TK$ and $KBR^{-1}Bq$ have to be substituted by

$$[\bar{A}_{1\Gamma\zeta}K^T(z,\zeta,t)]^T_{\zeta=L} br^{-1}(L)b^T[\bar{A}_{1\Gamma z}K(z,\zeta,t)]_{z=L}$$

and

$$[\overline{A}_{1\Gamma\zeta}K^T(z,\zeta,t)]^T_{\zeta=L}br^{-1}(L)b^Tq(z,t)$$

respectively. The optimal control force $u_L^o(t)$ may be generated by a ship screw or a thrust caused by water propulsion. However, the solution of the coupled Riccati equations (96a) is rather difficult. We therefore restrict ourselves to stationary solutions and assume time-constant system parameters (N_L = constant). The final time approaches infinity ($t_f \to \infty$). To assure the convergence of the performance functional (40b), we replace $u(z,t)$ by the difference $u(z,t) - u_d(z)$, where

$$u_d(z) \triangleq -(b^Tb)^{-1}b^TA_zx_d(z) = \left[-\mu \frac{\partial}{\partial z}\left(\gamma(z)\frac{\partial}{\partial z}\right) + a^2\frac{\partial^4}{\partial z^4}\right]w_d(z)$$

which can be obtained from Eq. (97) if the desired state is $x_d(z) = [w_d(z),0]^T$. In other words, a desired deflection profile $w_d(z)$ is given [for example, $w_d(z) = kz$], and $u_d(z) = k\overline{\mu}_pg$ can be obtained with the constant parameter k. Without loss of generality, we only consider the regulator problem where $x_d = 0$ and assume in the remainder of this section that $v(t) \equiv 0$, $d(z,t) \equiv 0$, which includes $q(z,t) \equiv 0$.

D. Approximate Solution of the Optimal Regulator Problem

Contrary to Example 2, the Riccati equations (96a) cannot be solved analytically since the parameter $\gamma(z)$ is space dependent and the solution of the associated eigenvalue problem would be rather difficult. However, the method of weighted residuals [54] can be applied if the orthogonal set of normalized basis functions $e_m(z)$ defined by

$$e_m^{IV}(z) = \lambda_m^4 e_m(z), \qquad e_m(0) = e_m''(0) = e_m''(L) = e_m'''(L) = 0 \tag{99a}$$

$$e_1(z) = \sqrt{\frac{3}{L}}\ \frac{z}{L}, \qquad \lambda_1 = 0$$

$$e_m(z) = \sqrt{\frac{2}{L}}\ \frac{\sin\lambda_m z\ \sinh\lambda_m L + \sinh\lambda_m z\ \sin\lambda_m L}{\sqrt{\sinh^2\lambda_m L - \sin^2\lambda_m L}} \qquad m=2,\ \ldots\ ,M \tag{99b}$$

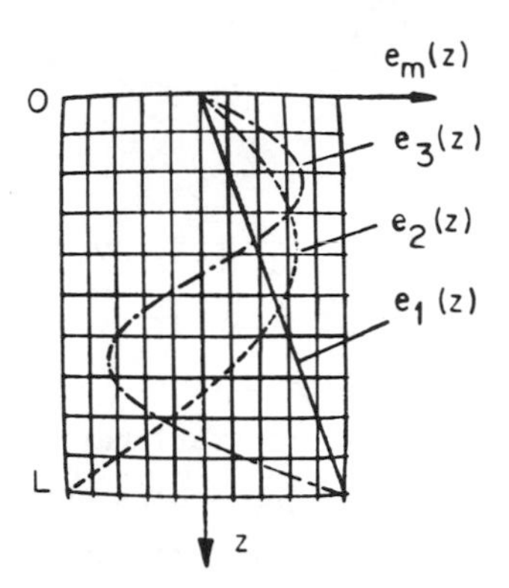

FIG. 8. The first three basis functions.

is chosen, where λ_m are solutions of $\tan \lambda_m L = \tanh \lambda_m L$ (see [11] for details). The first three basis functions are shown in Figure 8. Note that (99a) is not the eigenvalue problem of the considered system, but a reduced version of it.

The Galerkin approximation (50) of the Riccati matrix $K(z,\zeta)$ is now performed. This leads to an algebraic Riccati equation (similar to (52a), but with nondiagonal submatrices $\hat{K}_{i,j}$) which can only be solved numerically [12]. We prefer the Galerkin approximation method over others (finite differences, splines, or Walsh functions, for example), because of the following advantages:

Close relation to modal approximation

Rapid convergence (especially in elastic systems) if the basis functions are suitably chosen

A saving of computer time and storage (particularly if two-dimensional plates are considered)

The last point is obvious, since $K(z_1,z_2,\zeta_1,\zeta_2,t)$ depends on five independent variables which have to be discretized if, for example, finite difference methods are applied. Thus, a computer with a very large memory would be necessary. For our investigations only a relatively small digital computer, the EAI-Pacer 100, was available.

The same approximation method may be used for simulation of the closed-loop system. However, finite difference methods are as advantageous as the Galerkin approximation and will be applied here, since only two independent variables have to be considered. The following question arises: How many basis functions (99b) are necessary for the semianalytic approximate solution $\hat{K}(z,\zeta)$, and how small should the increments Δz and Δt of the variables z and t be chosen for the simulation of the controlled system? A detailed study of this problem leads to the following results [12]:

> M=4 basis functions $e_m(z)$ are sufficient for an approximation of high accuracy. For example, the increase in accuracy is less than 2% with M=6 basis functions, since no basis function of higher order is dominant in the ocean mining system considered.
>
> N=20 equal increments Δz are sufficient if finite differences are used for the simulation of the lateral deflection w of the controlled pipe string since, due to physical reasons, w(z,t) is a fairly smooth function.
>
> If explicit difference methods are applied, the choice of Δt depends on stability conditions and on the propagation velocity of flexural waves [2]. $\Delta t \leq 2s$ has been proven to satisfy these conditions.

The development of the approximate system equations is straightforward; therefore, we restrict ourselves to the presentation of several simulation results.

E. Simulation Results

The following dimensions are assumed for the simulated steel pipe:

Length of pipe, L,	= 5000 m
Outer diameter of pipe, $\emptyset$,	= 50 cm
Thickness of wall	= 1.5 cm
Flexural rigidity of pipe, α,	= $142 \cdot 10^6$ N m^2
Mass per unit length of pipe, μ_P,	= 173 kg/m
Mass per unit length of fluid, μ_F,	= 180 kg/m
Reduced mass due to buoyancy, $\bar{\mu}_P$,	= 150 kg/m
Coefficient of external damping, δ_E,	= 20 N sec/m^2

The following coefficients are varied through the specified range:

Fluid flow velocity, v_F,	= 0, ... ,10 m/sec
Longitudinal force at the end, N_L	= 0, ... ,10^6 N
Lumped mass at end of pipe, m_L,	= 0, ... ,10^5 kg

The other coefficients of the system equations, for example a, β_E, β_F, and γ, can be computed from these data.

First of all, we will study the natural dynamic behavior of the free system (u, v, d, u_L, N_L, m_L=0) with given initial conditions w(z,0) and $\dot{w}$(z,0) shown in Figure 9. These conditions occur if a constant distributed disturbance d=200 N/m acts on the pipe for a period of 1 min and is then removed. For the purpose of comparison, we retain these initial conditions for the remainder of this section. Observe the high-frequency lateral vibrations of the lower end of the hauling pipe. The reason for this can be deduced from Eq. (95a). With increasing z the tensile force f(z) decreases. In addition, the term $\beta_F\dot{w}'$ may produce destabilizing effects which increase with growing flow velocity v_F. Similar effects have been proven experimentally in [55]. The main bending stress also occurs in this part of the pipe, which can be seen in Figure 9c.

We will now investigate the influence of several weighting matrices on the dynamic behavior of the controlled hauling pipe. Assuming

$$q_{11}(z) = 0.1\ (m^3\ sec)^{-1}, q_{22}(z) = 0, \qquad r(z) = 10^{-2}\ m/N^2/sec$$

we obtain the results in Figure 10, where the transverse deflection $w(z,t_i)$, the velocity $\dot{w}(z,t_i)$, and the distributed control $u(z,t_i)$ have been plotted at several discrete points in time t_i. The controlled hauling pipe reaches the vertical position in approximately 50 sec (the free pipe needs 120 sec), but a maximum control force of 1500 N/m is necessary. If the space-dependent weighting

$$q_{11}(z) = 0.2\ \frac{z}{L}\ (m^3\ sec)^{-1}$$

is chosen, the deflection of the lower end of the pipe decreases due to the greater control force (Figure 11). This was expected from control theory. Both results are compared, together with the free motion, at the time t = 30 sec in Figure 12. The influence of space-independent but variable weighting q_{11} is demonstrated

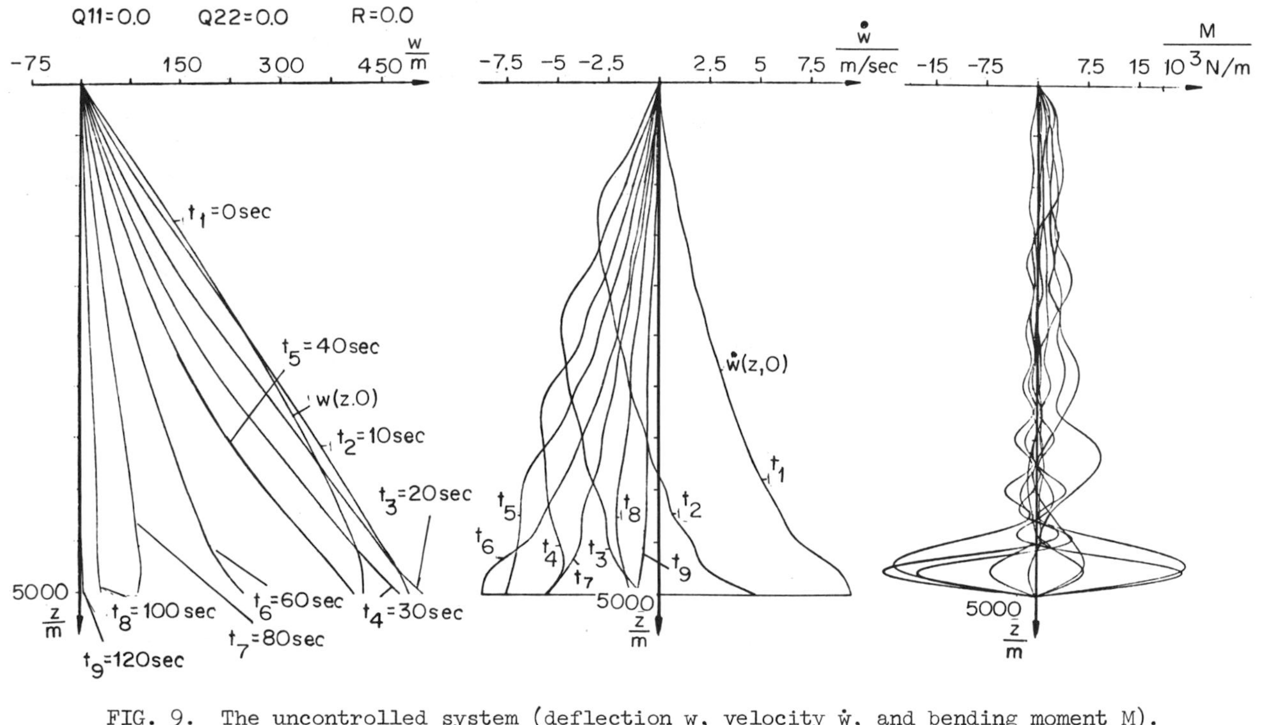

FIG. 9. The uncontrolled system (deflection w, velocity ẇ, and bending moment M).

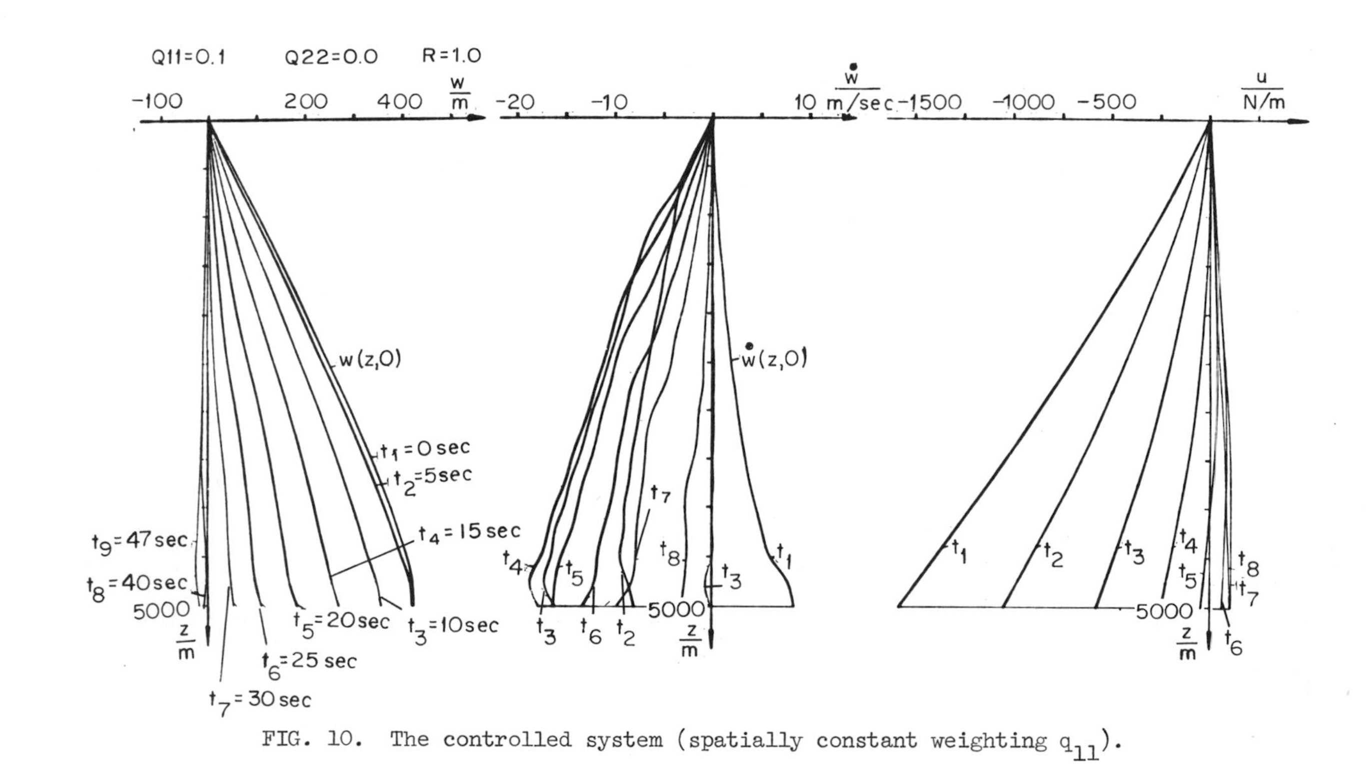

FIG. 10. The controlled system (spatially constant weighting q_{11}).

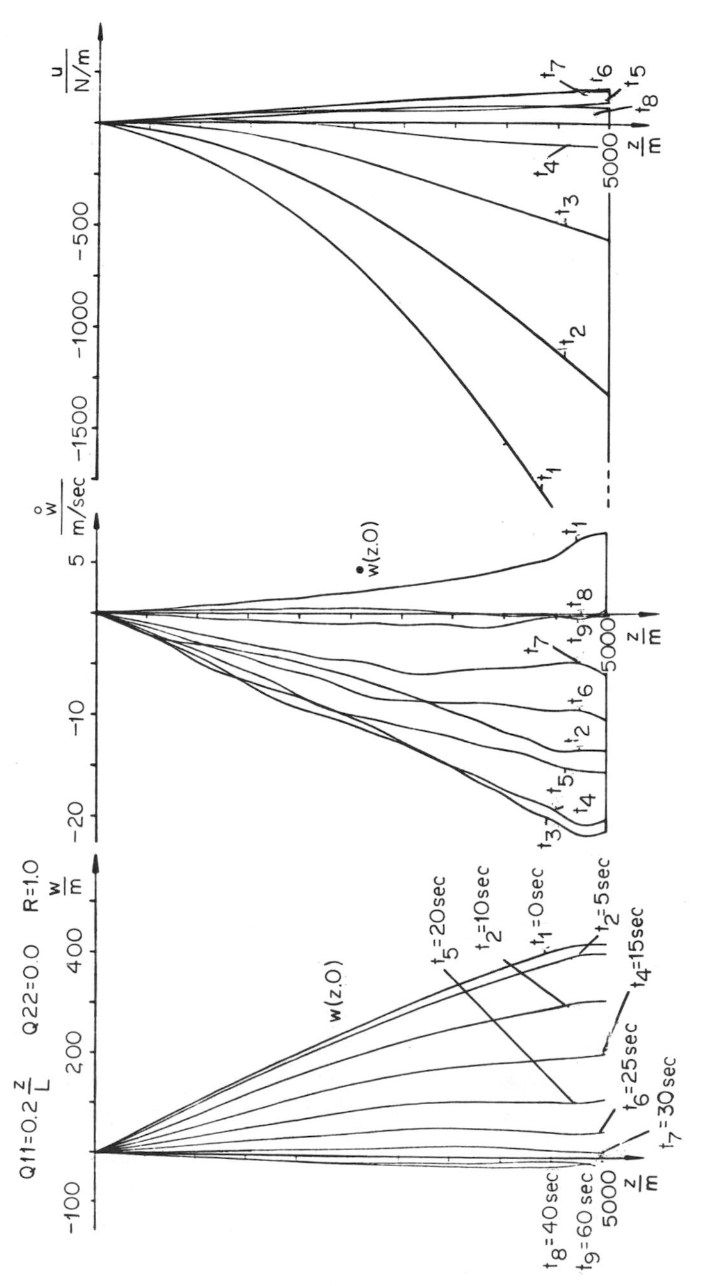

FIG. 11. The controlled system (space-dependent weighting q_{11}).

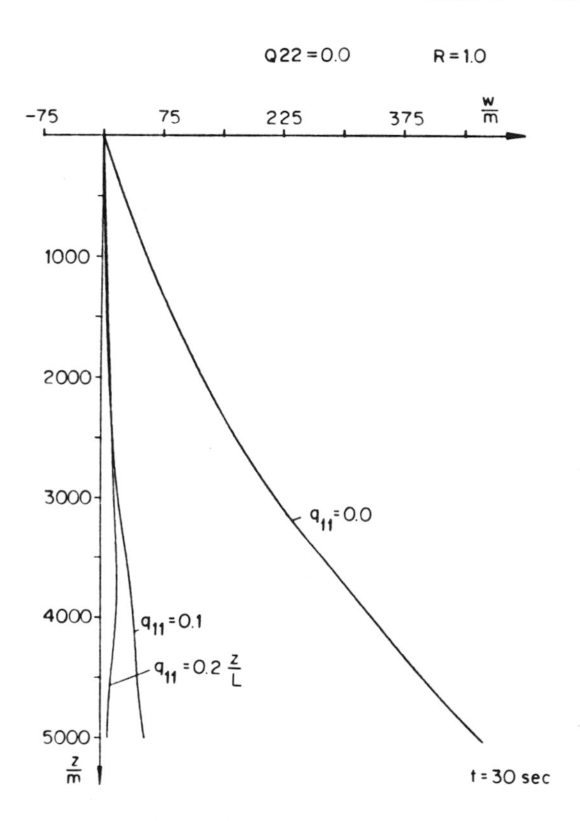

FIG. 12. Comparison of both regulators (q_{11} = 0.1 and q_{11} = 0.2 z/L).

in Figure 13 at the fixed time t = 30 sec and at the fixed point z = 5000 m = L. The high-frequency vibrations of the lower part of the pipe are not observed in the controlled case.

The final examples consider the fact that spatially continuous control forces cannot be implemented in real physical systems. We therefore assume lumped control forces $u_i(t)$ at equidistant points a_i, which can be approximated by "narrow" distributed loads. See Eq. (35), for example, with $\overline{\Omega}_i = (a_i - 1/2\,\Delta z,\ a_i + 1/2\,\Delta z)$, i = 1,2, ... ,I. Cases I = 10 and I = 5 have been simulated, as was the boundary control problem (I = 1), where the control force acts only at the end point z = L. Both the latter cases are shown in Figure 14. The excitation of the lateral vibrations of the lower

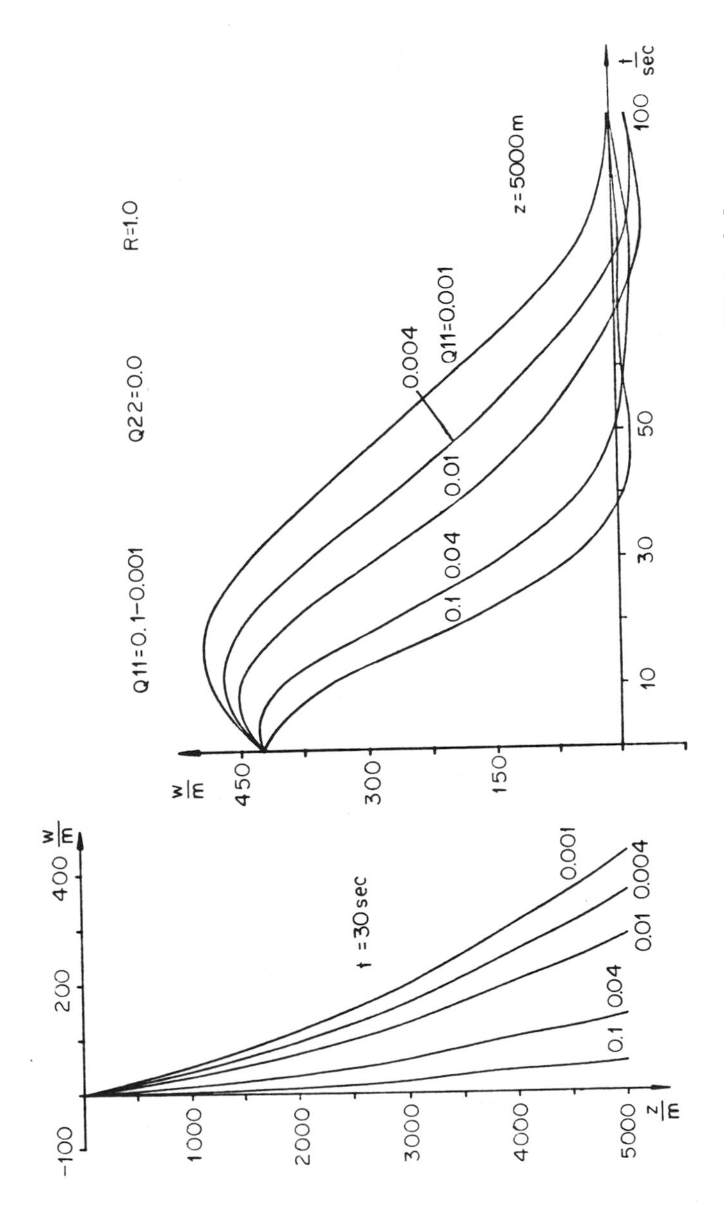

FIG. 13. Influence of different weighting factors q_{11} = 0.001 to 0.1.

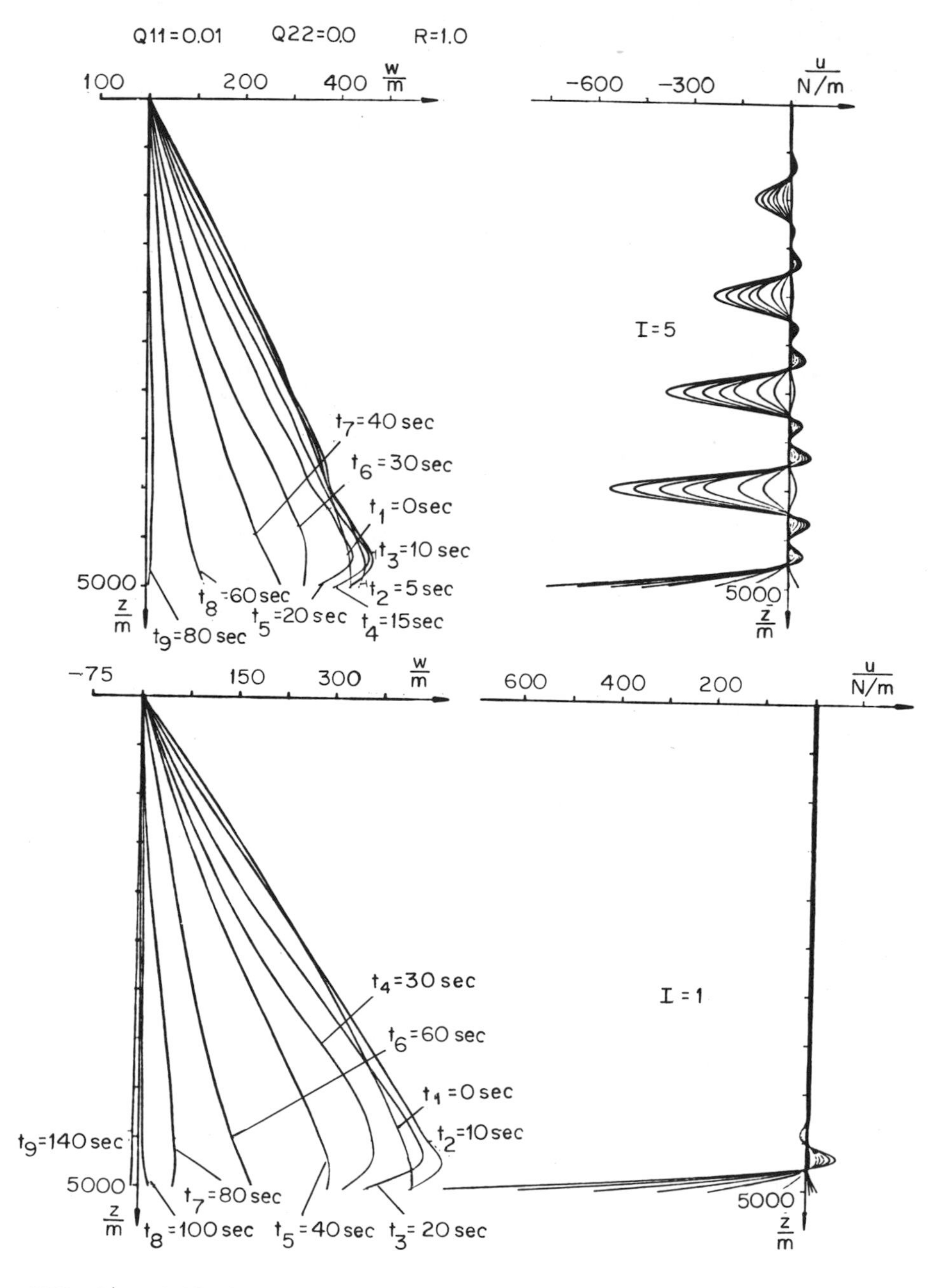

FIG. 14. Effect of I=5 and I=1 spatially concentrated control forces.

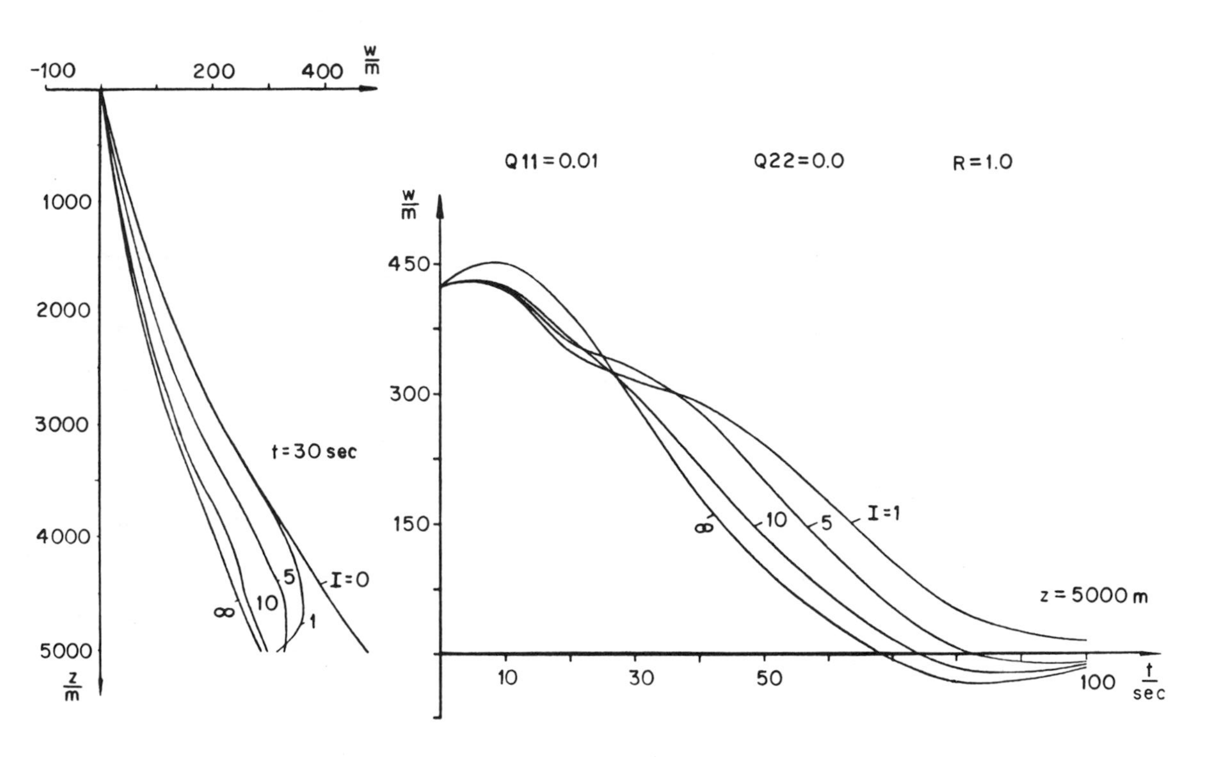

FIG. 15. Influence of I spatially concentrated control forces.

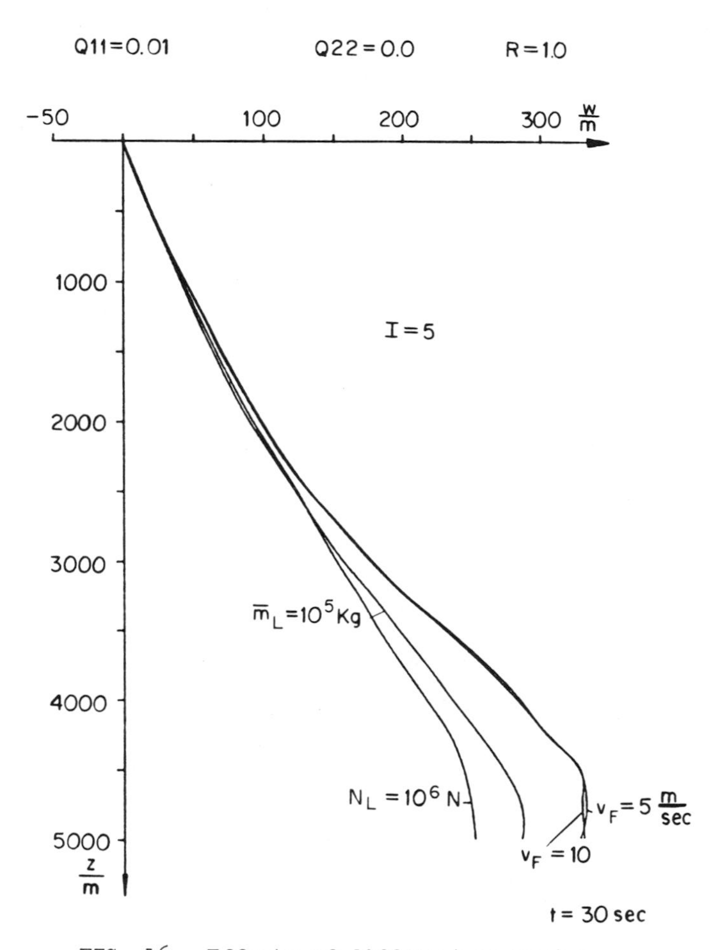

FIG. 16. Effects of different parameters.

end of the pipe occur again. This can also be seen in Figure 15, where the continuous time behavior of the end point z = 5000 m has been plotted. The results with I = 10 and I = ∞ (spatially continuous control) are given for comparison, as is the transverse deflection of the entire pipe at the time t = 30 sec.

Finally, Figure 16 demonstrates the effect of different parameters. The higher fluid flow velocity v_F = 10 m/sec (instead of 5 m/sec) only influences the vibration of the lower part of the pipe as discussed above. An additional longitudinal tensile force

$N_L = 10^6$ N avoids the increase of vibrations, as does an equivalent mass $m_L = 10^5$ kg. In the latter case, however, the inertia increases, and in both cases the static tensile stress increases.

Neither a lumped mass at the end of the pipe nor, as is sometimes proposed, a jacket of buoyant material along the pipe (reducing its weight) seem to be the best design. Rather, we propose a combination of both, since the lumped mass raises the dynamic bending stress at the lower part of the pipe and the static tensile stress along the pipe. The buoyant material should only reduce the weight at the upper part of the pipe (for example, in the interval $0 \le z \le \ell < L$), since a larger reduction of the weight would facilitate the excitation of lateral vibrations or would demand a very large end mass m_L. This compromise reduces the static tensile stress to the desired value σ_d as shown in Figure 17, and avoids unwelcome excitation of lateral vibrations.

Several simplification assumptions have been necessary to handle this problem. They may be removed by further research in this area. For example, state observers must be designed to estimate the position and the velocity of the hauling pipe from noise-free or noisy measurements. It would also be advantageous to take the ship dynamics into consideration and to solve time- or energy-minimal control problems, where the purpose is to reach a desired final position in

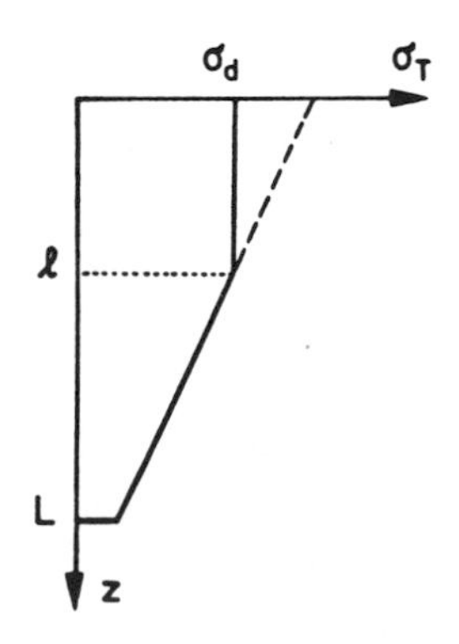

FIG. 17. Desired tensile stress.

a minimum time or to use a minimum amount of energy when the control forces are constrained.

Problems similar to those discussed in this section occur in mineral exploration on the sea floor and scientific drilling in the deep sea bed when the drill head has to be changed and the borehole later refound. Underwater cameras are used for the exploration of manganese modules. The control of the positioning of these cameras which are attached to the mining vessel using long elastic cables is still an unsolved problem.

The methods and results of this chapter can also be extended to other plants having industrial importance since lateral vibrations of axially moving tapes, bands, belts, chains, ropes, or cables are governed by equations comparable to those describing the hauling pipe with flowing fluid [54].

VI. CONCLUSIONS

The control of vibrating elastic systems, particularly the optimal feedback control of transverse vibrating beams and plates, is investigated from the engineering point of view. Based on a chosen state space description of these systems, the optimal tracking problem and implicit as well as explicit model-following problems are both formulated and partially solved. Some of the theoretical results obtained are applied to control the lateral vibrations of elastic hauling pipes for the elevation of minerals in ocean mining. A semianalytic method consisting of the Galerkin approximation (with respect to the space coordinates) and finite difference methods (with respect to time) gives the best approximate solution of the Riccati equation with smallest computational effort. A considerable improvement in the dynamic behavior of the controlled pipe is obtained.

The main section of this chapter is concerned with state reconstructors based on noise-free measurements, namely the distributed parameter observer design, since comparatively few publications have dealt with this problem. A basic concept of observer theory for systems with more than one distributed state is developed, and the design of full-order and reduced-order observers is proposed. A few examples serve to demonstrate the existence of such observers.

The results are not restricted to vibrating elastic systems; they can be extended to other multidimensional distributed systems. Further work in this area is necessary to prove the general existence of the defined observers and to reach a degree of theoretical maturity as in the distributed parameter filter theory. At the same time, it is the author's desire to stimulate the application of distributed parameter observers in different engineering fields and to initiate their industrial implementation [56,57].

ACKNOWLEDGMENTS

The author wishes to acknowledge motivation for his interest in the area of distributed parameter systems, and elastic systems in particular, stimulated by Professor Dr.-Ing. E. D. Gilles, director of the Institut für Systemdynamik und Regelungstechnik, Universität Stuttgart. Special thanks are also due to Dipl.-Ing. R. Locher, who carried out the numerical computations and the simulation of the hauling pipe.

Financial support was provided in part by the Deutsche Forschungsgemeinschaft (DFG).

REFERENCES

1. I. Szabó, Höhere Technische Mechanik, Springer-Verlag, Berlin, 1964.
2. L. Meirovitch, Analytical Methods in Vibrations, Macmillan, London, 1971.
3. S. G. Michlin, Lehrgang der Mathematischen Physik, Akademie-Verlag, Berlin, 1972.
4. H. Leipholz, ed., Instability of Continuous Systems, Springer-Verlag, Berlin, 1971.
5. M. P. Paidoussis, Dynamics of tubular cantilevers conveying fluid, J. Mech. Eng. Sci., 12: 85-103 (1970).
6. P. K. C. Wang, Stability analysis of elastic and aeroelastic systems via Lyapunov's direct method, J. Franklin Inst., 281: 51-72 (1966).

7. R. H. Plaut, On the optimal structural design for a nonconservative, elastic stability problem, JOTA, 7: 52-60 (1971).

8. F. Berkman and D. Karnopp, Complete response of distributed systems controlled by a finite number of linear feedback loops, Trans. ASME, J. Eng. Indust., 91: 1063-1068 (1969).

9. A. Bradshaw, Modal control of distributed-parameter vibratory systems, Int. J. Control, 19: 957-968 (1974).

10. M. Köhne, "Optimal feedback control of flexible mechanical systems," IFAC Symposium on the Control of Distributed Parameter Systems, Banff, Canada, paper no. 12-7 (1971).

11. M. Köhne, Modale Regelung des Förderrohres eines Tiefseebergbausystems, VDI/VDE-Aussprachetag, Frankfurt a.M., 1973.

12. M. Köhne, "Lineare optimale Regelung elastischer Systeme," Ph.D. thesis, Univ. of Stuttgart, 1975.

13. V. Komkov, Optimal Control Theory for the Damping of Vibrations of Simple Elastic Systems, Springer-Verlag, Berlin, 1972.

14. V. Komkov, Formulation of Pontryagin's maximality principle in a problem of structural mechanics, Int. J. Control, 17: 455-463 (1973).

15. R. S. Smith and E. L. S. Lum, Linear optimal theory applied to active structural bending control, J. Aircraft, 5: 479-485 (1968).

16. A. G. Butkowskii, Distributed Control Systems, American Elsevier, New York, 1969.

17. P. K. C. Wang, Control of Distributed Parameter Systems, in Advances in Control Systems, Vol. 1, Academic, New York, 1964, pp. 75-172.

18. J. L. Lions, Optimal Control of Systems Governed by Partial Differential Equations, Springer-Verlag, Berlin, 1971.

19. W. H. Yonkin, Optimal control of two distributed parameter systems via Lyapunov's direct method, J. Franklin Inst., 281: 51-72 (1966).

20. T. K. Sirazetdinov, Optimum control of elastic aircraft, Auto. Remote Contr., 27: 1139-1152 (1966).

21. J. Van de Vegte, Optimal and constrained optimal controls for vibrating beams, 11. JACC Atlanta, Georgia, paper 19-C, 1970, pp. 469-475.

22. J. Van de Vegte, The wave reflection matrix in beam vibration control, Trans. ASME, J. Dynamical Systems, Measurement and Control, 93: 94-101 (1971).

23. D. R. Vaughan, Application of distributed parameter concepts to dynamic analysis and control of bending vibrations, Trans. ASME, J. Basic Eng., 90: 157-166 (1968).

24. W. L. Brogan, Optimal control theory applied to systems described by partial differential equations, in Advances in Control Systems, Vol. 6, Academic, New York, 1968, pp. 221-316.

25. M. Köhne, Zustandsbeobachter für Systeme mit verteilten Parametern, VDI/VDE-Aussprachetag, Frankfurt a.M., 1975.

26. M. Köhne, Modellbildung und Simulation der Transversalschwingungen elastischer Föderrohre zur Mineralgewinnung aus der Tiefsee, Proceedings of the Symposium Simulation 1975, Zurich, Switzerland, 1975, pp. 228-234.

27. Y. Sakawa, A matrix Green's formula and optimal control of linear distributed-parameter systems, JOTA, 10: 290-299 (1972).

28. H. F. Vandevenne, Optimal control for a class of linear distributed parameter systems, Parts I and II, Revue A, XII: 131-146 (1970); XIII: 8-21 (1971).

29. G. A. Phillipson, Identification of Distributed Systems, American Elsevier, New York, 1971.

30. E. D. Gilles, Systeme mit verteilten Parametern – Einführung in die Regelungstheorie, Oldenbourg-Verlag, Munich, 1973.

31. C. Lanczos, Linear Differential Operators, Van Nostrand, London, 1964.

32. R. E. Goodson and R. E. Klein, A definition and some results for distributed system observability, IEEE Trans. Auto. Contr., AC-15: 165-174 (1970).

33. T. K. Yu and J. H. Seinfeld, Observability and optimal measurement location in linear distributed parameter systems, Int. J. Contr., 18: 785-799 (1973).

34. H. O. Fattorini, Controllability of Higher Order Linear Systems, in Mathematical Theory of Control (A. V. Balakrishnan and L. W. Neustadt, eds.), Academic, New York, 1967.

35. E. D. Gilles (ed.), Prozessanalyse und Prozessregelung. Fortgeschrittenen Studienprogramm der TU Berlin, Kurs-Nr. 31.2A (1970).

36. J. F. Creedon and A. G. Lindgren, Control of the optical surface of a thin, deformable primary mirror with application to an orbiting astronomical observatory, Automatica, 6: 643-660 (1970).

37. U. Knöpp, Zur optimalen Regelung von Systemen, die der Wellengleichung genügen, Regelungstechnik, 19: 535-538 (1971).

38. S. G. Tzafestas, Optimal distributed-parameter control using classical variational theory, Int. J. Contr., 12: 593-698 (1970).

39. E. Kreindler, On the linear optimal servo problem, Int. J. Contr., 9: 465-472 (1969).

40. C. A. Markland, Optimal model-following control-system synthesis techniques, Proc. IEE, 117: 623-627 (1970).

41. D. G. Luenberger, An introduction to observers, IEEE Trans. Auto. Contr., AC-16: 596-602 (1971).

42. Y. Ö. Yüksel and J. J. Bongiorno, Observers for linear multivariable systems with applications, IEEE Trans. Auto. Contr., AC-16: 603-613 (1971).

43. S. Kitamura, S. Sakairi, and M. Nishimura, Observer for distributed-parameter diffusion systems, Elect. Eng. Jap., 92: 142-149 (1972).

44. P. A. Orner and A. M. Foster, A design procedure for a class of distributed parameter control systems, Trans. ASME, J. Dynam. Syst., Meas. Contr., 93: 86-93 (1971).

45. M. B. Ajinkya, M. Köhne, H. F. Mäder, and W. H. Ray, The experimental implementation of a distributed parameter filter, Automatica, 11: in press.

46. M. Athans, Toward a practical theory for distributed parameter systems, IEEE Trans. Auto. Contr., AC-15: 245-247 (1970).

47. E. Gottzein and B. Lange, Magnetic suspension control systems for the MBB high speed train, Automatica, 11: 271-284 (1975).

48. E. D. Gilles and M. Zeitz, Modal simulation of a distributed parameter system, Simulation, 15: 179-188 (1970).

49. J. F. Wilson, Dynamic interactions between long, high speed trains of air cushion vehicles and their guideways, Trans. ASME, J. Dynam. Syst., Meas. Contr., 33: 16-24 (1971).

50. D. F. Wilkie, Dynamics, control and ride quality of a magnetically levitated high speed ground vehicle, Transp. Res., 6: 343-369 (1972).

51. G. Clauss, Probleme bei der Optimierung von Fördersystemen im Meeresbergbau, Erdoel-Erdgas-Zeitschrift, 86: 410-419 (1970).

52. J. E. Flipse, "An engineering approach to ocean mining," Offshore Technology Conference, Dallas, Texas, paper no. 1035, 1969.

53. D. Hody and J. O. Willums, Ein neues Verfahren zur Förderung von Manganerz-Knollen aus grossen Meerestiefen, Chemie-Ingenieur-Technik, 44: 1183-1188 (1972).

54. B. A. Finlayson and L. E. Scriven, The method of weighted residuals — A review, Appl. Mech. Rev., 19: 735-748 (1966).

55. H. S. Liu and C. D. Mote, Jr., "Vibration of pipes transporting high velocity fluids," Report of the Department of Mechanical Engineering, University of California, Berkeley, 1972.

56. M. Köhne, Zustandsbeobachter für Systeme mit verteilten Parametern - Theorie und Anwendung, VDI-Verlag, Düsseldorf, 1977.

57. M. Köhne, H. Schuler, and M. Zeitz, Einsatzmöglichkeiten von Beobachtern zur Messung von Zustandsgrössen verfahrenstechnischer Prozesse, in Fachberichte Messen - Steuern - Regeln (M. Syrbe and M. Thoma, eds.), Vol. 1, Springer-Verlag, Berlin, 1977, pp. 221-240.

Chapter 8

FEEDBACK STABILIZATION OF HYDROMAGNETIC EQUILIBRIA

P. K. C. Wang

Department of System Science
University of California
at Los Angeles
Los Angeles, California

G. Rodriguez

Jet Propulsion Laboratory
California Institute of Technology
Pasadena, California

I. INTRODUCTION

The possibility of stabilizing various types of instabilities in plasmas and hydromagnetic systems by means of feedback controls has been investigated both experimentally and theoretically during the past few years [1-3]. Recently, a number of experimental studies have been made on the feedback stabilization of various types of laboratory plasmas, in particular, those in thermonuclear devices such as the theta-pinch and the Tokamak [4,5]. In most of these studies the structural form of the feedback controls is determined in a heuristic manner. Consequently, the potentiality of feedback controls for plasma stabilization may not be fully utilized. Moreover, from the control-theoretic viewpoint, certain fundamental questions pertaining to stabilizability should be answered before one proceeds to implement any feedback controls.

The main objective here is to develop an applicable theory for the feedback stabilization of plasmas based on the single-fluid magnetohydrodynamic (MHD) equations. This involves the incorporation of the constraints imposed by the physical limitations on control and measurement into the mathematical formulation of the stabilization problem. Due to the spatially distributed nature of a plasma, it is difficult to implement the controls and sensors which are effective over the entire plasma. Consequently, we are limited to controls and sensors which are localized to certain portions of the plasma. In this chapter, an attempt will be made to answer the following basic questions associated with a given mathematical model of the plasma: (1) Do the localized controls affect all the unstable motions of the plasma? (2) Where should the controls and sensors be placed to achieve the most effective stabilization? (3) What is the simplest physically realizable form of stabilizing feedback controls for the given plasma model? Attention will be focused on the case of a perfectly conducting plasma with a sharp vacuum-plasma interface. The corresponding problems for the case of a resistive plasma will be discussed briefly.

II. MATHEMATICAL MODELS

Let Ω be an open bounded subset of the real three-dimensional Euclidean space R^3, with boundary $\partial\Omega$, representing the spatial domain of the plasma system. The plasma domain Ω_p is a proper subset of Ω such that its boundary $\partial\Omega_p$ has no points in common with $\partial\Omega$. Moreover, Ω_p is surrounded by a vacuum region Ω_v such that $\Omega_p \cap \Omega_v = \varphi$, $\Omega = \Omega_p \cup \Omega_v$, and $\partial\Omega_v = \partial\Omega \cup \partial\Omega_p$ (see Figure 1). In the physical system, $\partial\Omega_p$ may correspond to a rigid solid boundary (e.g., a glass

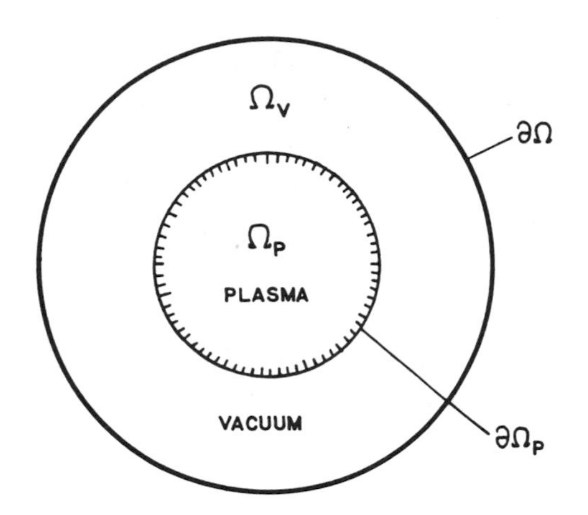

FIG. 1. Plasma and vacuum domains.

container wall) or the free boundary surface of the plasma. Let the control time interval $]0,T[$ be denoted by I_T, $T \le \infty$, and a point in Ω by $\underline{x} = (x_1,x_2,x_3)$. We shall consider a plasma whose motion is describable by the following single-fluid MHD equations [6]:

$$\frac{\partial \rho_m}{\partial t} + \underline{\nabla} \cdot (\rho_m \underline{V}) = 0 \qquad \text{(continuity)} \qquad (1)$$

$$\rho_m \frac{D\underline{V}}{Dt} = \underline{\nabla} p + \underline{J}_p \times \underline{B}_p \qquad \text{(momentum balance)} \qquad (2)$$

$$(\gamma - 1)^{-1} \rho_m^{\gamma} \frac{D}{Dt} (p \rho_m^{-\gamma}) = \eta \, |\underline{J}_p|^2 \qquad \text{(equation of state)} \qquad (3)$$

$$\left.\begin{aligned} &\frac{\partial \underline{B}_p}{\partial t} = -\underline{\nabla} \times \underline{E}_p &(4)\\ &\epsilon_o \frac{\partial \underline{E}_p}{\partial t} = \mu_o^{-1} \underline{\nabla} \times \underline{B}_p - \underline{J}_p &(5)\\ &\underline{\nabla} \cdot \underline{B}_p = 0 &(6) \end{aligned}\right\} \qquad \text{(Maxwell equations)}$$

$$\underline{E}_p + \underline{V} \times \underline{B}_p = \eta \underline{J}_p \qquad \text{(simplified Ohm's law)} \qquad (7)$$

where $D/Dt = (\partial/\partial t) + \underline{V} \cdot \underline{\nabla}$; $\rho_m(t,\underline{x})$, $\underline{V}(t,\underline{x})$, and $p(t,\underline{x})$ are the mass density, fluid velocity, and scalar pressure, respectively, of the plasma at time t and a point $\underline{x} \in \Omega_p$; $\underline{E}_p(t,\underline{x})$, $\underline{B}(t,\underline{x})$, and $\underline{J}_p(t,\underline{x})$ are the total electric and magnetic fields and the current density at $\underline{x}$ inside the plasma at t, respectively; ϵ_o, μ_o, and γ are constants corresponding to the permittivity and permeability of free space and the ratio of specific heats, respectively; and η is the plasma resistivity, which may depend on $\underline{x}$.

In the vacuum region Ω_v, the electric field $\underline{E}_v$ and magnetic field $\underline{B}_v$ due to both the plasma current and external current density $\underline{J}_v$ are governed by the Maxwell equations

$$\epsilon_o \frac{\partial \underline{E}_v}{\partial t} = \mu_o^{-1} \underline{\nabla} \times \underline{B}_v - \underline{J}_v \tag{8}$$

$$\frac{\partial \underline{B}_v}{\partial t} = -\underline{\nabla} \times \underline{E}_v \tag{9}$$

$$\underline{\nabla} \cdot \underline{B}_v = 0 \tag{10}$$

We shall use $\underline{J}_v$ as the manipulatable input or the control variable.

To simplify the foregoing mathematical model (1) to (10), we assume that the displacement currents $\epsilon_o \partial \underline{E}_p/\partial t$ and $\epsilon_o \partial \underline{E}_v/\partial t$ are negligible. Using Ohm's law (7), we can eliminate $\underline{E}_p$ in (4). Also, $\underline{J}_p$ in (2) can be eliminated by means of (5) with its left-hand side set to $\underline{0}$. Using elementary vector identities, we obtain the following simplified equations for the plasma system:

On Ω_p,

$$\frac{\partial}{\partial t}\begin{bmatrix} \rho_m \\ \underline{V} \\ p\rho_m^{-\gamma} \\ \underline{B}_p \end{bmatrix} = \begin{bmatrix} -\underline{\nabla} \cdot (\rho_m \underline{V}) \\ -\underline{V} \cdot \underline{\nabla}\underline{V} - \rho_m^{-1}[\underline{\nabla}(p + \mu_o^{-1}|\underline{B}_p|^2/2) - \mu_o^{-1}(\underline{B}_p \cdot \underline{\nabla})\underline{B}_p] \\ -\underline{V} \cdot \underline{\nabla}(p\rho_m^{-\gamma}) + \eta(\gamma - 1)\rho_m^{-\gamma}\mu_o^{-2}|\underline{\nabla} \times \underline{B}_p|^2 \\ \underline{\nabla} \times (\underline{V} \times \underline{B}_p) + \eta\mu_o^{-1}\nabla^2\underline{B}_p - \mu_o^{-1}\underline{\nabla}\eta \times (\underline{\nabla} \times \underline{B}_p) \end{bmatrix} \tag{11}$$

along with

$$\underline{\nabla} \cdot \underline{B}_p = 0 \tag{12}$$

and on Ω_v,

$$\underline{\nabla} \times \underline{B}_v = \mu_o \underline{J}_v, \qquad \underline{\nabla} \cdot \underline{B}_v = 0 \tag{13}$$

To complete the description of the foregoing simplified mathematical model, we need to specify the initial data at $t = 0$ along with appropriate boundary conditions at $\partial\Omega_v$ and $\partial\Omega_p$. The former is complete by specifying

$$\begin{aligned} \rho_m(0,\underline{x}) &= \rho_{mo}(\underline{x}) \\ \underline{V}(0,\underline{x}) &= \underline{V}_o(\underline{x}) \\ p(0,\underline{x}) &= p_o(\underline{x}) \\ \underline{B}_p(0,\underline{x}) &= \underline{B}_{po}(\underline{x}) \end{aligned} \tag{14}$$

In many plasma devices, the outer container wall is a conductor. For most purposes, $\partial\Omega$ or the outer boundary of Ω_v may be taken to be a perfectly conducting wall. Thus we have the following boundary condition for $\underline{B}_v$ at $\partial\Omega$:

$$\underline{n}(\underline{x}) \cdot \underline{B}_v(t,\underline{x}) = 0 \qquad \text{for} \quad \underline{x} \in \partial\Omega \quad \text{and} \quad t \geq 0 \tag{15}$$

where $\underline{n}(\underline{x})$ denotes the outward unit normal at $\underline{x} \in \partial\Omega$.

The boundary conditions at $\partial\Omega_p$ depend on the nature of the plasma under consideration. For a perfectly conducting plasma, it is meaningful to take $\partial\Omega_p$ as a free, sharp plasma-vacuum interface. In this case, we have the following free boundary conditions for $\underline{x} \in \partial\Omega_p$:

$$\underline{n}_p(\underline{x}) \cdot [\rho_m(t,\underline{x})\underline{V}(t,\underline{x})] = 0 \qquad \text{(no mass flow into } \Omega_v) \tag{16}$$

$$p(t,\underline{x}) + (\tfrac{1}{2}\mu_o)|\underline{B}_p(t,\underline{x})|^2 = (\tfrac{1}{2}\mu_o)|\underline{B}_v(t,\underline{x})|^2 \quad \text{(pressure balance)} \tag{17}$$

$$\underline{n}_p(\underline{x}) \cdot [\underline{B}_p(t,\underline{x}) - \underline{B}_v(t,\underline{x})] = 0 \tag{18}$$

$$\underline{n}_p(\underline{x}) \times [\underline{B}_p(t,\underline{x}) - \underline{B}_v(t,\underline{x})] = \mu_o \underline{J}_s \tag{19}$$

where $\underline{J}_s$ is a surface current density on $\partial\Omega_p$ and $\underline{n}_p(\underline{x})$ is the outward unit normal at $\underline{x} \in \partial\Omega_p$. Note that $\partial\Omega_p$ is a free boundary which may vary with time. In the case of a plasma with finite conductivity, a sharp plasma-vacuum interface does not exist. It is more realistic to consider a diffuse boundary. Here, we shall take $\partial\Omega_p$ to be a rigid, perfectly insulated wall. In a physical device, this type of wall permits separation between the control currents in Ω_v and the plasma. The boundary conditions for this case are given by

$$\underline{n}_p \cdot \underline{V}(t,\underline{x}) = 0 \tag{20}$$

$$\underline{B}_p(t,\underline{x}) = \underline{B}_v(t,\underline{x}) \tag{21}$$

for $\underline{x} \in \partial\Omega_p$ and $t \geq 0$. Now, the simplified equations for the plasma system are given by (11) to (13) with initial data (14) and boundary conditions (15) at $\partial\Omega$, and (16) to (19) or (20), (21) at $\partial\Omega_p$. Note that the control $\underline{J}_v$ affects the plasma indirectly through a "non-dynamic" system (13) and boundary conditions (17) to (19) or (21).

In the feedback control of plasmas, it is important to have instantaneous measurement of various pertinent plasma parameters which are usually time-dependent. In a plasma device, the direct observations or measurements of the plasma parameters are limited to the following forms:

1. Plasma mass density (ρ_m) measurements are limited to spatial averages over certain portions of Ω_p. For an electron-ion plasma, ρ_m is approximately equal to the ion mass density. If the temperature of the plasma is sufficiently low, local ion density measurements can be made using electric probes [7]. For high-temperature plasmas, ion density measurements based on Stark effect may be used.
2. Plasma fluid-velocity ($\underline{V}$) and displacement measurements are limited to local averages over the plasma boundary $\partial\Omega_p$ when it is a free surface. For a luminous plasma, optical sensors may be used for displacement measurements.
3. The direct measurement of pressure p in a physical device without disturbing the plasma is very difficult to accomplish in practice. However, it is possible to estimate p from temperature and mass density measurements by considering the plasma as a gas in quasistatic equilibrium.

4. The total magnetic field $\underline{B}_v$ in the vacuum region Ω_v can be measured by means of magnetic probes [8]. These measurements only provide spatial averages of $\underline{B}_v$ over subsets of Ω_v. Note that $\underline{B}_v$ is due to both the plasma and control current densities. In feedback control, it is necessary to extract the component of $\underline{B}_v$ due to the plasma only.

A. Static Equilibria

By a <u>static equilibrium</u> of the uncontrolled system (11) to (13) with boundary conditions (15) and (16) to (19) [or (20), (21)], we mean a time-independent solution of the system with $\underline{J}_v(t,\underline{x}) = \underline{0}$ and $\underline{V}(t,\underline{x}) = \underline{0}$ for all t and $\underline{x} \in \Omega_p$. We shall use the superscript "o" to denote quantities associated with a static equilibrium. The equations for determining the static equilibria are as follows:

On Ω_p,

$$\underline{\nabla}(p^o + \mu_o^{-1}|\underline{B}_p^o|^2/2) - \mu_o^{-1}(\underline{B}_p^o \cdot \underline{\nabla})\underline{B}_p^o = \underline{0} \tag{22}$$

$$\eta\nabla^2\underline{B}_p^o - \underline{\nabla}\eta \times (\underline{\nabla} \times \underline{B}_p^o) = \underline{0} \tag{23}$$

$$\eta\mu_o^{-2}|\underline{\nabla} \times \underline{B}_p^o|^2 = 0, \qquad \underline{\nabla} \cdot \underline{B}_p^o = 0 \tag{24}$$

and on Ω_v,

$$\underline{\nabla} \times \underline{B}_v^o = \mu_o\underline{J}_v^o, \qquad \underline{\nabla} \cdot \underline{B}_v^o = 0 \tag{25}$$

along with appropriate boundary conditions discussed earlier, where $\underline{J}_v^o$ corresponds to a given time-independent current density in the vacuum region.

First, we shall consider a perfectly conducting plasma ($\eta = 0$). In this case, (23) and the first equation in (24) are automatically satisfied. It is possible to seek static equilibria with a sharp plasma-vacuum interface $\partial\Omega_p$. We shall discuss a few static equilibria which are of importance in thermonuclear plasma devices.

1. Cylindrical Plasma Column

Consider a perfectly conducting cylindrical plasma column with radius R_1 and length ℓ, surrounded by a vacuum, which is enclosed by a

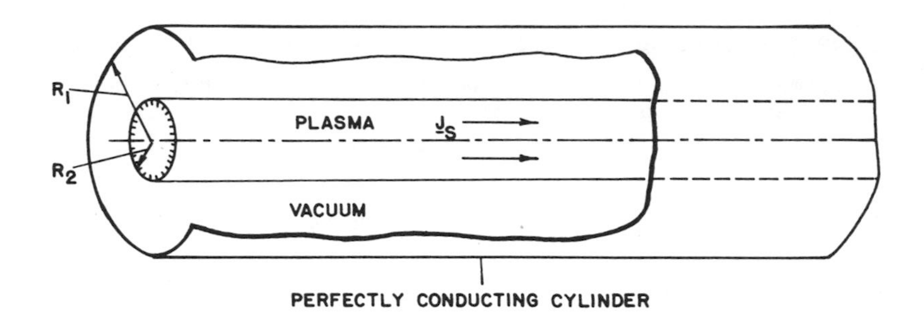

FIG. 2. A cylindrical plasma column.

perfectly conducting cylinder of radius R_2 (see Figure 2). The plasma has uniform mass density ρ_m^o and pressure p^o. If $\ell \gg R_2$, we may regard the plasma as infinite in length. In this case, the equilibrium magnetic fields $\underline{B}_p^o$ and $\underline{B}_v^o$ in the plasma and vacuum are given by

$$\begin{aligned} \underline{B}_p^o(\underline{x}) &= B_\theta^o b_i \underline{e}_z \\ \underline{B}_v^o(\underline{x}) &= B_\theta^o(R_1 r^{-1}\underline{e}_\theta + b_e \underline{e}_z) \end{aligned} \tag{26}$$

where $\underline{x} = (r,\theta,z)$ denotes the position vector in cylindrical coordinates, and $\underline{e}_\theta$ and $\underline{e}_z$ are unit vectors in the θ and z directions, respectively. The axial fields inside and outside the plasma are expressed as fractions b_i and b_e of the azimuthal field B_θ^o at the plasma surface, respectively. It is assumed that the plasma carries a sheet current $\underline{J}_s = J_{s\theta}^o \underline{e}_\theta + J_{sz}^o \underline{e}_z$ on its surface. Under the foregoing assumption, the configuration is in static equilibrium, provided that

$$\begin{aligned} \mu_o J_{s\theta}^o &= B_\theta^o(b_i - b_e) \\ \mu_o J_{sz}^o &= B_\theta^o \\ 2\mu_o p^o &= (B_\theta^o)^2(1 + b_e^2 - b_i^2) \end{aligned} \tag{27}$$

2. Toroidal Equilibria

An important class of thermonuclear plasma devices makes use of toroidal magnetic fields to confine the plasma. Recently, attention has been focused on the so-called Tokamak device in which a toroidal plasma current is used to generate a poloidal magnetic field. By combining this field with an external toroidal field, an improvement in plasma stability can be attained. In what follows, we shall discuss a few toroidal equilibria for perfectly conducting plasmas only.

First, consider a toroidal plasma in a Tokamak device as shown in Figure 3a. The plasma, contained in a conducting shell, carries a toroidal sheet current I_p on its surface $\partial\Omega_p$. Also, there is a toroidal magnetic field $\underline{B}_v$ in the vacuum region Ω_v. It can be shown

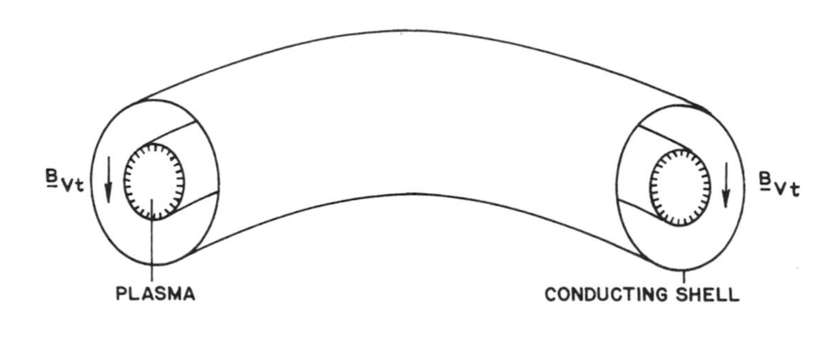

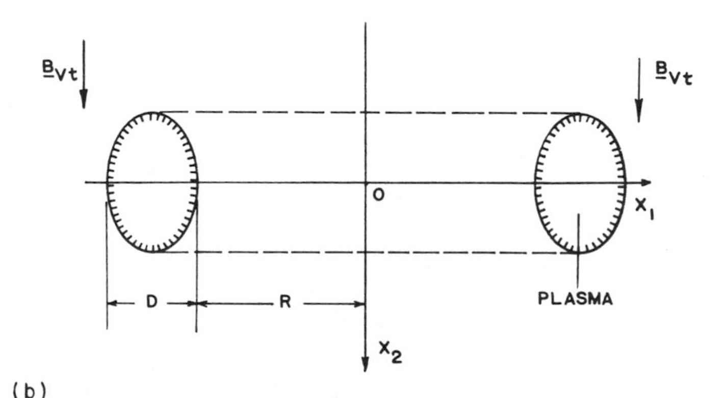

FIG. 3. (a) Cross section of a cutaway portion of plasma in a Tokamak device; (b) cross section of a toroidal plasma with a conductivity shell.

that there is no static equilibrium for the Tokamak without the conducting shell or any externally applied poloidal fields. To center the plasma, a static external vertical field $\underline{B}_{vt}$ is introduced to produce a net radial force on the plasma which counteracts the expanding force due to the plasma current I_p. The determination of static equilibria for the Tokamak is a free-boundary problem which cannot be solved analytically. However, by removing the conducting shell and replacing the plasma by two straight sections as shown in Figure 3b, the problem is reduced to a two-dimensional one whose solution can be expressed in explicit analytical form.

Let $\underline{B}^o_{vp}$ denote the equilibrium poloidal field in the vacuum region $\tilde{\Omega}_v \subset R^2$. This field satisfies

$$\underline{\nabla} \times \underline{B}^o_{vp} = \underline{0}, \qquad \underline{\nabla} \cdot \underline{B}^o_{vp} = 0 \qquad \text{on } \tilde{\Omega}_v \tag{28}$$

and $\underline{B}^o_{vp}(x_1,x_2) \to B^\infty_{vt}\underline{e}_{x_1}$ as $|\underline{x}| \to \infty$, where $\underline{x} = (x_1,x_2)$, $\underline{e}_{x_i}$ is the unit vector in the x_i direction, and B^∞_{vt} is a constant. Superimposed on $\underline{B}^o_{vp}$ there is a constant toroidal field $\underline{B}^o_{vt} = B^o_{vt}\underline{e}_{x_3}$ perpendicular to the (x_1,x_2)-plane. Inside the plasma domain $\tilde{\Omega}_p \subset R^2$, the poloidal field is zero since there is no current in the interior of $\tilde{\Omega}_p$. The toroidal field inside $\tilde{\Omega}_p$ is given by $\underline{B}^T_{po} = B^T_{po}\underline{e}_{x_3}$.

At the free boundary $\partial\tilde{\Omega}_p$, the poloidal field $\underline{B}^o_{vp}$ satisfies

$$\oint_{\Gamma_p} \underline{B}^o_{vp} \cdot d\ell = \mu_o I_p \tag{29}$$

where Γ_p is the contour along the boundary of $\tilde{\Omega}_v$. Note that (29) and the condition at infinity uniquely determine $\underline{B}^o_{vp}$.

Finally, we have the following pressure balance condition at the free boundary $\partial\tilde{\Omega}_p$:

$$p^o + (B^o_{pt})^2/2\mu_o = [|\underline{B}_{vp}|^2 + (B^o_{vt})^2]/2\mu_o \tag{30}$$

where p^o is the constant plasma pressure. Evidently, (30) implies that $|\underline{B}^o_{vp}|$ is a constant along Γ_p. Now, the free boundary problem is to find a domain $\tilde{\Omega}_p$ such that local pressure balance along Γ_p

can be maintained. This problem can be solved using conformal mapping. It has been shown [9] that for large aspect ratio $\alpha = (2R + D)/D$, the domain $\tilde{\Omega}_p$ consists of two circular regions; for small α, the circular region becomes D-shaped. Explicit expressions for Γ_p and the equilibrium quantities are given in [9]. The aforementioned results are obtained by introducing a two-dimensional approximation. The corresponding problem for the three-dimensional toroidal plasma with or without the conducting shell remains unsolved at this time.

Another interesting class of toroidal equilibria has been obtained recently by making use of the notion of "structural stability" for dynamical systems [10]. Roughly speaking, a dynamical system is structurally stable if the topological behavior of its state space trajectories is unaffected by small perturbations in the system equations. In the case of a toroidal plasma equilibrium, the magnetic field lines on the plasma surface are solutions of a differential equation. If we have structural stability, then the equilibrium magnetic field structure will not be destroyed under small perturbations of the system parameters, which is a desirable feature from the physical standpoint.

To fix ideas, consider a perfectly conducting plasma without a conducting shell as shown in Figure 4a. In the vacuum region Ω_v, the equilibrium magnetic field $\underline{B}_v^o$ is governed by (25). Hence there exists a scalar potential ψ such that $\underline{B}_v^o = \underline{\nabla}\psi$ and $\nabla^2\psi = 0$ in Ω_v. Assuming axial symmetry, this implies that the components of $\underline{B}_v^o$ are given by

$$B_{vr}^o = -r^{-1}(\partial\psi/\partial z), \qquad B_{vz}^o = r^{-1}(\partial\psi/\partial r), \qquad B_{v\varphi}^o = c/r \tag{31}$$

and ψ satisfies

$$(\partial^2\psi/\partial r^2) - r^{-1}(\partial\psi/\partial r) + (\partial^2\psi/\partial z^2) = 0 \tag{32}$$

On the plasma surface $\partial\Omega_p$, we require: (1) $|\underline{B}_v^o|^2/2\mu_o = p^o$ = constant; and (2) the normal component of $\underline{B}_v^o$ is continuous across $\partial\Omega_p$. It can be shown that Condition (1) implies

$$\underline{n}_p \cdot \underline{\nabla}\psi = \pm c(r^2/r_c^2 - 1)^{1/2} \quad \text{on } \partial\Omega_p \tag{33}$$

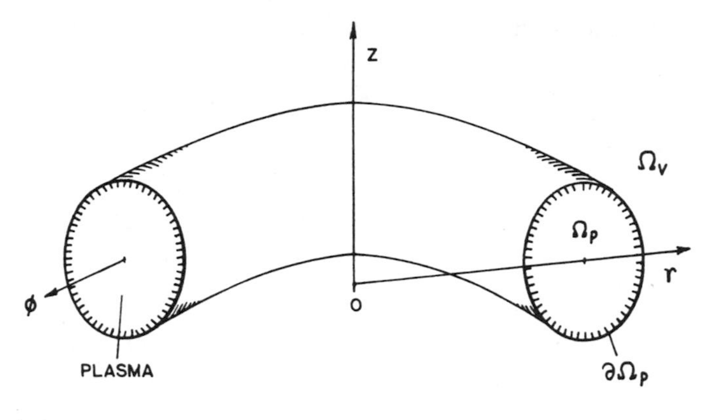

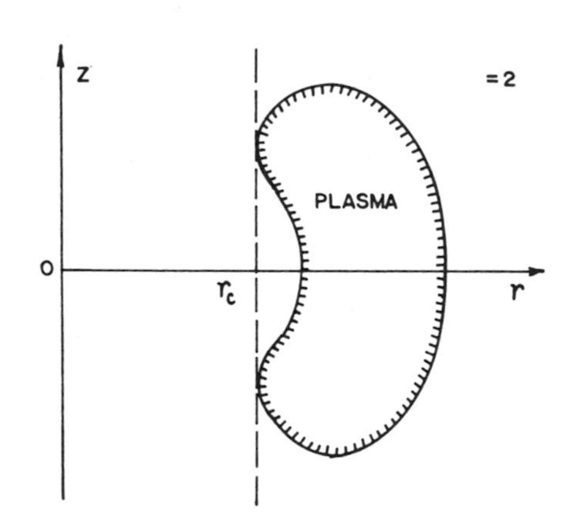

FIG. 4. (a) Cutaway portion of a toroidal plasma; (b) toroidal plasma with a kidney bean-shaped cross section.

where

$$r_c = |c|(2\mu_o p^o)^{-1/2} \tag{34}$$

Conditions (1) and (2) imply that $\underline{B}_v^o$ is tangent to $\partial\Omega_p$ or Ψ = constant on $\partial\Omega_p$. Since $p^o > 0$, Condition (1) also implies that $|\underline{B}_v^o| > 0$ on $\partial\Omega_p$. Consequently, the magnetic field lines form orbits on $\partial\Omega_p$. Moreover, the number of closed orbits around the torus is finite. On a closed orbit, $B^o_{vr} = B^o_{vz} = 0$, which implies that $\underline{n}_p \cdot \underline{\nabla}\psi = 0$ on $\partial\Omega_p$. Evidently, from (33), this holds if and only if $r = r_c$. We now require that the magnetic field lines on the torus be structurally stable. Using the results of Peixoto and Lefschetz [11], the number of closed orbits must be even for structural stability, or the poloidal cross section of the plasma must touch $r = r_c$ in an even number of points, and the remaining plasma must have $r > r_c$. For two closed orbits, the cross section has the kidney bean shape as shown in Figure 4b. Numerical results for such plasma cross sections are given in [10].

We now consider the static equilibria for the resistive case ($\eta > 0$). From the first equation in (24), if $\eta \neq 0$, then $\underline{\nabla} \times \underline{B}^o_p = \underline{0}$, or the equilibrium plasma current must be zero. Thus, (23) reduces to $\nabla^2 \underline{B}^o_p = \underline{0}$ or the equilibrium magnetic field in the plasma has the same structure as that in the vacuum region. Also, it is impossible to maintain a sharp plasma-vacuum interface at equilibrium. In the case where $\partial\Omega_p$ is a rigid, perfectly insulated wall, the plasma simply fills the domain Ω_p at equilibrium. When the plasma is only slightly resistive, we may ignore the effect of resistivity in the determination of the static equilibria. This is justified when the growth rates of the instabilities are much larger than the diffusive decay rates due to resistive effects.

B. Linearized Equations

In controlling the plasma motion about a given static equilibrium, it is useful to consider the linearized plasma equations. We shall now give the equations corresponding to (11) to (13) linearized about a given static equilibrium specified by ρ^o_m, $\underline{V}^o = \underline{0}$, p^o, $\underline{B}^o_p$, and $\underline{B}^o_v$. Let the perturbed quantities be defined as follows:

$$\begin{aligned} \rho_m &= \rho^o_m + \rho_{m1}, & \underline{V} &= \underline{V}^o + \underline{V}_1 = \underline{V}_1 \\ p &= p^o + p_1, & \underline{B}_p &= \underline{B}^o_p + \underline{B}_{p1} \\ \underline{B}_v &= \underline{B}^o_v + \underline{B}_{v1}, & \underline{J}_v &= \underline{J}^o_v + \underline{J}_{v1} \end{aligned} \tag{35}$$

Substituting (35) into (11) to (13) and eliminating the quadratic terms involving the perturbed variables leads to the following linearized equations:

On Ω_p,

$$\frac{\partial}{\partial t}\begin{bmatrix} \rho_{m1} \\ \underline{V}_1 \\ \\ p_1 \\ \underline{B}_{p1} \end{bmatrix} = \begin{bmatrix} -\underline{\nabla} \cdot (\rho_m^o \underline{V}_1) \\ -(\rho_m^o)^{-1}[\underline{\nabla}(p_1 + \mu_o^{-1}\underline{B}_p^o \cdot \underline{B}_{p1}) \\ \qquad + \mu_o^{-1}(\underline{B}_{p1} \cdot \underline{\nabla B}_p^o + \underline{B}_p^o \cdot \underline{\nabla B}_{p1})] \\ -\underline{V}_1 \cdot \underline{\nabla} p^o - \gamma p^o \underline{\nabla} \cdot \underline{V}_1 \\ \underline{\nabla} \times (\underline{V}_1 \times \underline{B}_p^o) + \eta\mu_o^{-1}\nabla^2 \underline{B}_{p1} - \mu_o^{-1}\underline{\nabla}\eta \times (\underline{\nabla} \times \underline{B}_{p1}) \end{bmatrix} \tag{36}$$

along with

$$\underline{\nabla} \cdot \underline{B}_{p1} = 0 \tag{37}$$

and on Ω_v,

$$\underline{\nabla} \times \underline{B}_{v1} = \mu_o \underline{J}_{v1}, \qquad \underline{\nabla} \cdot \underline{B}_{v1} = 0 \tag{38}$$

The linearized boundary conditions for the perturbed variables corresponding to (15) to (21) are given by

$$\underline{n}(\underline{x}) \cdot \underline{B}_{v1}(t,\underline{x}) = 0 \tag{15'}$$

$$\underline{n}_p(\underline{x}) \cdot [\rho_m^o(\underline{x})\underline{V}_1(t,\underline{x})] = 0 \tag{16'}$$

$$p_1(t,\underline{x}) + \mu_o^{-1}\underline{B}_p^o(\underline{x}) \cdot \underline{B}_{p1}(t,\underline{x}) = \mu_o^{-1}\underline{B}_v^o(\underline{x}) \cdot \underline{B}_{v1}(t,\underline{x}) \tag{17'}$$

$$\underline{n}_p(\underline{x}) \cdot [\underline{B}_{p1}(t,\underline{x}) - \underline{B}_{v1}(t,\underline{x})] = 0 \tag{18'}$$

$$\underline{n}_p(\underline{x}) \times [\underline{B}_{p1}(t,\underline{x}) - \underline{B}_{v1}(t,\underline{x})] = \mu_o \underline{J}_{s1} \tag{19'}$$

$$\underline{n}_p(\underline{x}) \cdot \underline{V}_1(t,\underline{x}) = 0 \tag{20'}$$

$$\underline{B}_{p1}(t,\underline{x}) = \underline{B}_{v1}(t,\underline{x}) \tag{21'}$$

In the special case of a perfectly conducting plasma ($\eta = 0$), (36) can be rewritten in a different form by expressing the small perturbations from a static equilibrium state in terms of the

Lagrangian displacement of a plasma fluid element. Let the initial position of such an element be denoted by $\underline{x}_o$ and the displaced position of this element at time t by $\underline{x}(t)$, where

$$\underline{x}(t) = \underline{x}_o + \underline{\xi}(t,\underline{x}_o), \qquad \underline{\xi}(0,\underline{x}_o) = \underline{0} \tag{39}$$

It can be shown [12] that the linearized equation for $\underline{\xi}$ has the form

$$\rho_m^o \, \partial^2\underline{\xi}/\partial t^2 = \underline{F}(\underline{\xi}) \tag{40}$$

with

$$\underline{F}(\underline{\xi}) = \underline{\nabla}[\xi \cdot \underline{\nabla}p^o + \gamma p^o \underline{\nabla} \cdot \underline{\xi}] + \underline{J}^o \times [\underline{\nabla} \times (\underline{\xi} \times \underline{B}_p^o)]$$
$$- \mu_o^{-1}\underline{B}_p^o \times \underline{\nabla} \times [\underline{\nabla} \times (\underline{\xi} \times \underline{B}_p^o)] \tag{41}$$

where

$$\underline{J}^o = \mu_o^{-1}(\underline{\nabla} \times \underline{B}_p^o) \tag{42}$$

Note that (40) and (41) are defined for $t > 0$ and any point $\underline{x}_o$ in the equilibrium plasma domain Ω_p^o. Assuming the existence of a free, sharp plasma-vacuum interface, any point on the perturbed plasma surface $\partial\Omega_{p1}$ can be written as $\underline{x} = \underline{x}_o + (\underline{n}_{po} \cdot \underline{\xi})\underline{n}_{po}$, where $\underline{n}_{po}$ is the outward unit normal at the point $\underline{x}_o$ on the equilibrium plasma surface $\partial\Omega_p^o$. For small displacement $\underline{\xi}$, we may use the first-order relations

$$p^o(\underline{x}) = p^o(\underline{x}_o) + (\underline{x} - \underline{x}_o) \cdot \underline{\nabla}p^o(\underline{x}_o)$$
$$= p^o(\underline{x}_o) + (\underline{n}_{po} \cdot \underline{\xi})\underline{n}_{po} \cdot \underline{\nabla}p^o(\underline{x}_o) \tag{43}$$

$$p_1(t,\underline{x}) = -\underline{\xi} \cdot \underline{\nabla}p^o(\underline{x}) - \gamma p^o(\underline{x})\underline{\nabla} \cdot \underline{\xi} \tag{44}$$

$$\underline{B}_p^o(\underline{x}) \cdot \underline{B}_{p1}(t,\underline{x}) = \underline{B}_p^o(\underline{x}_o) \cdot \underline{B}_{p1}(t,\underline{x}_o) \tag{45}$$

$$\underline{B}_{p1}(t,\underline{x}) = \underline{\nabla} \times [\underline{\xi}(t,\underline{x}_o) \times \underline{B}_p^o(\underline{x}_o)] \tag{46}$$

$$\underline{B}_v^o(\underline{x}) \cdot \underline{B}_{v1}(t,\underline{x}) = \underline{B}_v^o(\underline{x}_o) \cdot \underline{B}_{v1}(t,\underline{x}_o) \tag{47}$$

to obtain the following boundary conditions for (40):

$$\mu_o \gamma p^o(\underline{x}_o)[\underline{\nabla} \cdot \underline{\xi}(t,\underline{x}_o)] - \underline{B}_p^o(\underline{x}_o) \cdot \underline{B}_{p1}(t,\underline{x}_o)$$
$$+ \underline{B}_v^o(\underline{x}_o) \cdot \underline{B}_{v1}(t,\underline{x}_o) + \underline{\xi}(t,\underline{x}_o) \cdot \underline{\nabla}[(|\underline{B}_v^o(\underline{x}_o)|^2$$
$$- |\underline{B}_v^o(\underline{x}_o)|^2)/2] = 0 \tag{48}$$

$$\underline{n}_{po} \cdot \underline{B}_{v1}(t,\underline{x}_o) - \underline{n}_{po} \cdot [\underline{\nabla} \times (\underline{\xi}(t,\underline{x}_o) \times \underline{B}_v^o(\underline{x}_o))] = 0 \tag{49}$$

for $\underline{x}_o \in \partial\Omega_p^o$ and $t \geq 0$.

In the vacuum region Ω_v^o, the perturbed magnetic field $\underline{B}_{v1}$ can be written in terms of a vector potential $\underline{A}$ such that

$$\underline{B}_{v1} = \underline{\nabla} \times \underline{A} \tag{50}$$

and

$$\underline{\nabla} \times (\underline{\nabla} \times A) = \mu_o \underline{J}_{v1}, \qquad \underline{\nabla} \cdot \underline{A} = 0 \tag{51}$$

In terms of $\underline{A}$, (48) and (49) can be rewritten as

$$\mu_o \gamma p^o(\underline{x})[\underline{\nabla} \cdot \underline{\xi}(t,\underline{x})] + \underline{B}_v^o \cdot (\underline{\nabla} \times \underline{A}) - \underline{B}_p^o \cdot \underline{\nabla} \times [\underline{\xi}(t,\underline{x}) \times \underline{B}_p^o(\underline{x})]$$
$$+ \underline{\xi}(t,\underline{x}) \cdot \underline{\nabla}[(|\underline{B}_v^o(\underline{x})|^2 - |\underline{B}_p^o(\underline{x})|^2)/2] = 0 \tag{52}$$

$$[\underline{n}_p \cdot \underline{\xi}(t,\underline{x})]\underline{B}_v^o(\underline{x}) + \underline{n}_p \times \underline{A} = \underline{0} \tag{53}$$

defined for $t \geq 0$ and $\underline{x} \in \partial\Omega_p^o$ (for brevity, the subscript "o" for $\underline{x}$ and $\underline{n}_p$ has been dropped). Finally, the boundary condition at the outer perfectly conducting wall $\partial\Omega$ can be expressed in terms of $\underline{A}$:

$$\underline{n}(\underline{x}) \times \underline{A}(t,\underline{x}) = \underline{0} \qquad \text{for } t \geq 0,\ \underline{x} \in \Omega \tag{54}$$

Thus, for a perfectly conducting plasma with a sharp plasma-vacuum boundary, its perturbed motion about a given static equilibrium can be described approximately by the solution of the mixed initial boundary value problem (40) (defined on Ω_p^o) and (51) (defined on $\Omega_v^o = \Omega - \Omega_p^o$) along with boundary conditions (52), (53) at $\partial\Omega_p^o$ and (54) at $\partial\Omega$, and initial data at $t = 0$

$$\begin{aligned} &\underline{\xi}(0,\underline{x}) = \underline{\xi}_o(\underline{x}), \qquad \partial\underline{\xi}(0,\underline{x})/\partial t = \dot{\underline{\xi}}_o(\underline{x}), \qquad \underline{x} \in \Omega_p^o \\ &\underline{A}(0,\underline{x}) = \underline{A}_o(\underline{x}), \qquad \partial\underline{A}(0,\underline{x})/\partial t = \underline{A}_o(\underline{x}), \qquad \underline{x} \in \Omega_v^o \end{aligned} \tag{55}$$

This completes the formal description of the linearized equations for the plasma. These equations will be used later for developing a theory for the feedback stabilization of plasmas. In the following section, we shall consider first the mathematical problem of establishing the existence and uniqueness of solutions of the mixed initial boundary value problem (abbreviated hereafter as MIBVP) associated with the linearized equations.

C. Existence and Uniqueness of Solutions

First, we consider the MIBVP for (40) and (51) to (55). Use will be made of Lions' approach [13] to establish the existence and uniqueness of weak solutions by posing the problem in appropriate Sobolev spaces.

Let Σ be an arbitrary open set in R^3. The triple cartesian product of the complex Hilbert space $L_2(\Sigma)$ is denoted by $L_2^3(\Sigma)$, whose inner product is defined by

$$(\varphi,\overline{\varphi}') = \int_\Sigma \varphi\cdot\overline{\varphi}'\, d\Sigma = \int_\Sigma \sum_{i=1}^{3} \varphi_i\overline{\varphi}_i'\, d\Sigma \tag{56}$$

for any $\varphi = (\varphi_1,\varphi_2,\varphi_3)$, $\varphi' = (\varphi_1',\varphi_2',\varphi_3') \in L_2^3(\Sigma)$, where $(\bar{\cdot})$ denotes complex conjugation. We introduce the following closed subspace H of $L_2^3(\Omega_p^o) \times L_2^3(\Omega_v^o)$ defined by

$$H = \{\underline{u} = (\underline{u}_1,\underline{u}_2) \in L_2^3(\Omega_p^o) \times L_2^3(\Omega_v^o) : \underline{\nabla} \cdot \underline{u}_2 = 0\} \tag{57}$$

whose inner product is defined by

$$\langle \underline{u},\underline{u}'\rangle = (\underline{u}_1,\underline{u}_1')_1 + (\underline{u}_2,\underline{u}_2')_2 \tag{58}$$

where $(\cdot,\cdot)_1$ and $(\cdot,\cdot)_2$ denote the inner products for $L_2^3(\Omega_p^o)$ and $L_2^3(\Omega_v^o)$, respectively. Also we introduce the complex inner product space V_o defined by

$$\begin{aligned} V_o = \{\underline{u} = (\underline{u}_1,\underline{u}_2) \in (H^1(\Omega_p^o))^3 \times (H^1(\Omega_v^o))^3 : \\ \underline{\nabla} \cdot \underline{u}_2 = 0,\ (\underline{n}_p \cdot \underline{u}_1)\underline{B}_v^o + \underline{n}_p \times \underline{u}_2 = \underline{0} \text{ for } \underline{x} \in \partial\Omega_p \\ \underline{n} \times \underline{u}_2 = \underline{0} \text{ for } \underline{x} \in \partial\Omega\} \end{aligned} \tag{59}$$

with the inner product

$$((\underline{u},\underline{u}')) = \int_{\Omega_p^o} [\underline{u}_1 \cdot \overline{\underline{u}_1'} + \gamma p^o(\underline{\nabla} \cdot \underline{u}_1)(\underline{\nabla} \cdot \overline{\underline{u}_1'})$$

$$+ \mu_o^{-1}\underline{\nabla} \times (\underline{u}_1 \times \underline{B}_p^o) \cdot \underline{\nabla} \times (\overline{\underline{u}_1'} \times \underline{B}_p^o)]\, d\Omega_p^o$$

$$+ \mu_o^{-1} \int_{\Omega_p^o} (\underline{\nabla} \times \underline{u}_2) \cdot (\underline{\nabla} \times \overline{\underline{u}_2'})\, d\Omega_v^o \tag{60}$$

It can be verified [14] that V_o is a separable Hilbert space which is dense in H. Also, H is a dense subspace of V_o^*, the antidual of V_o. Note that it is possible to include the boundary conditions (53) and (54) into the definition of V_o, since the functions in the Sobolev space $H^1(\Omega_p^o)$ and $H^1(\Omega_v^o)$ are well-behaved at the boundaries $\partial\Omega_p^o$ and $\partial\Omega_v^o$, respectively (as established by the trace theorems [13]).

We now introduce the following sesquilinear continuous symmetric form on V_o:

$$a(\underline{u},\underline{u}') = \mu_o^{-1} \int_{\Omega_v^o} (\underline{\nabla} \times \underline{u}_2) \cdot (\underline{\nabla} \times \overline{\underline{u}_2'})\, d\Omega_v^o - \int_{\partial\Omega_p^o} (\underline{n}_p \cdot \underline{k}_o)$$

$$\times (\underline{n}_p \cdot \underline{u}_1)(\underline{n}_p \cdot \overline{\underline{u}_1'})\, d(\partial\Omega_p) + \int_{\Omega_p^o} [\gamma p^o(\underline{\nabla} \cdot \underline{u}_1)$$

$$\times (\underline{\nabla} \cdot \overline{\underline{u}_1'}) + \mu_o^{-1}\underline{\nabla} \times (\underline{u}_1 \times \underline{B}_p^o) \cdot \underline{\nabla} \times (\overline{\underline{u}_1'} \times \underline{B}_p^o)]\, d\Omega_p^o$$

$$- \int_{\Omega_p^o} [\underline{u}_1 \cdot \underline{J}^o \times \underline{\nabla} \times (\overline{\underline{u}_1'} \times \underline{B}_p^o) + \underline{u}_1 \cdot \underline{\nabla}(\overline{\underline{u}_1'} \cdot \underline{\nabla} p^o)]\, d\Omega_p^o \tag{61}$$

where $\underline{k}_o = \underline{\nabla}[(|\underline{B}_p^o|^2 - |\underline{B}_v^o|^2)/2\mu_o]$. Note that the surface integral in (61) is well-defined, since $\underline{u}_1 \in V_o$ implies $(\underline{n}_p \cdot \underline{u}_1) \in L_2(\partial\Omega_p^o)$ by the trace theorem [13]. It can be shown that $a(\cdot,\cdot)$ is V_o-coercive [relative to $L_2^3(\Omega_p^o)$], i.e., there exist real numbers λ_o, α with $\alpha > 0$ such that

$$a(\underline{u},\underline{u}) + \lambda_o\|\underline{u}_1\|^2 \geq \alpha|||\underline{u}|||^2 \tag{62}$$

for all $\underline{u} = (\underline{u}_1,\underline{u}_2) \in V_o$, where $\|\cdot\|$ and $|||\cdot|||$ denote the norms for $L_2^3(\Omega_p^o)$ and V_o, respectively.

In terms of the sesquilinear form a, we define a weak solution to the MIBVP (40), (51) to (55) as a function $\underline{u}(t) = (\underline{\xi}(t,\cdot),\underline{A}(t,\cdot))$, $t \in I_T$, which satisfies

$$(\rho_m^o \partial^2 \underline{\xi}/\partial t^2, \underline{\psi}_1)_1 + a(\underline{\xi},\underline{\psi}) = (\mu_o \underline{J}_{v1}, \underline{\psi}_2)_2 \tag{63}$$

for all $\underline{\psi} = (\underline{\psi}_1,\underline{\psi}_2) \in V_o$. Using Green's formula, (63) can be rewritten as:

$$(\rho_m^o \partial^2 \underline{\xi}/\partial t^2 - \underline{F}(\underline{\xi}), \underline{\psi}_1)_1 + (\underline{\nabla} \times \underline{\nabla} \times \underline{A} - \mu_o \underline{J}_{v1}, \psi_2)_2$$

$$= \int_{\partial\Omega_p^o} [(\underline{k}_o \cdot \underline{\xi}) + \gamma p^o(\underline{\nabla} \cdot \underline{\xi}) + \mu_o^{-1}\underline{B}_v^o \cdot \underline{\nabla} \times \underline{A}$$

$$- \mu_o^{-1}\underline{B}_p^o \cdot \underline{\nabla} \times (\underline{\xi} \times \underline{B}_p^o)](\underline{n}_p \cdot \overline{\underline{\psi}}_1)\, d(\partial\Omega_p^o) \tag{64}$$

for all $\underline{\psi} \in V_o$. The surface integral in (64) vanishes, implying that in this formulation (52) appears as a natural boundary condition. The boundary conditions (53) and (54) are included in the definition of V_o. The initial conditions are satisfied in the following sense:

$$\lim_{t\to 0} \|\underline{u}(t,\cdot) - \underline{u}_o(\cdot)\|_H + \|\dot{\underline{u}}(t,\cdot) - \dot{\underline{u}}_o(\cdot)\|_H = 0 \tag{65}$$

where $\|\cdot\|_H$ is the norm induced by the inner product (58), and $\underline{u}_o = (\underline{\xi}_o,\underline{A}_o)$, $\dot{\underline{u}}_o = (\dot{\underline{\xi}}_o,\dot{\underline{A}}_o)$, $\dot{u} = \partial\underline{u}/\partial t$.

Let $L_p(I_T;V)$, $1 \le p \le \infty$, denote the space of functions which are L_p with respect to t, and take their values in V for each fixed $t \in I_T$. We now give a theorem for the existence and uniqueness of weak solutions to the MIBVP.

<u>Theorem.</u> Let $\underline{u}_o \in V_o$, $\dot{\underline{u}}_o \in H$, and $\underline{J}_{v1} \in L_2(I_T;L_2^3(\Omega_v^o))$ be given. There then exists a unique weak solution $\underline{u} = (\underline{u}_1,\underline{u}_2) \triangleq (\underline{\xi},\underline{A})$ to the MIBVP (40), (51) to (55) with

$$\underline{u} \in L_\infty(I_T;V_o)$$

$$\dot{\underline{u}} \in L_\infty(I_T;H)$$

Moreover, the mapping $(\underline{J}_{v1},\underline{u}_o,\dot{\underline{u}}_o) \to (\underline{u},\dot{\underline{u}})$ is a linear continuous map of $L_2(I_T;L_2^3(\Omega_v^o)) \times V_o \times H \to L_\infty(I_T;V_o) \times L_\infty(I_T;H)$.

The foregoing theorem can be proved using Lions' approach [13]. We only outline the basic ideas here. The details are given in [14]. Since V_o is a separable Hilbert space, there exists a countable basis $\underline{\beta}^1, \underline{\beta}^2, \ldots, \underline{\beta}^m, \ldots$. To establish the existence of solutions to the MIBVP, we first construct a sequence of approximate solutions of the form

$$\underline{u}^m(t) = \sum_{i=1}^{m} \alpha_{im}(t)\underline{\beta}^i, \qquad m = 1,2, \ldots , \tag{66}$$

where $\underline{u}^m = (\underline{u}_1^m, \underline{u}_2^m)$ satisfies the following set of ordinary differential equations

$$\frac{d^2}{dt^2} (\underline{u}_1^m(t), \underline{\beta}_1^i)_1 + a(\underline{u}^m(t), \underline{\beta}^i) = (\mu_o \underline{J}_{v1}, \underline{\beta}_2^i)_2, \qquad 1 \le i \le m \tag{67}$$

where $\underline{\beta}^i = (\underline{\beta}_1^i, \underline{\beta}_2^i)$, with $\underline{\beta}_1^i$ and $\underline{\beta}_2^i$ defined on Ω_p^o and Ω_v^o, respectively. The initial conditions for (67) are given by

$$\underline{u}^m(0) = \underline{u}_o^m, \qquad \underline{\dot{u}}^m(0) = \underline{\dot{u}}_o^m \tag{68}$$

where $\underline{u}_o^m$ and $\underline{\dot{u}}_o^m$ are determined by

$$\underline{u}_o^m = \sum_{i=1}^{m} \alpha_{im}(0)\underline{\beta}^i, \qquad \underline{u}_o^m \to \underline{u}_o \quad \text{in } V_o \text{ as } m \to \infty$$

and

$$\underline{\dot{u}}_o^m = \sum_{i=1}^{m} \dot{\alpha}_{im}(0)\underline{\beta}^i, \qquad \underline{\dot{u}}_o^m \to \underline{\dot{u}}_o \quad \text{in } H \text{ as } m \to \infty$$

It can be shown that the sequence of approximate solutions converges weakly to a function in the proper space as the number of basis elements used in constructing the approximate solutions approaches infinity. The existence of a weak solution to the MIBVP can be established by showing that the limit function is indeed a weak solution. The uniqueness of a weak solution can be established by showing that the difference of two solutions with the same initial conditions is identically zero over the time interval I_T. The continuity of the mapping $(\underline{J}_{v1},\underline{u}_o,\dot{\underline{u}}_o) \to (\underline{u},\dot{\underline{u}})$ can be deduced from the a priori estimates of $\underline{u}^m(t)$.

For the case of a resistive plasma, the existence and uniqueness of weak solutions to the MIBVP corresponding to the linearized equations (36) to (38) along with appropriate boundary conditions in (15′) to (21′) can be established in a similar way. The details will be presented elsewhere.

D. Modal Representation of Solutions

In dealing with the linearized model of a plasma system, it is useful to express its motion as the superposition of its natural modes. For the plasma system discussed here, each natural mode is determined by the plasma properties and its magnetic field in the vacuum region. In the following, we shall consider only the case of a perfectly conducting plasma whose linearized equations about a given equilibrium are given by (40) to (42), (48), (49), and (55). The natural modes of this plasma system can be determined by solving the following boundary value problem: Find a scalar λ and a function $\underline{\varphi} = (\underline{\varphi}_1,\underline{\varphi}_2)$ such that

$$\begin{aligned} &\underline{F}(\underline{\varphi}_1)(\underline{x}) = -\lambda\underline{\varphi}_1(\underline{x}), && \underline{x} \in \Omega_p^o \\ &(\underline{\nabla} \times \underline{\nabla} \times \underline{\varphi}_2)(\underline{x}) = \underline{0}, \quad (\underline{\nabla} \cdot \underline{\varphi}_2)(\underline{x}) = 0, && \underline{x} \in \Omega_v^o \end{aligned} \tag{69}$$

subject to the boundary conditions

$$\begin{aligned} &-\mu_o \gamma p^o(\underline{\nabla} \cdot \underline{\varphi}_1) + \underline{B}_p^o \cdot \underline{\nabla} \times (\underline{\varphi}_1 \times \underline{B}_p^o) + \underline{\varphi}_1 \cdot \underline{\nabla}(|\underline{B}_p^o|^2 - |\underline{B}_v^o|^2)/2 \\ &\qquad -\underline{B}_v^o \cdot \underline{\nabla} \times \underline{\varphi}_2 = 0, \qquad \underline{x} \in \partial\Omega_p^o \end{aligned} \tag{70}$$

$$\underline{\nabla} \times \underline{\varphi}_2 + (\underline{n}_p \cdot \underline{\varphi}_1)\underline{B}_v^o = \underline{0}, \qquad \underline{x} \in \partial\Omega_p^o \tag{71}$$

$$\underline{n} \times \underline{\varphi}_2 = \underline{0}, \qquad \underline{x} \in \partial\Omega \tag{72}$$

A function $\underline{\varphi} = (\underline{\varphi}_1,\underline{\varphi}_2)$ is said to be a weak solution to the foregoing problem for some λ if and only if

$$a(\underline{\varphi},\underline{\psi}) = \lambda(\underline{\varphi}_1,\underline{\psi}_1)_1 \qquad \text{for all } \underline{\psi} = (\underline{\psi}_1,\underline{\psi}_2) \in V_o \tag{73}$$

where a is the sesquilinear form defined by (61). Using Green's formula, (73) can be rewritten as

$$[\underline{F}(\underline{\varphi}_1) + \lambda\underline{\varphi}_1,\underline{\psi}_1]_1 + (\underline{\nabla} \times \underline{\nabla} \times \underline{\varphi}_2,\underline{\psi}_2)_2$$

$$+ \int_{\partial\Omega_p^o} (\underline{n}_p \cdot \overline{\underline{\psi}}_1)[-\mu_o\gamma p^o(\underline{\nabla} \cdot \underline{\varphi}_1) + \underline{B}_p^o \cdot \underline{\nabla} \times (\underline{\varphi}_1 \times \underline{B}_p^o)$$

$$- \underline{B}_v^o \cdot \underline{\nabla} \times \underline{\varphi}_2 + \underline{\varphi}_1 \cdot \underline{\nabla}(|\underline{B}_p^o|^2 - |\underline{B}_v^o|^2)/2]\, d(\partial\Omega_p^o) = 0$$

$$\text{for all } \underline{\psi} \in V_o \qquad (74)$$

In view of (70), the last integral in (74) vanishes. Boundary conditions (71) and (72) are contained in the definition of V_o.

It can be shown [14] that the foregoing problem has an infinite sequence of weak solutions $\underline{\varphi}^m = (\underline{\varphi}_1^m,\underline{\varphi}_2^m)$, m = 1,2, ... , which is complete in a subspace of H defined by

$$H_s = \{\underline{u} = (\underline{u}_1,\underline{u}_2) \in H\colon \underline{\nabla} \times \underline{\nabla} \times \underline{u}_2 = \underline{0}\}$$

and $\{\underline{\varphi}_1^m\}$ is orthonormal and complete in $L_2^3(\Omega_p^o)$. Moreover, to each element $\underline{\varphi}_1^m$ in the sequence, there corresponds a real number $\lambda = \lambda_m$ which tends to infinity as $m \to \infty$. Consequently, for the case where ρ_m^o is a constant, the weak solutions of the MIBVP (40) to (42), (48), (55) can be expressed in the following form:

$$\underline{\xi}(t,\underline{x}) = \sum_{k=1}^{\infty} y_k(t)\underline{\varphi}_1^k(x), \qquad \underline{x} \in \Omega_p^o$$

$$\underline{A}(t,\underline{x}) = \sum_{k=1}^{\infty} y_k(t)\underline{\varphi}_2^k(\underline{x}), \qquad \underline{x} \in \Omega_v^o \qquad (75)$$

Substituting (75) into (63) and using (73) and the orthogonality property of the $\underline{\varphi}_1^k$'s, it can be deduced that the modal amplitudes y_k satisfy

$$(d^2y_k/dt^2) + \omega_k^2 y_k = (\rho_m^o)^{-1}(\underline{J}_{v1},\underline{\varphi}_2^k)_2, \qquad k = 1,2, \dots , \qquad (76)$$

where $\lambda_k = \rho_m^o\omega_k^2$.

III. FEEDBACK STABILIZATION

In this section, attention will first be focused on the fundamental question of stabilizability. The problem of determining the stabilizing feedback controls for a plasma will then be discussed with special reference to localized controls and noninteracting modal controls. The application of the results will be illustrated by an example. Throughout this section, results will be developed using the linearized equations (40) to (42), (48), (49), and (55) for a perfectly conducting plasma.

A. Stabilizability

A basic problem in feedback stabilization is to determine whether the controls affect all the unstable motions of the plasma under consideration. A closely related problem is that of controllability, which pertains to the question of whether all the plasma motions are affected by the controls. Clearly, a plasma system is stabilizable only if all its unstable motions are controllable. We shall therefore direct our attention to the controllability problem. Although this problem can be treated from a general viewpoint [14], we shall approach it by studying the effect of the controls on various natural modes of the plasma system. This can be accomplished by using Eq. (76) for the modal amplitudes $y_k(t)$.

Consider the right-hand side of (76) which, in view of (51), can be written as

$$(\rho_m^o)^{-1}(\underline{J}_{v1},\underline{\varphi}_2^k)_2 = (\rho_m^o\mu_o)^{-1} \int_{\Omega_v^o} (\underline{\nabla} \times \underline{B}_{v1}) \cdot \underline{\varphi}_2^k \, d\Omega_v^o \tag{77}$$

The perturbed magnetic field $\underline{B}_{v1}$ can be written as

$$\underline{B}_{v1} = \underline{B}_{v1}^p + \underline{B}_v^c \tag{78}$$

where $\underline{B}_{v1}^p$ and $\underline{B}_v^c$ are the perturbed fields due to the plasma and control currents, respectively. Since $\underline{\nabla} \times \underline{B}_{v1}^p = \underline{0}$, (77) can be rewritten as

$$(\rho_m^o)^{-1}(\underline{J}_{v1},\underline{\varphi}_2^k)_2 = (\rho_m^o\mu_o)^{-1}\int_{\Omega_v^o}(\underline{\nabla}\times\underline{B}_v^c)\cdot\underline{\varphi}_2^k\,d\Omega_v^o$$

$$= (\rho_m^o\mu_o)^{-1}\int_{\Omega_v^o}\nabla\cdot(\underline{B}_v^c\times\underline{\varphi}_2^k)\,d\Omega_v^o - \int_{\Omega_v^o}\underline{B}_v^c\cdot(\underline{\nabla}\times\underline{\varphi}_2^k)\,d\Omega_v^o$$

$$= (\rho_m^o\mu_o)^{-1}\int_{\partial\Omega_v^o}\underline{n}_v\cdot(\underline{B}_v^c\times\underline{\varphi}_2^k)\,d(\partial\Omega_v^o) - \int_{\Omega_v^o}\underline{B}_v^c\cdot(\underline{\nabla}\times\underline{\varphi}_2^k)\,d\Omega_v^o \quad (79)$$

where the vector identity $\underline{\nabla}\cdot(\underline{a}\times\underline{b}) = (\underline{b}\cdot\underline{\nabla}\times\underline{a}) - (\underline{a}\cdot\underline{\nabla}\times\underline{b})$ and the divergence theorem have been used, and $\underline{n}_v$ is the outward unit normal to $\partial\Omega_v^o = \partial\Omega_p^o \cup \partial\Omega$. Using the boundary conditions (71) and (72), the surface integral in (79) simplifies to

$$\int_{\partial\Omega_v^o}\underline{n}_v\cdot(\underline{B}_v^c\times\underline{\bar{\varphi}}_2^k)\,d(\partial\Omega_v^o)$$

$$= \int_{\partial\Omega_p^o}\underline{B}_v^c\cdot(\underline{n}_p\times\underline{\bar{\varphi}}_2^k)\,d(\partial\Omega_p^o) + \int_{\partial\Omega}\underline{B}_v^c\cdot(\underline{n}\times\underline{\bar{\varphi}}_2^k)\,d(\partial\Omega)$$

$$= -\int_{\partial\Omega_p^o}(\underline{B}_v^o\cdot\underline{B}_v^c)(\underline{n}_p\cdot\underline{\bar{\varphi}}_1^k)\,d(\partial\Omega_p^o) \quad (80)$$

Assuming that $\underline{J}_{v1} = \underline{0}$ in the vacuum region immediately adjacent to the plasma surface, we can replace $\underline{B}_v^c$ in (80) by $\underline{\nabla\Phi}_c$, where Φ_c is a magnetic scalar potential. Thus, (80) can be rewritten as

$$\int_{\partial\Omega_v^o}\underline{n}_v\cdot(\underline{B}_v^c\times\underline{\bar{\varphi}}_2^k)\,d(\partial\Omega_v^o)$$

$$= \int_{\partial\Omega_p^o}\Phi_c\underline{B}_v^o\cdot\underline{\nabla}(\underline{n}_p\cdot\underline{\bar{\varphi}}_1^k)\,d(\partial\Omega_p^o) - \int_{\partial\Omega_p^o}\underline{\nabla}\cdot[\Phi_c(\underline{n}_p\cdot\underline{\bar{\varphi}}_1^k)\underline{B}_v^o]\,d(\partial\Omega_p^o) \quad (81)$$

Substituting (81) into (79) leads to

$$(\rho_m^o)^{-1}(\underline{J}_{v1},\underline{\varphi}_2^k)_2 = (\rho_m^o\mu_o)^{-1}\int_{\partial\Omega_p^o}\Phi_c\underline{B}_v^o\cdot\underline{\nabla}(\underline{n}_p\cdot\underline{\bar{\varphi}}_1^k)\,d(\partial\Omega_p^o)$$

$$- \int_{\partial\Omega_p^o}\underline{\nabla}\cdot[\Phi_c(\underline{n}_p\cdot\underline{\bar{\varphi}}_1^k)\underline{B}_v^o]\,d(\partial\Omega_p^o) - \int_{\Omega_v^o}\underline{B}_v^o\cdot(\underline{\nabla}\times\underline{\bar{\varphi}}_2^k)\,d\Omega_v^o \quad (82)$$

Evidently, the k-mode is uncontrollable if $(\underline{J}_{v1}, \underline{\varphi}_2^k)_2 = 0$ for any $\underline{J}_{v1}$. From (82), we observe that there exist two classes of modes for which the right-hand side of (76) vanishes, namely when (i) $\underline{n}_p \cdot \underline{\varphi}_1^k = 0$ and (ii) $\underline{B}_v^o \cdot \underline{\nabla}(\underline{n}_p \cdot \underline{\varphi}_1^k) = 0$, but $(\underline{n}_p \cdot \underline{\varphi}_1^k) \neq 0$. It can be readily verified that under Condition (i) or (ii), $\underline{\nabla} \times \underline{\varphi}_2^k = \underline{0}$ on Ω_v^o so that the last integral in (82) vanishes. Also, if the control $\underline{J}_{v1}$ is effective over the entire surface $\partial\Omega_p^o$, then the second integral in (82) also vanishes. The first class of modes represents the "internal" modes which introduce no displacement perturbations at the plasma surface. The second class of modes represents the so-called "interchange" modes. Neither one of these two types of modes is controllable if the control is effective over the entire plasma surface. The underlying physical reason is that the magnetic field perturbations due to these modes are identically zero in the vacuum region. Consequently, they cannot interact with the control currents.

To overcome this difficulty, we observe that the interchange modes can be controlled using linear interaction of the control field with the equilibrium field by restricting the magnetic field due to the control currents to a portion of the plasma surface. To establish this result, let Γ denote a proper subset of $\partial\Omega_p^o$ over which the controls are effective. Then, for an interchange mode, (82) reduces to

$$(\rho_m^o)^{-1}(\underline{J}_{v1}, \underline{\varphi}_2^k)_2 = -(\rho_m^o \mu_o)^{-1} \int_\Gamma \underline{\nabla} \cdot [\Phi_c(\underline{n}_p \cdot \overline{\underline{\varphi}}_1^k)\underline{B}_v^o]\, d\Gamma$$
$$= -(\rho_m^o \mu_o)^{-1} \int_{C_\Gamma} \Phi_c(\underline{\nu} \cdot \underline{B}_v^o)(\underline{n}_p \cdot \overline{\underline{\varphi}}_1^k)\, d\ell \qquad (83)$$

where C_Γ is the boundary curve of Γ and $\underline{\nu}$ is the outward unit normal associated with C_Γ.

Note that since the magnetic field perturbations due to the interchange modes are identically zero in the vacuum region, restricting the control in this region (by imposing an additional boundary condition) changes neither the natural frequencies nor the mode shapes of the plasma. Hence, ω_k^2 and $\underline{\varphi}_1^k$ in (76) and (83) are still defined by the boundary value problem (69). However, the expansion

(75) is not suitable for evaluating the magnetic vector potential due to the control. Hence, this expansion cannot be used to evaluate the right-hand side of (76) given by (83) which contains the magnetic scalar potential Φ_c. Instead, Φ_c can be evaluated by means of an expansion which is independent of expansion (75) and which also takes into account the restriction of the control field to a portion of the plasma surface [14]. Fortunately, to establish that the interchange modes can be controlled, it is unnecessary to evaluate fully the right-hand side of (76) given by (83). We need only observe that the surface Γ and the control can be chosen in many ways so that the integral in (83) does not vanish. In particular, they can be chosen so that the effect of control on any given interchange mode is maximized in some sense. Certain practical aspects of implementing the restricted control fields are discussed in [14].

At this point we have established that the internal modes are uncontrollable; however, the interchange modes can be controlled by restricting or localizing the control fields to certain portions of the vacuum domain. Obviously, the plasma system is not stabilizable by means of current controls if any internal mode is unstable.

We now turn to the situation in which the control currents are spatially localized to certain subsets of Ω_v^o. In particular, the control current density $\underline{J}_{v1}$ corresponds to a finite number M of currents in Ω_v^o, which can be represented in the form:

$$\underline{J}_{v1}(t,x) = \sum_{j=1}^{M} \underline{W}_j(x) I_j(t) \tag{84}$$

where $\underline{W}_j$ is the spatial distribution of the jth current, I_j. In this case, (76) reduces to

$$d^2y_k/dt^2 + \omega_k^2 y_k = (\rho_m^o)^{-1} \sum_{j=1}^{M} (\underline{W}_j, \underline{\varphi}_2^k)_2 I_j(t), \qquad k = 1,2, \ldots , \tag{85}$$

To determine the conditions for an arbitrary finite number (say N) of unstable modes to be controllable, we extract from (85) the equations corresponding to the unstable modes only and write them in the following form:

$$d^2\underline{Y}/dt^2 + A\underline{Y} = B\underline{v} \tag{86}$$

where $\underline{Y}$ is an N-dimensional vector whose components are the unstable modal amplitudes; $\underline{v}$ is the control vector whose components are I_i; A is an N x N diagonal matrix with its diagonal elements being the negative ω_k^2's, and B is an N x M matrix whose (i,j)th element is $(\rho_m^o)^{-1}(\underline{W}_j,\underline{\varphi}_2^i)_2$.

In order to have controllability of all the unstable modes, the current distributions $\underline{W}_j$ and the number of currents M have to be chosen so that all the modes in (86) are affected by the controls. It is known [15] that this can be achieved if and only if the condition

$$\text{Rank } [B \mid AB \mid A^2B \mid \cdots \mid A^{N-1}B] = N \tag{87}$$

is satisfied. Assuming that both the internal and interchange modes are stable and the number of control currents is equal to the number of unstable modes, condition (87) can always be satisfied by choosing an appropriate set of W_j's. In fact, it can be shown [14] that condition (87) can be satisfied with a minimum number of control currents corresponding to the number of repeated eigenvalues of A. For the cylindrical plasma column considered in Section II, this minimum number is equal to two.

B. Stabilizing Feedback Controls

Having established that all the plasma motions which perturb the plasma surface can be controlled in principle, we proceed to determine the form of feedback controls for stabilization. This can be accomplished by considering the following functional:

$$Q(\underline{u},\dot{\underline{u}}) = (\rho_m^o\dot{\underline{u}}_1,\dot{\underline{u}}_1)_1 + a(\underline{u},\underline{u}) - (\underline{u}_2,\mu_o\underline{J}_{vl})_2 \tag{88}$$

where $\underline{u} = (\underline{u}_1,\underline{u}_2) = (\underline{\xi},\underline{A})$ and $\dot{\underline{u}} = \partial\underline{u}/\partial t$, and a is defined by (61). For $\underline{J}_{vl} = 0$, $Q(\underline{u},\dot{\underline{u}})$ is proportional to the total (kinetic plus potential) energy of the plasma system. To stabilize a given equilibrium of the system locally by means of feedback controls, it is sufficient to choose $\underline{J}_{vl}$ as a linear function of $\underline{u}$ such that Q is a constant of motion for the system and that the quadratic form

$$\Pi_{\underline{J}_{v1}}(\underline{u},\underline{u}) = a(\underline{u},\underline{u}) - (\underline{u}_2,\mu_o\underline{J}_{v1}(\underline{u}))_2 \quad (89)$$

is positive definite, i.e., there exists a real number $c > 0$ such that $\Pi_{\underline{J}_{v1}}(\underline{u},\underline{u}) \geq c\|\underline{u}\|_H^2$ for all $\underline{u} \in V_o$, where $\|\underline{u}\|_H$ is the norm of $\underline{u}$ induced by the inner product (58). A systematic way for examining the sign definiteness of $\Pi_{\underline{J}_{v1}}$ is to consider the problem of minimizing $\Pi_{\underline{J}_{v1}}(\underline{u},\underline{u})$ over all $\underline{u} \in V_o$ such that $(\underline{u}_1,\underline{u}_1)_1 = 1$. The solution $\tilde{\underline{u}}$ to this problem can be characterized by the boundary value problem

$$\Pi_{\underline{J}_{v1}}(\tilde{\underline{u}},\underline{\psi}) = \lambda_{\underline{J}_{v1}}(\tilde{\underline{u}}_1,\underline{\psi}_1)_1 \qquad \text{for all } \underline{\psi} \in V_o \quad (90)$$

where $\lambda_{\underline{J}_{v1}} = \Pi_{\underline{J}_{v1}}(\tilde{\underline{u}},\tilde{\underline{u}})$. For stability, the linear feedback control $\underline{J}_{v1}$ should be chosen so that $\Pi_{\underline{J}_{v1}}$ is positive. Note that $\tilde{\underline{u}}$ corresponds to a weak solution to the boundary value problem similar to that defined by (69) to (72) except for $\underline{\nabla} \times \underline{\nabla} \times \underline{u}_2 = \underline{J}_{v1}(\underline{u})$ on Ω_v^o. A simple form of feedback control is given by

$$\underline{J}_{v1}(\underline{u}) = -gG\underline{u}_2 \quad (91)$$

where G is any positive operator having the property that $G\underline{u}_2 \in L_2^3(\Omega_v^o)$ and $(\underline{u}_2,G\underline{u}_2)_2 \geq c(\underline{u}_2,\underline{u}_2)_2$ for all $\underline{u}_2$ such that $(\underline{u}_1,\underline{u}_2) \in V_o$ and for some $c > 0$; and g is a sufficiently large positive constant. This form of control corresponds to feeding back only the magnetic vector potential $\underline{A}$ in the vacuum region. Unfortunately, $\underline{A}$ is a nonphysical quantity which cannot be readily determined from the observable physical quantities.

We now consider the important case where the plasma system has a finite number of controllable unstable modes, and the control corresponds to a finite number of localized currents in Ω_v^o having a representation of the form (84). The feedback stabilization problem reduces to finding a stabilization feedback control for system (86) such that the stability of the originally stable modes remains unaltered.

First, let the number of control currents be equal to the number of unstable modes. In this case, a stabilizing feedback control is given by

$$\underline{v} = -g_1 B^T \underline{Y} \tag{92}$$

where $(\cdot)^T$ denotes transposition. The positive constant g_1 is chosen to be sufficiently large so that the symmetric matrix $P = (A + g_1 BB^T)$ is positive definite. This can be ensured by choosing g_1 such that $\omega^2_{min} + \lambda_{min} g_1 > 0$, where ω^2_{min} and λ_{min} are the minimum eigenvalues of A and BB^T, respectively. The stability of the trivial solution of (86) with feedback control (92) follows from the fact that $V(\underline{Y},\dot{\underline{Y}}) = (\dot{\underline{Y}}^T\dot{\underline{Y}} + \underline{Y}^T P\underline{Y})/2$ is a Lyapunov function for the system with $dV/dt = 0$ along any trajectory. Since all the nontrivial solutions of the stabilized feedback system are oscillatory and the originally stable modes are driven by the same controls, the possibility of resonance exists. This can be avoided by requiring that the eigenvalues of P be distinct from ω_i^2 corresponding to all originally stable modes. On the other hand, the resonance phenomenon can be avoided by using a feedback control of the form

$$\underline{v} = -g_1 B^T \underline{Y} - g_2 B^T \dot{\underline{Y}} \tag{93}$$

where the gain g_1 is as previously chosen and g_2 is positive. It is not difficult to show that the stability of the originally stable modes is unaffected by a feedback control in the form (92) or (93).

In the foregoing development we have assumed that the number of control currents is equal to the number of unstable modes. Identical results can be obtained for the case with two control currents only. Also, the effectiveness of the foregoing controls depends partially on the accuracy in estimating the unstable modal amplitudes from the sensor output data. Due to the localized nature of the sensors, they should be placed in such a way that the unstable modes are observable and the signal-to-noise ratio is maximized.

Finally, in the physical implementation of the stabilizing feedback controls, the sensors should be chosen such that all the unstable modal amplitudes can be estimated from the sensor output data or they are all observable. In the case of the interchange modes, the modal amplitudes are unobservable by means of magnetic probes placed in the vacuum region. Thus, nonmagnetic sensors such as capacitative or optical probes have to be used for feedback stabilization.

C. Noninteracting Modal Controls

The feedback controls (92) given in the previous section may lead to linear interaction of the stabilized modes. In other words, the differential equations for the modal amplitudes of the feedback system are not necessarily decoupled, since the matrix $(A + g_1 BB^T)$ may not be diagonal. In the absence of control ($g_1 = 0$), there is no interaction between the unstable modes (A is diagonal). It is of interest to find stabilizing controls which preserve this property of the uncontrolled system. This can be accomplished by means of a feedback control of the form

$$\underline{v} = -B^{-1}K\underline{Y} \quad (94)$$

where K is an $N \times N$ diagonal matrix to be determined. It has been shown that the control current distributions can always be chosen so that B^{-1} exists [16]. Substituting (94) into (86) leads to

$$d^2\underline{Y}/dt^2 + (A + K)\underline{Y} = \underline{0} \quad (95)$$

Since $(A + K)$ is diagonal, each mode can be stabilized individually by choosing the corresponding diagonal element of K, which implies feeding back the amplitude of that mode only.

D. Application to Cylindrical Plasma Column

To illustrate the application of some of the foregoing results, we again consider the cylindrical plasma column described in Section II.A with length $\ell = 2\pi$. The control $\underline{J}_{v1}$ is taken to be a current sheet

$$\underline{g}(\underline{x}) = \mu_o \underline{J}_{v1}(\underline{x}) = g_\theta(\theta,z)\underline{e}_\theta + g_z(\theta,z)\underline{e}_z \tag{96}$$

which flows on the cylindrical surface

$$\Gamma_c = \{\underline{x} = (r,\theta,z): r + R,\ 0 < \theta < 2\pi,\ 0 < z < 2\pi\}$$

where R is the radius at which the surface is located. The functions g_θ and g_z satisfy the condition $\underline{\nabla}_{\theta,z} \cdot \underline{g} = R^{-1}(\partial g_\theta/\partial\theta) + (\partial g_z/\partial z) = 0$, but they can be specified arbitrarily otherwise. In particular, they can be specified to be zero on a subset of Γ_c.

With this type of control, the boundary value problem (90) can be reformulated as

$$\begin{aligned}
&\underline{F}(\underline{u}_1) = -\lambda\underline{u}_1, \qquad \underline{x} \in \Omega_p^o \\
&\nabla^2\Psi_i = 0, \qquad \underline{x} \in \Sigma_i,\ i = 1,2, \\
&\underline{n}_c \cdot \underline{\nabla}(\Psi_2 - \Psi_1) = 0, \qquad \underline{n}_c \times \underline{\nabla}(\Psi_2 - \Psi_1) = \underline{g}, \qquad \underline{x} \in \Gamma_c \\
&\underline{n}_p \cdot \underline{\nabla}\Psi_1 = \underline{B}_v^o \cdot \underline{\nabla}(\underline{n}_p \cdot \underline{u}_1), \qquad \underline{x} \in \partial\Omega_p^o \\
&-\mu_o\gamma p^o(\underline{\nabla} \cdot \underline{u}_1) + \underline{B}_p^o \cdot \underline{\nabla} \times (\underline{u}_1 \times \underline{B}_p^o) + \underline{u}_1 \cdot \underline{\nabla}[(|\underline{B}_p^o|^2 - |\underline{B}_v^o|^2)/2] \\
&\qquad - \underline{B}_v^o \cdot \underline{\nabla}\Psi_1 = 0, \qquad \underline{x} \in \partial\Omega_p^o \\
&\underline{n} \cdot \underline{\nabla}\Psi_2 = 0, \qquad \underline{x} \in \partial\Omega
\end{aligned} \tag{97}$$

where $\underline{n}_c$ is the outward unit normal at Γ_c, and Ψ_1 and Ψ_2 are scalar magnetic potentials defined on the regions

$$\Sigma_1 = \{\underline{x} \in \Omega_v^o: R_1 < r < R\}$$

$$\Sigma_2 = \{\underline{x} \in \Omega_v^o: R < r < R_2\}$$

The normalization $(\underline{u}_1,\underline{u}_1)_1 = 1$ is assumed. For every $\underline{g}$, this problem admits an infinite sequence of nontrivial solutions. To each solution there corresponds a value of the parameter λ. For stability, $\underline{g}$ should be chosen so that the smallest value of λ is positive.

The functions $\underline{u}_1$, $\underline{\Psi}_1$, $\underline{\Psi}_2$, and $\underline{g}$ in (97) are expressible [17] as

$$\underline{u}_1(\underline{x}) = \sum_{k=-\infty}^{\infty} \sum_{m=-\infty}^{\infty} c_1^{k,m} \underline{h}_1^{k,m}(r) \exp(jm\theta + jkz), \qquad \underline{x} \in \Omega_p^o$$

$$\Psi_1(\underline{x}) = \sum_{k=-\infty}^{\infty} \sum_{m=-\infty}^{\infty} \psi_1^{k,m}(r) \exp(jm\theta + jkz), \qquad \underline{x} \in \Sigma_1 \tag{98}$$

$$\Psi_2(\underline{x}) = \sum_{k=-\infty}^{\infty} \sum_{m=-\infty}^{\infty} d_2^{k,m} \psi_2^{k,m}(r) \exp(jm\theta + jkz), \qquad \underline{x} \in \Sigma_2$$

$$\underline{g}(\underline{x}) = \sum_{k=-\infty}^{\infty} \sum_{m=-\infty}^{\infty} g^{k,m}(0, jk, -jm/R)^T \exp(jm\theta + jkz), \qquad \underline{x} \in \Gamma_c$$

where

$$\underline{h}_1^{k,m}(r) = [-j(k^2 - \omega_s^2) I_m'(\alpha r)/k\alpha,\ m(k^2 - \omega_s^2) I_m(\alpha r)/k\alpha^2 r,\ I_m(\alpha r)]^T \Big/ \|h_1^{k,m}\|$$

$$\psi_1^{k,m}(r) = d_1^{k,m} I_m(kr) + q_1^{k,m} K_m(kr) \tag{99}$$

$$\psi_2^{k,m}(r) = K_m'(kR_2) I_m(kr) - I_m'(kR_2) K_m(kr)$$

with

$$\alpha^2 = (k^2 - \omega_s^2)(k^2 - \omega_H^2)/(k^2 - \omega_s^2 - \omega_H^2), \qquad \rho_m^o \omega^2 = \lambda,$$

$$\omega_s^2 = \omega_{k,m}^2 / c_s^2, \qquad \omega_H^2 = \omega_{k,m}^2 / c_H^2$$

$$c_s^2 = \gamma p^o / \rho_m^o, \qquad c_H^2 = (B_\theta^o b_i)/\mu_o \rho_m^o$$

$$\|\underline{h}_1^{k,m}\|^2 = 4\pi^2 \int_0^R \underline{h}_1^{k,m}(r) \cdot \overline{\underline{h}}_1^{k,m}(r)\, dr$$

I_m and K_m are the Bessel functions of order m of the first and second kind, respectively, and $'$ denotes differentiation with respect to r. The constants $d_1^{k,m}$, $d_2^{k,m}$, and $q_1^{k,m}$ are given by

$$\begin{bmatrix} d_1^{k,m} \\ \hline q_1^{k,m} \end{bmatrix} = \left[\begin{array}{c|c} q_{11}^{k,m} & q_{12}^{k,m} \\ \hline q_{21}^{k,m} & q_{22}^{k,m} \end{array}\right] \begin{bmatrix} g^{k,m} \\ \hline jB_\theta^o(b_e + m/kR_1)(\underline{n}_p \cdot \underline{h}_1^{k,m}) \end{bmatrix}, \tag{100}$$

$$d_2^{k,m} = k[d_1^{k,m} I_m'(kR) + q_1^{k,m} K_m'(kR)]/\psi_2'(R)$$

where

$$q_{11}^{k,m} = k\psi_2^{k,m}(R) I_m'(kR) - \psi_2'(R) I_m(kR)$$

$$q_{12}^{k,m} = k\psi_2^{k,m}(R) K_m'(kR) - \psi_2'(R) K_m(kR)$$

$$q_{21}^{k,m} = I_m'(kR_1)$$

$$q_{22}^{k,m} = K_m'(kR_1)$$

$$\psi_2'(R) = d\psi_2^{k,m}(r)/dr\Big|_{r=R}$$

Note that (98) to (100) determine the solution to the boundary value problem (97) in terms of $\underline{g}$. Substituting this solution into the pressure balance condition in (97), we obtain the equation

$$\frac{\alpha I_m(\alpha R_1)}{I_m'(\alpha R_1)} \left\{ \frac{\mu_o \rho_m^o \omega_{k,m}^2}{(k^2 - \omega_s^2)} - (b_i B_\theta^o)^2 \right\} (\underline{n}_p \cdot \underline{h}_1^{k,m}) - (B_\theta^o)^2 (\underline{n}_p \cdot \underline{h}_1^{k,m})/R_1$$

$$= jB_\theta^o (kb_e + m/R_1)[d_1^{k,m} I_m(kR_1) + q_1^{k,m} K_m(kR_1)] \tag{101}$$

whose roots $\omega_{k,m}^2$ are the eigenvalues depending on $\underline{g}$. If a specific feedback control is substituted into (100), $d_1^{k,m}$ and $q_1^{k,m}$ are determined in terms of the plasma surface displacement $(\underline{n}_p \cdot \underline{h}_1^{k,m})$. If (100) is then substituted into (101), a dispersion relation is obtained. We note that the foregoing expressions are valid only for nonzero k and arbitrary m. Similar results can be obtained for $k = 0$ and arbitrary m.

Although the method just outlined can be used to determine the eigenvalues in terms of any arbitrary physically realizable control of the form (96), it leads to Eq. (101), which can only be solved numerically. Thus, no general explicit expressions for the controls can be derived. On the other hand, explicit expressions [such as (94)] can be derived by the method described in Section III.C.

The controls (94) can be determined by evaluating the matrix B in terms of the eigenfunctions which correspond to a given equilibrium. We assume that the equilibrium described earlier is modified to include an internal helical current $\underline{J}_o$ with an arbitrary radial distribution and an axial magnetic field which is much larger than the azimuthal field. Furthermore, we simulate the toroidal geometry by letting the length of the column be $2\pi R$, where R denotes the major radius of the torus.

The eigenfunctions corresponding to this equilibrium have the form

$$\underline{\varphi}_1^{m,n}(r)\exp(jm\theta - jnz/R) \quad \text{in the plasma} \tag{102}$$

$$\underline{\varphi}_2^{m,n}(r)\exp(jm\theta - jnz/R) \quad \text{in the vacuum} \tag{103}$$

where m and n are the wave numbers in the θ and z directions, respectively. It is known [18] that there are equilibrium current distributions for which only the $m = \pm 1$ modes are unstable. For these unstable modes, $\underline{\varphi}_1^{m,n}(r)$ and $\underline{\varphi}_2^{m,n}(r)$ in (102) and (103) are given approximately by

$$\underline{\varphi}_1^{m,n}(r) = (c_1^{m,n}, jc_1^{m,n}, 0)^T \quad \text{in the plasma}$$

$$\underline{\varphi}_2^{m,n}(r) = \begin{bmatrix} -jc_2^{m,n}\partial u^{m,n}(r)/\partial r - jmd_2^{m,n}v^{m,n}(r)/r \\ mc_2^{m,n}u^{m,n}(r)/r - d_2^{m,n}\partial v^{m,n}(r)/\partial r \\ -nc_2^{m,n}u^{m,n}(r)/R \end{bmatrix} \quad \text{in the vacuum}$$

where

$$u^{m,n}(r) = K_m(kR_2)I_m(kr) - I_m(kR_2)K_m(kr)$$

$$v^{m,n}(r) = K_m'(kR_2)I_m(kr) - I_m'(kR_2)K_m(kr)$$

$$k = -n/R, \qquad c_1^{m,n} = (4\pi^2 R_1^2 R)^{-1/2}, \qquad c_2^{m,n} = c_1^{m,n}B_\theta^o/ku^{m,n}(R_1) \tag{104}$$

$$d_2^{m,n} = -c_1^{m,n}B_\theta^o R(m - nq)/nR_1 v'(R_1), \qquad q = B_z^o R_1/B_\theta^o R$$

and

$$v'(R_1) = dv^{m,n}(r)/dr\Big|_{r=R_1}$$

The eigenvalues $\omega^2_{m,n}$ which correspond to these eigenfunctions are given approximately by

$$\omega^2_{m,n} = \frac{-2(B^o_\theta)^2}{\mu_o \rho^o_m R_1^2} [(1 - nq) - (1 - nq)^2/(1 - R_1^2/R_2^2)] \qquad (105)$$

where q is as in (104) a ratio related to the pitch of the equilibrium magnetic lines at the plasma surface. We suppose that $1/2 < q < 1$ so that only the $m = n = \pm 1$ modes are unstable. For convenience, we take the real and imaginary parts of the complex functions in (102) and (103). Thus, we obtain two unstable real modes

$$\left.\begin{aligned} \underline{\psi}^1_1(r) &= [c_1^{m,n} \sin(\theta - qz),\ c_1^{m,n} \cos(\theta - qz),\ 0]^T \\ \underline{\psi}^2_1(r) &= [c_1^{m,n} \cos(\theta - qz),\ -c_1^{m,n} \sin(\theta - qz),\ 0]^T \end{aligned}\right\} \text{ in the plasma}$$

and in the vacuum

$$\underline{\psi}^1_2(r) = \begin{bmatrix} [-c_2^{m,n}u'(r) - md_2^{m,n}v(r)]r^{-1} \cos(\theta - qz) \\ [mc_2^{m,n}r^{-1}u(r) - d_2v'(r)] \sin(\theta - qz) \\ -nR^{-1}c_2^{m,n}u(r) \sin(\theta - qz) \end{bmatrix}$$

$$\underline{\psi}^2_2(r) = \begin{bmatrix} [c_2^{m,n}u'(r) - r^{-1}md_2^{m,n}v(r)] \sin(\theta - qz) \\ [mr^{-1}c_2^{m,n}u(r) - d_2^{m,n}v'(r)] \cos(\theta - qz) \\ -nR^{-1}c_2^{m,n}u(r) \sin(\theta - qz) \end{bmatrix}$$

where the functions $v^{m,n}$ and $u^{m,n}$ are defined in (104) (the superscripts m and n have been omitted for brevity).

We suppose that the control currents flow in two helical strips, and that each strip follows a crest of one of the modes. The axis of the strips can be described by the equations

$$\theta_1 - qz_1 = \pi/2$$

$$\theta_2 - qz_2 = 0$$

where $\underline{x}_1 = (\alpha, \theta_1, z_1)$ and $\underline{x}_2 = (\alpha, \theta_2, z_2)$ are points on the axis of the first and second strips, respectively, and α denotes the radius at which the strips are located. The matrix B in (92) can be expressed approximately as

$$B = \beta \begin{bmatrix} 1 & 0 \\ 0 & 1 \end{bmatrix}$$

where

$$\beta = \Delta[c_2 u(\alpha)(qm\alpha^{-1} - nR^{-1}) - q\ d_2 v'(\alpha)]/(1 + q^2)^{1/2}$$

and Δ is the surface area of the strips. The two-dimensional control vector consisting of the currents I_1 and I_2 can be obtained by substituting this expression for B into (92). A sufficient condition for stability is

$$g_1 > -\omega^2_{min}/\lambda_{min}$$

where

$$\omega^2_{min} = \frac{-(B^o_\theta)^2}{2\mu_o\rho^o_m R_1^2}(1 - R_1^2/R_2^2)$$

and $\lambda_{min} = \beta^2$. It is also possible to obtain the following estimates for the required magnitudes of the control currents in terms of the magnitudes of the modal amplitudes

$$(\underline{v}^T\underline{v}) \simeq \omega^2_{min}(\underline{Y}^T\underline{Y})^{1/2}/|\beta|$$

IV. CONCLUSIONS

In this chapter, we have considered certain fundamental aspects of the feedback stabilization problem for plasmas describable by the single-fluid MHD equations. Since the main results are developed using the linearized MHD equations for a perfectly conducting plasma with sharp plasma-vacuum interface, they are valid only when the plasma perturbations about a given equilibrium and the resistive effects are sufficiently small. A more complete study necessitates the use of nonlinear models such as the one given by (11) to (13) with boundary conditions (15) and (16) to (19) [or (20), (21)]. One may investigate the possibility of using linear and nonlinear feedback controls to achieve global stability.

An important aspect of the plasma stabilization problem which is not considered here is the problem of optimal feedback stabilization, namely that it is desirable to choose a feedback control belonging to a given class of stabilizing feedback controls such that certain performance criteria are optimized. For example, in high-current plasma devices such as the theta-pinch and the Tokamak, it is important to seek feedback controls which minimize the control energy. Also, in the case of a resistive plasma describable by (11) to (13) with boundary conditions (15), (20), and (21), one may minimize a performance functional which involves the control energy and penalizes high values of the plasma density near the inner solid wall. Thus, the optimal control tends to confine the plasma toward the central portion of the plasma domain. Some results in this area have been obtained recently. They will be presented elsewhere by the authors.

ACKNOWLEDGMENTS

The work of the first author was performed during his visit with the Institut de Recherche d'Informatique et d'Automatique, Rocquencourt, Le Chesnay, France, in 1975. Their hospitalities are greatly appreciated. Also, he wishes to acknowledge the support provided by an AFOSR Grant No. 74-2662.

REFERENCES

1. T. K. Chu and H. W. Hendel (eds.), Feedback and Dynamic Control of Plasmas, American Institute of Physics, New York, 1970.

2. K. I. Thomassen, Feedback stabilization in plasmas: A review, Nucl. Fusion, 11: 175-186 (1971).

3. P. K. C. Wang, Feedback stabilization of distributive systems with applications to plasma stabilization, in Instability of Continuous Systems (H. Leipholz, ed.), Springer-Verlag, Berlin, 1971, pp. 228-237.

4. Progress Report, LASL Controlled Thermonuclear Research Program, Los Alamos Scientific Laboratory, Los Alamos, New Mexico, Report no. LA-5656-PR, July 1974.

5. Yu. P. Ladikov and Yu. I. Samoilenko, Magnetic feedback stabilization in a Tokamak, Sov. Phys.-Tech. Phys., vol. 17, no. 10, 1644-1650 (1973).

6. N. A. Krall and A. W. Trivelpiece, Principles of Plasma Physics, McGraw-Hill, New York, 1973.

7. F. F. Chen, Electric probes, in Plasma Diagnostic Techniques (R. H. Huddlestone and S. L. Leonard, eds.), Academic, New York, 1965, pp. 113-199.

8. R. H. Lovberg, Magnetic probes, in Plasma Diagnostic Techniques (R. H. Huddlestone and S. L. Leonard, eds.), Academic, New York, 1965, pp. 69-112.

9. H. Grad, A. Kadish, and D. C. Stevens, A free boundary Tokamak equilibrium, Comm. Pure Appl. Math., XXVII: 39-57 (1974).

10. B. K. Harrison, R. W. Bass, et al., Uniqueness of the Topolotron design relative to structural stability, Proc. Utah Acad. Sci., Arts Lett., vol. 50, no. 2, 19-26 (1973).

11. S. Lefschetz, Differential Equations; Geometric Theory, Interscience, New York, 1957.

12. I. B. Bernstein, et al., An energy principle for hydromagnetic stability problems, Proc. Royal Soc., 244: 17-40 (1958).

13. J. L. Lions, Optimal Control of Systems Governed by Partial Differential Equations, Springer-Verlag, Berlin, 1971.

14. G. Rodriguez, "Feedback stabilization of plasmas," School of Engineering and Applied Science, Univ. of California, Los Angeles, Report no. UCLA-ENG-7401, Jan. 1974.

15. L. A. Zadeh and C. A. Desoer, Linear System Theory, McGraw-Hill, New York, 1963.

16. G. Rodriguez and P. K. C. Wang, "Modal feedback stabilization of hydromagnetic equilibria," School of Engineering and Applied Science, Univ. of California, Los Angeles, Report no. UCLA-ENG-7285, Sept. 1972.

17. R. J. Taylor, The influence of an axial magnetic field on the stability of a constricted gas discharge, Proc. Phys. Soc. (London), B-70: 1049-1063 (1957).

18. V. D. Shafranov, Hydromagnetic stability of a current-carrying pinch in a strong magnetic field, Sov. Phys.-Tech. Phys., vol. 15, no. 2, 175-183 (1970).

Chapter 9

IDENTIFICATION OF PETROLEUM RESERVOIR PROPERTIES

John H. Seinfeld
Department of Chemical Engineering
California Institute of Technology
Pasadena, California

Wen H. Chen*
Gulf Research and Development Company
Pittsburgh, Pennsylvania

I. INTRODUCTION

Estimation of subsurface properties of the earth gives rise to a number of challenging distributed parameter identification problems. One such problem is the identification of petroleum reservoir properties on the basis of data obtained during the course of producing oil. The dynamic behavior of a petroleum reservoir, i.e., the time-dependent flow of the fluids contained within the reservoir, is described in general by a coupled set of nonlinear partial differential

*Current affiliation: Production Research Department, Gulf Science and Technology Company, Pittsburgh, Pennsylvania.

equations. The physical properties of the reservoir are represented by parameters that appear in the partial differential equations. These properties, such as the porosity of the rock which contains the fluids, are not generally uniform throughout the reservoir. Thus, the reservoir physical properties to be identified will vary with their position in the reservoir. In addition, there is a class of parameters potentially to be identified which represent the ease with which the various fluids flow. These parameters are nonlinear functions of the state variables of the reservoir. In short, then, the identification of petroleum reservoir properties leads to distributed parameter identification problems in which the parameters to be estimated depend on the spatial variables of the system and, in some cases, on the state variables themselves.

II. MATHEMATICAL DESCRIPTION OF PETROLEUM RESERVOIRS

A cross section and an aerial view of an idealized petroleum reservoir are shown in Figure 1. In this reservoir gas, oil, and liquid

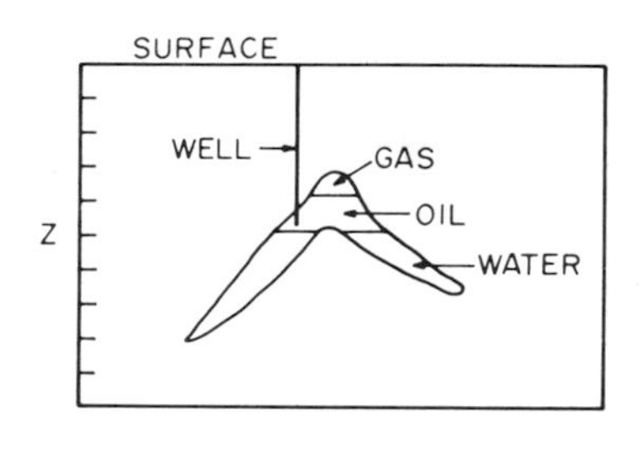

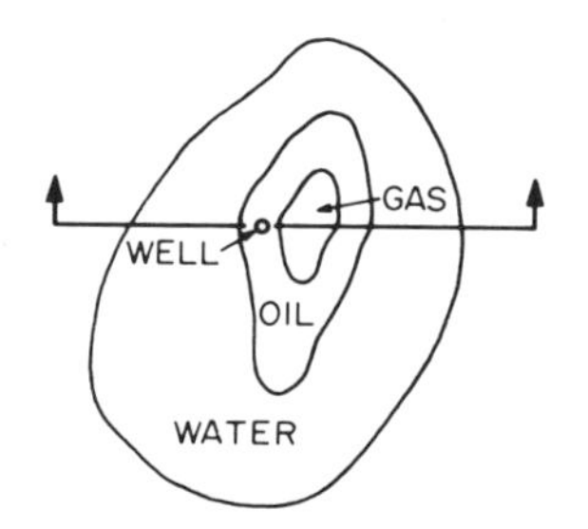

FIG. 1. A typical petroleum reservoir.

are trapped in an underground structure. Both the top and bottom of the structure are bounded by impermeable rock. The fluids are contained in porous rock and are usually segregated by gravity. The pressure in the formation is often several thousand pounds per square inch. When oil is withdrawn from the well, the pressure is reduced and the fluids expand. Gas may come out of solution with the water interface rising and the gas interface falling. In the aerial view the oil region becomes smaller as the gas and water displace the oil.

Consequently, the flow process is complicated. It involves one, two, or three phases simultaneously flowing in three dimensions. Because of the large amount of surface area in porous media, capillary forces may be significant. To solve the flow problem, principles of thermodynamics in addition to material and momentum balances are required. For some recovery processes where the temperature changes, an energy balance is also required. Three processes are naturally available to drive the oil through the reservoir to the well. They are: volume expansion of reservoir fluids and rock, gas coming out of solution, and water influx from surrounding rock. In one reservoir, any combination of these may be present.

Conceptually, the water-oil interface may assume either of the between the oil and water occurs. In this zone the water saturation

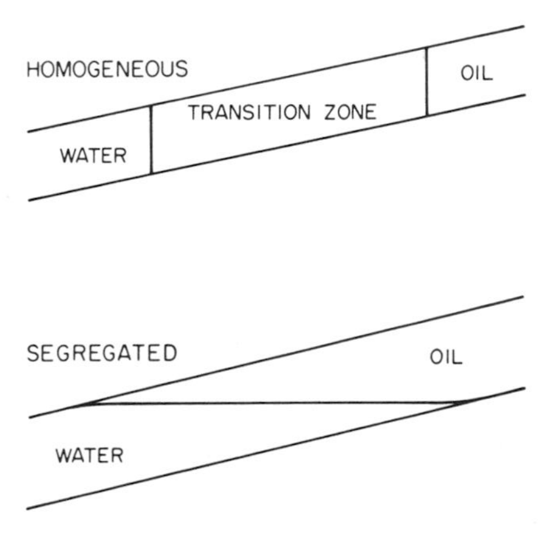

FIG. 2. Two-phase flow.

varies from 1.0 to 0, and as the water moves up the formation, the transition zone becomes larger. In segregated flow, a well-defined water-oil interface is assumed to exist. On each side of the interface the single phase can be considered to be flowing separately, and flow equations appropriate for a single phase can be used. For the purposes of the present report we confine our attention to water-oil systems. Our results can be readily extended to include gas-containing reservoirs.

The mathematical model for the reservoir consists of a set of partial differential equations (PDEs) obtained from material balances on the water and oil phases. Data describing the fluids and the formation enter the equations as parameters. The initial or discovery conditions of pressure and oil and water saturation must be known, and the production or injection schedule for each well as a function of time must be specified. The model predicts the oil and water saturation and pressure distribution in the reservoir as a function of time.

A material balance on a differential volume element of the porous medium yields the following equations for the oil and water phases (in a region of homogeneous flow):

$$\frac{\partial}{\partial t}(\rho_o S_o \varphi) = -\nabla \cdot \rho_o \underline{v}_o - q_o \tag{1}$$

$$\frac{\partial}{\partial t}(\rho_w S_w \varphi) = -\nabla \cdot \rho_w \underline{v}_w - q_w \tag{2}$$

where

ρ_o, ρ_w = mass densities of oil and water

S_o, S_w = oil and water saturations (fraction of available pore volume filled with oil and water; i.e., $S_o + S_w = 1$)

φ = porosity (volume fraction of voids in the porous medium)

q_o, q_w = rate of removal (production) of oil and water from wells

$\underline{v}_o$, $\underline{v}_w$ = flow velocities of oil and water

The flow velocities can be obtained in terms of the pressure p by Darcy's law. For the flow of a single phase (say oil) through a porous medium, Darcy's law is [1]

$$\underline{v} = -\frac{1}{\mu}\underline{K} \cdot (\nabla p + \rho \underline{g}) \tag{3}$$

where μ is the viscosity of the fluid, $\underline{g}$ is the gravitational acceleration vector, and $\underline{K}$ is a second-order tensor (3 x 3 matrix) of parameters called permeabilities. The permeabilities are not in principle fundamental properties of the rock (as, for example, is the porosity); rather they are defined by Eq. (3). Thus, if one measures the flow induced by a known pressure change, ∇p, then the appropriate permeability is that value which satisfies (3). Since (3) is generally taken as a valid relationship for flow in porous media, $\underline{K}$ is often considered as a fundamental property of the rock. $\underline{K}$ is usually taken to be diagonal with elements k_x, k_y, k_z.

Because we are interested in the simultaneous flow of two phases (oil and water), Darcy's law must be extended to describe the flow of each phase. The velocities of oil and water in two-phase flow are given by

$$\underline{v}_o = -\frac{k_{ro}}{\mu_o}\underline{K} \cdot (\nabla p + \rho_o \underline{g}) \tag{4}$$

$$\underline{v}_w = -\frac{k_{rw}}{\mu_w}\underline{K} \cdot (\nabla p + \rho_w \underline{g}) \tag{5}$$

where the pressure in both phases at a point is assumed to be the same; thus, capillary forces have been neglected.†

The parameters k_{ro} and k_{rw} in Eqs. (4) and (5) are the relative permeabilities for oil and water, respectively. These relative permeabilities are functions of saturation and are usually found empirically in laboratory studies of oil-water flow in samples of the porous medium. Typical curves of k_{ro} and k_{rw} as functions of S_o are shown in Figure 3. The relative permeability k_{ro} is defined as the

---

†In general, p_o and p_w differ because of surface tension between the two phases. For the purpose of introducing the equations governing petroleum reservoirs, we will neglect here the difference between p_o and p_w, the so-called capillary pressure. In Section IV we will present the basic equations for an oil-water reservoir in which capillary pressure effects are included.

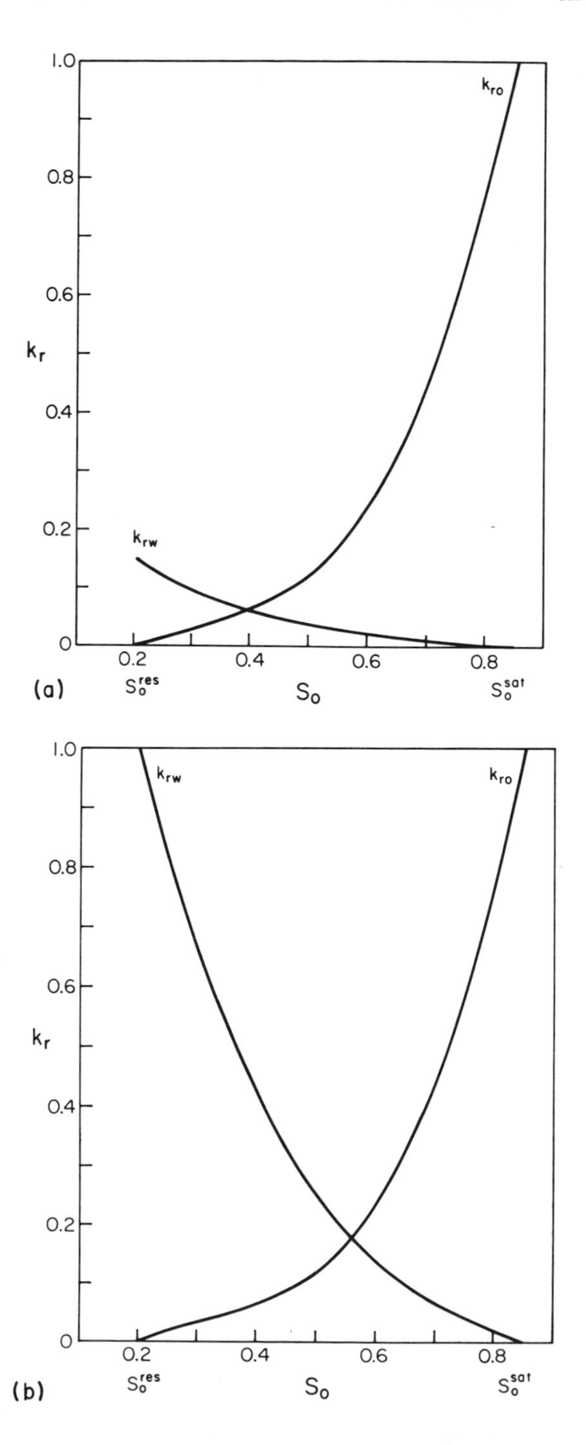
1.0
0.8
0.6
k_r
0.4
0.2
0
k_{ro}
k_{rw}
0.2
0.4
0.6
0.8
(a)
S_o^{res}
S_o
S_o^{sat}
1.0
0.8
0.6
k_r
0.4
0.2
0
k_{rw}
k_{ro}
0.2
0.4
0.6
0.8
(b)
S_o^{res}
S_o
S_o^{sat}

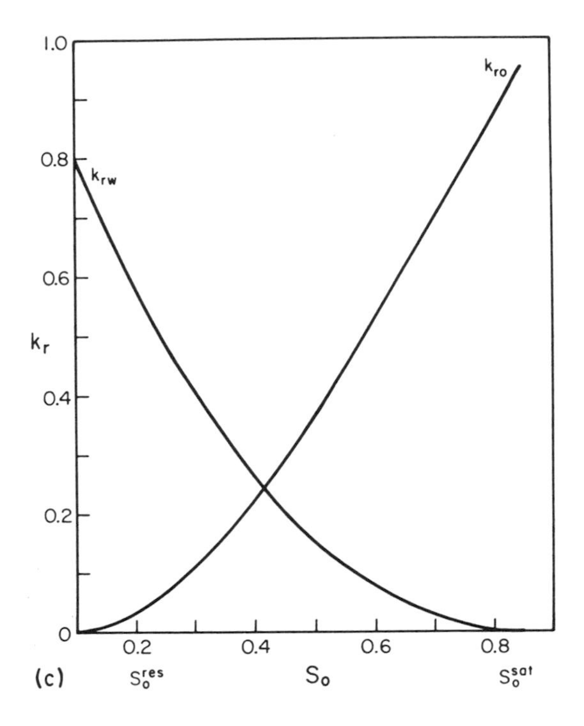

FIG. 3. (Left and above) Relative permeabilities: (a) Nolen and Berry [19]; (b) the curves of [19] with k_{rw} divided by 0.15; (c) Hagoort [18].

ratio of the permeability for the flow of oil in a two-phase system to the permeability for flow in a single-phase system. Outside of a two-phase region and for segregated flow, Eqs. (4) and (5) can be used, but the relative permeability is 1.0 for the phase present and zero for the other. For a two-phase system, therefore, the permeability for each component is different and depends on the saturation conditions. Thus, if $k_{ro} = 0.8$ at $S_o = 0.6$, the permeability is 80% of what it would be if only oil were present. It is seen from Figure 3 that $k_{ro} = 0$ at $S_o = S_o^{res}$, when S_o^{res} is the residual oil saturation that remains in the porous medium because of capillary forces. Likewise, S_w^c represents the water saturation that remains in the formation. (S_w^c is called the <u>connate</u> water saturation.)

It may be difficult to obtain a core sample from the reservoir in which to determine the relative permeability/saturation relationships such as shown in Figure 3. As an alternative, relative permeability curves determined for rock of similar composition are used. It has generally not been possible to judge to what extent the selected relative permeability curves are applicable to the given reservoir.

We now develop the basic equations governing the simultaneous flow of oil and water in a porous medium. For simplicity, let us assume that the flow is horizontal.† Equations (1), (2), (4), and (5) yield

$$\frac{\partial}{\partial t}(\rho_o S_o \varphi) = \nabla \cdot \left[\frac{\rho_o k_{ro}}{\mu_o} \underline{K} \cdot \nabla p\right] - q_o \tag{6}$$

$$\frac{\partial}{\partial t}(\rho_w S_w \varphi) = \nabla \cdot \left[\frac{\rho_w k_{rw}}{\mu_w} \underline{K} \cdot \nabla p\right] - q_w \tag{7}$$

where ∇ is now taken to be the two-dimensional gradient operator. The left-hand sides of Eqs. (6) and (7) can be expanded to give

$$\frac{\partial}{\partial t}(\rho_o S_o \varphi) = \rho_o \varphi S_o (c_r + c_o) \frac{\partial p}{\partial t} + \rho_o \varphi \frac{\partial S_o}{\partial t} \tag{8}$$

$$\frac{\partial}{\partial t}(\rho_w S_w \varphi) = \rho_w \varphi (c_r + c_w)(1 - S_o) \frac{\partial p}{\partial t} - \rho_w \varphi \frac{\partial S_o}{\partial t} \tag{9}$$

where

$$c_r = \text{rock compressibility} = \frac{1}{\varphi}\frac{\partial \varphi}{\partial p}$$

$$c_o = \text{oil compressibility} = \frac{1}{\rho_o}\frac{\partial \rho_o}{\partial p}$$

$$c_w = \text{water compressibility} = \frac{1}{\rho_w}\frac{\partial \rho_w}{\partial p}$$

---

†The lateral extent of a reservoir is often considerably greater than its thickness. For this reason, the flow of fluid can usually be taken as nearly horizontal, and gradients of properties in the vertical direction need not be included explicitly in the description.

Thus, Eqs. (6) to (9) become

$$\rho_o \varphi S_o (c_r + c_o) \frac{\partial p}{\partial t} + \rho_o \varphi \frac{\partial S_o}{\partial t} = \nabla \cdot \left[\frac{\rho_o k_{ro}}{\mu_o} \underline{K} \cdot \nabla p \right] - q_o \tag{10}$$

$$\rho_w \varphi (c_r + c_w)(1 - S_o) \frac{\partial p}{\partial t} - \rho_w \varphi \frac{\partial S_o}{\partial t} = \nabla \cdot \left[\frac{\rho_w k_{rw}}{\mu_w} \underline{K} \cdot \nabla p \right] - q_w \tag{11}$$

If the thickness of the reservoir, h, varies with horizontal position, the variables in Eqs. (10) and (11) are usually interpreted as the average values at a point (x,y) integrated over the thickness of the reservoir. In that case, Eqs. (10) and (11) become

$$\rho_o \varphi h S_o (c_r + c_o) \frac{\partial p}{\partial t} + \rho_o \varphi h \frac{\partial S_o}{\partial t} = \nabla \cdot \left[\frac{\rho_o k_{ro}}{\mu_o} (\underline{K}h) \cdot \nabla p \right] - q_o' \tag{12}$$

$$\rho_w \varphi h (c_r + c_w)(1 - S_o) \frac{\partial p}{\partial t} - \rho_w \varphi h \frac{\partial S_o}{\partial t} = \nabla \cdot \left[\frac{\rho_w k_{rw}}{\mu_w} (\underline{K}h) \cdot \nabla p \right] - q_w' \tag{13}$$

where q_o' and q_w' are the production rates at position (x,y).

The equations describing the flow of fluid in a reservoir become considerably simplified if only a single fluid is present. Conservation of mass applied to a volume element containing only a single fluid in the absence of sources or sinks yields (we drop the subscripts o and w for convenience)

$$\frac{\partial}{\partial t} (\rho \varphi h) = -\nabla \cdot (h \rho \underline{v}) \tag{14}$$

Introducing Darcy's law into Eq. (14), we obtain

$$h\varphi \frac{\partial \rho}{\partial t} = \nabla \cdot \left(\frac{h \rho \underline{K}}{\mu} \cdot \nabla p \right) \tag{15}$$

The density ρ is assumed to be a function of pressure only. Thus, Eq. (15) becomes

$$h\varphi c \rho \frac{\partial p}{\partial t} = \rho \nabla \cdot \left(\frac{\underline{K}h}{\mu} \cdot \nabla p \right) + c\rho \left(\frac{\underline{K}h}{\mu} \cdot \nabla p \right) \cdot \nabla p \tag{16}$$

If the fluid motion is assumed to be slow as to yield small pressure gradients, then the second term on the right-hand side of Eq. (16) may be neglected in comparison with the first. As a result, we obtain the following equation for p

$$\varphi ch \frac{\partial p}{\partial t} = \nabla \cdot \left(\frac{\underline{K}h}{\mu} \cdot \nabla p \right) \tag{17}$$

The parameter combinations φch and $\underline{K}h/\mu$ are called the storage m and transmissibility $\underline{\theta}$, respectively.

In order to complete the mathematical model of the physical reservoir, we need to prescribe boundary and initial conditions. The condition at the boundary $\partial\Omega$ of the domain Ω of the reservoir depends on what adjoins the reservoir. If the reservoir under consideration is surrounded by impermeable rock, then the appropriate condition is zero flow of fluid normal to the boundary; this condition leads, through the use of Darcy's law, to

$$\underline{\nabla} p(x,y,t) \cdot \underline{n} = 0, \qquad x,y \in \partial\Omega \tag{18}$$

where $\underline{n}$ is the unit vector at (x,y) normal to the boundary. Alternatively, if the reservoir is surrounded by a large aquifer, the pressure of which remains constant, then the boundary conditions are

$$p(x,y,t) = p_o, \qquad x,y \in \partial\Omega \tag{19}$$

We shall assume that the location of the boundary is known. For the case in which the boundary position is to be estimated the reader is referred to [2].

The initial condition for Eq. (17), describing the state of the reservoir at the beginning of the oil flow, is that the fluid is at rest with a uniform pressure throughout the horizontal domain Ω of the reservoir. The value of this initial pressure p_o will be assumed known, giving

$$p(x,y,0) = p_o, \qquad x,y \in \Omega \tag{20}$$

The description of the producing reservoir is completed by including the producing wells. The wells opened in the reservoir will be assumed to traverse the whole of the thickness h. The flow in their vicinity may then be approximated by a two-dimensional flow in the horizontal plane. The boundary condition at a well is the mathematical statement of the fact that the total flow of oil toward the well at any time must account for its current production rate. The wells may be approximated by point sinks with strength $q_i(t)$. Thus, Eq. (17) becomes

$$\varphi ch \frac{\partial p}{\partial t} = \nabla \cdot \left(\frac{\underline{K}h}{\mu} \cdot \nabla p \right) - \sum_{i=1}^{M} q_i(t)\, \delta(x - x_i)\, \delta(y - y_i) \qquad (21)$$

where it is assumed that there are M wells in the reservoir.

III. THE PARAMETER IDENTIFICATION PROBLEM

The study of petroleum reservoir identification has been almost exclusively concerned with single-phase (oil) reservoirs. Two-phase identification problems, on the other hand, have received little attention to date in spite of their great practical importance. The reason for this lack of attention lies in the troublesome nonlinearity of the governing equations and in the associated computational difficulties in their numerical solution. The major portion of the present report is devoted to the development of techniques for the identification of multiphase reservoirs. In this section we formulate the identification problems associated with both single and multiphase reservoirs.

A. Single-phase Reservoirs

It is necessary to know the distribution of the transmissibility $\underline{\theta}(x,y) = \underline{K}h/\mu$ and the storage $m(x,y) = \varphi ch$ to accurately model a given two-dimensional single-phase oil reservoir. These functions are never known completely a priori, and must therefore be estimated from the available information about the reservoir. This information is usually in the form of records of production rate histories for the different wells and the pressure measurements at different time instants at some or all of the wells. In addition, the analysis of core samples provides information about $\underline{\theta}$ and m at the well locations.

Geological information about the type of the reservoir and information about the extent of the reservoir, including the location of the boundary, are also usually available.

A generally used method of utilizing the pressure and production history records involves seeking functions $\underline{\theta}(x,y)$ and $m(x,y)$ which yield model pressures that match the measurements. This procedure of parameter estimation is commonly referred to in the petroleum industry as <u>history matching</u>.

The problem of distributed parameter (DP) estimation can be approached conceptually in two different ways, depending on the stage at which a finite-dimensional approximation to the infinite-dimensional distributed system is introduced. On the one hand, the estimation problem can be posed in infinite dimensions, an algorithm derived, and at the end it is implemented by a suitable finite-dimensional approximation, such as finite differences. On the other hand, the original infinite-dimensional system is approximated at the beginning by a finite-dimensional model through finite differencing or other suitable methods. Subsequently, an appropriate algorithm is derived for the finite-dimensional minimization problem. In this chapter we shall illustrate both approaches. In this section, where we consider single-phase problems, we will employ approximation at the beginning. In Section IV we will preserve the infinite-dimensional nature of the problem when treating two-phase estimation problems.

We shall use a finite difference scheme employing a uniform spatial grid covering the domain of the reservoir for the purpose of the finite-dimensional approximation. This reduces the original PDE to a set of coupled ordinary differential equations (ODEs) for the oil pressure at the grid points. These will be further approximated by differencing in time, using constant time steps, to yield a set of discrete algebraic equations, which will be taken as the model for the system. The minimization will be carried out subject to these model equations, with respect to the grid point values of the rock properties $\underline{\theta}$ and m.

The most detailed description of the unknown rock properties is obtained by allowing $\underline{\theta}$ and m to vary independently at each block of the spatial grid used in the finite differencing. For accurate modeling, a fine spatial grid is essential; this would lead to a large number of unknowns compared to the available data. Thus, while minimizing the modeling error, this approach entails a great deal of uncertainty, or may even lead to a situation in which the number of unknowns exceeds the number of data points, resulting in nonuniqueness of the solution to the parameter estimation problem.

The problem of nonuniqueness can be dealt with by reduction of the number of unknowns, by injection of additional information, or by a combination of the two. The reduction of the unknowns can be effected by parameterization, i.e., the introduction of a set of fewer parameters along with a mapping which uniquely determines the values of $\underline{\theta}$ and m at all the grid points corresponding to a given value for this parameter set. Zonation, the assumption that the reservoir properties are uniform in each of a certain number of zones, is an example of parameterization. The number of zones can be as large as the number of grid blocks utilized in the numerical solution of Eq. (21). Employing a small number of zones enables a rapid solution to the parameter estimation problem but introduces considerable modeling error by forcing the parameters to be constant over arbitrarily defined regions of the reservoir. If the zones correspond to the grid used for numerical solution, modeling error is minimized, but the problem may be ill-conditioned. Examples of zonation approaches can be found in [3-9].

The single-phase identification problem in one dimension can be stated generally as follows: Determine the estimates of m(x) and θ(x) to minimize the difference between the predicted and observed pressures at each of the M wells

$$J = \sum_{i=1}^{R} \sum_{j=1}^{M} [p(x_j,t_i) - p^{obs}(x_j,t_i)]^2 \sigma_{j,i}^{-2} \qquad (22)$$

where $\sigma_{j,i}^2$ is the variance of the error in the observations, subject to the model describing the reservoir pressure, Eq. (21), and appropriate boundary conditions.

Ordinarily Eq. (21) will be solved on a grid of, say, N interior points, so that p(x,t) is replaced by the N-vector $\underline{p}(t)$ and m(x) and $\theta(x)$ are replaced by the L-vector $\underline{\pi}$. Note that the maximum value of L is 2N if both m(x) and $\theta(x)$ at each grid point are taken as unknown. L may be smaller if either m(x) or $\theta(x)$ is known or has a specified functional dependence on x. Equation (21) can then be written in discrete time form as

$$\underline{G}\underline{p}_{i+1} = \underline{H}\underline{p}_i + \underline{q}_i, \qquad i = 1,2, \ldots ,T \tag{23}$$

where $\underline{p}_{i+1}$ represents the ith time step in the numerical solution of Eq. (21), i.e., $\underline{p}_{i+1} = \underline{p}(i\,\Delta t)$. The matrix $\underline{G}$ generally includes the porosity, and the matrix $\underline{H}$ depends on the permeability. For generality, we simply indicate that $\underline{G} = \underline{G}(\underline{\pi})$ and $\underline{H} = \underline{H}(\underline{\pi})$. We note that at each observation location we assume that we have R data points over the time period of interest. The number of time increments in the numerical solution of Eq. (21) is T, where $T \geq R$. Thus, in general, data are not available at every time step corresponding to the numerical solution. The pressure $\underline{p}_i$ is an N-vector, with elements indexed according to the grid point. The M measurement locations will be indexed as $j_1, j_2, \ldots, j_M$, as shown below.

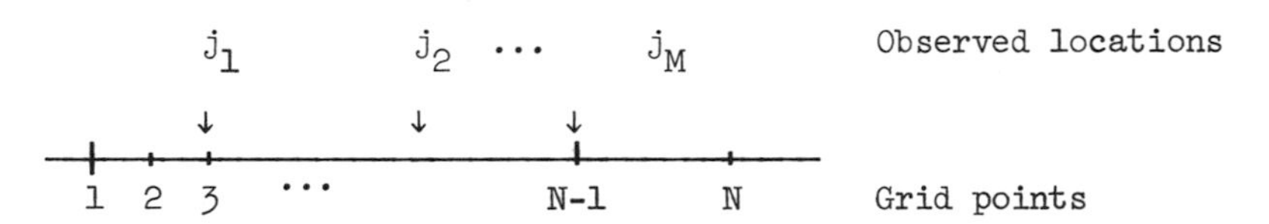

The time points at which data are available will be indexed as $i_1, i_2, \ldots, i_R$. We will assume for simplicity that data are taken at the same times at each location.

If the number of unknowns, L, exceeds the total number of observations, I = MR, the problem is ill-determined.† Even if $I > MMR$, it still may be virtually impossible to obtain the value of $\underline{\pi}$ which minimizes J because of the inherent insensitivity of J to one or more

---

†Note that the maximum value of I is MT if R = T.

of the elements of $\underline{\pi}$. History-matching problems are consistently plagued by nonunique estimates, the values of which depend on the initial guess and the efficiency of the numerical minimization routine. For fixed values of M and R the inherently ill-determined nature of the identification problem can be alleviated only by decreasing the number of unknown parameters L, for example, through zonation, or by introducing additional information into the problem. Thus, one may consider reformulating a given problem by reducing the number of unknowns, by introducing constraints on the estimates, or by modifying the performance index.

The most efficient way to introduce additional information into the problem is to specify certain statistical properties of the unknown parameters in the following way. We define an objective function

$$J = J_p + J_2 \tag{24}$$

where J_p is the customary objective function based on the difference between the observed and calculated pressures defined by Eq. (22), and

$$J_2 = (\underline{\pi} - \bar{\underline{\pi}})^T \underline{P}_o^{-1} (\underline{\pi} - \bar{\underline{\pi}}) \tag{25}$$

where $\bar{\underline{\pi}} = E\{\underline{\pi}\}$ and $\underline{P}_o = E\{(\underline{\pi} - \bar{\underline{\pi}})(\underline{\pi} - \bar{\underline{\pi}})^T\}$.

Implicit in this formulation is the assumption that the unknown property vector $\underline{\pi}$ can be described by an a priori probability distribution, at least with respect to its mean and variance. The estimation is then defined by the minimization of the composite index J. The term J_2 is a Bayesian-like term penalizing the weighted deviations of the parameters from their prior mean values. Minimization of J_p results in the utilization of the information in the observations; J_2 introduces the prior information about the parameters into the estimation procedure. By adding J_2 we require that the parameters follow some preconceived pattern. This requirement reduces the statistical uncertainty at the cost of increasing the residual observation error. The relative weight in the two terms,

J_p, J_2, is determined by the variance of the observation error, $\sigma^2_{j,i}$, and by $\underline{P}_o^{-1}$. When little is known a priori about the rock properties, $\underline{P}_o$ will have large diagonal elements, and the elements of $\underline{P}_o^{-1}$ will be small so that the term J_2 will be given a small weight.

Regardless of how the problem is formulated, it is necessary to employ the most efficient numerical minimization algorithm available in minimizing J to obtain the estimate of $\underline{\pi}$.

There are essentially two classes of numerical minimization algorithms which are available for inclusion in a history-matching algorithm:

1. First-order methods

2. Second-order methods

In first-order gradient algorithms the correction $\delta\underline{\pi}^i$ depends only on $\partial J/\partial\underline{\pi}(J_{\underline{\pi}})$. Thus, the corrections are largest for the most sensitive parameters. The first step in the implementation of a first-order gradient method is the calculation of $J_{\underline{\pi}}$. Let us write Eq. (22) in the form corresponding to (23) and consider only the pressure term J_p with equal errors in all data (adding a factor of 1/2)

$$J = \frac{1}{2} \sum_{m=1}^{R} \sum_{n=1}^{M} (p_{i_m,j_n} - p^o_{mn})^2 \tag{26}$$

To compute $J_{\underline{\pi}}$ we can differentiate Eq. (26) to give

$$J_{\underline{\pi}} = \sum_{m=1}^{R} \sum_{n=1}^{M} \left(\frac{\partial p_{i_m,j_n}}{\partial \underline{\pi}} \right) (p_{i_m,j_n} - p^o_{mn}) \tag{27}$$

The sensitivity vector $(\partial p_{i_m,j_n}/\partial\underline{\pi})$ can be computed from the full $N \times L$ sensitivity matrix $(\partial\underline{p}_i/\partial\underline{\pi})$. This approach is prohibitively costly in terms of computing time. An efficient means of computing $J_{\underline{\pi}}$ lies in the use of optimal control theory [10-13]. Let us define a new performance index by

$$J = \frac{1}{2} \sum_{m=1}^{R} \sum_{n=1}^{M} (p_{i_m,j_n} - p^o_{mn})^2 + \sum_{i=1}^{T} \underline{\lambda}_i^T[\underline{\underline{G}}\underline{p}_{i+1} - \underline{\underline{H}}\underline{p}_i - \underline{q}_i] \tag{28}$$

where $\underline{\lambda}_i$, i = 1,2, ... ,T is a sequence of arbitrary N-vectors. We introduce a perturbation $\delta\underline{\pi}$ in $\underline{\pi}$, inducing a perturbation $\delta\bar{J}$ in $\bar{J}$,

$$\delta\bar{J} = \sum_{m=1}^{R}\sum_{n=1}^{M} (p_{i_m,j_n} - p^o_{mn})\, \delta p_{i_m,j_n} + \sum_{i=1}^{T} \underline{\lambda}_i^T[\underline{G}\, \delta\underline{p}_{i+1} + \delta\underline{G}\underline{p}_{i+1} - \underline{H}\, \delta\underline{p}_i - \delta\underline{H}\underline{p}_i] \tag{29}$$

which can be rewritten as

$$\delta\bar{J} = \sum_{i=2}^{T}\left[\underline{\lambda}_{i-1}^T\underline{G} - \underline{\lambda}_i^T\underline{H} + \sum_{n=1}^{M}\sum_{m=1}^{R} (p_{i_m,j_n} - p^o_{mn})\, \delta_{i,i_m}\underline{e}_{j_n}^T\right]\delta\underline{p}_i + \underline{\lambda}_T^T\underline{G}\, \delta\underline{p}_{T+1} + \sum_{i=1}^{T}\underline{\lambda}_i^T[\delta\underline{G}\underline{p}_{i+1} - \delta\underline{H}\underline{p}_i] + \sum_{n=1}^{M}(p_{i_R,j_n} - p^o_{Rn})\, \delta_{T+1,i_R}\underline{e}_{j_n}^T\, \delta\underline{p}_{T+1} \tag{30}$$

where $\underline{e}_{j_n}$ is an N-dimensional vector with elements δ_{ℓ,j_n}, the Kronecker delta. If we choose $\underline{\lambda}_i$ to satisfy

$$\underline{G}^T\underline{\lambda}_{i-1} = \underline{H}^T\underline{\lambda}_i - \sum_{n=1}^{M}\sum_{m=1}^{R}(p_{i_m,j_n} - p^o_{mn})\, \delta_{i,i_m}\underline{e}_{j_n}, \qquad i = T, T-1, \ldots ,2 \tag{31}$$

$$\underline{G}^T\underline{\lambda}_T = -\sum_{n=1}^{M}(p_{i_R,j_n} - p^o_{Rn})\, \delta_{T+1,i_R}\underline{e}_{j_n} \tag{32}$$

then

$$\delta\bar{J} = \sum_{i=1}^{T}\underline{\lambda}_i^T[\delta\underline{G}\underline{p}_{i+1} - \delta\underline{H}\underline{p}_i] \tag{33}$$

or

$$\bar{J}_{\underline{\pi}} = \sum_{i=1}^{T}\underline{\lambda}_i^T\left[\frac{\partial\underline{G}}{\partial\underline{\pi}}\, p_{i+1} - \frac{\partial\underline{H}}{\partial\underline{\pi}}\, \underline{p}_i\right] \tag{34}$$

We note that as long as Eq. (23) is satisfied at any point in the iteration, $J = \bar{J}$, and (34) therefore gives the expression for $J_{\underline{\pi}}$.

In second-order methods the correction $\delta\underline{\pi}^i$ depends on the second derivative matrix $J_{\underline{\pi}\underline{\pi}}$.

The second derivative of J with respect to $\underline{\pi}$ is

$$J_{\underline{\pi}\underline{\pi}} = \sum_{m=1}^{R} \sum_{n=1}^{M} \left[\left(\frac{\partial^2 p_{i_m,j_n}}{\partial\underline{\pi}\partial\underline{\pi}} \right)(p_{i_m,j_n} - p_{mn}^o) + \left(\frac{\partial p_{i_m,j_n}}{\partial\underline{\pi}} \right)\left(\frac{\partial p_{i_m,j_n}}{\partial\underline{\pi}} \right)^T \right] \tag{35}$$

Second-order methods employ $J_{\underline{\pi}\underline{\pi}}$ or an approximation thereto. In the Newton-Raphson method, one step in the iteration is

$$\underline{\pi}^{i+1} = \underline{\pi}^i - (J_{\underline{\pi}\underline{\pi}})^{i^{-1}}(J_{\underline{\pi}})^i \tag{36}$$

Rapid convergence is obtained if $J_{\underline{\pi}\underline{\pi}}$ is positive definite; however, the computation of $J_{\underline{\pi}\underline{\pi}}$ is very time-consuming. For that reason, the Newton-Raphson method is seldom used for parameter estimation problems.

The Gauss-Newton method is an approximation to the fully second-order Newton-Raphson method which maintains the convergence characteristics of the latter in the vicinity of the minimum. If it is assumed that $\underline{\pi}^i$ is close to the true value, then $(p_{i_m,j_n} - p_{mn}^o)$ can be considered as small, and we may approximate $J_{\underline{\pi}\underline{\pi}}$ by

$$J_{\underline{\pi}\underline{\pi}} \cong \sum_{m=1}^{R} \sum_{n=1}^{M} \left(\frac{\partial p_{i_m,j_n}}{\partial\underline{\pi}} \right)\left(\frac{\partial p_{i_m,j_n}}{\partial\underline{\pi}} \right)^T \tag{37}$$

The updated estimate $\underline{\pi}^{i+1}$ is obtained from $\underline{\pi}^i$ using Eqs. (35) and (36). As we noted earlier, the sensitivity matrix $\partial\underline{p}/\partial\underline{\pi}$ can be obtained by integration of the NL sensitivity equations. Again, however, there is an alternate way to compute this vector based on adjoint equations. We note also that only the derivatives at the measurement times and locations, denoted by i_m and j_n, are required in Eq. (37).

Our objective is to compute the vectors in Eq. (37). Consider a variation $\delta\underline{\pi}$. The first-order variational equations are

$$\underline{G}\ \delta p_{i+1} = -\delta \underline{\underline{G}} \underline{p}_{i+1} + \delta \underline{\underline{H}} \underline{p}_i + \underline{H}\ \delta \underline{p}_i, \qquad i = 1,2, \ldots ,T \tag{38}$$

with

$$\delta \underline{p}_1 = 0 \tag{39}$$

since the initial pressure is assumed known. Let us multiply Eq. (38) by a sequence of arbitrary N-dimensional vectors $\underline{\lambda}_i$ and sum from i = 1 to any one of the values of i corresponding to a grid point, $i_m - 1$, where m can assume values from 1 to R. We obtain

$$\sum_{i=1}^{i_m-1} \underline{\lambda}_i^T \underline{G}\ \delta \underline{p}_{i+1} = - \sum_{i=1}^{i_m-1} \underline{\lambda}_i^T\ \delta \underline{\underline{G}} \underline{p}_{i+1} + \sum_{i=1}^{i_m-1} \underline{\lambda}_i^T\ \delta \underline{\underline{H}} \underline{p}_i + \sum_{i=1}^{i_m-1} \underline{\lambda}_i^T \underline{H}\ \delta \underline{p}_i, \qquad m = 1,2, \ldots ,R \tag{40}$$

Rearranging Eq. (40) yields

$$\sum_{i=2}^{i_m-1} [\underline{\lambda}_{i-1}^T \underline{G} - \underline{\lambda}_i^T \underline{H}]\ \delta \underline{p}_i = - \sum_{i=1}^{i_m-1} \underline{\lambda}_i^T\ \delta \underline{\underline{G}} \underline{p}_{i+1} + \sum_{i=1}^{i_m-1} \underline{\lambda}_i^T\ \delta \underline{\underline{H}} \underline{p}_i - \underline{\lambda}_{i_m-1}^T \underline{G}\ \delta \underline{p}_{i_m}, \qquad m = 1,2, \ldots ,R \tag{41}$$

Let us choose the sequence $\underline{\lambda}_i$, $i = 1,2, \ldots ,i_m-1$, to satisfy

$$\underline{G}^T \underline{\lambda}_{i-1} = \underline{H}^T \underline{\lambda}_i, \qquad i = i_m-1,\ i_m-2, \ldots ,2 \tag{42}$$

$$\underline{G}^T \underline{\lambda}_{i_m-1} = \underline{e}_{j_n}, \qquad n = 1,2, \ldots ,M \tag{43}$$

Using Eqs. (42) and (43), (41) becomes

$$\delta p_{i_m, j_n} = - \sum_{i=1}^{i_m-1} \underline{\lambda}_i^T [\delta \underline{\underline{G}} \underline{p}_{i+1} - \delta \underline{\underline{H}} \underline{p}_i] \tag{44}$$

If we let $\delta \underline{G} = \underline{G}_{\underline{\pi}}\ \delta \underline{\pi}$ and $\delta \underline{H} = \underline{H}_{\underline{\pi}}\ \delta \underline{\pi}$, Eq. (44) becomes

$$\frac{\partial p_{i_m, j_n}}{\partial \pi_\ell} = - \sum_{i=1}^{i_m - 1} \underline{\lambda}_i^T \left[\frac{\partial \underline{G}}{\partial \pi_\ell} \underline{p}_{i+1} - \frac{\partial \underline{H}}{\partial \pi_\ell} \underline{p}_i \right], \tag{45}$$

$$m = 1,2, \ldots ,R \qquad n = 1,2, \ldots ,M$$

Equation (45) can be used to compute the sensitivity needed in (37). Note that it is necessary to solve the adjoint system (42) and (43) separately for each measurement time, i_m, $m = 1,2, \ldots ,R$, and for each measurement location j_n, $n = 1,2, \ldots ,M$. Thus, the derivative (45) at each i_m and j_n must be calculated separately and independently of all the others.

In this subsection we have briefly reviewed the problem of parameter estimation for single-phase reservoirs. We have presented the problem based on a finite-dimensional approximation at the beginning, in which the original PDE is replaced by a difference equation model before the estimation problem is considered. We have shown how conventional numerical minimization methods can be employed in conjunction with adjoint equation approaches for calculating the necessary derivatives of the objective function and the pressures with respect to the unknown parameters. The use of pseudo-Bayesian estimation as a means of reducing the ill-determinacy in reservoir parameter estimation problems was discussed. For a thorough exposition of this technique the reader is referred to Gavalas, Shah, and Seinfeld [14].

We have not discussed the important problem of determining covariances of the estimated parameters. An early work aimed at this problem was that of Dixon et al. [15]. A recent comprehensive treatment of the determination of covariances of parameters estimated in history matching is that of Shah, Gavalas, and Seinfeld [16].

B. Two-phase Reservoirs

The identification problem for two-phase reservoirs involves m, θ_x, and θ_y as in the single-phase case and may in addition require the identification of the two functions $k_{ro}(S_o)$ and $k_{rw}(S_o)$. It is clear that, in general, enough data will not be available to estimate all of these quantities simultaneously. The type of data available for a reservoir from which both oil and water are produced is pressure measurements and/or oil and water flow rates at the wells.

Before proceeding to the application of optimal control theory to two-phase identification problems, it is useful to consider the dynamics of a hypothetical oil-water reservoir to determine the degree of sensitivity of observed quantities to variations in the parameters which one might wish to estimate on the basis of those observations. For this purpose let us consider a circular reservoir with a centrally-located producing well as depicted in Figure 4. The reservoir is surrounded by an aquifer (water-containing rock) at pressure p_s and residual oil saturation S_o^{res}. The initial pressure and saturation distributions in the reservoir are known. Let us assume that they are p_s and $S_o(r,0) = f(r)$, respectively. At $t = 0$ the pressure at the well is reduced to γp_s $(\gamma < 1)$, and, as a result, fluid flows to the well and is removed from the reservoir. If we assume that the reservoir thickness h is constant, and that ρ_o, ρ_w, μ_o, μ_w, and φ are approximately constant over the pressure range considered, Eqs. (12) and (13) become

$$\varphi S_o(c_r + c_o)\frac{\partial p}{\partial t} + \varphi\frac{\partial S_o}{\partial t} = \frac{1}{r\mu_o}\frac{\partial}{\partial r}\left[rkk_{ro}\frac{\partial p}{\partial r}\right] \tag{46}$$

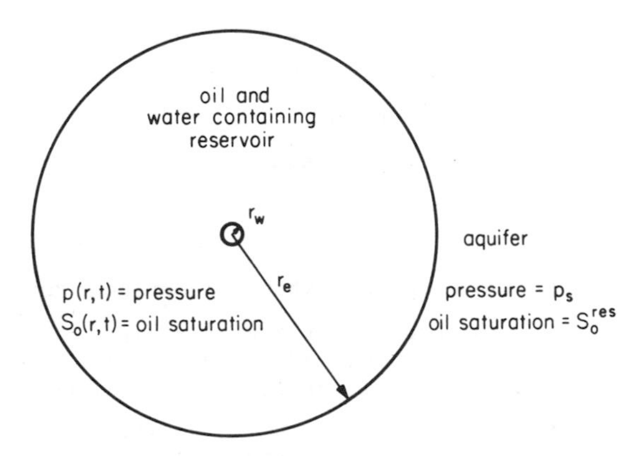

FIG. 4. Circular reservoir with central well.

$$\varphi\,(1 - S_o)(c_r + c_w)\,\frac{\partial p}{\partial t} + \varphi\,\frac{\partial S_o}{\partial t} = \frac{1}{r\mu_w}\,\frac{\partial}{\partial r}\left[rkk_{rw}\,\frac{\partial p}{\partial r}\right] \tag{47}$$

Initial and boundary conditions are†

$$p(r,0) = p_s \qquad t = 0,\ r_w \le r \le r_e \tag{48}$$

$$S_o(r,0) = f(r) \qquad t = 0,\ r_w \le r \le r_e \tag{49}$$

$$p(r_e,t) = p_s \qquad r = r_e,\ t \ge 0 \tag{50}$$

$$p(r_w,t) = \gamma p_s \qquad r = r_w,\ t > 0 \tag{51}$$

where $f(r)$ is the initial oil saturation profile in the reservoir. For example, for a region of oil surrounded by a region of water with boundary at r_o:

$$f(r) = \begin{cases} S_o^{sat}, & r_w \le r < r_o \\ S_o^{res}, & r_o \le r \le r_e \end{cases} \tag{52}$$

For simplicity we take φ and k to be independent of r. In that case variations of k and φ are not independent since both equations can be divided by φ to produce the single parameter k/φ. It is then clear that k/φ simply controls the time scale of the changes in p and S_o. Thus, the oil and water production rates will depend directly on k/φ, and it should be relatively easy to estimate k/φ on the basis of production data. If k and φ are each functions of r, the ability to estimate $k(r)$ and $\varphi(r)$ independently solely on the basis of oil and water production data will probably be rather poor.

---

†The oil and water production rates q_o' and q_w' do not enter Eqs. (46) and (47) since the removal occurs at the inner boundary r_w. Because we have specified the pressure at the well as γp_s, the flow rates of oil and water cannot be specified. These flow rates can be computed from the oil and water fluxes at $r = r_w$. If we had chosen to specify the flow rates at the well, then, conversely, the pressure at the well must be determined from the solution.

Our prime objective in considering the reservoir of Figure 4 is to determine the effect of changes in the relative permeability curves on the response of the system. The three sets of relative permeability curves given in Figure 3 are used. Three sets of initial saturation distributions f(r) are used: one that varies approximately linearly from the well to the external boundary of the reservoir (the "0 days" curve in Figure 5a); one that is semilinear (the "0 days" curve in Figure 5d); and one that approximates a step function (the "0 days" curve in Figure 5g). The other parameters used in solving Eqs. (46) and (47) are:

$c_o = 4.5 \times 10^{-5}\ \text{atm}^{-1}$ $\qquad$ $p_s = 1500$ psi

$c_r = 0$ $\qquad$ $r_e = 1000$ ft

$c_w = 2 \times 10^{-4}\ \text{atm}^{-1}$ $\qquad$ $r_w = 0.25$ ft

$k = 1$ darcy $\qquad$ $\gamma = 0.8$

$\varphi = 0.2$ $\qquad$ $\mu_o = \mu_w = 0.7$ centipoise

The results of the simulation are presented in Figures 5 to 7. Figures 5a-i show oil saturation S_o versus r/r_e at different times. Figures 6a-c depict pressure versus r/r_e at different times, corresponding to the initial saturation profiles of Figures 5a, d, and g, respectively, and to the relative permeability curves of Figure 3a. Pressure curves are not included for all the cases, since they are qualitatively similar. Figures 7a-c show the principal results of interest, namely the total oil produced as a function of time. The oil production is expressed in Figures 7a-c in percent of producible oil in place. Figures 7a-c correspond to the three different initial saturation conditions, i.e., the "0 days" curves of Figures 5a, 5d, and 5g, respectively. The curves labeled a, b, and c on each of Figures 7a-c correspond to the relative permeability curves of Figures 3a-c, respectively.

There are two items of interest about the pressure-time curves of Figures 6a-c. One is the characteristic time for pressure relaxation from the initial conditions. For a single-phase system, the characteristic time for pressure change is

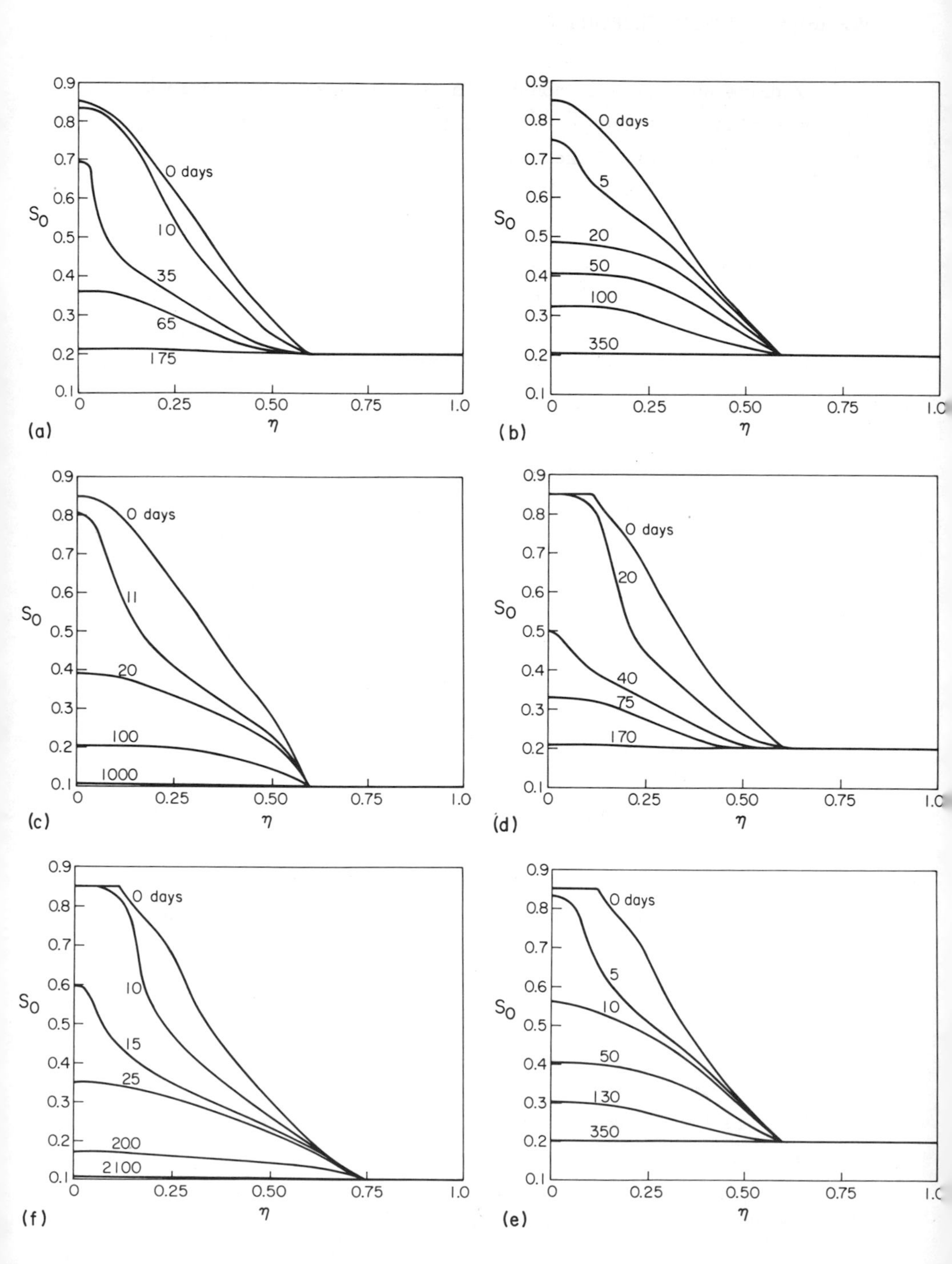

(a)
0 days
10
35
65
175
(b)
0 days
5
20
50
100
350
(c)
0 days
11
20
100
1000
(d)
0 days
20
40
75
170
(f)
0 days
10
15
25
200
2100
(e)
0 days
5
10
50
130
350
S_0
η

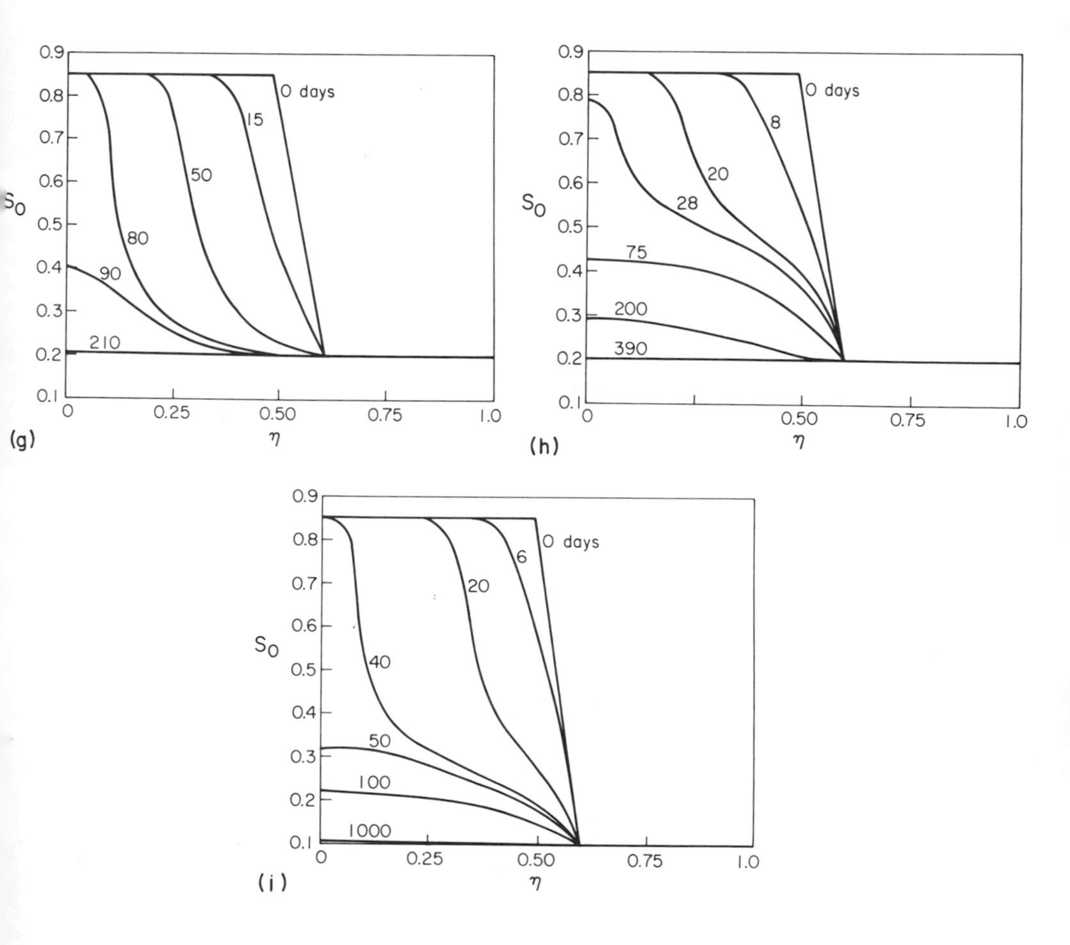

FIG. 5. Oil saturation as a function of radial position at various times for the reservoir shown in Figure 4. (a) Linear initial saturation profile and relative permeabilities from Figure 3a; (b) linear initial saturation profile and relative permeabilities from Figure 3b; (c) linear initial saturation profile and relative permeabilities from Figure 3c; (d) semilinear initial saturation profile and relative permeabilities from Figure 3a; (e) semilinear initial saturation profile and relative permeabilities from Figure 3b; (f) semilinear initial saturation profile and relative permeabilities from Figure 3c; (g) step-function initial saturation profile and relative permeabilities for Figure 3a; (h) step-function initial saturation profile and relative permeabilities for Figure 3b; (i) step-function initial saturation profile and relative permeabilities for Figure 3c.

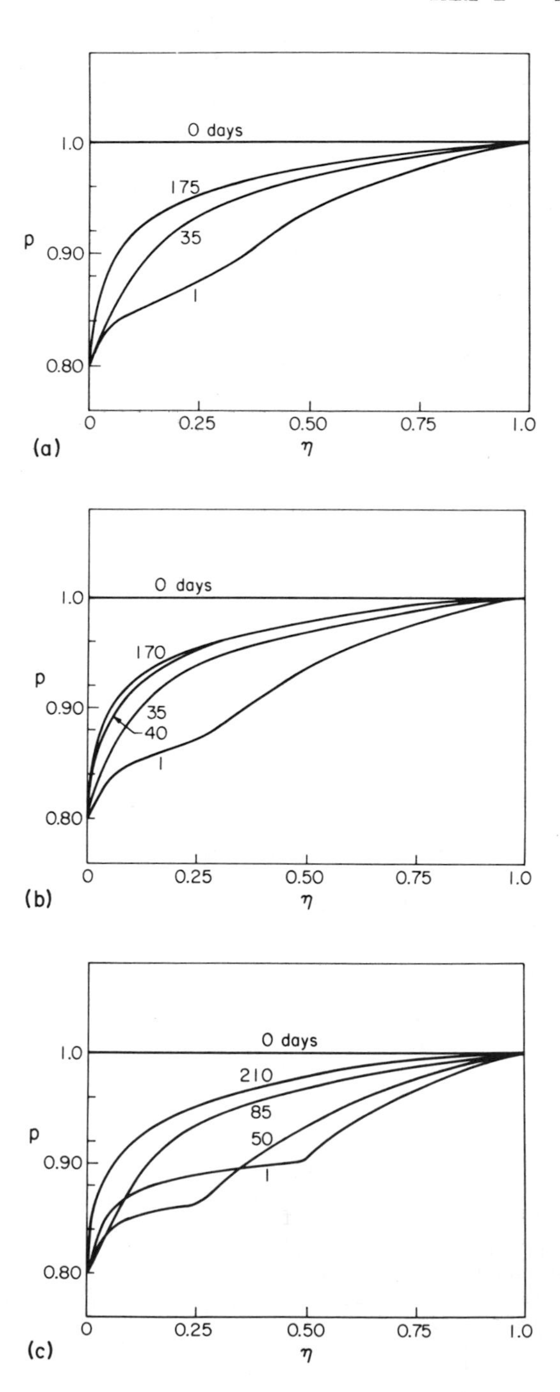
0 days
1.0
175
35
1
p
0.90
0.80
0
0.25
0.50
0.75
1.0
(a)
η
0 days
1.0
170
35
40
1
p
0.90
0.80
0
0.25
0.50
0.75
1.0
(b)
η
0 days
1.0
210
85
50
1
p
0.90
0.80
0
0.25
0.50
0.75
1.0
(c)
η

$$\tau = r_e^2(c_r + c_o)\varphi\mu_o/k$$

For the parameters given above, $\tau \cong 7$ hours. Thus, the "1 day" pressure curves in Figures 6a-c have changed appreciably from the initial curves. From the expression for τ we also see that, as noted above, τ depends directly on k/φ. The other item of note with respect to the pressure curves is the development of a kink or bend in the curves, most noticeable in Figure 6c. These bends are due to the front between the oil and water in the system, and they progress in toward the well bore as the oil is produced. As the oil is produced, the pressure at a point in the reservoir rises since the compressibility of the water replacing the oil is less than that of the oil. This compressibility difference also leads to the rise in pressure near the well bore as the oil is produced.

Analysis of Figure 7a-c is of interest to determine the extent to which information about the relative permeability curves can be deduced from production data. The first item to note is the value of k_{rw} at S_o^{res}. These values at 1.0, 0.8, and 0.15 for the curves in Figures 3b, c, and a, respectively. Since initially oil saturates the center of the reservoir and water saturates the outer regions, the speed with which the water can flow in toward the center and replace the oil flowing up the well will affect the oil production rate. This effect is illustrated in Figure 7a-c, where early in the life of the reservoir the rate of production (i.e., the slope of the production curves) is greatest for the relative permeability curves in Figure 3b and least for those in Figure 3a.

As S_o decreases with time, the rate at which the oil can flow decreases, and the magnitude of k_{rw} at S_o^{res} becomes less important than in the early life of the reservoir. Indeed, after S_o has

---

FIG. 6. Pressure as a function of radial position at various times for the reservoir of Figure 4. (a) Linear initial saturation profile and relative permeabilities of Figure 3a (corresponds to Figure 5a); (b) semilinear initial saturation profile and relative permeabilities of Figure 3a (corresponds to Figure 5d); (c) step-function initial saturation profile and relative permeabilities of Figure 3a (corresponds to Figure 5g).

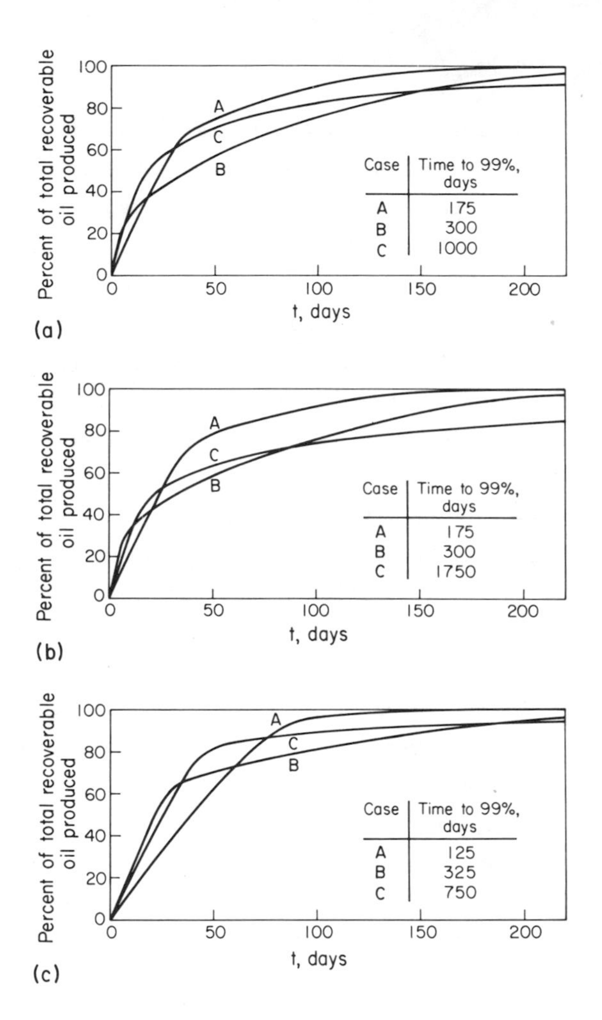

FIG. 7. Effect of relative permeabilities the total oil produced. (a) Linear initial saturation profile; (b) semilinear initial saturation profile; (c) step-function initial saturation profile.

decreased below the value of S_o at which $k_{ro} = k_{rw}$, the water is able to flow more readily than the oil. These crossover points are seen in Figure 3a-c to be at $S_o = 0.4$, 0.56, and 0.42, respectively. By comparison of the saturation-time curves of Figure 5 with the production-time curves of Figure 7 it can be seen that the change in slope of the production-time curve in each case occurs when S_o at $r = r_w$ is equal to the value at which $k_{ro} = k_{rw}$. If we call the time at which this bend in the production-time curves occurs as τ_B (e.g., in Figure 7a, curve c, $\tau_B \cong 45$ days), τ_B reflects the time when the saturation at which $k_{ro} = k_{rw}$ is reached. Thus, τ_B can be used to determine one important feature of k_{ro} and k_{rw}.

Another property of the production curves which should be sensitive to k_{ro} and k_{rw} is the total time required to produce a certain quantity of oil, say 99% of the producible oil in place. At first it might seem that the larger the value of k_{ro} at low S_o, the faster the reservoir will be completely produced. This is one, but not the only, consideration. In the cases in which the initial production rate is high, the oil at lower saturations farther out in the reservoir cannot "keep up with" the faster flowing higher saturation oil closer in toward the well bore, so that a low, flat saturation profile tends to develop in the reservoir (e.g., Figure 5h). When the initial production rate is not so high, this oil farther out can keep up more easily with the higher saturation oil closer to the well bore, and steep saturation profiles tend to develop (Figure 5g). Thus, when breakthrough does occur, there is much more oil left to produce (at relatively slow rates) in cases where initial production was fast, and the time required for 99% production may be much longer than for the case in which the initial oil production was slow. This is the explanation for why in Figures 7a-c there is a longer time to complete production in case b than in case a, even though the k_{ro} curves are identical.[†] (Since case c has a different value of S_o^{res} than cases a and b and thus more total oil to produce, it is not directly comparable to cases a and b.)

[†]A note of explanation is necessary for the cases in which f(r) is a step function. Equations (46) and (47) predict that the oil would be completely produced at the instant water breakthrough occurs. This sharp break is not visible in the presented results because the necessity of using finite difference approximations results in "numerical diffusion" which tends to spread out the oil saturation profiles, as seen in Figures 5g-i.

In summary, certain features of the relative permeability/saturation relationships are reflected in the production curves for the hypothetical single-well reservoir considered here. Thus, it is feasible to attempt to estimate certain aspects of k_{ro} and k_{rw} from such data. In Section IV, therefore, we develop an algorithm for the estimation of k_{ro} and k_{rw} from production data for oil-water reservoirs.

IV. PARAMETER IDENTIFICATION IN TWO-PHASE RESERVOIRS

In this section we apply optimal control theory to the problem of parameter identification in two-phase (oil-water) reservoirs. The parameter identification problem can be divided into two parts: (1) estimation of the absolute permeability (and porosity), and (2) estimation of the relative permeabilities, $k_{ro}(S_o)$ and $k_{rw}(S_o)$. The main purpose of this section is to present the algorithms for estimating the parameters and to discuss their use. Since the algorithms have not yet been applied to either simulated or actual reservoirs, no results are presented here.

We will consider a two-dimensional reservoir in cylindrical coordinates (see Figure 8). The basic equations, including capillary pressure effects, are [17]

$$\frac{\partial}{\partial t}(\rho_o S_o \varphi) = \frac{\partial}{\partial z}\left[k_z \frac{k_{ro}\rho_o}{\mu_o}\left(\frac{\partial p_o}{\partial z} - \rho_o g\right)\right] + \frac{1}{r}\frac{\partial}{\partial r}\left[rk_r \frac{k_{ro}\rho_o}{\mu_o}\frac{\partial p_o}{\partial r}\right] \quad (53)$$

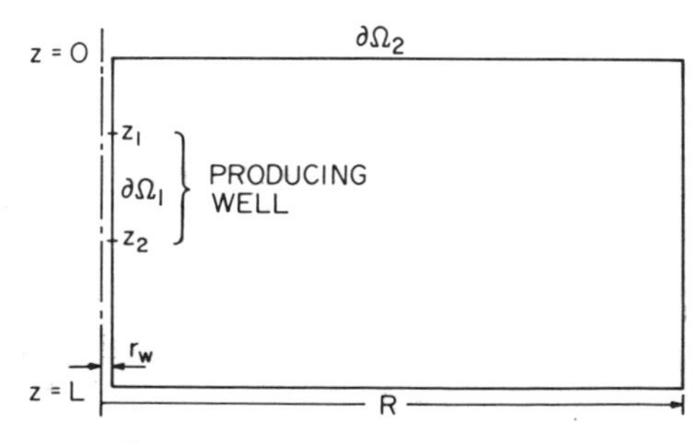

FIG. 8. Vertical section through a radially symmetric reservoir.

$$\frac{\partial}{\partial t}(\rho_w S_w \varphi) = \frac{\partial}{\partial z}\left[k_z \frac{k_{rw}\rho_w}{\mu_w}\left(\frac{\partial p_w}{\partial z} - \rho_w g\right)\right] + \frac{1}{r}\frac{\partial}{\partial r}\left[rk_r \frac{k_{rw}\rho_w}{\mu_w}\frac{\partial p_w}{\partial r}\right] \quad (54)$$

$$S_o + S_w = 1 \quad (55)$$

$$p_c = p_o - p_w = f(S_w) \quad (56)$$

subject the boundary conditions on $\partial\Omega_2$ (the reservoir boundary excluding the well)

$$\frac{\partial p_\ell}{\partial z} - \rho_\ell g = 0, \qquad z = 0, L, \; r_w \le r \le R \; (\ell = o, w) \quad (57)$$

$$\frac{\partial p_\ell}{\partial r} = 0, \qquad r = R, \quad 0 \le z \le L \quad (58)$$

$$r = r_w, \quad 0 \le z < z_1, \; z_2 < z \le L$$

and on $\partial\Omega_1$ (the well)

Before water breakthrough:

$$\frac{\partial p_o}{\partial z} = \rho_o g, \qquad r = r_w \quad (59)$$

$$p_o = g(t) \qquad \text{at a given depth on } \partial\Omega_1 \quad (60)$$

After breakthrough:

$$\frac{\partial p}{\partial z} = \bar{\rho} g \qquad p = p_o = p_w \text{ at } r_w \quad (61)$$

$$p = \bar{g}(t) \qquad \text{at a given depth on } \partial\Omega_1 \quad (62)$$

We therefore assume that a reference pressure $g(t)$ [and $\bar{g}(t)$ after breakthrough] at the well are known. This allows calculation of the pressure at the well as a function of z by integration of (59) and (61) over the height of the well. Since the pressure at the well is assumed known at any vertical height z, the oil and water flow rates can be calculated at any height z as a function of time. Thus, we can predict the total oil and water flow rates at any time without the need of specifying the total oil flow rate.

A. Absolute Permeability

The history-matching problem we consider here is to estimate k_z to minimize J. We assume that the other physical constants are known and, for simplicity, that φ, μ_o, ρ_o, μ_w, and ρ_w are constants. The development for the estimation of k_r and φ is analogous to that for k_z. The performance index is given by

$$J = \int_{z_1}^{z_2} \int_0^T \left(q_o^{obs} - 2\pi r k_r \frac{k_{ro}\rho_o}{\mu_o} \frac{\partial p_o}{\partial r} \bigg|_{r_w} \right)^2 + \left(q_w^{obs} - 2\pi r k_r \frac{k_{rw}\rho_w}{\mu_w} \frac{\partial p_w}{\partial r} \bigg|_{r_w} \right)^2 dt\, dz \tag{63}$$

where T is the time period over which data are available. We note that $q_w = 0$ before breakthrough.

1. First Variation of State Equations

The change in p_o, p_w, S_w, and S_o resulting from a perturbation in k_z, δk_z, is given by the solution of

$$\frac{\partial \delta S_o}{\partial t} = \frac{\partial}{\partial z}\left[\delta k_z \alpha_o \left(\frac{\partial p_o}{\partial z} - \rho_o g \right) + k_z \beta_o \left(\frac{\partial p_o}{\partial z} - \rho_o g \right) \frac{dk_{ro}}{dS_w} \delta S_w + k_z \alpha_o \frac{\partial \delta p_o}{\partial z} \right] + \frac{1}{r}\frac{\partial}{\partial r}\left[r k_r \beta_o \frac{\partial p_o}{\partial r} \frac{dk_{ro}}{dS_w} \delta S_w + r k_r \alpha_o \frac{\partial \delta p_o}{\partial r} \right] \tag{64}$$

$$\frac{\partial \delta S_w}{\partial t} = \frac{\partial}{\partial z}\left[\delta k_z \alpha_w \left(\frac{\partial p_w}{\partial z} - \rho_w g \right) + k_z \beta_w \left(\frac{\partial p_w}{\partial z} - \rho_w g \right) \frac{dk_{rw}}{dS_w} \delta S_w + k_z \alpha_w \frac{\partial \delta p_w}{\partial z} \right] + \frac{1}{r}\frac{\partial}{\partial r}\left[r k_r \beta_w \frac{\partial p_w}{\partial r} \frac{dk_{rw}}{dS_w} \delta S_w + r k_r \alpha_w \frac{\partial \delta p_w}{\partial r} \right] \tag{65}$$

where

$$\alpha_o = \frac{k_{ro}}{\mu_o \varphi} \qquad \alpha_w = \frac{k_{rw}}{\mu_w \varphi}$$

$$\beta_o = \frac{1}{\mu_o \varphi} \qquad \beta_w = \frac{1}{\mu_w \varphi}$$

2. Introduction of Adjoint Variables

We now introduce the adjoint variables $\psi_1(r,z,t)$ and $\psi_2(r,z,t)$ and multiply Eq. (64) by ψ_1 and Eq. (65) by ψ_2 and add to give

$$\frac{\partial(\psi_1 \delta S_o)}{\partial t} + \frac{\partial(\psi_2 \delta S_w)}{\partial t} = \frac{\partial \psi_1}{\partial t}\,\delta S_o + \frac{\partial \psi_2}{\partial t}\,\delta S_w$$

$$+ \psi_1 \frac{\partial}{\partial z}\Big[\delta k_z \alpha_o \Big(\frac{\partial p_o}{\partial z} - \rho_o g\Big) + k_z \beta_o \Big(\frac{\partial p_o}{\partial z} - \rho_o g\Big)\frac{dk_{ro}}{dS_w}\,\delta S_w$$

$$+ k_z \alpha_o \frac{\partial \delta p_o}{\partial z}\Big] + \psi_1 \frac{1}{r}\frac{\partial}{\partial r}\Big[r k_r \beta_o \frac{\partial p_o}{\partial r}\frac{dk_{ro}}{dS_w}\,\delta S_w + r k_r \alpha_o \frac{\partial \delta p_o}{\partial r}\Big]$$

$$+ \psi_2 \frac{\partial}{\partial z}\Big[\delta k_z \alpha_w \Big(\frac{\partial p_w}{\partial z} - \rho_w g\Big) + k_z \beta_w \Big(\frac{\partial p_w}{\partial z} - \rho_w g\Big)\frac{dk_{rw}}{dS_w}\,\delta S_w$$

$$+ k_z \alpha_w \frac{\partial \delta p_w}{\partial z}\Big] + \psi_2 \frac{1}{r}\frac{\partial}{\partial r}\Big[r k_r \beta_w \frac{\partial p_w}{\partial r}\frac{dk_{rw}}{dS_w}\,\delta S_w + r k_r \alpha_w \frac{\partial \delta p_w}{\partial r}\Big] \qquad (66)$$

Using the following relations,

$$\psi_1 \frac{\partial}{\partial z}\Big[\delta k_z \alpha_o \Big(\frac{\partial p_o}{\partial z} - \rho_o g\Big)\Big] = \frac{\partial}{\partial z}\Big[\psi_1 \alpha_o \Big(\frac{\partial p_o}{\partial z} - \rho_o g\Big)\delta k_z\Big]$$

$$- \frac{\partial \psi_1}{\partial z}\alpha_o \Big(\frac{\partial p_o}{\partial z} - \rho_o g\Big)\delta k_z$$

$$\psi_1 \frac{\partial}{\partial z}\Big[k_z \beta_o \Big(\frac{\partial p_o}{\partial z} - \rho_o g\Big)\frac{dk_{ro}}{dS_w}\,\delta S_w\Big] = \frac{\partial}{\partial z}\Big[\psi_1 k_z \beta_o \Big(\frac{\partial p_o}{\partial z} - \rho_o g\Big)$$

$$\times \frac{dk_{ro}}{dS_w}\,\delta S_w\Big] - \frac{\partial \psi_1}{\partial z} k_z \beta_o \Big(\frac{\partial p_o}{\partial z} - \rho_o g\Big)\frac{dk_{ro}}{dS_w}\,\delta S_w$$

$$\psi_1 \frac{\partial}{\partial z}\Big[k_z \alpha_o \frac{\partial \delta p_o}{\partial z}\Big] = \frac{\partial}{\partial z}\Big[\psi_1 k_z \alpha_o \frac{\partial \delta p_o}{\partial z}\Big] - \frac{\partial}{\partial z}\Big[k_z \alpha_o \frac{\partial \psi_1}{\partial z}\,\delta p_o\Big]$$

$$+ \frac{\partial}{\partial z}\Big[k_z \alpha_o \frac{\partial \psi_1}{\partial z}\Big]\delta p_o$$

$$\psi_1 \frac{1}{r}\frac{\partial}{\partial r}\left[rk_r\beta_o \frac{\partial p_o}{\partial r}\frac{dk_{ro}}{dS_w}\delta S_w\right] = \frac{\partial}{\partial r}\left[\psi_1 k_r\beta_o \frac{\partial p_o}{\partial r}\frac{dk_{ro}}{dS_w}\delta S_w\right]$$

$$- \frac{\partial}{\partial r}\left(\frac{\psi_1}{r}\right) rk_r\beta_o \frac{\partial p_o}{\partial r}\frac{dk_{ro}}{dS_w}\delta S_w$$

$$\psi_1 \frac{1}{r}\frac{\partial}{\partial r}\left[rk_r\alpha_o \frac{\partial \delta p_o}{\partial r}\right] = \frac{\partial}{\partial r}\left[\psi_1 k_r\alpha_o \frac{\partial \delta p_o}{\partial r}\right] - \frac{\partial}{\partial r}\left[rk_r\alpha_o \frac{\partial}{\partial r}\left(\frac{\psi_1}{r}\right)\delta p_o\right]$$

$$+ \frac{\partial}{\partial r}\left[rk_r\alpha_o \frac{\partial}{\partial r}\left(\frac{\psi_1}{r}\right)\right]\delta p_o$$

with the analogous ones for ψ_2, Eq. (66) becomes

$$\frac{\partial(\psi_1\delta S_o)}{\partial t} + \frac{\partial(\psi_2\delta S_w)}{\partial t} = \frac{\partial\psi_1}{\partial t}\delta S_o + \frac{\partial\psi_2}{\partial t}\delta S_w$$

$$- \left[\frac{\partial\psi_1}{\partial z}\alpha_o\left(\frac{\partial p_o}{\partial z} - \rho_o g\right) + \frac{\partial\psi_2}{\partial z}\alpha_w\left(\frac{\partial p_w}{\partial z} - \rho_w g\right)\right]\delta k_z$$

$$- \frac{\partial\psi_1}{\partial z} k_z\beta_o\left(\frac{\partial p_o}{\partial z} - \rho_o g\right)\frac{dk_{ro}}{dS_w}\delta S_w - \frac{\partial}{\partial r}\left(\frac{\psi_1}{r}\right) rk_r\beta_o \frac{\partial p_o}{\partial r}\frac{dk_{ro}}{dS_w}\delta S_w$$

$$- \frac{\partial\psi_2}{\partial z} k_z\beta_w\left(\frac{\partial p_w}{\partial z} - \rho_w g\right)\frac{dk_{rw}}{dS_w}\delta S_w - \frac{\partial}{\partial r}\left(\frac{\psi_2}{r}\right) rk_r\beta_w \frac{\partial p_w}{\partial r}\frac{dk_{rw}}{dS_w}\delta S_w$$

$$+ \frac{\partial}{\partial z}\left(k_z\alpha_o \frac{\partial\psi_1}{\partial z}\right)\delta p_o + \frac{\partial}{\partial r}\left[rk_r\alpha_o \frac{\partial}{\partial r}\left(\frac{\psi_1}{r}\right)\right]\delta p_o$$

$$+ \frac{\partial}{\partial z}\left(k_z\alpha_w \frac{\partial\psi_2}{\partial z}\right)\delta p_w + \frac{\partial}{\partial r}\left[rk_r\alpha_w \frac{\partial}{\partial r}\left(\frac{\psi_2}{r}\right)\right]\delta p_w$$

$$+ \frac{\partial}{\partial z}\left[\psi_1\alpha_o\left(\frac{\partial p_o}{\partial z} - \rho_o g\right)\delta k_z + \psi_2\alpha_w\left(\frac{\partial p_w}{\partial z} - \rho_w g\right)\delta k_z\right.$$

$$+ \psi_1 k_z\beta_o\left(\frac{\partial p_o}{\partial z} - \rho_o g\right)\frac{dk_{ro}}{dS_w}\delta S_w$$

$$+ \psi_2 k_z\beta_w\left(\frac{\partial p_w}{\partial z} - \rho_w g\right)\frac{dk_{rw}}{dS_w}\delta S_w - k_z\alpha_o \frac{\partial\psi_1}{\partial z}\delta p_o$$

$$+ \psi_1 k_z \alpha_o \frac{\partial \delta p_o}{\partial z} - k_z \alpha_w \frac{\partial \psi_2}{\partial z} \delta p_w + \psi_2 k_z \alpha_w \frac{\partial \delta p_w}{\partial z} \Big]$$

$$+ \frac{\partial}{\partial r}\Big[\psi_1 k_r \beta_o \frac{\partial p_o}{\partial r} \frac{dk_{ro}}{dS_w} \delta S_w + \psi_2 k_r \beta_w \frac{\partial p_w}{\partial r} \frac{dk_{rw}}{dS_w} \delta S_w$$

$$- r k_r \alpha_o \frac{\partial}{\partial r}\left(\frac{\psi_1}{r}\right) \delta p_o + \psi_1 k_r \alpha_o \frac{\partial \delta p_o}{\partial r} - r k_r \alpha_w \frac{\partial}{\partial r}\left(\frac{\psi_2}{r}\right) \delta p_w$$

$$+ \psi_2 k_r \alpha_w \frac{\partial \delta p_w}{\partial r}\Big] \tag{67}$$

From Eqs. (55) and (56) we have

$$\delta S_o = - \delta S_w \tag{68}$$

and

$$\delta p_w = \delta p_o - \delta p_c = \delta p_o - \frac{dp_c}{dS_w} \delta S_w \tag{69}$$

Thus, we will treat the two dependent variables henceforth as p_o and S_w. Substituting Eqs. (68) and (69) into (67), we obtain

$$\frac{\partial(\psi_2 - \psi_1)\ \delta S_w}{\partial t} = \frac{\partial(\psi_2 - \psi_1)}{\partial t}\ \delta S_w - \Big[\frac{\partial \psi_1}{\partial z} \alpha_o \left(\frac{\partial p_o}{\partial z} - \rho_o g\right)$$

$$+ \frac{\partial \psi_2}{\partial z} \alpha_w \left(\frac{\partial p_w}{\partial z} - \rho_w g\right)\Big] \delta k_z$$

$$- \Big\{\frac{\partial \psi_1}{\partial z} k_z \beta_o \left(\frac{\partial p_o}{\partial z} - \rho_o g\right) \frac{dk_{ro}}{dS_w} + \frac{\partial}{\partial r}\left(\frac{1}{r}\psi_1\right) r k_r \beta_o \frac{\partial p_o}{\partial r} \frac{dk_{ro}}{dS_w}$$

$$+ \frac{\partial \psi_2}{\partial z} k_z \beta_w \left(\frac{\partial p_w}{\partial z} - \rho_w g\right) \frac{dk_{rw}}{dS_w}$$

$$+ \frac{\partial}{\partial r}\left(\frac{1}{r}\psi_2\right) r k_r \beta_w \frac{\partial p_w}{\partial r} \frac{dk_{ro}}{dS_w} + \frac{dp_c}{dS_w} \frac{\partial}{\partial z}\left(k_z \alpha_w \frac{\partial \psi_2}{\partial z}\right)$$

$$+ \frac{dp_c}{dS_w} \frac{\partial}{\partial r}\Big[r k_r \alpha_w \frac{\partial}{\partial r}\left(\frac{1}{r}\psi_2\right)\Big]\Big\} \delta S_w$$

$$
+ \left\{ \frac{\partial}{\partial z}\left(k_z \alpha_o \frac{\partial \psi_1}{\partial z}\right) + \frac{\partial}{\partial r}\left[r k_r \alpha_o \frac{\partial}{\partial r}\left(\frac{1}{r}\psi_1\right)\right] + \frac{\partial}{\partial z}\left(k_z \alpha_w \frac{\partial \psi_2}{\partial z}\right)\right.
$$

$$
\left. + \frac{\partial}{\partial r}\left[r k_r \alpha_w \frac{\partial}{\partial r}\left(\frac{1}{r}\psi_2\right)\right]\right\} \delta p_o
$$

$$
+ \frac{\partial}{\partial z}\left[\psi_1 \alpha_o \left(\frac{\partial p_o}{\partial z} - \rho_o g\right) \delta k_z + \psi_2 \alpha_w \left(\frac{\partial p_w}{\partial z} - \rho_w g\right) \delta k_z \right.
$$

$$
+ \psi_1 k_z \beta_o \left(\frac{\partial p_o}{\partial z} \; \rho_o g\right) \frac{dk_{ro}}{dS_w} \delta S_w
$$

$$
+ \psi_2 k_z \beta_w \left(\frac{\partial p_w}{\partial z} - \rho_w g\right) \frac{dk_{rw}}{dS_w} \delta S_w + k_z \alpha_w \frac{\partial \psi_2}{\partial z} \frac{dp_c}{dS_w} \delta S_w
$$

$$
- \psi_2 k_z \alpha_w \frac{dp_c}{dS_w} \frac{\partial \delta S_w}{\partial z}
$$

$$
\left. - k_z \alpha_o \frac{\partial \psi_1}{\partial z} \delta p_o - k_z \alpha_w \frac{\partial \psi_2}{\partial z} \delta p_o + \psi_1 k_z \alpha_o \frac{\partial \delta p_o}{\partial z} + \psi_2 k_z \alpha_w \frac{\partial \delta p_o}{\partial z} \right]
$$

$$
+ \frac{\partial}{\partial r}\left[\psi_1 k_r \beta_o \frac{\partial p_o}{\partial r} \frac{dk_{ro}}{dS_w} \delta S_w + \psi_2 k_r \beta_w \frac{\partial p_w}{\partial r} \frac{dk_{ro}}{dS_w} \delta S_w \right.
$$

$$
+ r k_r \alpha_w \frac{\partial}{\partial r}\left(\frac{1}{r}\psi_2\right) \frac{dp_c}{dS_w} \; \partial S_w
$$

$$
- \psi_2 k_r \alpha_w \frac{dp_c}{dS_w} \frac{\partial \delta S_w}{\partial r} - r k_r \alpha_o \frac{\partial}{\partial r}\left(\frac{1}{r}\psi_1\right) \delta p_o
$$

$$
- r k_r \alpha_w \frac{\partial}{\partial r}\left(\frac{1}{r}\psi_2\right) \delta p_o + \psi_1 k_r \alpha_o \frac{\partial \delta p_o}{\partial r}
$$

$$
\left. + \psi_2 k_r \alpha_w \frac{\partial \delta p_o}{\partial r} \right] \tag{70}
$$

Integrating both sides of Eq. (70) over the reservoir domain and over the time interval [0,T] and using the condition

$$
\delta S_w(r,z,0) = 0
$$

we obtain

$$\int_0^L\int_0^R (\psi_2 - \psi_1)\ \delta S_w \Big|_{t=T} dr\ dz = \int_0^T\int_0^L\int_0^R \Big\{ \frac{\partial(\psi_2 - \psi_1)}{\partial t} - k_z\beta_o\Big(\frac{\partial p_o}{\partial z}$$

$$- \rho_o g\Big)\frac{dk_{ro}}{dS_w}\frac{\partial\psi_1}{\partial z} - rk_r\beta_o\frac{\partial p_o}{\partial r}\frac{dk_{ro}}{dS_w}\frac{\partial}{\partial r}\Big(\frac{1}{r}\psi_1\Big) - k_z\beta_w\Big(\frac{\partial p_w}{\partial z}$$

$$- \rho_w g\Big)\frac{dk_{rw}}{dS_w}\frac{\partial\psi_2}{\partial z} - rk_r\beta_w\frac{\partial p_w}{\partial r}\frac{dk_{rw}}{dS_w}\frac{\partial}{\partial r}\Big(\frac{1}{r}\psi_2\Big)$$

$$- \frac{dp_c}{dS_w}\frac{\partial}{\partial z}\Big(k_z\alpha_w\frac{\partial\psi_2}{\partial z}\Big) - \frac{dp_c}{dS_w}\frac{\partial}{\partial r}\Big[rk_r\alpha_w\frac{\partial}{\partial r}\Big(\frac{1}{r}\psi_2\Big)\Big]\Big\}\ \delta S_w\ dr\ dz\ dt$$

$$+ \int_0^T\int_0^L\int_0^R \Big\{\frac{\partial}{\partial z}\Big(k_z\alpha_o\frac{\partial\psi_1}{\partial z}\Big) + \frac{\partial}{\partial r}\Big[rk_r\alpha_o\frac{\partial}{\partial r}\Big(\frac{1}{r}\psi_1\Big)\Big] + \frac{\partial}{\partial r}\Big(k_z\alpha_w\frac{\partial\psi_2}{\partial z}\Big)$$

$$+ \frac{\partial}{\partial r}\Big[rk_r\alpha_w\frac{\partial}{\partial r}\Big(\frac{1}{r}\psi_2\Big)\Big]\Big\}\ \delta p_o\ dr\ dz\ dt$$

$$+ \int_0^L\int_0^R\Big\{\int_0^T\Big[\frac{\partial\psi_1}{\partial z}\alpha_o\Big(\frac{\partial p_o}{\partial z} - \rho_o g\Big) + \frac{\partial\psi_2}{\partial z}\alpha_w\Big(\frac{\partial p_w}{\partial z} - \rho_w g\Big)\Big]dt\Big\}\ \delta k_z\ dr\,dz$$

$$+ \int_0^T\int_0^R\Big[\psi_1\alpha_o\Big(\frac{\partial p_o}{\partial z} - \rho_o g\Big) + \psi_2\alpha_w\Big(\frac{\partial p_w}{\partial z} - \rho_w g\Big)\Big]\ \delta k_z\ \Big|_{z=0}^{z=L} dr\ dt$$

$$+ \int_0^T\int_0^R\Big\{\Big[\psi_1 k_z\beta_o\Big(\frac{\partial p_o}{\partial z} - \rho_o g\Big)\frac{dk_{ro}}{dS_w} + \psi_2 k_z\beta_w\Big(\frac{\partial p_w}{\partial z} - \rho_w g\Big)\frac{dk_{rw}}{dS_w}$$

$$+ k_z\alpha_w\frac{\partial\psi_2}{\partial z}\frac{dp_c}{dS_w}\Big]\ \delta S_w$$

$$- \psi_2 k_z\alpha_w\frac{dp_c}{dS_w}\frac{\partial\delta S_w}{\partial z}\Big\}\ \Big|_{z=0}^{z=L} dr\ dt$$

$$+ \int_0^T\int_0^R\Big\{\Big[k_z\alpha_o\frac{\partial\psi_1}{\partial z} + k_z\alpha_w\frac{\partial\psi_2}{\partial z}\Big]\ \delta p_o - \Big[\psi_1 k_z\alpha_o + \psi_2 k_z\alpha_w\Big]$$

$$\times\frac{\partial\delta p_o}{\partial z}\Big\}\ \Big|_{z=0}^{z=L} dr\ dt$$

$$+ \int_0^T \int_0^L \left\{ \left[\psi_1 k_r \beta_o \frac{\partial p_o}{\partial r} \frac{dk_{ro}}{dS_w} + \psi_2 k_r \beta_w \frac{\partial p_w}{\partial r} \frac{dk_{rw}}{dS_w} + r k_r \alpha_w \frac{\partial}{\partial r} \right. \right.$$

$$\left. \left. \times \left(\frac{1}{r} \psi_2 \right) \frac{dp_c}{dS_w} \right] \delta S_w - \psi_2 k_r \alpha_w \frac{dp_c}{dS_w} \frac{\partial \delta S_w}{\partial r} \right\} \Bigg|_{r=r_w}^{r=R} dz\ dt$$

$$- \int_0^T \int_0^L \left\{ \left[r k_r \alpha_o \frac{\partial}{\partial r} \left(\frac{1}{r} \psi_1 \right) + r k_r \alpha_w \frac{\partial}{\partial r} \left(\frac{1}{r} \psi_2 \right) \right] \delta p_o \right.$$

$$\left. - \left[\psi_1 k_r \alpha_o + \psi_2 k_r \alpha_w \right] \frac{\partial \delta p_o}{\partial r} \right\} \Bigg|_{r=r_w}^{r=R} dz\ dt \tag{71}$$

3. First Variation of J

The change in J, δJ, resulting from the perturbation in k_z, is given by

$$\delta J = -2 \int_0^T \int_{z_1}^{z_2} \left[\left(q_o^{obs} - 2\pi r k_r \alpha_o \frac{\partial p_o}{\partial r} \right) \left(2\pi r k_r \alpha_o \frac{\partial \delta p_o}{\partial r} \right. \right.$$

$$\left. + 2\pi r k_r \beta_o \frac{\partial p_o}{\partial r} \frac{dk_{ro}}{dS_w} \delta S_w \right)$$

$$+ \left(q_w^{obs} - 2\pi r k_r \alpha_w \frac{\partial p_w}{\partial r} \right) \left(2\pi r k_r \alpha_w \frac{\partial \delta p_w}{\partial r} \right.$$

$$\left. \left. + 2\pi r k_r \beta_w \frac{\partial p_w}{\partial r} \frac{dk_{rw}}{dS_w} \delta S_w \right) \right] \Bigg|_{r=r_w} dz\ dt \tag{72}$$

Combining Eqs. (71) and (72), we obtain

$$\int_0^L \int_0^R (\psi_2 - \psi_1)\, \delta S_w \Bigg|_{t=T} dr\ dz + \delta J = \int_0^T \int_0^L \int_0^R \left\{ \frac{\partial (\psi_2 - \psi_1)}{\partial t} - k_z \beta_o \right.$$

$$\times \left(\frac{\partial p_o}{\partial z} - \rho_o g \right) \frac{dk_{ro}}{dS_w} \frac{\partial \psi_1}{\partial z} - r k_r \beta_o \frac{\partial p_o}{\partial r} \frac{dk_{ro}}{dS_w} \frac{\partial}{\partial r} \left(\frac{1}{r} \psi_1 \right)$$

$$- k_z \beta_w \left(\frac{\partial p_w}{\partial z} - \rho_w g \right) \frac{dk_{rw}}{dS_w} \frac{\partial \psi_2}{\partial z} - r k_r \beta_w \frac{\partial p_w}{\partial r} \frac{dk_{rw}}{dS_w} \frac{\partial}{\partial r} \left(\frac{1}{r} \psi_2 \right)$$

$$- \frac{dp_c}{dS_w} \frac{\partial}{\partial z}\left(k_z \alpha_w \frac{\partial \psi_2}{\partial z}\right) - \frac{dp_c}{dS_w} \frac{\partial}{\partial r}\left[r k_r \alpha_w \frac{\partial}{\partial r}\left(\frac{1}{r} \psi_2\right)\right]\Bigg\} \delta S_w \, dr \, dz \, dt$$

$$+ \int_0^T \int_0^L \int_0^R \Bigg\{ \frac{\partial}{\partial z}\left(k_z \alpha_o \frac{\partial \psi_1}{\partial z}\right) + \frac{\partial}{\partial z}\left[r k_r \alpha_o \frac{\partial}{\partial r}\left(\frac{1}{r} \psi_1\right)\right]$$

$$+ \frac{\partial}{\partial z}\left(k_z \alpha_w \frac{\partial \psi_2}{\partial z}\right)$$

$$+ \frac{\partial}{\partial r}\left[r k_r \alpha_w \frac{\partial}{\partial r}\left(\frac{1}{r} \psi_2\right)\right]\Bigg\} \delta p_o \, dr \, dz \, dt$$

$$- \int_0^L \int_0^R \Bigg\{ \int_0^T \Bigg[\frac{\partial \psi_1}{\partial z} \alpha_o \left(\frac{\partial p_o}{\partial z} - \rho_o g\right) + \frac{\partial \psi_2}{\partial z} \alpha_w \left(\frac{\partial p_w}{\partial z} - \rho_w g\right)$$

$$- \left(\psi_1 \alpha_o \left(\frac{\partial p_o}{\partial z} - \rho_o g\right) + \psi_2 \alpha_w \left(\frac{\partial p_w}{\partial z} - \rho_w g\right)\right) \delta(z - L)$$

$$+ \left(\psi_1 \alpha_o \left(\frac{\partial p_o}{\partial z} - \rho_o g\right) + \psi_2 \alpha_w \left(\frac{\partial p_w}{\partial z} - \rho_w g\right)\right) \delta(z) \Bigg] dt \Bigg\} \delta k_z \, dr \, dz$$

$$+ \int_0^T \int_0^R \Bigg\{ \Bigg[\psi_1 k_z \beta_o \left(\frac{\partial p_o}{\partial z} - \rho_o g\right) \frac{dk_{ro}}{dS_w} + \psi_2 k_z \beta_w \left(\frac{\partial p_w}{\partial z} - \rho_w g\right) \frac{dk_{rw}}{dS_w}$$

$$+ k_z \alpha_w \frac{\partial \psi_2}{\partial z} \frac{dp_c}{dS_w} \Bigg] \delta S_w$$

$$- \psi_2 k_z \alpha_w \frac{dp_c}{dS_w} \frac{\partial \delta S_w}{\partial z} \Bigg\} \Bigg|_{z=0}^{z=L} dr \, dt$$

$$- \int_0^T \int_0^R \Bigg\{ \Bigg[k_z \alpha_o \frac{\partial \psi_1}{\partial z} + k_z \alpha_w \frac{\partial \psi_2}{\partial z} \Bigg] \delta p_o - \Big[\psi_1 k_z \alpha_o + \psi_2 k_z \alpha_w \Big]$$

$$\times \frac{\partial \delta p_o}{\partial z} \Bigg\} \Bigg|_{z=0}^{z=L} dr \, dt$$

$$+ \int_0^T \int_0^L \Bigg\{ \Bigg[\psi_1 k_r \beta_o \frac{\partial p_o}{\partial r} \frac{dk_{ro}}{dS_w} + \psi_2 k_r \beta_w \frac{\partial p_w}{\partial r} \frac{dk_{rw}}{dS_w}$$

$$+ rk_r\alpha_w \frac{\partial}{\partial r}\left(\frac{1}{r}\psi_2\right)\frac{dp_c}{dS_w}\Big]\delta S_w - \psi_2 k_r\alpha_w \frac{dp_c}{dS_w}\frac{\partial\delta S_w}{\partial r}\Big\}\Big|_{r=R} dz\,dt$$

$$-\int_0^T\int_0^{z_1}\Big\{\Big[\psi_1 k_r\beta_o \frac{\partial p_o}{\partial r} + \psi_2 k_r\beta_w \frac{\partial p_w}{\partial r}\frac{dk_{rw}}{dS_w} + rk_r\alpha_w \frac{\partial}{\partial r}\left(\frac{1}{r}\psi_2\right)$$

$$\times \frac{dp_c}{dS_w}\Big]\delta S_w - \psi_2 k_r\alpha_w \frac{dp_c}{dS_w}\frac{\partial\delta S_w}{\partial r}\Big\}\Big|_{r=r_w} dz\,dt$$

$$-\int_0^T\int_{z_2}^L \Big\{\Big[\psi_1 k_r\beta_o \frac{\partial p_o}{\partial r}\frac{dk_{ro}}{dS_w} + \psi_2 k_r\beta_w \frac{\partial p_w}{\partial r}\frac{dk_{rw}}{dS_w} + rk_r\alpha_w \frac{\partial}{\partial r}.$$

$$\times \left(\frac{1}{r}\psi_2\right)\frac{dp_c}{dS_w}\Big]\delta S_w - \psi_2 k_r\alpha_w \frac{dp_c}{dS_w}\frac{\partial\delta S_w}{\partial r}\Big\}\Big|_{r=r_w} dz\,dt$$

$$-\int_0^T\int_{z_1}^{z_2}\Big\{\Big[\psi_1 k_r\beta_o \frac{\partial p_o}{\partial r}\frac{dk_{ro}}{dS_w} + \psi_2 k_r\beta_w \frac{\partial p_w}{\partial r}\frac{dk_{ro}}{dS_w} + rk_r\alpha_w \frac{\partial}{\partial r}$$

$$\times \left(\frac{1}{r}\psi_2\right)\frac{dp_c}{dS_w} + 4\pi rk_r\left(q_o^{obs} - 2\pi rk_r \frac{k_{ro}\rho_o}{\mu_o}\frac{\partial p_o}{\partial r}\right)\frac{\rho_o}{\mu_o}\frac{\partial p_o}{\partial r}\frac{dk_{ro}}{dS_w}$$

$$+ 4\pi rk_r\left(q_w^{obs} - 2\pi rk_r \frac{k_{rw}\rho_w}{\mu_w}\frac{\partial p_w}{\partial r}\right)\frac{\rho_w}{\mu_w}\frac{\partial p_w}{\partial r}\frac{dk_{ro}}{dS_w}\Big]\delta S_w$$

$$-\Big[\psi_2 k_r\alpha_w \frac{dp_c}{dS_w} + 4\pi rk_r\left(q_w^{obs} - 2\pi rk_r \frac{k_{rw}\rho_w}{\mu_w}\frac{\partial p_w}{\partial r}\right)\frac{k_{rw}\rho_w}{\mu_w}\frac{dp_c}{dS_w}\Big]$$

$$\times \frac{\partial\delta S_w}{\partial r}\Big\}\Big|_{r=r_w} dz\,dt$$

$$-\int_0^T\int_0^L \Big\{\Big[rk_r\alpha_o \frac{\partial}{\partial r}\left(\frac{1}{r}\psi_1\right) + rk_r\alpha_w \frac{\partial}{\partial r}\left(\frac{1}{r}\psi_2\right)\Big]\delta p_o$$

$$-\Big[\psi_1 k_r\alpha_o + \psi_2 k_r\alpha_w\Big]\frac{\partial\delta p_o}{\partial r}\Big\}\Big|_{r=R} dz\,dt$$

$$+\int_0^T\int_0^{z_1}\left\{\left[rk_r\alpha_o\frac{\partial}{\partial r}\left(\frac{1}{r}\psi_1\right)+rk_r\alpha_w\frac{\partial}{\partial r}\left(\frac{1}{r}\psi_2\right)\right]\delta p_o\right.$$

$$\left.-\left[\psi_1k_r\alpha_o+\psi_2k_r\alpha_w\right]\frac{\partial\delta p_o}{\partial r}\right\}\Big|_{r=r_w}dz\ dt$$

$$+\int_0^T\int_{z_2}^{L}\left\{\left[rk_r\alpha_o\frac{\partial}{\partial r}\left(\frac{1}{r}\psi_1\right)+rk_r\alpha_w\frac{\partial}{\partial r}\left(\frac{1}{r}\psi_2\right)\right]\delta p_o\right.$$

$$\left.-\left[\psi_1k_r\alpha_o+\psi_2k_r\alpha_w\right]\frac{\partial\delta p_o}{\partial r}\right\}\Big|_{r=r_w}dz\ dt$$

$$+\int_0^T\int_{z_1}^{z_2}\left\{\left[rk_r\alpha_o\frac{\partial}{\partial r}\left(\frac{1}{r}\psi_1\right)+rk_r\alpha_w\frac{\partial}{\partial r}\left(\frac{1}{r}\psi_2\right)\right]\delta p_o\right.$$

$$-\left[\psi_1k_r\alpha_o+\psi_2k_r\alpha_w\right.$$

$$+4\pi rk_r\left(q_o^{obs}-2\pi rk_r\frac{k_{ro}\rho_o}{\mu_o}\frac{\partial p_o}{\partial r}\right)\frac{k_{ro}\rho_o}{\mu_o}$$

$$\left.\left.+4\pi rk_r\left(q_w^{obs}-2\pi rk_r\frac{k_{rw}\rho_w}{\mu_w}\frac{\partial p_w}{\partial r}\right)\frac{k_{rw}\rho_w}{\mu_w}\right]\frac{\partial\delta p_o}{\partial r}\right\}\Big|_{r=r_w}dz\,dt \qquad (73)$$

From the boundary conditions we obtain the relations

$$\frac{\partial\delta p_\ell}{\partial r}=0 \qquad \text{on } \partial\Omega_2 \quad (\ell=o,w) \qquad (74)$$

$$\left.\begin{aligned}\frac{\partial\delta p_o}{\partial z}&=0\\ \delta p_o&=0\end{aligned}\right\} \quad \text{on } \partial\Omega_1 \text{ before breakthrough} \qquad (75)$$

$$\frac{\partial\delta p_o}{\partial z}=0,$$

$$\frac{\partial \delta p_w}{\partial z} = 0 \quad \text{on } \partial\Omega_1 \text{ after breakthrough} \tag{76}$$

$$\delta p_o = \delta p_w = 0$$

Substituting Eqs. (74) to (76) into Eq. (73) gives

$$\int_0^L\int_0^R (\psi_2 - \psi_1)\,\delta S_w \Big|_{t=T} dr\,dz + \delta J = \int_0^T\int_0^L\int_0^R \Big\{ \frac{\partial(\psi_2 - \psi_1)}{\partial t}$$

$$- k_z\beta_o\left(\frac{\partial p_o}{\partial z} - \rho_o g\right)\frac{dk_{ro}}{dS_w}\frac{\partial \psi_1}{\partial z}$$

$$- rk_r\beta_o\frac{\partial p_o}{\partial r}\frac{dk_{ro}}{dS_w}\frac{\partial}{\partial r}\left(\frac{1}{r}\psi_1\right) - k_z\beta_w\left(\frac{\partial p_w}{\partial z} - \rho_w g\right)\frac{dk_{rw}}{dS_w}\frac{\partial \psi_2}{\partial z}$$

$$- rk_r\beta_w\frac{\partial p_w}{\partial r}\frac{dk_{rw}}{dS_w}\frac{\partial}{\partial r}\left(\frac{1}{r}\psi_2\right)$$

$$- \frac{dp_c}{dS_w}\frac{\partial}{\partial z}\left(k_z\alpha_w\frac{\partial \psi_2}{\partial z}\right) - \frac{dp_c}{dS_w}\frac{\partial}{\partial r}\left[rk_r\alpha_w\frac{\partial}{\partial r}\left(\frac{1}{r}\psi_2\right)\right]\Big\}\,\delta S_w\,dr\,dz\,dt$$

$$+ \int_0^T\int_0^L\int_0^R \Big\{\frac{\partial}{\partial z}\left(k_z\alpha_o\frac{\partial \psi_1}{\partial z}\right) + \frac{\partial}{\partial r}\left[rk_r\alpha_o\frac{\partial}{\partial r}\left(\frac{1}{r}\psi_1\right)\right]$$

$$+ \frac{\partial}{\partial z}\left(k_z\alpha_w\frac{\partial \psi_2}{\partial z}\right) + \frac{\partial}{\partial r}\left[rk_r\alpha_w\frac{\partial}{\partial r}\left(\frac{1}{r}\psi_2\right)\right]\Big\}\,\delta p_o\,dr\,dz\,dt$$

$$- \int_0^L\int_0^R \Big\{\int_0^T \Big[\frac{\partial \psi_1}{\partial z}\alpha_o\left(\frac{\partial p_o}{\partial z} - \rho_o g\right) + \frac{\partial \psi_2}{\partial z}\alpha_w\left(\frac{\partial p_w}{\partial z} - \rho_w g\right)$$

$$- \left(\psi_1\alpha_o\left(\frac{\partial p_o}{\partial z} - \rho_o g\right) + \psi_2\alpha_w\left(\frac{\partial p_w}{\partial z} - \rho_w g\right)\right)\delta(z - L)$$

$$+ \left(\psi_1\alpha_o\left(\frac{\partial p_o}{\partial z} - \rho_o g\right) + \psi_2\alpha_w\left(\frac{\partial p_w}{\partial z} - \rho_w g\right)\right)\delta(z)\Big]\,dt\Big\}\,\delta k_z\,dr\,dz$$

$$+ \int_0^T\int_0^R k_z\alpha_w\frac{\partial \psi_2}{\partial z}\frac{dp_c}{dS_w}\,\delta S_w\Big|_{z=0}^{z=L} dr\,dt - \int_0^T\int_0^R \Big[k_z\alpha_o\frac{\partial \psi_1}{\partial z}$$

$$+ k_z\alpha_w \frac{\partial\psi_2}{\partial z}\Big] \delta p_o \Big|_{z=0}^{z=L} dr\ dt$$

$$+ \int_0^T\int_0^L rk_r\alpha_w \frac{\partial}{\partial r}\left(\frac{1}{r}\psi_2\right)\frac{dp_c}{dS_w}\delta S_w \Big|_{r=R} dz\ dt$$

$$- \int_0^T\int_0^{z_1} rk_r\alpha_w \frac{\partial}{\partial r}\left(\frac{1}{r}\psi_2\right)\frac{dp_c}{dS_w}\delta S_w \Big|_{r=r_w} dz\ dt$$

$$- \int_0^T\int_{z_2}^L rk_r\alpha_w \frac{\partial}{\partial r}\left(\frac{1}{r}\psi_2\right)\frac{dp_c}{dS_w}\delta S_w \Big|_{r=r_w} dz\ dt$$

$$+ \int_0^T\int_{z_1}^{z_2} \Big[\psi_2 k_r\alpha_w \frac{dp_c}{dS_w} + 4\pi rk_r\left(q_w^{obs} - 2\pi rk_r \frac{k_{rw}\rho_w}{\mu_w}\frac{\partial p_w}{\partial r}\right)$$

$$\times \frac{k_{rw}\rho_w}{\mu_w}\frac{dp_c}{dS_w}\Big] \frac{\partial\delta S_w}{\partial r}\Big|_{r=r_w} dz\ dt$$

$$- \int_0^T\int_0^L \Big[rk_r\alpha_o \frac{\partial}{\partial r}\left(\frac{1}{r}\psi_1\right) + rk_r\alpha_w \frac{\partial}{\partial r}\left(\frac{1}{r}\psi_2\right)\Big] \delta p_o \Big|_{r=R} dz\ dt$$

$$+ \int_0^T\int_0^{z_1} \Big[rk_r\alpha_o \frac{\partial}{\partial r}\left(\frac{1}{r}\psi_1\right) + rk_r\alpha_w \frac{\partial}{\partial r}\left(\frac{1}{r}\psi_2\right)\Big] \delta p_o \Big|_{r=r_w} dz\ dt$$

$$+ \int_0^T\int_{z_2}^L \Big[\psi_1 k_r\alpha_o \frac{\partial}{\partial r}\left(\frac{1}{r}\psi_1\right) + rk_r\alpha_w \frac{\partial}{\partial r}\left(\frac{1}{r}\psi_2\right)\Big] \delta p_o \Big|_{r=r_w} dz\ dt$$

$$- \int_0^T\int_{z_1}^{z_2} \Big[\psi_1 k_r\alpha_o + \psi_2 k_r\alpha_w + 4\pi rk_r\left(q_o^{obs} - 2\pi rk_r \frac{k_{ro}\rho_o}{\mu_o}\frac{\partial p_o}{\partial r}\right)\frac{k_{ro}\rho_o}{\mu_o}$$

$$+ 4\pi rk_r\left(q_w^{obs} - 2\pi rk_r \frac{k_{rw}\rho_w}{\mu_w}\frac{\partial p_w}{\partial r}\right)\frac{k_{rw}\rho_w}{\mu_w}\Big] \frac{\partial\delta p_o}{\partial r}\Big|_{r=r_w} dz\, dt \quad (77)$$

4. Adjoint Equations

The objective is to define ψ_1 and ψ_2 in such a way that a relation between δJ and δk_z is obtained from Eq. (77). We therefore specify that ψ_1 and ψ_2 be governed by

$$\begin{aligned}\frac{\partial(\psi_2 - \psi_1)}{\partial t} = {} & k_z\beta_o\left(\frac{\partial p_o}{\partial z} - \rho_o g\right)\frac{dk_{ro}}{dS_w}\frac{\partial\psi_1}{\partial z} \\ & + rk_r\beta_o \frac{\partial p_o}{\partial r}\frac{dk_{ro}}{dS_w}\frac{\partial}{\partial r}\left(\frac{\psi_1}{r}\right) \\ & + k_z\beta_w\left(\frac{\partial p_w}{\partial z} - \rho_w g\right)\frac{dk_{rw}}{dS_w}\frac{\partial\psi_2}{\partial z} \\ & + rk_r\beta_w \frac{\partial p_w}{\partial r}\frac{dk_{rw}}{dS_w}\frac{\partial}{\partial r}\left(\frac{\psi_2}{r}\right) \\ & + \frac{dp_c}{dS_w}\frac{\partial}{\partial z}\left(k_z\alpha_w\frac{\partial\psi_2}{\partial z}\right) + \frac{dp_c}{dS_w}\frac{\partial}{\partial r} \\ & \times\left[rk_r\alpha_w\frac{\partial}{\partial r}\left(\frac{\psi_2}{r}\right)\right]\end{aligned} \tag{78}$$

and

$$\begin{aligned}& \frac{\partial}{\partial z}\left(k_z\alpha_o\frac{\partial\psi_1}{\partial z}\right) + \frac{\partial}{\partial r}\left[rk_r\alpha_o\frac{\partial}{\partial r}\left(\frac{\psi_1}{r}\right)\right] \\ & \quad + \frac{\partial}{\partial z}\left(k_z\alpha_w\frac{\partial\psi_2}{\partial z}\right) + \frac{\partial}{\partial r}\left[rk_r\alpha_w\frac{\partial}{\partial r}\left(\frac{\psi_2}{r}\right)\right] = 0\end{aligned} \tag{79}$$

Equation (78) is solved subject to the final condition

$$\psi_2 - \psi_1 = 0, \qquad t = T \tag{80}$$

The boundary conditions for Eqs. (78) and (79) are chosen such that the last 10 terms on the right-hand side of Eq. (77) vanish. Because the variations of δS_w and δp_o are arbitrary on $\partial\Omega_2$ and the variations of $\partial\delta S_w/\partial r$ and $\partial\delta p_o/\partial r$ are arbitrary on $\partial\Omega_1$, we obtain the conditions

$$\left.\begin{aligned} \frac{\partial}{\partial r}\left(\frac{\psi_1}{\psi}\right) &= 0 \\ \frac{\partial}{\partial r}\left(\frac{\psi_2}{r}\right) &= 0, \end{aligned}\right\} \quad \begin{aligned} &\text{at } r = r_w,\ 0 \le z < z_1,\ z_2 < z \le L \\ &r = R, \end{aligned} \tag{81}$$

$$\frac{\partial \psi_1}{\partial z} = \frac{\partial \psi_2}{\partial z} = 0, \qquad z = 0, L \tag{82}$$

$$\left.\begin{aligned} \psi_1 k_r \alpha_o + 4\pi r k_r \frac{k_{ro}\rho_o}{\mu_o}\left(q_o^{obs} - 2\pi r k_r \frac{k_{ro}\rho_o}{\mu_o}\frac{\partial p_o}{\partial r}\right) &= 0 \\ \psi_2 k_r \alpha_w + 4\pi r k_r \frac{k_{rw}\rho_w}{\mu_w}\left(q_w^{obs} - 2\pi r k_r \frac{k_{rw}\rho_w}{\mu_w}\frac{\partial p_w}{\partial r}\right) &= 0 \end{aligned}\right. \quad \text{on } \partial\Omega_1 \tag{83}$$

5. Necessary Conditions for Optimality

Substituting Eqs. (78) to (83) into Eq. (77), we obtain

$$\delta J = -\int_0^L\int_0^R \left\{ \int_0^T \left[\frac{\partial \psi_1}{\partial z}\alpha_o\left(\frac{\partial p_o}{\partial z} - \rho_o g\right) + \frac{\partial \psi_2}{\partial z}\alpha_w\left(\frac{\partial p_w}{\partial z} - \rho_w g\right)\right] dt\right\}$$
$$\delta k_z \, dr\, dz \tag{84}$$

Note that we have used Eq. (57) to arrive at (84). From (84), we conclude that the necessary condition for optimality is

$$\int_0^T \left[\frac{\partial \psi_1}{\partial z}\alpha_o\left(\frac{\partial p_o}{\partial z} - \rho_o g\right) + \frac{\partial \psi_2}{\partial z}\alpha_w\left(\frac{\partial p_w}{\partial z} - \rho_w g\right)\right] dt = 0 \tag{85}$$

To place the adjoint equations in the form of the state equations, let $\Phi_1 = \psi_2 - \psi_1$ and $\Phi_1 + \Phi_2 = 0$. Then the adjoint equations, (78) and (79), become

$$\frac{\partial \Phi_1}{\partial t} = k_z \beta_o\left(\frac{\partial p_o}{\partial z} - \rho_o g\right)\frac{dk_{ro}}{dS_w}\frac{\partial \psi_1}{\partial z} + r k_r \beta_o \frac{\partial p_o}{\partial r}\frac{dk_{ro}}{dS_w}\frac{\partial}{\partial r}\left(\frac{1}{r}\psi_1\right)$$
$$+ k_z \beta_w\left(\frac{\partial p_w}{\partial z} - \rho_w g\right)\frac{dk_{rw}}{dS_w}\frac{\partial \psi_2}{\partial z}$$
$$+ r k_r \beta_w \frac{\partial p_w}{\partial r}\frac{dk_{ro}}{dS_w}\frac{\partial}{\partial r}\left(\frac{1}{r}\psi_2\right) + \frac{dp_o}{dS_w}\frac{\partial}{\partial z}\left(k_z \alpha_w \frac{\partial \psi_2}{\partial z}\right)$$

$$+ \frac{dp_c}{dS_w} \frac{\partial}{\partial r}\left[rk_r\alpha_w \frac{\partial}{\partial r}\left(\frac{1}{r}\psi_2\right)\right] \tag{86}$$

and

$$\frac{\partial \Phi_2}{\partial t} = \frac{\partial}{\partial z}\left(k_z\alpha_o \frac{\partial \psi_1}{\partial z}\right) - k_z\beta_o\left(\frac{\partial p_o}{\partial z} - \rho_o g\right)\frac{dk_{ro}}{dS_w}\frac{\partial \psi_1}{\partial z}$$

$$+ \frac{\partial}{\partial r}\left[rk_r\alpha_o \frac{\partial}{\partial r}\left(\frac{1}{r}\psi_1\right)\right]$$

$$- rk_r\beta_o \frac{\partial p_o}{\partial r}\frac{dk_{ro}}{dS_w}\frac{\partial}{\partial r}\left(\frac{1}{r}\psi_1\right) + \left(1 - \frac{dp_c}{dS_w}\right)\frac{\partial}{\partial z}\left(k_z\alpha_w \frac{\partial \psi_2}{\partial z}\right)$$

$$- k_z\beta_w\left(\frac{\partial p_w}{\partial z} - \rho_w g\right)\frac{dk_{ro}}{dS_w}\frac{\partial \psi_2}{\partial z}$$

$$+ \left(1 - \frac{dp_c}{dS_w}\right)\frac{\partial}{\partial r}\left[rk_r\alpha_w \frac{\partial}{\partial r}\left(\frac{1}{r}\psi_2\right)\right] - rk_r\beta_w \frac{\partial p_w}{\partial r}\frac{dk_{rw}}{dS_w}$$

$$\times \frac{\partial}{\partial r}\left(\frac{1}{r}\psi_2\right) \tag{87}$$

B. Relative Permeability

In this section we consider the problem of estimating the relative permeabilities $k_{ro}(S_w)$ and $k_{rw}(S_w)$, to minimize J given by Eq. (63). We note that k_{ro} and k_{rw} are unknown functions of the state variable S_w. This type of problem seems to be rather unique in identification problems. We assume that the absolute permeabilities are known and, as before, that φ, μ_o, ρ_o, μ_w, and ρ_w are constant. We develop an algorithm for estimating k_{ro} and k_{rw} based on optimal control theory.

1. First Variation of State Equations

Perturbations in k_{ro}, $\delta\bar{k}_{ro}$, and in k_{rw}, $\delta\bar{k}_{rw}$, are given by

$$\delta\bar{k}_{ro} = \bar{k}_{ro}(\bar{S}_w) - k_{ro}(S_w) \tag{88}$$

$$\delta\bar{k}_{rw} = \bar{k}_{rw}(\bar{S}_w) - k_{rw}(S_w) \tag{89}$$

where $\bar{S}_w$ is the solution of Eqs. (53) to (62) with $\bar{k}_{ro}$ and $\bar{k}_{rw}$, and S_w is the solution of (53) to (62) with k_{ro} and k_{rw}.

For a first-order approximation, $\delta\bar{k}_{ro}$ and $\delta\bar{k}_{rw}$ can be expressed as

$$\delta\bar{k}_{ro} = k_{ro}(\bar{S}_w) + \delta k_{ro}(\bar{S}_w) - k_{ro}(S)_w$$

$$\simeq k_{ro}(S_w) + \frac{dk_{ro}}{dS_w}\,\delta S_w + \delta k_{ro}(S_w) - k_{ro}(S_w)$$

$$= \delta k_{ro}(S_w) + \frac{dk_{ro}}{dS_w}\,\delta S_w \tag{90}$$

and

$$\delta\bar{k}_{rw} \simeq \delta k_{rw}(S_w) + \frac{dk_{rw}}{dS_w}\,\delta S_w \tag{91}$$

where $\delta S_w = \bar{S}_w - S_w$.

Keeping Eqs. (90) and (91) in mind, the change in p_o, p_w, S_w, and S_o resulting from a perturbation in k_{ro} and a perturbation in k_{rw} is given by the solution of

$$\frac{\partial \delta S_o}{\partial t} = \frac{\partial}{\partial z}\left[\alpha_{zo}\left(\delta k_{ro} + \frac{dk_{ro}}{dS_w}\,\delta S_w\right)\left(\frac{\partial p_o}{\partial z} - \rho_o g\right) + \alpha_{zo}k_{ro}\frac{\partial \delta p_o}{\partial z}\right]$$

$$+ \frac{1}{r}\frac{\partial}{\partial r}\left[r\alpha_{ro}\left(\delta k_{ro} + \frac{dk_{ro}}{dS_w}\,\delta S_w\right)\frac{\partial p_o}{\partial r} + r\alpha_{ro}k_{ro}\frac{\partial \delta p_o}{\partial r}\right] \tag{92}$$

and

$$\frac{\partial \delta S_w}{\partial t} = \frac{\partial}{\partial z}\left[\alpha_{zw}\left(\delta k_{rw} + \frac{dk_{rw}}{dS_w}\,\delta S_w\right)\left(\frac{\partial p_w}{\partial z} - \rho_w g\right) + \alpha_{zw}k_{rw}\frac{\partial \delta p_w}{\partial z}\right]$$

$$+ \frac{1}{r}\frac{\partial}{\partial r}\,r\left[\alpha_{rw}\left(\delta k_{rw} + \frac{dk_{rw}}{dS_w}\,\delta S_w\right)\frac{\partial p_w}{\partial r} + r\alpha_{rw}k_{rw}\frac{\partial \delta p_w}{\partial r}\right] \tag{93}$$

where $\alpha_{zo} = k_z/\mu_o\varphi$, $\alpha_{zw} = k_z/\mu_w\varphi$, $\alpha_{ro} = k_r/\mu_o\varphi$, and $\alpha_{rw} = k_r/\mu_w\varphi$.

2. Introduction of Adjoint Variables

We now introduce the adjoint variables $\psi_1(r,z,t)$ and $\psi_2(r,z,t)$ and multiply Eq. (92) by ψ_1 and Eq. (93) by ψ_2 and add to give

$$\frac{\partial(\psi_1 \delta S_o)}{\partial t} + \frac{\partial(\psi_2 \delta S_w)}{\partial t} = \frac{\partial \psi_1}{\partial t}\delta S_o + \frac{\partial \psi_2}{\partial t}\delta S_w$$

$$+ \psi_1 \frac{\partial}{\partial z}\left[\alpha_{zo}\left(\delta k_{ro} + \frac{dk_{ro}}{dS_w}\delta S_w\right)\left(\frac{\partial p_o}{\partial z} - \rho_o g\right) + \alpha_{zo}k_{ro}\frac{\partial \delta p_o}{\partial z}\right]$$

$$+ \psi_1 \frac{1}{r}\frac{\partial}{\partial r}\left[r\alpha_{ro}\left(\delta k_{ro} + \frac{dk_{ro}}{dS_w}\delta S_w\right)\frac{\partial p_o}{\partial r} + r\alpha_{ro}k_{ro}\frac{\partial \delta p_o}{\partial r}\right]$$

$$+ \psi_2 \frac{\partial}{\partial z}\left[\alpha_{zw}\left(\delta k_{rw} + \frac{dk_{rw}}{dS_w}\delta S_w\right)\left(\frac{\partial p_w}{\partial z} - \rho_w g\right) + \alpha_{zw}k_{rw}\frac{\partial \delta p_w}{\partial z}\right]$$

$$+ \psi_2 \frac{1}{r}\frac{\partial}{\partial r}\left[r\alpha_{rw}\left(\delta k_{rw} + \frac{dk_{rw}}{dS_w}\delta S_w\right)\frac{\partial p_w}{\partial r} + r\alpha_{rw}k_{rw}\frac{\partial \delta p_w}{\partial r}\right] \qquad (94)$$

Integrating by parts and using Eqs. (68) and (69), (94) becomes

$$\frac{\partial(\psi_2 - \psi_1)\delta S_w}{\partial t} = \frac{\partial(\psi_2 - \psi_1)}{\partial t}\delta S_w - \left[\frac{\partial \psi_1}{\partial z}\alpha_{zo}\left(\frac{\partial p_o}{\partial z} - \rho_o g\right) + \frac{\partial}{\partial r}\left(\frac{\psi_1}{r}\right)r\alpha_{ro}\right.$$

$$\left.\times \frac{\partial p_o}{\partial r}\right]\delta k_{ro} - \left[\frac{\partial \psi_2}{\partial z}\alpha_{zw}\left(\frac{\partial p_w}{\partial z} - \rho_w g\right) + \frac{\partial}{\partial r}\left(\frac{\psi_2}{r}\right)r\alpha_{rw}\frac{\partial p_w}{\partial r}\right]\delta k_{rw}$$

$$- \left\{\frac{\partial \psi_1}{\partial z}\alpha_{zo}\left(\frac{\partial p_o}{\partial z} - \rho_o g\right)\frac{dk_{ro}}{dS_w} + \frac{\partial}{\partial r}\left(\frac{\psi_1}{r}\right)r\alpha_{ro}\frac{\partial p_o}{\partial r}\frac{dk_{ro}}{dS_w}\right.$$

$$+ \frac{\partial \psi_2}{\partial z}\alpha_{zw}\left(\frac{\partial p_w}{\partial z} - \rho_w g\right)\frac{dk_{rw}}{dS_w} + \frac{\partial}{\partial r}\left(\frac{\psi_2}{r}\right)r\alpha_{rw}\frac{\partial p_w}{\partial r}\frac{dk_{rw}}{dS_w}$$

$$\left. + \frac{dp_c}{dS_w}\frac{\partial}{\partial z}\left(\alpha_{zw}k_{rw}\frac{\partial \psi_2}{\partial z}\right) + \frac{dp_c}{dS_w}\frac{\partial}{\partial r}\left[r\alpha_{rw}k_{rw}\frac{\partial}{\partial r}\left(\frac{\psi_2}{r}\right)\right]\right\}\delta S_w$$

$$+ \left\{\frac{\partial}{\partial z}\left(\alpha_{zo}k_{ro}\frac{\partial \psi_1}{\partial z}\right) + \frac{\partial}{\partial r}\left[r\alpha_{ro}k_{ro}\frac{\partial}{\partial r}\left(\frac{\psi_1}{r}\right)\right]\right.$$

$$+ \frac{\partial}{\partial z}\left[\alpha_{zo}k_{rw}\frac{\partial\psi_2}{\partial z}\right] + \frac{\partial}{\partial r}\left[r\alpha_{rw}k_{rw}\frac{\partial}{\partial r}\left(\frac{\psi_2}{r}\right)\right]\Bigg\}\,\delta p_o$$

$$+ \frac{\partial}{\partial z}\Bigg[\psi_1\alpha_{zo}\left(\frac{\partial p_o}{\partial z} - \rho_o g\right)\delta k_{ro} + \psi_2\alpha_{zw}\left(\frac{\partial p_w}{\partial z} - \rho_w g\right)\delta k_{rw}$$

$$+ \psi_1\alpha_{zo}\left(\frac{\partial p_o}{\partial z} - \rho_o g\right)\frac{dk_{ro}}{dS_w}\,\delta S_w$$

$$+ \psi_2\alpha_{zw}\left(\frac{\partial p_w}{\partial z} - \rho_w g\right)\frac{dk_{rw}}{dS_w}\,\delta S_w + \alpha_{zw}k_{rw}\frac{\partial\psi_2}{\partial z}\frac{dp_c}{dS_w}\,\delta S_w$$

$$- \psi_2\alpha_{zw}k_{rw}\frac{\partial}{\partial z}\left(\frac{dp_c}{dS_w}\right)\delta S_w$$

$$- \psi_2\alpha_{zw}k_{rw}\frac{dp_c}{dS_w}\frac{\partial\delta S_w}{\partial z} - \alpha_{zo}k_{ro}\frac{\partial\psi_1}{\partial z}\,\delta p_o - \alpha_{zw}k_{rw}\frac{\partial\psi_2}{\partial z}\,\delta p_o$$

$$+ \psi_1\alpha_{zo}k_{ro}\frac{\partial\delta p_o}{\partial z} + \psi_2\alpha_{zw}k_{rw}\frac{\partial\delta p_o}{\partial z}\Bigg]$$

$$+ \frac{\partial}{\partial r}\Bigg[\psi_1\alpha_{ro}\frac{\partial p_o}{\partial r}\,\delta k_{ro} + \psi_2\alpha_{rw}\frac{\partial p_w}{\partial r}\,\delta k_{rw} + \psi_1\alpha_{ro}\frac{\partial p_o}{\partial r}\frac{dk_{ro}}{dS_w}\,\delta S_w$$

$$+ \psi_2\alpha_{rw}\frac{\partial p_w}{\partial r}\frac{dk_{rw}}{dS_w}\,\delta S_w + r\alpha_{rw}k_{rw}\frac{\partial}{\partial r}\left(\frac{\psi_2}{r}\right)\frac{dp_c}{dS_w}\,\delta S_w$$

$$- \psi_2\alpha_{rw}k_{rw}\frac{\partial}{\partial r}\left(\frac{dp_c}{dS_w}\right)\delta S_w$$

$$- \psi_2\alpha_{rw}k_{rw}\frac{dp_c}{dS_w}\frac{\partial\delta S_w}{\partial r} - r\alpha_{ro}k_{ro}\frac{\partial}{\partial r}\left(\frac{\psi_1}{r}\right)\delta p_o$$

$$- r\alpha_{rw}k_{rw}\frac{\partial}{\partial r}\left(\frac{1}{r}\psi_2\right)\delta p_o$$

$$+ \psi_1\alpha_{ro}k_{ro}\frac{\partial\delta p_o}{\partial r} + \psi_2\alpha_{rw}k_{rw}\frac{\partial\delta p_o}{\partial r}\Bigg] \tag{95}$$

Integrating both sides of Eq. (95) over the reservoir domain and over the time interval [0,T] and using the condition $\delta S_w(r,z,0) = 0$, we obtain

$$\int_0^L\int_0^R (\psi_2 - \psi_1)\ \delta S_w \Big|_{t=T} dr\ dz = \int_0^T\int_0^L\int_0^R \Big\{ \frac{\partial(\psi_2 - \psi_1)}{\partial t}$$

$$- \alpha_{zo}\left(\frac{\partial p_o}{\partial r} - \rho_o g\right)\frac{dk_{ro}}{dS_w}\frac{\partial \psi_1}{\partial z} - r\alpha_{ro}\frac{\partial p_o}{\partial r}\frac{dk_{ro}}{dS_w}\frac{\partial}{\partial r}\left(\frac{\psi_1}{r}\right)$$

$$- \alpha_{zw}\left(\frac{\partial p_w}{\partial z} - \rho_w g\right)\frac{dk_{rw}}{dS_w}\frac{\partial \psi_2}{\partial z} - r\alpha_{rw}\frac{\partial p_w}{\partial r}\frac{dk_{rw}}{dS_w}\frac{\partial}{\partial r}\left(\frac{\psi_2}{r}\right)$$

$$- \frac{dp_c}{dS_w}\frac{\partial}{\partial z}\left(\alpha_{zw}k_{rw}\frac{\partial \psi_2}{\partial z}\right)$$

$$- \frac{dp_c}{dS_w}\frac{\partial}{\partial r}\left[r\alpha_{rw}k_{rw}\frac{\partial}{\partial r}\left(\frac{\psi_2}{r}\right)\right]\Big\}\ \delta S_w\ dr\ dz\ dt$$

$$+ \int_0^T\int_0^L\int_0^R \Big\{\frac{\partial}{\partial z}\left(\alpha_{zo}k_{ro}\frac{\partial \psi_1}{\partial z}\right) + \frac{\partial}{\partial r}\left[r\alpha_{ro}k_{ro}\frac{\partial}{\partial r}\left(\frac{\psi_1}{r}\right)\right]$$

$$+ \frac{\partial}{\partial z}\left[\alpha_{zw}k_{rw}\frac{\partial \psi_2}{\partial z}\right] + \frac{\partial}{\partial r}\left[r\alpha_{rw}k_{rw}\frac{\partial}{\partial r}\left(\frac{\psi_2}{r}\right)\right]\Big\}\ \delta p_o\ dr\ dz\ dt$$

$$- \int_0^T\int_0^L\int_0^R \left[\frac{\partial \psi_1}{\partial z}\alpha_{zo}\left(\frac{\partial p_o}{\partial z} - \rho_o g\right) + \frac{\partial}{\partial r}\left(\frac{\psi_1}{r}\right)r\alpha_{ro}\frac{\partial p_o}{\partial r}\right]\delta k_{ro}\ dr\ dz\ dt$$

$$- \int_0^T\int_0^L\int_0^R \left[\frac{\partial \psi_2}{\partial z}\alpha_{zw}\left(\frac{\partial p_w}{\partial z} - \rho_w g\right) + \frac{\partial}{\partial r}\left(\frac{\psi_2}{r}\right)r\alpha_{rw}\frac{\partial p_w}{\partial r}\right]\delta k_{rw}\ dr\ dz\ dt$$

$$+ \int_0^T\int_0^R \psi_1\alpha_{zo}\left(\frac{\partial p_o}{\partial z} - \rho_o g\right)\delta k_{ro}\Big|_{z=0}^{z=L} dr\ dt$$

$$+ \int_0^T\int_0^R \psi_2\alpha_{zw}\left(\frac{\partial p_w}{\partial z} - \rho_w g\right)\delta k_{rw}\Big|_{z=0}^{z=L} dr\ dt$$

$$+ \int_0^T\int_0^R \Big\{\Big[\psi_1\alpha_{zo}\left(\frac{\partial p_o}{\partial z} - \rho_o g\right)\frac{dk_{ro}}{dS_w} + \psi_2\alpha_{zw}\left(\frac{\partial p_w}{\partial z} - \rho_w g\right)\frac{dk_{rw}}{dS_w}$$

$$+ \alpha_{zw}k_{rw} \frac{\partial\psi_2}{\partial z}\frac{dp_c}{dS_w} - \psi_2\alpha_{zw}k_{rw}\frac{\partial}{\partial z}\left(\frac{dp_c}{dS_w}\right)\Big]\delta S_w$$

$$- \psi_2\alpha_{zw}k_{rw}\frac{dp_c}{dS_w}\frac{\partial\delta S_w}{\partial z}\Big\}\Big|_{z=0}^{z=L} dr\ dt$$

$$- \int_0^T\int_0^R \Big\{\Big[\alpha_{zo}k_{ro}\frac{\partial\psi_1}{\partial z} + \alpha_{zw}k_{rw}\frac{\partial\psi_2}{\partial z}\Big]\delta p_o$$

$$- \Big[\psi_1\alpha_{zo}k_{ro} + \psi_2\alpha_{zw}k_{rw}\Big]\frac{\partial\delta p_o}{\partial z}\Big\}\Big|_{z=0}^{z=L} dr\ dt$$

$$+ \int_0^T\int_0^L \psi_1\alpha_{ro}\frac{\partial p_o}{\partial r}\delta k_{ro}\Big|_{r=r_w}^{r=R} dz\ dt$$

$$+ \int_0^T\int_0^L \psi_2\alpha_{rw}\frac{\partial p_w}{\partial r}\delta k_{rw}\Big|_{r=r_w}^{r=R} dz\ dt$$

$$+ \int_0^T\int_0^L \Big\{\Big[\psi_1\alpha_{ro}\frac{\partial p_o}{\partial r}\frac{dk_{ro}}{dS_w} + \psi_2\alpha_{rw}\frac{\partial p_w}{\partial r}\frac{dk_{rw}}{dS_w} + r\alpha_{rw}k_{rw}\frac{\partial}{\partial r}\left(\frac{\psi_2}{r}\right)\frac{dp_c}{dS_w}$$

$$- \psi_2\alpha_{rw}k_{rw}\frac{\partial}{\partial r}\left(\frac{dp_c}{dS_w}\right)\Big]\delta S_w - \psi_2\alpha_{rw}k_{rw}\frac{dp_c}{dS_w}\frac{\partial\delta S_w}{\partial r}\Big\}\Big|_{r=r_w}^{r=R} dz\ dt$$

$$- \int_0^T\int_0^L \Big\{\Big[r\alpha_{ro}k_{ro}\frac{\partial}{\partial r}\left(\frac{\psi_1}{r}\right) + r\alpha_{rw}k_{rw}\frac{\partial}{\partial r}\left(\frac{\psi_2}{r}\right)\Big]\delta p_o$$

$$- \Big[\psi_1\alpha_{ro}k_{ro} + \psi_2\alpha_{rw}k_{rw}\Big]\frac{\partial\delta p_o}{\partial r}\Big\}\Big|_{r=r_w}^{r=R} dz\ dt \qquad (96)$$

3. First Variation of J

The change in J, δJ, resulting from the perturbation in k_{ro} and k_{rw} is given by

$$\delta J = -2\int_0^T\int_{z_1}^{z_2}\Big\{\left(q_o^{obs} - 2\pi r\alpha_{ro}k_{ro}\frac{\partial p_o}{\partial r}\right)\Big[2\pi r\alpha_{ro}k_{ro}\frac{\partial\delta p_o}{\partial r}$$

$$+ 2\pi r\alpha_{ro}\frac{\partial p_o}{\partial r}\left(\delta k_{ro} + \frac{dk_{ro}}{dS_w}\delta S_w\right)\Big]$$

$$+ \left(q_w^{obs} - 2\pi r\alpha_{rw}k_{rw}\frac{\partial p_w}{\partial r}\right)\Big[2\pi r\alpha_{rw}k_{rw}\frac{\partial\delta p_w}{\partial r}$$

$$+ 2\pi r\alpha_{rw}\frac{\partial p_w}{\partial r}\left(\delta k_{rw} + \frac{dk_{rw}}{dS_w}\delta S_w\right)\Big]\Big\}\Big|_{r=r_w} dz\ dt \qquad (97)$$

Combining Eqs. (96) and (97) and using the boundary conditions give

$$\int_0^L\int_0^R (\psi_2 - \psi_1)\delta S_w\Big|_{t=T} dr\ dz + \delta J = \int_0^T\int_0^L\int_0^R \Big\{\frac{\partial(\psi_2 - \psi_1)}{\partial t}$$

$$- \alpha_{zo}\left(\frac{\partial p_o}{\partial z} - \rho_o g\right)\frac{dk_{ro}}{dS_w}\frac{\partial\psi_1}{\partial z}$$

$$- r\alpha_{ro}\frac{\partial p_o}{\partial r}\frac{dk_{ro}}{dS_w}\frac{\partial}{\partial r}\left(\frac{\psi_1}{r}\right) - \alpha_{zw}\left(\frac{\partial p_w}{\partial z} - \rho_w g\right)\frac{dk_{rw}}{dS_w}\frac{\partial\psi_2}{\partial z}$$

$$- r\alpha_{rw}\frac{\partial p_w}{\partial r}\frac{dk_{ro}}{dS_w}\frac{\partial}{\partial r}\left(\frac{\psi_2}{r}\right)$$

$$- \frac{dp_c}{dS_w}\frac{\partial}{\partial z}\left(\alpha_{zw}k_{rw}\frac{\partial\psi_2}{\partial z}\right) - \frac{dp_c}{dS_w}\frac{\partial}{\partial r}\Big[r\alpha_{rw}k_{rw}\frac{\partial}{\partial r}\left(\frac{\psi_2}{r}\right)\Big]\Big\}\delta S_w\ dr\ dz\ dt$$

$$+ \int_0^T\int_0^L\int_0^R \Big\{\frac{\partial}{\partial z}\left(\alpha_{zo}k_{ro}\frac{\partial\psi_1}{\partial z}\right) + \frac{\partial}{\partial r}\Big[r\alpha_{ro}k_{ro}\frac{\partial}{\partial r}\left(\frac{\psi_1}{r}\right)\Big]$$

$$+ \frac{\partial}{\partial z}\left(\alpha_{zw}k_{rw}\frac{\partial\psi_2}{\partial z}\right) + \frac{\partial}{\partial r}\Big[r\alpha_{rw}k_{rw}\frac{\partial}{\partial r}\left(\frac{\psi_2}{r}\right)\Big]\Big\}\delta p_o\ dr\ dz\ dt$$

$$- \int_0^T\int_0^L\int_0^R \Big[\frac{\partial\psi_1}{\partial z}\alpha_{zo}\left(\frac{\partial p_o}{\partial z} - \rho_o g\right) + \frac{\partial}{\partial r}\left(\frac{\psi_1}{r}\right)r\alpha_{ro}\frac{\partial p_o}{\partial r}$$

$$+ \psi_1\alpha_{ro}\frac{\partial p_o}{\partial r}\delta(r - r_w)$$

$$+ 4\pi r\alpha_{ro}\left(q_o^{obs} - 2\pi r\alpha_{ro}k_{ro}\frac{\partial p_o}{\partial r}\right)\frac{\partial p_o}{\partial r}\delta(r - r_w)\Big]\delta k_{ro}\ dr\ dz\ dt$$

$$- \int_0^T \int_0^L \int_0^R \left[\frac{\partial \psi_2}{\partial z} \alpha_{zw} \left(\frac{\partial p_w}{\partial z} - \rho_w g \right) + \frac{\partial}{\partial r} \left(\frac{\psi_2}{r} \right) r \alpha_{rw} \frac{\partial p_w}{\partial r} \right.$$

$$+ \psi_2 \alpha_{rw} \frac{\partial p_w}{\partial r} \delta(r - r_w)$$

$$\left. + 4\pi r \alpha_{rw} \left(q_w^{obs} - 2\pi r \alpha_{rw} k_{rw} \frac{\partial p_w}{\partial r} \right) \frac{\partial p_w}{\partial r} \delta(r - r_w) \right] \delta k_{rw} \, dr \, dz \, dt$$

$$+ \int_0^T \int_0^R \alpha_{rw} \frac{\partial \psi_2}{\partial z} \frac{dp_c}{dS_w} \delta S_w \Big|_{z=0}^{z=L} dr \, dt - \int_0^T \int_0^R \left[\alpha_{zo} k_{ro} \frac{\partial \psi_1}{\partial z} \right.$$

$$\left. + \alpha_{zw} k_{rw} \frac{\partial \psi_2}{\partial z} \right] \partial p_o \Big|_{z=0}^{z=L} dr \, dt$$

$$+ \int_0^T \int_0^L r \alpha_{rw} k_{rw} \frac{\partial}{\partial r} \left(\frac{\psi_2}{r} \right) \frac{dp_c}{dS_w} \delta S_w \Big|_{r=R} dz \, dt$$

$$- \int_0^T \int_0^{z_1} r \alpha_{rw} k_{rw} \frac{\partial}{\partial r} \left(\frac{\psi_2}{r} \right) \frac{dp_c}{dS_w} \delta S_w \Big|_{r=r_w} dz \, dt$$

$$- \int_0^T \int_{z_2}^L r \alpha_{rw} k_{rw} \frac{\partial}{\partial r} \left(\frac{\psi_2}{r} \right) \frac{dp_c}{dS_w} \delta S_w \Big|_{r=r_w} dz \, dt$$

$$+ \int_0^T \int_{z_1}^{z_2} \left[\psi_2 \alpha_{rw} k_{rw} \frac{dp_c}{dS_w} + 4\pi r \alpha_{rw} k_{rw} \left(q_w^{obs} \right. \right.$$

$$\left. \left. - 2\pi r \alpha_{rw} k_{rw} \frac{\partial p_w}{\partial r} \right) \frac{dp_c}{dS_w} \right] \frac{\partial \delta S_w}{\partial r} \Big|_{r=r_w} dz \, dt$$

$$- \int_0^T \int_0^L \left[r \alpha_{ro} k_{ro} \frac{\partial}{\partial r} \left(\frac{\psi_1}{r} \right) + r \alpha_{rw} k_{rw} \frac{\partial}{\partial r} \left(\frac{\psi_2}{r} \right) \right] \delta p_o \Big|_{r=R} dz \, dt$$

$$+ \int_0^T \int_0^{z_1} \left[r \alpha_{ro} k_{ro} \frac{\partial}{\partial r} \left(\frac{\psi_1}{r} \right) + r \alpha_{rw} k_{rw} \frac{\partial}{\partial r} \left(\frac{\psi_2}{r} \right) \right] \delta p_o \Big|_{r=r_w} dz \, dt$$

$$+ \int_0^T \int_{z_2}^L \left[r \alpha_{ro} k_{ro} \frac{\partial}{\partial r} \left(\frac{\psi_1}{r} \right) + r \alpha_{rw} k_{rw} \frac{\partial}{\partial r} \left(\frac{\psi_2}{r} \right) \right] \delta p_o \Big|_{r=r_w} dz \, dt$$

$$- \int_0^T \int_{z_2}^{z_1} \Big[\psi_1 \alpha_{ro} k_{ro} + \psi_2 \alpha_{rw} k_{rw} + 4\pi r \alpha_{ro} k_{ro} \Big(q_n^{obs} - 2\pi r \alpha_{ro} k_{ro} \frac{\partial p_o}{\partial r} \Big)$$

$$+ 4\pi r \alpha_{rw} k_{rw} \Big(q_w^{obs} - 2\pi r \alpha_{rw} k_{rw} \frac{\partial p_w}{\partial r} \Big) \Big] \frac{\partial \delta p_o}{\partial r} \Big|_{r=r_w} dz\, dt \qquad (98)$$

4. Adjoint Equations

The objective is to define ψ_1 and ψ_2 in such a way that a relation between δJ and δk_{ro} and δk_{rw} is obtained from Eq. (98). We therefore specify that ψ_1 and ψ_2 be governed by

$$\frac{\partial(\psi_2 - \psi_1)}{\partial t} = \alpha_{zo} \Big(\frac{\partial p_o}{\partial z} - \rho_o g \Big) \frac{dk_{ro}}{dS_w} \frac{\partial \psi_1}{\partial z} + r\alpha_{ro} \frac{\partial p_o}{\partial r} \frac{dk_{ro}}{dS_w} \frac{\partial}{\partial r} \Big(\frac{\psi_1}{r} \Big)$$

$$+ \alpha_{zw} \Big(\frac{\partial p_w}{\partial z} - \rho_w g \Big) \frac{dk_{rw}}{dS_w} \frac{\partial \psi_2}{\partial z} + r\alpha_{rw} \frac{\partial p_w}{\partial r} \frac{dk_{rw}}{dS_w} \frac{\partial}{\partial r} \Big(\frac{\psi_2}{r} \Big)$$

$$+ \frac{dp_c}{dS_w} \frac{\partial}{\partial z} \Big(\alpha_{zw} k_{rw} \frac{\partial \psi_2}{\partial z} \Big) + \frac{dp_c}{dS_w} \frac{\partial}{\partial r} \Big[r\alpha_{rw} k_{rw} \frac{\partial}{\partial r} \Big(\frac{\psi_2}{r} \Big) \Big] \qquad (99)$$

and

$$\frac{\partial}{\partial z} \Big(\alpha_{zo} k_{ro} \frac{\partial \psi_1}{\partial z} \Big) + \frac{\partial}{\partial r} \Big[r\alpha_{ro} k_{ro} \frac{\partial}{\partial r} \Big(\frac{\psi_1}{r} \Big) \Big]$$

$$+ \frac{\partial}{\partial z} \Big(\alpha_{zw} k_{rw} \frac{\partial \psi_2}{\partial z} \Big) + \frac{\partial}{\partial r} \Big[r\alpha_{rw} k_{rw} \frac{\partial}{\partial r} \Big(\frac{\psi_2}{r} \Big) \Big] = 0 \qquad (100)$$

Equation (99) is solved subject to the final condition

$$\psi_2 - \psi_1 = 0, \qquad t = T \qquad (101)$$

The boundary conditions for Eqs. (99) and (100) are chosen such that the last ten terms on the right-hand side of Eq. (98) vanish. Because the variations of δS_w and δp_o are arbitrary on $\partial\Omega_2$, and the variations of $\partial \delta S_w / \partial r$ and $\partial \delta p_o / \partial r$ are arbitrary on $\partial\Omega_1$, we obtain the conditions

$$\frac{\partial \psi_1}{\partial z} = \frac{\partial \psi_2}{\partial z} = 0, \qquad z = 0, L \tag{102}$$

$$\left.\begin{aligned} &\frac{\partial}{\partial r}\left(\frac{\psi_1}{r}\right) = 0, \\ &\frac{\partial}{\partial r}\left(\frac{\psi_2}{r}\right) = 0, \end{aligned}\right\} \quad r = R \tag{103}$$

$$\left.\begin{aligned} &\frac{\partial}{\partial r}\left(\frac{\psi_1}{r}\right) = 0, \\ &\frac{\partial}{\partial r}\left(\frac{\psi_2}{r}\right) = 0, \end{aligned}\right\} \quad r = r_w,\ 0 \le z < z_1,\ z_2 < z \le L \tag{104}$$

$$\left.\begin{aligned} &\psi_1 \alpha_{ro} k_{ro} + 4\pi r \alpha_{ro} k_{ro}\left(q_o^{obs} - 2\pi r \alpha_{ro} k_{ro} \frac{\partial p_o}{\partial r}\right) = 0 \\ &\psi_2 \alpha_{rw} k_{rw} + 4\pi r \alpha_{rw} k_{rw}\left(q_w^{obs} - 2\pi r \alpha_{rw} k_{rw} \frac{\partial p_w}{\partial r}\right) = 0 \end{aligned}\right\} \quad \text{on } \partial\Omega_1 \tag{105}$$

Substituting Eqs. (99) to (105) into (98), we obtain

$$\begin{aligned} \delta J = &- \int_0^T \int_0^L \int_0^R \left[\frac{\partial \psi_1}{\partial z} \alpha_{zo} \left(\frac{\partial p_o}{\partial z} - \rho_o g \right) + \frac{\partial}{\partial r}\left(\frac{\psi_1}{r}\right) \right. \\ &\left. \times\ r \alpha_{ro} \frac{\partial p_o}{\partial r} \right] \delta k_{ro}\ dr\ dz\ dt \\ &- \int_0^T \int_0^L \int_0^R \left[\frac{\partial \psi_2}{\partial z} \alpha_{zw} \left(\frac{\partial p_w}{\partial z} - \rho_w g \right) + \frac{\partial}{\partial r}\left(\frac{\psi_2}{r}\right) \right. \\ &\left. \times\ r \alpha_{rw} \frac{\partial p_w}{\partial r} \right] \delta k_{rw}\ dr\ dz\ dt \end{aligned} \tag{106}$$

From Eq. (106) we conclude that the necessary conditions for optimality are

$$\frac{\partial \psi_1}{\partial z} \alpha_{zo} \left(\frac{\partial p_o}{\partial z} - \rho_o g \right) + \frac{\partial}{\partial r}\left(\frac{\psi_1}{r}\right) r \alpha_{ro} \frac{\partial p_o}{\partial r} = 0 \tag{107}$$

$$\frac{\partial \psi_2}{\partial z} \alpha_{zw} \left(\frac{\partial p_w}{\partial z} - \rho_w g \right) + \frac{\partial}{\partial r}\left(\frac{\psi_2}{r}\right) r \alpha_{rw} \frac{\partial p_w}{\partial r} = 0 \tag{108}$$

It is interesting to note that the adjoint equations for the estimation of relative permeabilities, which depend on the state variables, and for the estimation of absolute permeability, which is a function of the spatial variables, are the same for this system.

Because δk_{ro} and δk_{rw} are functions of the state variables, the left-hand sides of Eqs. (107) and (108) are no longer the gradient of J with respect to k_{ro} and k_{rw}, respectively. However, if we assume an algebraic form containing unknown constant parameters for the functions $k_{ro}(S_w)$ and $k_{rw}(S_w)$, for example $k_{ro}(S_w) = k_{ro}(S_w, a_1, \ldots, a_n)$ and $k_{rw}(S_w) = k_{rw}(S_w, b_1, \ldots, b_n)$, then the gradients of J with respect to those parameters can be calculated as follows:

$$\frac{\partial J}{\partial a_i} = -\int_0^T\int_0^L\int_0^R \left[\frac{\partial \psi_1}{\partial z} \alpha_{zo} \left(\frac{\partial p_o}{\partial z} - \rho_o g \right) + \frac{\partial}{\partial r}\left(\frac{\psi_1}{r}\right) r\alpha_{ro} \frac{\partial p_o}{\partial r} \right] \frac{\partial k_{ro}}{\partial a_i}\, dr\, dz\, dt, \qquad i = 1, \ldots, n \tag{109}$$

$$\frac{\partial J}{\partial b_i} = -\int_0^T\int_0^L\int_0^R \left[\frac{\partial \psi_2}{\partial z} \alpha_{zw} \left(\frac{\partial p_w}{\partial z} - \rho_w g \right) + \frac{\partial}{\partial r}\left(\frac{\psi_2}{r}\right) r\alpha_{rw} \frac{\partial p_w}{\partial r} \right] \frac{\partial k_{rw}}{\partial b_i}\, dr\, dz\, dt, \qquad i = 1, \ldots, n \tag{110}$$

Thus, the gradient method developed in the previous section for determining absolute permeability can be applied to estimate the relative permeabilities.

V. CONCLUSIONS

In this chapter we have defined the general problems arising in the identification of petroleum reservoir properties. This class of problems is characterized by

1. Distributed parameter models (linear and nonlinear)
2. Spatially-dependent parameters
3. State variable-dependent parameters
4. Nonuniqueness (more unknown parameters than data)

Algorithms based on optimal control theory for obtaining parameter estimates have been developed. The major remaining problems are related to improving the parameter identifiability and obtaining estimates for the accuracy of the parameter estimates.

It is necessary to develop systematic means of reducing the number of unknown parameters by eliminating or consolidating those to which the data are insensitive. Then methods must be developed for estimating the accuracy of parameters determined from history matching.

REFERENCES

1. R. E. Collins, Flow of Fluids Through Porous Materials, Holt Rheinhold, New York, 1961.
2. W. H. Chen and J. H. Seinfeld, Estimation of the location of the boundary of a petroleum reservoir, Soc. Petrol. Eng. J., 15: 19-38 (1975).
3. P. Jacquard and C. Jain, Permeability distributions from field pressure data, Soc. Petrol. Eng. J., 5: 281-294 (1965).
4. H. O. Jahns, A rapid method for obtaining a two-dimensional reservoir description from well pressure response data, Soc. Petrol. Eng. J., 6: 315-327 (1966).
5. K. H. Coats, J. R. Dempsey, and J. H. Henderson, A new technique for determining reservoir description from field performance data, Soc. Petrol. Eng. J., 10: 66-74 (1970).
6. G. E. Slater and D. J. Durrer, Adjustment of reservoir simulator models to match field performance, Soc. Petrol. Eng. J., 11: 295-305 (1971).
7. R. W. Veatch, Jr. and G. W. Thomas, "A direct approach for history matching," 46th Annual Fall Meeting of the Society of Petroleum Engineers, SPE Paper 3515, Oct. 3-6, 1971.
8. L. K. Thomas, L. J. Hellums, and G. M. Reheis, A nonlinear automatic history matching technique for reservoir simulation models, Soc. Petrol. Eng. J., 12: 508-514 (1972).
9. R. D. Carter, L. F. Kemp, Jr., A. C. Pierce, and D. L. Williams, Performance matching with constraints, Soc. Petrol. Eng. J., 14: 187-196 (1974).
10. W. H. Chen, G. R. Gavalas, J. H. Seinfeld, and M. L. Wasserman, A new algorithm for automatic history matching, Soc. Petrol. Eng. J., 14: 593-608 (1974).
11. G. Chavent, M. Dupuy, and P. Lemonnier, History matching by use of optimal control theory, Soc. Petrol. Eng. J., 15: 74-86 (1975).

12. M. L. Wasserman, A. S. Emanuel, and J. H. Seinfeld, Practical applications of optimal control theory to history matching multiphase simulator models, Soc. Petrol. Eng. J., 15: 347-355 (1975).

13. E. L. Dougherty and D. Khairklah, "History matching of gas simulation models using optimal control theory," 45th Annual California Regional Meeting of the Society of Petroleum Engineers, SPE Paper 5371, April 2-4, 1975.

14. G. R. Gavalas, P. C. Shah, and J. H. Seinfeld, Reservoir history matching by Bayesian estimation, Soc. Petrol. Eng. J., 16: 337-350 (1976).

15. T. N. Dixon, J. H. Seinfeld, R. A. Startzman, and W. H. Chen, "Reliability of reservoir parameters from history matched drill stem tests," Third Numerical Simulation of Reservoir Performance Symposium of the Society of Petroleum Engineers, Houston, Texas, Jan. 10-12, 1973, SPE Paper 4282.

16. P. C. Shah, G. R. Gavalas, and J. H. Seinfeld, The structure of history matching problems, Soc. Petrol. Eng. J., submitted for publication.

17. A. Settari and K. Aziz, A computer model for two-phase coning simulation, Soc. Petrol. Eng. J., 14: 221-236 (1974).

18. J. Hagoort, Displacement stability of water drives in water-west connate-water-bearing reservoirs, Soc. Petrol. Eng. J., 14: 70 (1974).

19. J. S. Nolen and D. W. Berry, Tests of stability and time step sensitivity of semi-implicit reservoir simulation techniques, Soc. Petrol. Eng. J., 12: 253-266 (1972).

Chapter 10

IDENTIFICATION OF PARAMETERS IN DISTRIBUTED CHEMICAL REACTORS

Bruno A. J. Van den Bosch

Instituut voor Chemie-ingenieurstechniek
Katholieke Universiteit te Leuven
Leuven, Belgium

I. INTRODUCTION

The problem of identification of parameters in mathematical models is frequently encountered in various fields of science and engineering. In addition, in chemical engineering practice it often happens that lack of experimental data for some of the parameters necessitates the use of a numerical technique for estimating the parameters from a series of data for measurable variables of the system. Due to the nature of many chemical engineering processes, models describing their behavior often consist of sets of partial differential equations. These processes are commonly called distributed parameter systems, although actually the state variables are distributed. Their models often contain unknown parameters related to the reaction kinetics or to the transport mechanisms.

In this chapter several aspects of the parameter estimation problem for chemical engineering distributed parameter systems will be discussed. It will be assumed that the model describing the system is completely known from conservation principles, chemical kinetics mechanisms, and transport phenomena principles. Only the case of nonsequential estimation is considered here. This means that all data are collected before the analysis is undertaken. Attention will be given only to the most general techniques for solving the parameter identification problem. Special methods such as frequency response and moment matching have limited applicability and are described elsewhere [1].

A concise presentation of the parameter estimation problem is given in Figure 1. A process is considered modeled by a set of state equations giving the relations between the state variables $\underline{x}$, the independent variables $\underline{t}$, and the parameters $\underline{a}$. Only distributed

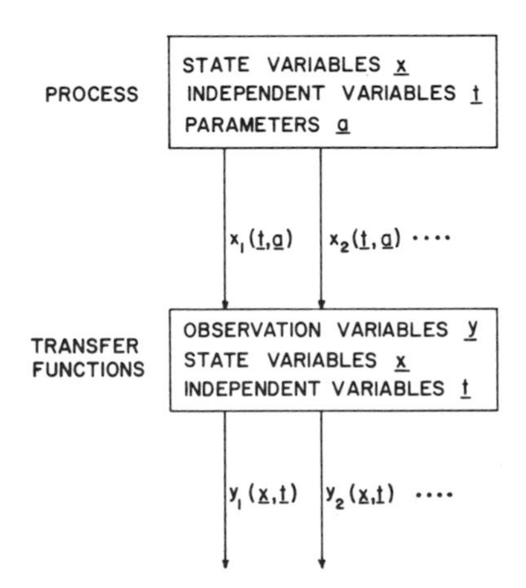

FIG. 1. Schematic presentation of a system.

parameter systems modeled by partial differential equations will be considered. The state variables $\underline{x}$ can be considered as outputs of the process and serve at the same time as inputs to an observation system. In the latter system a set of transfer functions relates the state variables $\underline{x}$ to a series of observation variables $\underline{y}$. These variables are explicit functions of the independent variables and implicit functions of the parameters through dependence upon the state variables. The entire set of state equations (eventually including the required boundary conditions) and observation equations forms the model of the considered process. In addition to this model, a certain amount of data is obtained from experiments. These data represent the observation variables at a series of observation points, but are subject to measurement errors. The deviations between the data and the values predicted by the model for a chosen set of parameters should be small. This matching of the data to the given model determines a parameter estimation or identification problem. The optimal matching is based upon the principles of statistics, determining an objective function based upon the residuals which has to be minimized with respect to the unknown parameters.

It will be assumed in this chapter that the form of the objective function is completely known. The estimation problem stated in this way belongs to the field of numerical analysis. Serious numerical complications arise in distributed parameter systems because of the implicit dependence of the objective functions upon the parameters and the difficulties involved in solving partial differential equations numerically. Techniques which have been proposed make use of the gradient methods for the minimization of an objective function or of the quasilinearization method for solving differential equations. These methods will be discussed, as well as a more recent technique based upon the principles of orthogonal collocation, a method combining the advantages of both finite differences and weighted residuals for solving differential equations. The example of an isothermal tubular reactor in laminar flow illustrates the principles of the technique.

The estimates obtained by minimization of an objective function derived from statistics only partially solve the parameter identification problem. An analysis of the reliability of the obtained results completes the solution of the problem. This is achieved by the construction of confidence regions in the parameter space for a given confidence level. Sections VI to VIII are concerned with problems related to this aspect of the estimation of parameters.

II. GRADIENT METHODS

Before discussing the several techniques for parameter identification, the most general form of the system presented schematically in Figure 1 will be considered. A set of n state equations is considered, giving the relations between the n state variables $\underline{x}$, the S independent variables $\underline{t}$, and the P parameters $\underline{a}$ in a region Ω

$$\underline{L}(\underline{x},\underline{t},\underline{a}) = \underline{0} \quad \text{in } \Omega \tag{1}$$

In chemical engineering applications, the model often consists of a set of partial differential equations (PDEs). This is the case for tubular reactors with radial effects, since the temperature and the concentrations of the different reactants vary with respect to both radial and axial position. Another class of problems concerns catalyst deactivation, where the time enters into the problem as an independent variable in addition to the spatial coordinates. Unknown parameters arise from the reaction kinetics such as the frequency factor and the activation energy as well as the reaction order or a deactivation constant characterizing the catalyst poisoning or coking. Also, several constants related to the heat or mass transport mechanisms may be unknown due to lack of experimental information for the processes as they occur in real practice.

The model described by the state equations (1) is completed by a series of appropriate boundary conditions

$$\underline{d}(\underline{x},\underline{t},\underline{a}) = \underline{0} \quad \text{at } \partial\Omega \tag{2}$$

the number and the form of which depend upon the type of the state equations (1). As can be seen, the unknown parameters may enter the boundary conditions.

The explicit form of the solution of Eqs. (1) and (2)

$$\underline{x} = \underline{g}(\underline{t},\underline{a}) \tag{3}$$

is usually unknown, and $\underline{x}$ can only be found by use of numerical techniques for solving differential equations.

The observation of the process is characterized by a series of transfer functions relating the m observation variables $\underline{y}$ to the state variables $\underline{x}$ and the independent variables $\underline{t}$

$$\underline{y} = \underline{h}(\underline{x},\underline{t}) \tag{4}$$

Measurements of the variables $\underline{y}$ are performed at a series of R observation points $\underline{t}_r$. Due to measurement errors the obtained data $\underline{\eta}_r$ differ from the exact values of $\underline{y}$, and the deviations in each point are represented by the vector of residuals $\underline{e}_r$. These errors originate from the presence of noise and imperfections in the measuring instruments. The data are matched to the predicted values of $\underline{y}$ by minimizing an objective function derived from statistics for known distributions of the residuals. Assuming independent measurements in the observation points $\underline{t}_r$ with normally distributed errors with mean value zero and known variance-covariance matrix $\underline{M}_r$, the principle of maximum likelihood leads to the objective function [2]

$$\Psi = \sum_{r=1}^{R} \underline{e}_r^T \underline{M}_r^{-1} \underline{e}_r = \sum_{r=1}^{R} (\underline{y}_r - \underline{\eta}_r)^T \underline{M}_r^{-1} (\underline{y}_r - \underline{\eta}_r) \tag{5}$$

where

$$\underline{M}_r = E(\underline{e}_r \underline{e}_r^T) \tag{6}$$

The symbol E denotes the statistical concept expectation.

If the errors of the different observation variables are independent with a same variance σ_r^2, the moment matrix becomes

$$\underline{M}_r = \sigma_r^2 \underline{I} \tag{7}$$

leading to a weighted least squares criterion

$$\Psi = \sum_{r=1}^{R} \sigma_r^{-2} \underline{e}_r^T \underline{e}_r \tag{8}$$

The least squares criterion is obtained if the variances have the same value at all observation points

$$\sigma_r^2 = \sigma^2 \qquad \text{for } r = 1,R \tag{9}$$

resulting in the objective function

$$\Psi = \sum_{r=1}^{R} \underline{e}_r^T \underline{e}_r \tag{10}$$

The maximum likelihood estimates of $\underline{a}$ are found by minimization of the appropriate objective function Ψ with respect to $\underline{a}$. The dependence of Ψ upon $\underline{a}$ is implicit: Ψ contains $\underline{y}$ which is a function of $\underline{x}$ from Eq. (4), and $\underline{x}$ in turn depends upon $\underline{a}$ through the state equations (1) and the boundary conditions (2).

A large number of numerical techniques for iterative minimization of an objective function have been developed [2]. An important class of methods are the gradient methods for which the calculation of the gradient of Ψ with respect to $\underline{a}$ at each iteration is typical. They correspond to the general form

$$\underline{a}_{k+1} = \underline{a}_k - \rho_k \underline{K}_k \left(\frac{\partial \Psi}{\partial \underline{a}} \right)_k \tag{11}$$

where $(\partial\Psi/\partial\underline{a})$ is the gradient vector of the objective function. Different methods arise from the choice of ρ and $\underline{K}$ at each iteration.

The steepest descent method is the simplest form since it uses a unity matrix $\underline{I}$ for $\underline{K}$ and requires only the calculation of the gradient vector. Much more complex is the Newton-Raphson procedure, in which the matrix $\underline{K}$ is the inverse of the Hessian matrix

$$\underline{H} = \frac{\partial^2 \Psi}{\partial \underline{a}^2} \tag{12}$$

This method is known to have excellent convergence properties in its region of convergence, but the actual calculation of all second derivatives involved in the Hessian matrix $\underline{H}$ becomes a time-consuming task. As a compromise and in order to limit the number of calculations involved, different methods have been proposed using a matrix

$\underline{K}$ approximating the inverse of $\underline{H}$. From an extensive collection of computational results it was found that of all methods investigated, the Gauss-Newton method and the Marquardt method ranked first [3]. Concerning the choice of the step length ρ at each iteration, there appeared to be no need to locate the minimum precisely, the only remaining limitation for ρ thus being the decrease of the objective function Ψ. The superiority of the Gauss-Newton method in terms of its speed and region of convergence has been confirmed by another study [4].

Since the objective function Ψ depends upon the parameters $\underline{a}$ through the state variables $\underline{x}$ in the case of models consisting of differential equations, the knowledge of the sensitivity coefficients

$$\lambda_{j,d} = \frac{\partial x_j}{\partial a_d} \qquad \text{for } j = 1,n \text{ and } d = 1,P \tag{13}$$

is required for the calculation of the gradient vector $(\partial\Psi/\partial\underline{a})$. Their calculation is tedious for systems modeled by ordinary differential equations (ODEs), and it complicates the use of the gradient methods even more for distributed parameter systems (DPS). The sensitivity coefficients are found as the solution of the sensitivity equations obtained from the derivation of the state equations (1) with respect to the parameters and interchanging the order of differentiation following

$$\frac{\partial}{\partial a_d}\left(\frac{\partial x_j}{\partial t_s}\right) = \frac{\partial}{\partial t_s}(\lambda_{j,d}) \tag{14}$$

The corresponding boundary conditions are found in a similar manner from Eq. (2). This leads to a total number of $(n \times P)$ sensitivity equations, the solution of which can be found numerically by simultaneous integration with the n state equations (1). The disadvantage of this approach is that an eventually large number of complex coupled differential equations (DEs) has to be solved at each iteration of the minimization procedure. This may become extremely time-consuming and, together with the convergence problem of the minimization algorithm, it seriously restricts the applicability of this approach, even for small values of n and P. The numerical difficulties have been demonstrated for a system modeled by ODEs. An industrial example involving seven parameters concerning the oxidation of methanol in a packed bed reactor was found to be very time-consuming [5]. The difficulties will of course increase for DPS since the required computation time for each repetitive integration may be very long. This calls for alternative approaches of parameter identification in systems modeled by PDEs.

III. QUASILINEARIZATION

A method has been developed for systems modeled by ODEs based upon a redefinition of the minimization problem. It has been successfully applied to examples of different complexity. A similar approach could be used for DPS if it is possible to reduce each PDE to a series of ODEs by applying finite differences [6]. In order to use this second estimation technique, it is necessary that the system can be modeled by a set of first-order ODEs

$$\underline{\dot{x}} = \underline{f}(\underline{x},\underline{a},t) \tag{15}$$

containing only one independent variable, t. Eventually, any ODE of higher order appearing in the model can be replaced by an equivalent set of first-order equations.

In addition to the set of state equations (15), a series of DEs can be obtained by putting the derivatives of the unknown but constant parameters with respect to t equal to zero:

$$\underline{\dot{a}} = \underline{0} \tag{16}$$

The missing boundary conditions are obtained by the requirement that the objective function is minimized with respect to the parameters:

$$\frac{\partial \Psi}{\partial \underline{a}} = \underline{0} \tag{17}$$

This restates the estimation problem as a multipoint boundary value problem consisting of a set of (n + P) differential equations (15) and (16) with the corresponding boundary conditions (2) and (17). Any numerical technique suitable for integration of these differential equations could be used to solve the estimation problem stated in this way. The quasilinearization method, however, treats this type of problem fairly easily and, consequently, it has been widely used. This explains why quasilinearization is commonly considered to be a classical method for parameter identification, although in a strict sense the problem is actually solved by the reformulation as a multipoint boundary value problem, with quasilinearization one of several ways to handle the new problem.

Basically, quasilinearization uses the Newton-Raphson formula for functional equations in order to replace the nonlinear ODEs by linear equations. This is obtained by use of the first and second terms in the Taylor's series expansion [7]. The original equations are then solved recursively by a series of linear DEs. These can be solved by the superposition principle, the particular and the homogeneous solutions being obtained routinely by initial value techniques such as the well-known Runge-Kutta method. The advantage of this approach is the quadratic convergence of the Newton-Raphson formula, provided that it converges. Therefore, it is important to have good initial approximations which could be obtained from the physical situation or from engineering intuition.

In the situation considered here, the set of (n + P) Eqs. (15) and (16) can be summarized as

$$\dot{\underline{q}} = \underline{F}(\underline{q},t) \tag{18}$$

where $\underline{q}$ is a vector composed of $\underline{x}$ and $\underline{a}$. The first n equations are given by Eq. (15), whereas the last P equations of (18) correspond to (16). The boundary conditions are given by Eqs. (2) and (17).

The following iterative scheme is based upon the Newton-Raphson formula:

$$\dot{\underline{q}}_{k+1} = \underline{F}(\underline{q}_k,t) + \underline{J}_k(\underline{q}_{k+1} - \underline{q}_k) \qquad \text{for } k = 0,1, \ldots \tag{19}$$

where $\underline{J}$ is the Jacobi matrix

$$\underline{J} = \begin{bmatrix} \frac{\partial F_1}{\partial q_1} & \frac{\partial F_1}{\partial q_2} & \cdots & \frac{\partial F_1}{\partial q_{n+P}} \\ \frac{\partial F_2}{\partial q_1} & \frac{\partial F_2}{\partial q_2} & \cdots & \frac{\partial F_2}{\partial q_{n+P}} \\ \vdots & & \vdots & \\ \frac{\partial F_{n+P}}{\partial q_1} & \cdots & \cdots & \frac{\partial F_{n+P}}{\partial q_{n+P}} \end{bmatrix} \tag{20}$$

The solution of Eq. (19) for each iteration is found by use of the superposition principle

$$\underline{q}_{k+1} = \underline{q}_{p,k+1} + \underline{Q}_{h,k+1}\underline{c} \tag{21}$$

where $\underline{q}_{p,k+1}$ is a particular solution of Eq. (19) with arbitrary initial conditions such as

$$\underline{q}_{p,k+1}(0) = [0 \quad 0 \quad \ldots\ldots \quad 0]^T \tag{22}$$

The homogeneous solutions appearing as columns in the square matrix $\underline{Q}$ are obtained from the initial value problems

$$\dot{\underline{q}}_{k+1} = \underline{J}_k \underline{q}_{k+1} \qquad \text{for } k = 0,1, \ldots \tag{23}$$

with the initial conditions

$$\underline{Q}_{h,k+1}(0) = \underline{I} \tag{24}$$

where $\underline{I}$ is the unity matrix. The $(n + P)$ integration constants are represented by the vector $\underline{c}$. If some of the initial conditions are given explicitly it is possible to reduce the number of homogeneous solutions required by an appropriate choice of the initial conditions (22) [7], but the most general situation is treated here.

By use of the initial conditions (22) and (24) and the general expression (21) of the solution, we find that

$$\underline{q}_{k+1}(0) = \underline{Ic} \tag{25}$$

This means that because of the composition of $\underline{q}$ from $\underline{x}$ and $\underline{a}$, the last P components of $\underline{c}$ are equal to the initial values of the parameters $\underline{a}$. Since these parameters are constant, they equal the last P components of $\underline{c}$ summarized by $\underline{c}_2$ for all t. The boundary condition (17) can be replaced by

$$\frac{\partial \Psi}{\partial \underline{c}_2} = \underline{0} \tag{26}$$

where Ψ is a function of $\underline{c}$ only after the substitution of (21). The n remaining equations for $\underline{c}$ are found by the substitution of (21) in the boundary conditions (2).

The entire approach can now be summarized as an iterative procedure of two steps starting from an initial approximation of $\underline{x}$ and $\underline{a}$. In the first step of each iteration a set of (n + P) linear ODEs (19) and (23) is solved by an initial value technique such as the Runge-Kutta method for one set (22) and (n + P) sets (24) of initial conditions, respectively. In the second step a series of (n + P) algebraic equations is solved with respect to the unknown coefficients $\underline{c}$.

The proposed procedure may lead to serious complications. A convergence problem is inherently connected with the Newton-Raphson scheme (19), restricting the approach to problems with small n and P. The convergence properties may be improved by a data perturbation technique [8]; new data are generated falling between the actual data and the data predicted from the current solution. In this way a sequence of perturbed problems must be solved. Numerical difficulties may also be encountered in both steps of each iteration. The solution of large sets of initial-value problems may be impeded by computer storage limitations, although excellent initial value routines exist. Serious difficulties can arise in the second step when solving the set of algebraic equations for $\underline{c}$ by nonlinear regression techniques. These methods are indeed iterative, and thus time-consuming; moreover, the regression matrix is often ill-conditioned.

In order to limit the computation time in the latter step, the observation equations (4) could be linearized in correspondence with the linearization of the state equations in order to obtain a linear regression problem from Eqs. (2) and (26). This simplification, however, turns out to deteriorate the convergence of the procedure. In order to obtain convergence it is necessary to modify this approach by adding a variable scalar matrix to the regression matrix as shown in detail in a study concerning the aforementioned example of methanol oxidation involving eight parameters [9]. The study also found this modified quasilinearization method to be superior to the gradient method. It was shown, however, that the region of convergence is very small, whereas the required computation time increases rapidly with the number of parameters. This réstricts the applicability of the method in real practice.

These difficulties encountered when estimating parameters occurring in ODEs by the quasilinearization method may be expected in an even more pronounced way when applying the procedure to DPS. The original set of PDEs must be replaced by a set of ODEs by the use of finite differences. Even if the number of parameters is small, the applicability of the method is greatly restricted due to the fact that the total number of ODEs obtained from finite differences will be significant.

IV. ORTHOGONAL COLLOCATION

Both the gradient methods and the quasilinearization technique have been widely used in estimating parameters in ODEs because of their generality. An important part of the computation time required in both methods goes to the repetitive solution of a set of DEs by classical numerical techniques such as finite differences or the Runge-Kutta method. Recent developments in numerical analysis for chemical engineering models, however, emphasize the efficiency of weighted residuals methods for solving the DEs. These methods make use of trial functions expansions in order to approximate the solutions of the DEs, and the unknown expansion coefficients can be determined by several criteria making the residuals of the DEs equal to zero in an average sense [10]. The method is flexible because of the free choice of the trial functions and the criterion. For an appropriate choice it turns out to give accurate solutions, but from a computational point of view it is less convenient for automatic treatment on computers than the method of finite differences. A discrete version of the method of weighted residuals, however, namely the collocation method applied in terms of the unknown values of the state variables, called ordinates, at a series of points can be easily treated on computers. The method actually replaces each DE by a series of algebraic equations at the collocation points, the derivatives being replaced by weighted averages of the ordinates [11]. The weights are found by making the formulas exact for the derivatives of a series of polynomials of increasing degree, the number of which corresponds to the number of collocation points. The resulting set of algebraic equations must be solved with respect to the unknown ordinates, which is usually done by applying the Newton-Raphson formula. Proof has been given of the accuracy of the collocation method when selecting the collocation points as the roots of orthogonal polynomials. Applications of "orthogonal collocation" is actually equivalent to the use of optimal quadrature formulas for the evaluation of the integrals arising in the Galerkin method, one of the classical weighted residuals methods, which has been used extensively because of its accuracy and convergence [10].

The principle of replacing DEs by algebraic equations, as involved in the collocation method, may offer an interesting alternative for the identification of parameters, since it highly simplifies the form of the objective function which must be minimized with respect to the unknown parameters. The first proposal to apply this principle made use of the solution of a low-order collocation technique [6]. A similar idea for orthogonal polynomial expansions for parameter estimation has been reported elsewhere [12]. In both cases the identification was performed by matching the coefficients of the low-order expansions to the data. A general treatment of the possibilities offered by the orthogonal collocation method applied in terms of the ordinates has been presented for approximations of arbitrary order [13]; this procedure will be reformulated here.

The orthogonal collocation technique uses polynomial approximations for the solutions of the differential equations. Each state variable is replaced by a polynomial with respect to the independent variables in the considered region

$$x_j(\underline{a},\underline{t}) = \sum_{u=1}^{u_f} \cdots \sum_{i=1}^{i_f} \cdots \sum_{v=1}^{v_f} b_{j,u,\ldots,i,\ldots v}(\underline{a}) \times t_1^u \times \cdots \times t_s^i \times \cdots \times t_S^v \quad \text{for } j = 1,n \tag{27}$$

In this way the dependence upon the independent variables $\underline{t}$ is separated from the dependence upon the parameters $\underline{a}$, the latter one entering through the coefficients. For each state variable x_j the knowledge of M coefficients $b_{j,u,\ldots,i,\ldots,v}$ is required, where

$$M = u_f \times \cdots \times i_f \times \cdots \times v_f \tag{28}$$

These coefficients are summarized by a vector $\underline{b}_j$, all vectors $\underline{b}_j$ composing a vector $\underline{b}$ with (n x M) components. The knowledge of $\underline{b}_j$ is equivalent to knowledge of the values of $\underline{x}_j$ in M collocation points. These values are summarized by the vector of ordinates $\underline{z}_j$, all vectors defining a vector $\underline{z}$.

Substitution of the approximations (27) in the objective function Ψ defined by Eqs. (5) or (8) leads to a function $\Psi[\underline{b}(\underline{a})]$ which is equivalent to a function $\Psi[\underline{z}(\underline{a})]$. Minimization of Ψ with respect to all components of either $\underline{b}$ or $\underline{z}$ can be considered as a sufficient condition for the requirement (17). This can be performed by the classical gradient methods. Very often this turns out to be a problem of linear regression with respect to the components of $\underline{b}$ occurring in Ψ.

Using the approximations (27), the parameters $\underline{a}$ can then be estimated from the state equations (1). The requirement expressed by this equation will be satisfied in an average sense by minimizing

$$\Phi = \int_\Omega \underline{L}^T \underline{W} \underline{L} \, d\Omega \tag{29}$$

where Ω is the entire region of observation and $\underline{W}$ is a positive definite weighting matrix chosen as

$$\underline{W} = w(\underline{t})\underline{I} = w(t_1) \times \cdots \times w(t_s) \times \cdots \times w(t_S)\underline{I} \tag{30}$$

The simplest choice of the individual weighting functions is given by

$$w(t_s) = 1 \tag{31}$$

Since the approximations (27) are substituted in the objective function (29), the quadrature should be performed by a formula, the accuracy of which corresponds to the accuracy of Eq. (27). When the choice of the weighting functions $w(t_s)$ is limited to

$$w(t_s) = (1 - t_s)^{\alpha_s} t_s^{\beta_s} \tag{32}$$

with both exponents nonnegative and small, this accuracy is obtained by use of the Lobatto quadrature formulas [11] for all intervals reduced to [0,1]:

$$\Phi \simeq \sum_{c=1}^{M} \underline{L}_c^T \underline{W}_c \underline{L}_c \tag{33}$$

where the summation applies to all quadrature nodes, the index c referring to these points. A particular point $\underline{t}_c$ is determined by $t_{1,u}, \ldots, t_{s,i}, \ldots, t_{S,v}$, and the summation stands for

$$\sum_{c=1}^{M} = \sum_{u=1}^{u_f} \cdots \sum_{i=1}^{u_f} \cdots \sum_{v=1}^{v_f} \tag{34}$$

The optimal quadrature nodes are determined as the zeros of orthogonal polynomials with respect to the weighting functions

$$w(t_s) = (1 - t_s)^{\alpha_s + 1} t_s^{\beta_s + 1} \tag{35}$$

The weighting matrix $\underline{W}_c$ at each node is given by

$$\underline{W}_c = \underline{I}\omega_{1,u} \times \cdots \times \omega_{s,i} \times \cdots \times \omega_{S,v} \tag{36}$$

where the constant $\omega_{s,i}$ is the ith Lobatto weight with respect to the sth independent variable, and all constants can be obtained from straightforward algorithms.

The quadrature formula (33) requires the knowledge of the state equations $\underline{L}_c$ in all quadrature nodes. Because of the optimal choice of the nodes from orthogonal polynomials with respect to the weighting functions (35), the derivatives occurring in the state equations at the nodes can be replaced by the orthogonal collocation formulas which are still of the same accuracy as the underlying approximations (27). The weights used in calculating the derivatives are given by easily obtainable square matrices [11] indicated by $\underline{A}_s$ and $\underline{B}_s$, respectively, for the first and second derivatives with respect to t_s reduced to the interval [0,1]. The formulas are

$$\left.\frac{\partial x_j}{\partial t_s}\right|_{\underline{t}=\underline{t}_c} \simeq \sum_{\epsilon=1}^{i_f} A_{s,i,\epsilon} x_{j,u,\ldots,\epsilon,\ldots,v} \tag{37}$$

and

$$\left.\frac{\partial^2 x_j}{\partial t_s^2}\right|_{\underline{t}=\underline{t}_c} \simeq \sum_{\epsilon=1}^{i_f} B_{s,i,\epsilon} x_{j,u,\ldots,\epsilon,\ldots,v} \tag{38}$$

The combined use of the quadrature formula (33) and of the collocation equations based upon (37) and (38) leads to an objective function which can be more easily minimized with respect to all unknowns, namely the P parameters $\underline{a}$ and the (M x n) ordinates $\underline{z}$ of the state variables, because of the explicit dependence. The total number of unknowns surpasses the number (M x n) of squares occurring in the objective function (33) and seems to lead to an infinite number of solutions. The unknowns $\underline{a}$ and $\underline{z}$, however, are subject to a series of constraints given by the boundary conditions (2) and the requirement of minimization of Ψ with respect to $\underline{z}$. The total number of constraints should thus at least equal P.

The method presented has been shown to be efficient in solving parameter identification problems of variable complexity [13,14].

The main advantages are the gain of computation time, which can be enormous in some cases, as well as the generality. The speed of the method actually follows from the drastic reduction of the number of evaluations of the functions occurring in the differential equations. This number may be very large when using classical techniques such as the Runge-Kutta method or finite differences involved in the estimation techniques described in the previous sections. The orthog onal collocation approach actually limits the number of evaluations corresponding to the limited number of values of $\underline{x}$ considered in the objective function Ψ. The obtainable accuracy will of course be limited by the approximations (27). However, in view of the measurement errors occurring in the observation equations (4), polynomials of an appropriate degree frequently offer a sufficiently accurate representation within the accuracy of the data. If, however, the obtained value of the objective function Ψ minimized with respect to the coefficients of the chosen approximation turns out to be unacceptably large, the accuracy may be improved by subdividing the domain of observation Ω in a set of subdomains by a series of grid points. The orthogonal collocation principle can then be used in each subdomain with low-degree polynomials. The approximations in each subdomain have to be matched at the boundaries and are called "spline functions." The total number of function evaluations in the latter case will still be limited in comparison with the number required when applying the gradient or quasilinearization method.

V. ISOTHERMAL LAMINAR-FLOW TUBULAR REACTOR: AN EXAMPLE

The number of problems of parameter identification in DPS discussed in the chemical engineering literature is limited. This is certainly due to the poor convergence and long computation time involved in both the gradient methods and the quasilinearization method. Catalyst deactivation problems in tubular reactors have been treated by replacement of the original model by a series of ODEs [15] as well as by the generally applicable orthogonal collocation method [14]. In addition, a comparison of the three methods described in the previous sections, based upon three examples, has been reported [6]. A typical example discussed in the literature is the estimation of a kinetic parameter in a tubular reactor, which has been solved by the steepest descent method [16] and the orthogonal collocation method [13]. The last application will be discussed in detail here for the reactor depicted in Figure 2.

The dimensionless equations for an isothermal laminar-flow tubular reactor with the axial diffusion negligible in comparison with the radial diffusion will be considered [16]

$$(1 - t_2^2)\frac{\partial x}{\partial t_1} = 0.1\left(\frac{\partial^2 x}{\partial t_2^2} + \frac{1}{t_2}\frac{\partial x}{\partial t_2}\right) - ax \tag{39}$$

with the initial and boundary conditions

$$x(0,t_2) = 1 \tag{40}$$

$$\frac{\partial x}{\partial t_2} = 0 \qquad \text{for } t_2 = 0 \text{ and } t_2 = 1 \tag{41}$$

The state variable is the dimensionless concentration x, whereas the reduced axial and radial positions in the reactor serve as independent variables t_1 and t_2. The problem consists in the estimation of the kinetic parameter a from measurements of the concentration x. Values of x are actually generated for a parameter value a = 1 within the t_1-interval [0,1] [17].

Two different cases are considered following the form of the observation equation (4). First, it is assumed that the wall concentration is measured down the length of the reactor:

$$y(t_1) = x(t_1,1) \tag{42}$$

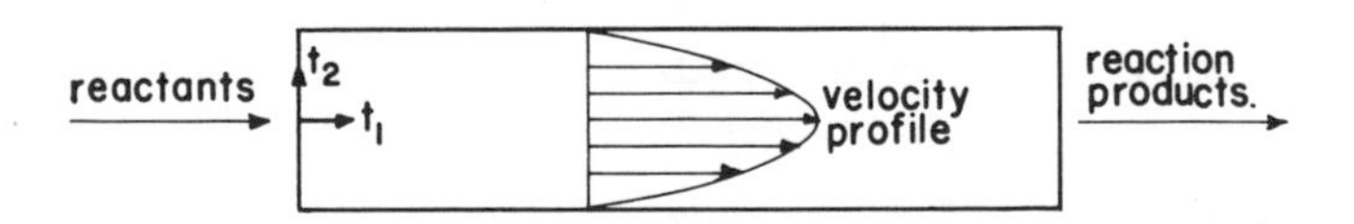

Fig. 2. Laminar-flow tubular reactor.

In the second case the integral average effluent conversion based on the volumetric flow rate is measured:

$$y = 4 \int_0^1 t_2(1 - t_2^2)x(1,t_2)\, dt_2 \tag{43}$$

Both problems have been solved by the steepest descent method, and accurate estimates for a have been found for starting values within the interval [0,2] [16].

When using orthogonal collocation the state equation (39) is replaced by a series of collocation equations

$$\begin{aligned} L_{u,v} = {} & (1 - t_{2,v}^2) \sum_{\mu=1}^{u_f} A_{1,u,\mu} z_{\mu,v} \\ & - 0.1\left(\sum_{\nu=1}^{v_f} B_{2,v,\nu} z_{u,\nu} + \frac{1}{t_{2,v}} \sum_{\nu=1}^{v_f} A_{2,v,\nu} z_{u,\nu} \right) \\ & + a z_{u,v} = 0 \qquad \text{for } u = 1,u_f \text{ and } v = 1,v_f \end{aligned} \tag{44}$$

The objective function Φ of Eq. (29) with the weighting functions defined by Eq. (31) becomes by Lobatto quadrature

$$\Phi = \sum_{u=1}^{u_f} \sum_{v=1}^{v_f} \omega_{1,u} \omega_{2,v} L_{u,v}^2 \tag{45}$$

which has to be minimized with respect to a and all $x_{u,v}$. The boundary conditions (40) and (41) lead to the following linear constraints:

$$z_{1,v} = 1 \qquad \text{for } v = 1,v_f \tag{46}$$

$$\sum_{\nu=1}^{v_f} A_{2,1,\nu} z_{u,\nu} = 0 \qquad \text{for } u = 2,u_f \tag{47}$$

$$\sum_{\nu=1}^{v_f} A_{2,v_f,\nu} z_{u,\nu} = 0 \qquad \text{for } u = 2,u_f \tag{48}$$

Additional constraints are obtained from the observation equations. Equation (42) gives

$$y_u = z_{u,v_f} \qquad \text{for } u = 2,u_f \tag{49}$$

Actually, only a value $u_f = 3$ will be considered, since in that case accurate values for y_2 and y_3 are found in the literature [17]

$$y_2 = 0.244944 \tag{50}$$

$$y_3 = 0.096416 \tag{51}$$

In the second example the following constraint is derived from Eq. (43):

$$y = 4 \sum_{v=1}^{v_f} \omega_{2,v} t_{2,v}(1 - t_{2,v}^2) z_{u_f,v} \tag{52}$$

with the value

$$y = 0.1378 \tag{53}$$

The total number of linear constraints is N, where

$$N = v_f + 3(u_f - 1) \tag{54}$$

in the first case, and

$$N = v_f + 2(u_f - 1) + 1 \tag{55}$$

in the second case. The number of collocation equations is given by

$$M = u_f \times v_f \tag{56}$$

which are all linear with respect to the vector of ordinates $\underline{z}$.

The minimization of the objective function (45) with respect to a and $\underline{z}$ subject to the constraints (46) to (48) and (43) or (52) can be performed by a one-dimensional search with respect to a. For each value of a the problem is indeed completely linear with respect to $\underline{z}$, allowing the use of linear regression techniques for the estimation of $\underline{z}$. The corresponding value of Φ can then be considered as a function of the parameter a only, and the one-dimensional search intends to minimize this value.

Special situations arise in the first example when $u_f = v_f = 3$ and in the second example when $u_f = 2$ and $v_f = 3$, since then the number of ordinates M given by Eq. (56) equals the number of constraints N given by (54) and (55), respectively. This means that all ordinates can be found directly from the constraints, and the objective function (45) can be easily minimized with respect to the only remaining unknown a, since all $L_{u,v}$ are linear with respect to this parameter. The results for a are 1.59 and 0.64, respectively. These values have poor accuracy due to the low-degree approximations with respect to t_1 and t_2, but they are obtained without any convergence problem.

The accuracy can be improved by increasing the degree of the polynomial approximations. For the example of wall measurements, the estimation of a for $u_f = 3$ and $v_f = 4$ gives a value a = 0.95, which is already much more accurate than the first estimate, 1.59. The corresponding values of the state variable at the center line of the reactor at the three collocation points are found to be 1.0, 0.57, and 0.25, as compared with the accurate values 1.0, 0.56, and 0.26 [17]. As shown in the case of ODEs the number of accurate figures obtainable by this method is actually unlimited when error-free data are available [14]. In the case of PDEs, however, it is necessary to increase the degree of the approximations with respect to different independent variables simultaneously, otherwise the accuracy remains restricted by the degree of the approximation with respect to one or more of these variables. This will be clarified by the results presented in connection with the example treated in Section VII.

In the case of effluent conversion measurements, and with the low-order approximations corresponding to $u_f = v_f = 4$, a remarkably accurate value a = 1.001 is found. The accuracy of this low-order approximation is sufficient in view of the observation errors occurring in practice. An increase of u_f and v_f could lead to more accurate results comparable to the very accurate estimates obtained by the method of steepest descent [16], but the corresponding increase in computation time would not pay off in realistic situations. The computation time is indeed determined by the repetitive solution for different values of a of a set of linear equations with respect to (M - N) components of $\underline{z}$. This number, and consequently the computation time, increases with increasing values of u_f and v_f.

VI. THE JOINT CONFIDENCE REGION

The minimization of an objective function with respect to the parameters $\underline{a}$ only partially solves the parameter identification problem. An estimate of the reliability should always be associated with the obtained results. This is important in view of the fact that parameters estimated from a set of experimental data often do not have the stature of true theoretical constants. Consequently, the use of the estimates beyond the range of the data could lead to erroneous conclusions.

The reliability of the parameter estimates is closely connected with the concept of observability of a system. A system is said to be observable if it is possible to recover the system state at any point from the output data. Analytical results concerning observability can be obtained only for linear systems [16] and consequently are of limited value. It is clear, however, that it will be possible to recover the system states if parameters can be estimated from the observations, assuming that the model solutions are unique functions of the parameters. The convergence of any algorithm for parameter estimation could thus be considered as a more general condition of observability [16].

Convergence of an algorithm is actually indicated by the confidence region. This region can be constructed based upon a comparison of the residual value of the objective function Ψ with an estimate of the value arising from random errors. For linear models the knowledge of the positive definite variance-covariance matrix $\underline{U}$ of the parameters, defined as

$$\underline{U} = E[(\underline{a} - \underline{a}_o)(\underline{a} - \underline{a}_o)^T] \tag{57}$$

where $\underline{a}_o$ is the vector of the obtained parameter estimates, suffices to construct a joint confidence region of these estimates for an arbitrary confidence level by using the properties of the F-distribution [18]. The confidence regions are hyperellipsoids in the P-dimensional parameter space.

The same approach can be used for nonlinear models after linearization of the system equations, the results being useful as far as linearization holds. The main problem thus consists in the calculation of $\underline{U}$.

For initial-value systems a rigorous development has been presented [2], leading to the following expression for the variance-covariance matrix $\underline{U}$:

$$\underline{U}^{-1} = \sum_{r=1}^{R} \underline{D}_r^T \underline{G}_r^t \underline{M}_r^{-1} \underline{G}_r \underline{D}_r \tag{58}$$

obtained by linearization of the objective function (5). The matrices $\underline{G}_r$ and $\underline{D}_r$ are defined by the following expressions calculated at the minimum of the objective function Ψ:

$$|G_{q,j}(t_r)| = \frac{\partial}{\partial x_j} h_q[\underline{x}(\underline{a}_o, t_r), t_r] \tag{59}$$

$$|D_{j,d}(t_r)| = \frac{\partial}{\partial a_d} x_j(\underline{a}_o, t_r) \tag{60}$$

Comparison of Eqs. (13) and (60) shows that $\underline{D}_r$ is the matrix of the sensitivity coefficients at t_r calculated for $\underline{a}_o$. Since the observability of the system stated in terms of convergence of the parameter estimation algorithm depends upon the existence of $\underline{U}$, the nonsingularity of the matrix of Eq. (58) is a necessary condition for observability. As can be seen, this condition depends upon the number and location of the observation points [6].

The same equation (58) could be used for DPS. The main problem consists in the calculation of the sensitivity coefficients at all observation points. This requires the simultaneous integration of n state equations (1) and (P x n) sensitivity equations obtained by differentiation of Eq. (1) using (14). The required values of the sensitivity coefficients are immediately available when using the gradient method, whereas after application of quasilinearization it is necessary to set up an integration scheme for solving the state and sensitivity equations.

The completely different approach used in the estimation of parameters following the principles of orthogonal collocation suggests an adapted procedure for the calculation of $\underline{U}$. Straightforward formulas are derived by linearization of the objective functions (5) and (33) in a manner similar to the preceding analysis.

When considering $\underline{x}$ as a function of the coefficients $\underline{b}$ given by Eq. (27) instead of the parameters $\underline{a}$, the variance-covariance matrix $\underline{P}$ of the coefficients $\underline{b}$ can be found in a manner similar to $\underline{U}$ in Eq. (58). The matrix is defined as

$$\underline{P} = E[(\underline{b} - \underline{b}_o)(\underline{b} - \underline{b}_o)^T] \tag{61}$$

and can be found from

$$\underline{P}^{-1} = \sum_{r=1}^{R} \underline{C}_r^T \underline{G}_r^T \underline{M}_r^{-1} \underline{G}_r \underline{C}_r \tag{62}$$

where $\underline{G}_r$ is defined by Eq. (59) and $\underline{C}_r$, similar to $\underline{D}_r$ in (60), is found from

$$|c_{j,d}(\underline{t}_r)| = \frac{\partial}{\partial b_d} x_j(\underline{b}_o, \underline{t}_r) \tag{63}$$

Because of Eq. (27), each row of $\underline{C}$ contains only M nonzero elements for a total number of (n x M).

Very often the matrix $\underline{P}$ resulting from Eq. (62) will be a block-diagonal matrix with square matrices of order M as diagonal elements, corresponding to

$$\underline{P}_j = E[(\underline{b}_j - \underline{b}_{j,o})(\underline{b}_j - \underline{b}_{j,o})^T] \tag{64}$$

where each $\underline{P}_j$ is the variance-covariance matrix of the coefficients $\underline{b}_j$ of the jth state variable.

The relation between the ordinates $\underline{z}$ and the coefficients $\underline{b}$ as given by Eq. (27) can be represented as

$$\underline{z} = \underline{T}^T \underline{b} \tag{65}$$

where $\underline{T}$ is a block-diagonal matrix with all diagonal elements equal to square matrices of order M, the elements of which are obtained by multiplication of appropriate powers of the abscissas of the collocation points. The variance-covariance matrix of $\underline{z}$ defined as

$$\underline{V} = E[(\underline{z} - \underline{z}_o)(\underline{z} - \underline{z}_o)^T] \tag{66}$$

from Eqs. (65) and (61) is found to be

$$\underline{V} = \underline{T}^T \underline{PT} \tag{67}$$

If $\underline{P}$ is a block-diagonal matrix with diagonal elements $\underline{P}_j$ defined by Eq. (64), $\underline{V}$ also has the same form with diagonal elements $\underline{V}_j$ defined as the variance-covariance matrix for the ordinates of the jth state variable

$$\underline{V}_j = E[(\underline{z}_j - \underline{z}_{j,o})(\underline{z}_j - \underline{z}_{j,o})^T] \tag{68}$$

Finally, linearization of the objective function (33) leads to the variance-covariance matrix $\underline{U}$ defined by Eq. (57) by a completely analogous treatment as before [2]:

$$\underline{RUR} = \underline{S} \tag{69}$$

Here

$$\underline{R} = \sum_{c=1}^{M} \underline{E}_c^T \underline{W}_c \underline{E}_c \tag{70}$$

is a symmetric matrix with $\underline{E}_c$ the matrix of the sensitivity coefficients of the collocation equations $\underline{L}_c$ at the collocation point $\underline{t}_c$

$$|E_{c,j,d}| = \frac{\partial L_{j,c}}{\partial a_d} \qquad \text{for } j = 1,n \text{ and } d = 1,P \tag{71}$$

Furthermore,

$$\underline{S} = \sum_{c=1}^{M} \sum_{c'=1}^{M} \underline{E}_c^T \underline{W}_c E[(\underline{L}_c - \underline{L}_{c,o})(\underline{L}_c - \underline{L}_{c,o})^T] \underline{W}_{c'} \underline{E}_{c'} \tag{72}$$

is a symmetric matrix, the central part of the last equation denoting the expectation of the product of the variations of the collocation equation points c and c′ with respect to variations of $\underline{z}$. Thus,

$$\underline{L}_c - \underline{L}_{c,o} = \underline{Z}_c^T(\underline{z} - \underline{z}_o) \tag{73}$$

where

$$\underline{Z}_{c,w,j} = \frac{\partial L_{j,c}}{\partial z_w} \quad \text{for } j = 1,n \text{ and } w = 1,M \times n \tag{74}$$

This results in

$$\underline{S} = \sum_{c=1}^{M} \sum_{c'=1}^{M} \underline{E}_c^T \underline{W}_c \underline{Z}_c^T \underline{V} \underline{Z}_{c'} \underline{W}_{c'} \underline{E}_{c'} \tag{75}$$

If the matrix $\underline{V}$ is a block-diagonal matrix with diagonal elements $\underline{V}_j$, the central part of Eq. (75) can be replaced by

$$\underline{Z}_c^T \underline{V} \underline{Z}_{c'} = \sum_{j=1}^{n} \underline{Z}_{c,j}^T \underline{V}_j \underline{Z}_{c',j} \tag{76}$$

where each $\underline{Z}_{c,j}$ is a matrix with elements

$$|Z_{c,j,c',j'}| = \frac{\partial L_{j',c}}{\partial z_{j,c'}} \quad \text{for } j' = 1,n \text{ and } c' = 1,M \tag{77}$$

where $z_{j,c'}$ is the ordinate of the jth state variable in the collocation point $\underline{t}_{c'}$.

Although these matrix manipulations seem complex, all calculations involved are straightforward and do not require iterations at any step of the procedure. An immediate calculation of the matrix $\underline{U}$ is thus possible as soon as the parameter estimates $\underline{a}_o$ are obtained.

VII. HEAT EQUATION: AN EXAMPLE

The transient behavior of a slab in an environment at constant temperature is considered. Although this is not an example of a distributed chemical reactor, this system is also described by a PDE and the estimation of the parameter determining the heat flux is of a nature similar to the identification problem in chemical reactors. Moreover, the form of the state equation and of the observation equation allows a simple and clear illustration of the confidence analysis described in the previous section. Transient profiles in the slab are depicted in Figure 3.

The heat equation [6]

$$\frac{dx}{dt_1} = a\,\frac{d^2x}{dt_2^2} \tag{78}$$

models the transient behavior of a slab where the dimensionless temperature x is a function of the dimensionless variables, time t_1 and position t_2. The parameter a characterizes the heat transfer. The boundary conditions are

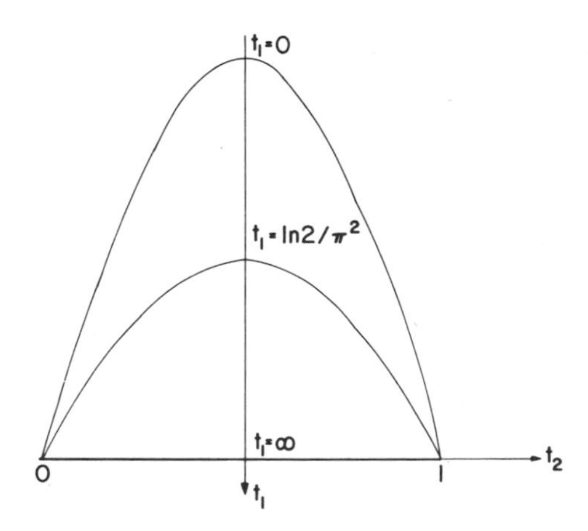

FIG. 3. Transient temperature profiles in a slab in uniform environment.

$$x(0,t_2) = \sin \pi t_2 \qquad \text{for } 0 \le t_2 \le 1 \tag{79}$$

and

$$x(t_1,0) = x(t_1,1) = 0 \qquad \text{for } 0 \le t_1 \le 0.2 \tag{80}$$

The observation equation is

$$y = x \tag{81}$$

and observations are made at a series of grid points which are equidistant with respect to both independent variables. The number of values of t_1 at which observations are made is R_1, where the last point is at $t_1' = 1$, whereas for t_2, R_2 inner values are considered.

If the parameter value $a = 1$ is considered, the solution of the state equation (78) with the boundary conditions (79) and (80) is given by

$$x = \exp(-a\pi^2 t_1) \sin \pi t_2 \tag{82}$$

and can be used to generate the data.

The orthogonal collocation procedure replaces the state equation (78) by a series of collocation equations

$$\frac{1}{0.2} \sum_{\mu=1}^{u_f} A_{1,u,\mu} z_{\mu,v} = a \sum_{\nu=1}^{v_f} B_{2,v,\nu} z_{u,\nu} \qquad \text{for } u = 1,u_f \text{ and } v = 1,v_f \tag{83}$$

written more simply as

$$\tau_{u,v} = a\theta_{u,v} \tag{84}$$

The values of the ordinates $z_{u,v}$ are derived from an approximation of x using the data for Eq. (81) and the boundary conditions (79) and (80). The parameter a is then estimated by minimization of

$$\Phi = \sum_{u=1}^{u_f} \sum_{v=1}^{v_f} \omega_{1,u} \omega_{2,v} L^2_{u,v} \tag{85}$$

Putting the derivative of Φ with respect to a equal to zero leads to the result

$$a = \frac{\sum_{u=1}^{u_f} \sum_{v=1}^{v_f} \omega_{1,u}\omega_{2,v}\tau_{u,v}\theta_{u,v}}{\sum_{u=1}^{u_f} \sum_{v=1}^{v_f} \omega_{1,u}\omega_{2,v}\theta_{u,v}^2} \tag{86}$$

Table 1 shows the absolute errors of the estimates obtained from Eq. (86) for variables u_f and v_f, assuming that accurate data for all ordinates $z_{u,v}$ are available. It is clear that in this situation the estimate can be made as accurate as desired by an appropriate increase of u_f and v_f. In real practice, however, the data are subject to error and the obtainable accuracy of the estimates is limited. It is therefore important to give an idea of the reliability of the obtained results.

Two different methods are possible to construct a confidence region for the estimate given by Eq. (86). The first method consists in performing a sufficiently large number of experiments, followed by the calculation of the mean and the standard deviation of all estimates. This procedure is purely theoretical, since the repetition

TABLE 1

Absolute Errors of the Estimate for Variables u_f and v_f

v_f	$u_f = 3$	$u_f = 4$	$u_f = 5$	$u_f = 6$	$u_f = 7$
3	2×10^{-1}	2×10^{-1}	2×10^{-1}	2×10^{-1}	2×10^{-1}
4	2×10^{-1}	2×10^{-1}	2×10^{-1}	2×10^{-1}	2×10^{-1}
5	6×10^{-2}	3×10^{-3}	2×10^{-3}	1×10^{-3}	1×10^{-3}
6	5×10^{-2}	3×10^{-3}	1×10^{-3}	1×10^{-3}	1×10^{-3}
7	5×10^{-2}	2×10^{-3}	3×10^{-5}	1×10^{-6}	1×10^{-6}
8	5×10^{-2}	2×10^{-3}	3×10^{-5}	8×10^{-7}	5×10^{-7}

of a large number of experiments would be a time- and money-consuming operation. For the example considered here, however, it can be done easily because of the generation of the data from Eq. (82). Normally distributed random errors with mean zero and variable standard deviation

$$\sigma_r(x) = \sigma x_r \tag{87}$$

with $\sigma = 0.09$ are imposed. Because of the boundary conditions (79) and (80), the following approximation is accepted:

$$x = \sin(\pi t_2) + t_2(1 - t_2) \sum_{u=1}^{u_f-1} \sum_{v=2}^{v_f-1} b_{u,v} t_1^u t_2^{v-2} \tag{88}$$

A new variable

$$X \equiv \frac{x - \sin(\pi t_2)}{t_2(1 - t_2)} = \sum_{u=1}^{u_f-1} \sum_{v=2}^{v_f-1} b_{u,v} t_1^u t_2^{v-2} \tag{89}$$

is introduced in order to allow the estimation of the coefficients $b_{u,v}$ from a weighted least squares objective function (5) for the corresponding data

$$\chi_r = \frac{\eta_r - \sin(\pi t_{2,r})}{t_{2,r}(1 - t_{2,r})} \tag{90}$$

with standard deviation

$$\sigma_r(X) = \frac{\sigma_r(x)}{t_{2,r}(1 - t_{2,r})} \tag{91}$$

The degree of approximation $u_f = v_f = 5$ was found to be appropriate for the imposed value $R_2 = 9$ and variable R_1.

The results obtained from the previous analysis are summarized in Table 2 for thirty experiments. It can be seen that the obtained estimates are remarkably accurate both in terms of the mean and the standard deviation. From the decrease of the values of the standard deviation for increasing value R_1, the trivial conclusion follows

TABLE 2

Mean Values and Standard Deviations of a_o

R_1	Mean	SDE	Steepest descent[a]	Quasilinearization[a]
10	1.0066	0.048	1.076	0.973
20	1.016	0.036	1.020	0.983
100	1.020	0.013	1.026	0.981
200	1.022	0.010	1.017	0.979

[a]Results obtained from [7].

that the accuracy increases with the number of data. The mean, however, is hardly influenced. Results reported elsewhere [6] obtained by use of the steepest descent method as well as by quasilinearization are included in Table 2 for comparison. The latter results, however, concern a single experiment, but their accuracy seems to be in agreement with the results of orthogonal collocation. In the case R_2 = 20, the method of steepest descent led to a mean 1.013 with corresponding standard deviation 0.0097 [6], indicating a smaller dispersion than for the orthogonal collocation results.

The main advantage of the method of orthogonal collocation as compared to steepest descent and quasilinearization is its convergence. The estimation of the parameter a from Eq. (86) does not require an initial estimate, making the region of convergence infinitely large. Moreover, it hardly requires any computation time, since iterative calculations are avoided. The method of steepest descent and quasilinearization, however, have a limited region of convergence and are time-consuming, even for the simple example considered here, due to their repetitive integrations [6].

A second way to perform the analysis of the reliability of the obtained parameter estimates is the use of the procedure developed in the previous section. This is the only feasible way in practice when only one experiment is performed. For the example considered here the analysis becomes simple because several of the matrices involved contain only one element. In order to further simplify the formulas, a constant value of the standard deviation of the errors $\sigma_r = 0.02$ is imposed upon all data generated from Eq. (82) as well as upon the boundary conditions (79) and (80), calling for a polynomial approximation

$$x = \sum_{u=1}^{u_f} \sum_{v=1}^{v_f} b_{u,v} t_1^{u-1} t_2^{v-1} \tag{92}$$

the coefficients of which are estimated from the least squares objective function (10).

The analysis has been performed for $R_1 = 20$ and $R_2 = 9$ and the orders of approximation $u_f = v_f = 5$. For a total number of thirty experiments, a mean value $a_o = 0.9775$ with a corresponding standard deviation 0.023 was obtained. It is the purpose of the confidence analysis to estimate the latter value when data from only one experiment are available.

The variance of the parameter a can be calculated from Eq. (69), where $\underline{R}$, $\underline{U}$, and $\underline{S}$ now contain only one element, leading to

$$\sigma^2(a) = \underline{U} = \underline{S}/\underline{R}^2 \tag{93}$$

with $\underline{R}$ and $\underline{S}$ defined by Eqs. (70) and (75). The different matrices involved in the calculation of $\underline{R}$ and $\underline{S}$ are:

(i) From Eqs. (71) and (83)

$$\underline{E}_c = \underline{E}_{u,v} = -\theta_{u,v} \tag{94}$$

(ii) $\underline{W}_c = \underline{W}_{u,v} = \omega_{1,u}\omega_{2,v}$ (95)

(iii) From Eqs. (77) and (83) the matrices $\underline{Z}_c$ are column vectors with elements

$$|Z_{c,c'}| = \frac{\partial L_{u,v}}{\partial z_{u',v'}} = \delta_{v,v'} A_{1,u,u'} - a\, \delta_{u,u'} B_{2,v,v'} \tag{96}$$

where δ is the Kronecker delta.

(iv) From Eqs. (59) and (81)

$$\underline{G}_r = 1 \tag{97}$$

(v) The moment matrix

$$\underline{M}_r = \sigma_r^2 = \sigma^2 \tag{98}$$

(vi) From Eq. (63)

$$\underline{C}_r^T = \begin{bmatrix} t_{i,r_1}^{0} \times t_{2,r_2}^{0} \\ t_{1,r_1}^{1} \times t_{2,r_2}^{0} \\ \vdots \\ t_{1,r_1}^{u_f-1} \times t_{2,r_2}^{0} \\ t_{1,r_1}^{0} \times t_{2,r_2}^{1} \\ \vdots \\ t_{1,r_1}^{u_f-1} \times t_{2,r_2}^{v_f-1} \end{bmatrix} \tag{99}$$

(vii) From Eq. (62)

$$\underline{P} = \sigma^2 \left(\sum_{r_1=1}^{R_1+1} \sum_{r_2=1}^{R_2+2} \underline{C}_r^T \underline{C}_r \right)^{-1} \tag{100}$$

Actual calculations resulted in a series of values $\sigma(a)$ with mean 0.026 and standard deviation 0.005, which is in excellent agreement with the true value $\sigma(a) = 0.023$.

The final conclusion is that it is possible to obtain accurate estimates by use of the orthogonal collocation technique. As soon as the parameter is estimated, it is easy to obtain an estimate of its standard deviation. In this way it is possible to connect any confidence interval with the parameter value.

VIII. NONLINEAR CONFIDENCE REGIONS

In the previous sections the construction of the confidence region for linear systems has been discussed. Most systems, however, are nonlinear, and the joint confidence region based upon linearization of the system equations may give a distorted picture of the true situation.

The joint confidence region generally has to be constructed from the equation

$$\Psi(\underline{a}) = \Psi(\underline{a}_o)\left[1 + \frac{P}{m \times R - P} F_{1-\gamma}(P, m \times R - P)\right] \tag{101}$$

requiring the appropriate value of the F-distribution for a given probability level $(1 - \gamma)$ [18]. For linear systems Eq. (101) leads to the joint confidence hyperellipsoid in parameter space for which the knowledge of the variance-covariance matrix $\underline{U}$ as calculated in the previous sections is sufficient.

For nonlinear systems the distribution properties are not known and the probability level $(1 - \gamma)$ holds only approximately. The major problem, however, is that the construction of the joint confidence region becomes very tedious. Calculations performed for a system of ODEs containing four parameters required an enormous amount of computation time not acceptable in real practice [19]. The joint confidence region was found to be strongly unsymmetrical and very difficult to present.

It is obvious that because of computer time restrictions the construction of the exact joint confidence region is not possible for DPS. Only the joint confidence hyperellipsoid based upon linearization can be constructed following the procedure outlined in the previous sections for the calculation of the variance-covariance matrix of the parameter estimates. This is a useful result, since in most applications great accuracy in estimating the confidence region is not needed. It is more important to be warned when the obtained parameter estimates are meaningless. Such a warning may be expected from the results obtained by linearization. Indeed, the unfortunate situation where large deviations invalidating the linearization would result in a small joint confidence region is very unlikely [2]. A small joint confidence region may be expected to occur when the deviations are small, a situation for which the linearization is valid.

IX. CONCLUSIONS

Three methods have been presented for solving parameter identification problems in DPS which are not subject to any a priori restrictions with respect to the state and observation equations nor with respect to the number of parameters. Restrictions may only arise from problems of convergence or of accuracy encountered when actually applying the different methods.

In view of the convergence of a method, both the region of convergence and the rate of convergence are important. With respect to the region of convergence the gradient methods as well as the quasilinearization technique usually require good starting values for the parameters in order to obtain convergence. Deviations allowed from the exact estimates are very small when several parameters are involved. Moreover, the computation time becomes very large due to the repetitive and time-consuming numerical integrations of an eventually large set of differential equations, the number of which becomes significantly large when applying quasilinearization since this technique assumes the possibility of reducing each PDE to a series of ODEs. Although both methods can be successfully applied to relatively simple problems involving only one state variable and one parameter, as demonstrated by the examples, the difficulties in terms of the required accuracy of the initial estimates as well as of computation time increase drastically with the dimension of $\underline{x}$ and $\underline{a}$, eventually impeding the applicability of both techniques. In such situations the only remaining alternative is the use of the orthogonal collocation approach. An important gain in computation time is possible by this method, since it actually limits the generation of values of $\underline{x}$ relating to the number of data available. The region of convergence is also significantly enlarged, and the problem of convergence is even completely avoided for systems which are linear with respect to the parameters, as shown by Section VII. The two advantages concerning convergence become very clear when a large number of parameters is involved. This has been demonstrated for the estimation of seven kinetic parameters from laboratory data concerning the isomerization of pentane in a packed bed reactor subject to catalyst deactivation [14]. The method becomes almost a thousand times faster than the gradient method, which is hardly useful in similar situations.

The second important concept concerning the applicability of a method is the accuracy of the estimates. Whenever the gradient method and the quasilinearization technique can be used, as is the case for the examples treated in this chapter, they turn out to give fairly accurate estimates. The polynomial approximations, however, involved in the application of orthogonal collocation limit the obtainable accuracy within the adequacy of the preset approximations in representing the actual profiles. The accuracy is good in most situations, and difficulties which sometimes arise may be avoided by the use of an appropriate number of subdomains of the entire observation domain. Moreover, it is possible to calculate the joint confidence hyperellipsoid connected with the obtained estimates fairly easily. This gives an easy way of estimating the reliability of the results. The application described in this chapter turns out to be remarkably accurate.

ACKNOWLEDGMENTS

The work presented in this chapter was made possible by a fellowship of "Aspirant van het Nationaal Fonds voor Wetenschappelijk Onderzoek" (Belgium). Our thanks are due to Professor L. Hellinckx for his encouragement and assistance, and to Dr. L. Hosten and Ir. B. Bossaert for their most helpful discussions. Finally, the obliging cooperation of Ing. F. Bierwerts is gratefully acknowledged.

NOTATION

$\underline{a}$	vector of parameters
$\underline{A}$	collocation matrix for the first derivative
$\underline{b}$	vector of coefficients
$\underline{B}$	collocation matrix for the second derivative
$\underline{c}$	vector of integration constants
$\underline{C}$	matrix of sensitivity coefficients for $\underline{b}$
$\underline{d}$	vector of boundary conditions
$\underline{D}$	matrix of sensitivity coefficients for $\underline{a}$
$\underline{e}$	vector of residuals
E	expectation
$\underline{E}$	matrix of sensitivity coefficients of $\underline{L}$
$\underline{f}$	vector of functions of $\underline{x}$, $\underline{t}$, and $\underline{a}$
F	F-distribution
$\underline{F}$	vector of functions of $\underline{q}$ and $\underline{t}$
$\underline{g}$	vector of functions of $\underline{t}$ and $\underline{a}$
$\underline{G}$	matrix derived from the observation equations
$\underline{h}$	vector of observation functions
$\underline{H}$	Hessian matrix
i	variable exponent of polynomial approximation
$\underline{I}$	unity matrix
$\underline{J}$	Jacobi matrix
$\underline{K}$	matrix characterizing the gradient method
$\underline{L}$	vector of state equations
m	number of observation variables
M	total number of collocation points
$\underline{M}$	variance-covariance matrix of $\underline{y}$
n	number of state variables
N	number of constraints
P	number of parameters
$\underline{P}$	variance-covariance matrix of $\underline{b}$
$\underline{q}$	vector of $\underline{q}$ and $\underline{a}$

$\underline{Q}$	matrix of homogeneous solutions
R	number of observation points
$\underline{R}$	symmetric matrix for the determination of $\underline{U}$
S	number of independent variables
$\underline{S}$	symmetric matrix for the determination of $\underline{U}$
$\underline{t}$	vector of independent variables
$\underline{T}$	block-diagonal matrix based upon the collocation points
u	variable exponent of polynomial approximation
$\underline{U}$	variance-covariance matrix of $\underline{a}$
v	variable exponent of polynomial approximation
$\underline{V}$	variance-covariance matrix of $\underline{z}$
w	weighting function
$\underline{W}$	weighting matrix
$\underline{x}$	vector of state variables
X	reduced state variable
$\underline{y}$	vector of observation variables
$\underline{z}$	vector of all ordinates
$\underline{Z}$	matrix for the dependence of $\underline{L}$ upon $\underline{z}$
α	exponent of weighting function w
β	exponent of weighting function w
γ	probability level
δ	Kronecker delta
$\underline{\eta}$	data for $\underline{y}$
θ	collocation group
λ	sensitivity coefficient
ρ	step length
σ	standard deviation
τ	collocation group
Φ	objective function based upon $\underline{L}$
χ	data for X
Ψ	objective function based upon $\underline{e}$
ω	Lobatto weight
Ω	domain of observation

Subscripts

c, c' collocation point
d index of parameter
f final value
h homogeneous solution
i index of collocation point of t_s
j, j' state variable
k iteration number
o estimated value
p particular solution
q index of observation variable
r observation point
s index of independent variable
u, u' index of collocation point of t_1
v, v' index of collocation point of t_S
w index of components of $\underline{z}$

ϵ index of collocation point of t_s
μ index of collocation point of t_1
ν index of collocation point of t_S

Superscript

T transpose

REFERENCES

1. J. H. Seinfeld and L. Lapidus, in Process Modeling, Estimation, and Identification, Prentice-Hall, Englewood Cliffs, New Jersey, 1974.

2. H. H. Rosenbrock and C. Storey, in Computational Techniques for Chemical Engineers, Pergamon Press, New York, 1966.

3. Y. Bard, Comparison of gradient methods for the solution of nonlinear parameter estimation problems, Siam. J. Numer. Anal., 7: 157 (1970).

4. A. J. Berger, Comparing sensitivity methods for estimating parameters in nonlinear ordinary differential equations, AIChE J., 19: 365 (1973).

5. G. Emig, H. Hofmann, and H. Friedrich, Ermittlung kinetischer Parameter aus integralen Labordaten, Proceedings of the Fifth European Congress on Chemical Reaction Engineering, Amsterdam, The Netherlands, 1972.

6. J. H. Seinfeld and W. H. Chen, Estimation of parameters in partial differential equations from noisy experimental data, Chem. Eng. Sci., 26: 753 (1971).

7. E. S. Lee, in Quasilinearization and Invariant Imbedding, Academic Press, New York, 1968.

8. J. K. Donnelly and D. Quon, Identification of parameters in systems of ordinary differential equations using quasilinearization and data perturbation, Can. J. Chem. Eng., 48: 114 (1970).

9. G. Emig and M. Köppner, Die Methode der Quasilinearisierung zur Ermittlung kinetischer Parameter, Chem. Eng. Sci., 29: 2339 (1974).

10. B. A. Finlayson, in The Method of Weighted Residuals and Variational Principles, Academic Press, New York, 1972.

11. J. V. Villadsen, in Selected Approximation Methods for Chemical Engineers, Danmarks tekniske Højskole, Lyngby, Denmark, 1970.

12. R. Tanner, Estimating kinetic rate constants using orthogonal polynomials and Picard's iteration method, Ind. Eng. Chem. Fundam., 11: 1 (1972).

13. B. A. J. Van den Bosch and L. J. Hellinckx, A new method for the estimation of parameters in differential equations, AIChE J., 20: 250 (1974).

14. B. A. J. Van den Bosch, F. Jonckheere, and L. J. Hellinckx, Rapid estimation of parameters from multiresponse data, Proceedings of the Third Symposium on the Use of Computers in Chemical Engineering, Gliwice, Poland, 1974.

15. G. R. Gavalas, G. C. Hsu, and J. H. Seinfeld, Estimation of catalyst deactivation parameters from operating reactor data, Chem. Eng. Jl., 4: 77 (1972).

16. J. H. Seinfeld, Identification of parameters in partial differential equations, Chem. Eng. Sci., 24: 65 (1969).

17. L. Lapidus, in Digital Computation for Chemical Engineers, McGraw-Hill, New York, 1962.

18. N. R. Draper and H. Smith, in Applied Regression Analysis, Wiley, New York, 1966.

19. G. Emig and L. H. Hosten, On the reliability of parameter estimates in a set of simultaneous nonlinear differential equations, Chem. Eng. Sci., 29: 475 (1974).

INDEX